Computational Methods in Applied Sciences

Volume 36

Series editor

Eugenio Oñate, Barcelona, Spain

More information about this series at http://www.springer.com/series/6899

David Greiner · Blas Galván
Jacques Périaux · Nicolas Gauger
Kyriakos Giannakoglou · Gabriel Winter
Editors

Advances in Evolutionary and Deterministic Methods for Design, Optimization and Control in Engineering and Sciences

Springer

Editors
David Greiner
Blas Galván
Gabriel Winter
SIANI
Universidad de Las Palmas
de Gran Canaria
Las Palmas de Gran Canaria
Spain

Jacques Périaux
University of Jyväskylä
Jyväskylä
Finland

and

CIMNE
Universitat Politècnica de Catalunya
Barcelona
Spain

Nicolas Gauger
RWTH Aachen University
Aachen
Germany

Kyriakos Giannakoglou
National Technical University of Athens
Athens
Greece

ISSN 1871-3033
ISBN 978-3-319-11540-5 ISBN 978-3-319-11541-2 (eBook)
DOI 10.1007/978-3-319-11541-2

Library of Congress Control Number: 2014953590

Springer Cham Heidelberg New York Dordrecht London

Printed on acid-free paper

Springer is part of Springer Science+Business Media (www.springer.com)

Preface

EUROGEN 2013, the tenth in the international series of Conferences was organized by the Institute of Intelligent Systems and Numerical Applications in Engineering (SIANI) of Universidad de Las Palmas de Gran Canaria (ULPGC), Spain, in association with the European Community on Computational Methods in Applied Sciences—ECCOMAS (Thematic Conference series) and the European Research Community on Flow, Turbulence and Combustion—ERCOFTAC (Special Group of Interest), and took place in the Industrial and Civil Engineering School at the Tafira Campus of ULPGC in October 7–9, 2013.

This event gathered experts from Universities, Research Institutions and Industries developing or applying evolutionary and deterministic methods in design optimization with emphasis on industrial and societal applications.

Additionally to the traditional themes and applications in the Series, EUROGEN 2013 focused particularly on:

- Intelligent systems for multidisciplinary design optimization (MDO) problems based on multi-hybridized software
- Adjoint-based and one-shot methods
- Uncertainty quantification and optimization
- Multidisciplinary design optimization
- Applications of game theory to industrial optimization problems
- Applications in structural and civil engineering optimum design
- Offshore, coastal and marine applications
- Surrogate models-based optimization methods in aerodynamic design
- Neural networks applied to logistic transport

Among the 75 presentations of the EUROGEN 2013 Conference, 34 extended full papers were selected for publication in this volume after peer-review by members of the European Scientific Program Committee and classified in the following sections:

- Theoretical and Numerical Methods and Tools for Optimization:
  - Theoretical Methods and Tools
  - Numerical Methods and Tools
- Engineering Design and Societal Applications:
  - Turbo machinery
  - Structures, Materials and Civil Engineering
  - Aeronautics and Astronautics
  - Societal Applications
  - Electrical and Electronics Applications

This volume presents up-to-date material on the state of the art in Evolutionary and Deterministic Methods for Design, Optimization and Control with Applications to Industrial and Societal Problems from Europe, Asia, North and South America.

The Scientific Organizing Committee and the Local Organizing Committee acknowledge the sponsorship of the following organizations through financial support or/and assistance during the development of the event: Cátedra Endesa-Red, European Community on Computational Methods in Applied Sciences (ECCOMAS), European Research Community on Flow, Turbulence and Combustion (ERCOFTAC), European Community (EC), Research Project E-CAERO, Center for Numerical Methods in Engineering (CIMNE), Plataforma Oceánica de Canarias (PLOCAN), CEANI Division—Institute of Intelligent Systems and Numerical Applications in Engineering (SIANI)—Universidad de Las Palmas de Gran Canaria (ULPGC), Gobierno de Canarias.

The two Committees above are grateful to all the members of the European Scientific Committee, the European Technical Committee and the International Corresponding members.

Special thanks are also addressed to Dr. Dietrich Knoerzer, Project Officer EC DG Aeronautics who initialized the EUROGEN series through the INGENET Networking EC project and has been continuously participating in this two-year frequency event since 1995.

Finally, the editors acknowledge Nathalie Jacobs, Springer, and Eugenio Oñate for the interest to this series in publishing the most representative scientific and industrial material presented in the EUROGEN 2013 ECCOMAS Thematic Conference in the Springer—ECCOMAS Series entitled: *Computational Methods in Applied Sciences.*

David Greiner
Blas Galván
Jacques Périaux
Nicolas Gauger
Kyriakos Giannakoglou
Gabriel Winter

Contents

Part I
Theoretical and Numerical Methods and Tools for Optimization: Theoretical Methods and Tools

Chapter 1
Multi-objective Evolutionary Algorithms in Real-World Applications: Some Recent Results and Current Challenges

Carlos A. Coello Coello

Abstract This chapter provides a short overview of the most significant research work that has been conducted regarding the solution of computationally expensive multi-objective optimization problems. The approaches that are briefly discussed include problem approximation, function approximation (i.e., surrogates) and evolutionary approximation (i.e., clustering and fitness inheritance). Additionally, the use of alternative approaches such as cultural algorithms, small population sizes and hybrids that use a few solutions (generated with optimizers that sacrifice diversity for the sake of a faster convergence) to reconstruct the Pareto front with powerful local search engines are also briefly discussed. In the final part of the chapter, some topics that (from the author's perspective) deserve more research, are provided.

Keywords Evolutionary algorithms · Multi-objective optimization · Metaheuristics

1.1 Introduction

In real-world applications, most problems have several (often conflicting) objectives that we aim to optimize at the same time. Such problems are called "multi-objective" and their solution gives rise to a set of solutions that represent the best possible trade-offs among all the objectives (i.e., the so-called *Pareto optimal set*). The image of the Pareto optimal set (i.e., the objective function values corresponding to this set) forms to so-called *Pareto front* of the multi-objective optimization problem being solved.

Starting in the mid-1980s, Evolutionary Algorithms (EAs) have become a popular search engine to solve multi-objective optimization problems, mainly because of their ease of use, and wide applicability (i.e., they require little domain-specific information to operate) [11, 15].

The author acknowledges the financial support obtained through a "Cátedra Marcos Moshinsky".

C.A. Coello Coello (✉)
CINVESTAV-IPN (Evolutionary Computation Group), 07360 Mexico, D.F., Mexico
e-mail: ccoello@cs.cinvestav.mx

D. Greiner et al. (eds.), *Advances in Evolutionary and Deterministic Methods for Design, Optimization and Control in Engineering and Sciences*, Computational Methods in Applied Sciences 36, DOI 10.1007/978-3-319-11541-2_1

Modern multi-objective evolutionary algorithms (MOEAs) consist of two main components:

1. A selection mechanism that is normally (but not necessarily) based on Pareto optimality. Performance indicators can also be used for selecting solutions in a population and that has been, indeed, a relatively popular research trend in recent years [3].
2. A density estimator, which is responsible for producing different elements of the Pareto optimal set in a single run of a MOEA. Different options are available for this mechanism, such as: fitness sharing [16], entropy [82], clustering [79], adaptive grids [31] and crowding [17], among others.

Additionally, all modern MOEAs are *elitist*, which means that they retain the non-dominated solutions generated at each iteration, so that at the end of a run, the user can have the globablly nondominated solutions that had been produced. Elitism is normally implemented through the use of an external archive, but the use of the main population for this purpose is also possible.

In spite of their popularity, one of the main limitations of MOEAs, when used for solving real-world problems, is their high computational cost, which is associated to the relatively high number of objective function evaluations that most current MOEAs require [62]. Although there are several remarkable efforts in this regard, several challenges still lie ahead, and the purpose of this chapter is precisely to review some of the most representative research that has been conducted in this area.

The remainder of this chapter is organized as follows. In Sect. 1.2, we present basic concepts related to multi-objective optimization. Then, in Sect. 1.3, we discuss the main schemes that have been proposed for dealing with expensive multi-objective optimization problems. In Sect. 1.4, we explore other ideas that havel also been used for dealing with real-world applications having objective functions that are computationally expensive. Section 1.5, provides some potential paths for future research in this area. Finally, the conclusions of this chapter are presented in Sect. 1.6.

1.2 Basic Concepts

We are interested in solving problems of the type[1]:

$$\text{minimize}\, \mathbf{f}(\mathbf{x}) := [f_1(\mathbf{x}), f_2(\mathbf{x}), \ldots, f_k(\mathbf{x})] \tag{1.1}$$

subject to:

$$g_i(\mathbf{x}) \leq 0 \quad i = 1, 2, \ldots, m \tag{1.2}$$

$$h_i(\mathbf{x}) = 0 \quad i = 1, 2, \ldots, p \tag{1.3}$$

[1] Without loss of generality, we will assume only minimization problems.

where $\mathbf{x} = [x_1, x_2, \ldots, x_n]^T$ is the vector of decision variables, $f_i : \mathbb{R}^n \rightarrow \mathbb{R}, i = 1, ..., k$ are the objective functions and $g_i, h_j : \mathbb{R}^n \rightarrow \mathbb{R}, i = 1, ..., m, j = 1, ..., p$ are the constraint functions of the problem.

To describe the concept of optimality in which we are interested, we will introduce next a few definitions.

Definition 1 Given two vectors $\mathbf{x}, \mathbf{y} \in \mathbb{R}^k$, we say that $\mathbf{x} \leq \mathbf{y}$ if $x_i \leq y_i$ for $i = 1, ..., k$, and that **x dominates y** (denoted by $\mathbf{x} \prec \mathbf{y}$) if $\mathbf{x} \leq \mathbf{y}$ and $\mathbf{x} \neq \mathbf{y}$.

Definition 2 We say that a vector of decision variables $\mathbf{x} \in \mathscr{X} \subset \mathbb{R}^n$ is **non-dominated** with respect to $\mathscr{X}$, if there does not exist another $\mathbf{x}' \in \mathscr{X}$ such that $\mathbf{f}(\mathbf{x}') \prec \mathbf{f}(\mathbf{x})$.

Definition 3 We say that a vector of decision variables $\mathbf{x}^* \in \mathscr{F} \subset \mathbb{R}^n$ ($\mathscr{F}$ is the feasible region) is **Pareto-optimal** if it is nondominated with respect to $\mathscr{F}$.

Definition 4 The **Pareto Optimal Set** $\mathscr{P}^*$ is defined by:

$$\mathscr{P}^* = \{\mathbf{x} \in \mathscr{F} | \mathbf{x} \text{ is Pareto-optimal}\}$$

Definition 5 The **Pareto Front** $\mathscr{P}\mathscr{F}^*$ is defined by:

$$\mathscr{P}\mathscr{F}^* = \{\mathbf{f}(\mathbf{x}) \in \mathbb{R}^k | \mathbf{x} \in \mathscr{P}^*\}$$

Therefore, we wish to determine the Pareto optimal set from the set $\mathscr{F}$ of all the decision variable vectors that satisfy (1.2) and (1.3). In practice, however, not all the Pareto optimal set is normally desirable or even achievable.

1.3 Dealing with Expensive Problems

In general, MOEAs can be unaffordable for an application when:

- The evaluation of the fitness functions is computationally expensive (e.g., it takes several hours).
- The total number of fitness function evaluations that can be performed is limited (e.g., we only have a certain computational budget available).

According to [29], there are three main schemes that can be used to deal with expensive problems:

Problem approximation: In this case, the idea is to replace the original (expensive) statement of the problem by another one which is easier (and less expensive) to solve.

Functional approximation: In this case, instead of using the original objective function(s) (which is/are expensive to evaluate), an alternative expression(s) is adopted. The new expression(s) is built based on the previous data obtained from evaluating the real objective function(s). The models that are obtained from the data that is currently available are called *meta-models* or *surrogates*.

Evolutionary approximation: This is an approach that is specific to EAs, and that aims to save fitness function evaluations by estimating the fitness of an individual using information from other (similar) individuals. The two main approaches in this class are: fitness inheritance and clustering.

Next, we will provide a short discussion of each of these schemes, as well as some real-world problems in which they have been adopted.

1.3.1 Use of Problem Approximation

As indicated before, in this case, the idea is to replace the original problem by another one which is easier to solve. This sort of approach has been relatively popular in aeronautical/aerospace engineering, in which complex Computational Fluid Dynamics (CFD), Computational Aero-Acoustics (CAA) and Computational Structural Mechanics (CSM) are adopted. When using such tools, it is possible to approximate the original problem by using different resolutions in the flow or structural simulation, adopting either coarse or fine grids. For CFD simulations is also possible to rely on Euler flows instead of (the more expensive) Navier-Stokes flow simulations.

An example of this sort of approach is the work of Lee et al. [41, 42]. In this case, the authors applied the HAPMOEA (Hierarchical Asynchronous Parallel Multi-Objective Evolutionary Algorithm) [24] to the robust design optimization of an ONERA M6 wing shape. The authors considered uncertainties in the design environment, related to the flow Mach number, and the Taguchi method was used to transform the problem into one with two objectives to be minimized: (1) the mean value of an objective function with respect to variability of the operating conditions, and (2) the variance of the objective function of each solution candidate, with respect to its mean value. HAPMOEA uses an evolution strategy as its search engine, incorporating the concept of Covariance Matrix Adaptation (CMA). It also incorporates a distance-dependent mutation operator, and a hierarchical set of CFD models (varying the grid resolution of the solver). Small populations are evolved using fine mesh CFD solutions in order to exploit the search space, while large populations are evolved with coarse mesh CFD solutions for exploring the search space. Good solutions from the coarse mesh populations (in which evaluations have a low computational cost) are transferred to the fine mesh populations (in which evaluations are computationally expensive).

For more information on this topic, the interested reader must refer to: [6, 62, 68]

1.3.2 Use of Functional Approximation

The use of meta-models or surrogate models has been very popular in engineering. In order to build a meta-model, a set of data points that lie on the local neighborhood of the design is required. The accuracy of the meta-model relies on the number of samples provided (from the real objective function evaluations), as well as on the accuracy of the model that is used to approximate the objective functions. Such an approximate model must also have a low computational cost, since it will be evaluated many times during the search.

There are several techniques available for constructing surrogate models, from which the main ones are [62]: response surface methods, Gaussian processes (or Kriging), radial basis functions, artificial neural networks and support vector machines.

An example of this sort of approach is the work of Voutchkov et al. [81], in which the Nondominated Sorting Genetic Algorithm-II (NSGA-II) [17] was used to perform a robust structural design of a simplified jet engine model. The aim was to find the best jet engine structural configuration that minimized: the variation of reacting forces under a range of external loads, the mass for the engine and the engine's fuel consumption. The evaluation of the structural response was done in parallel by means of finite element simulations. The authors adopted a kriging based response surface method in order to reduce the computational time required to solve this problem. Four objectives were minimized: (1) standard deviation of the internal reaction forces, (2) mean value of the internal reaction forces, (3) engine's mass, and (4) mean value of the specific fuel consumption. The first two objectives were computed over 200 external load variations. Due to the many combinations of loads and finite element thicknesses, the multi-objective optimization problem would have taken on the order of one year of computational time on a single 1 GHZ CPU, if no effort had been made to perform a more efficient search. When using the surrogate model that they report, combined with parallel processing, the total optimization time was reduced to about 26 h, in a cluster with 30 cores.

For more information on this topic, the interested reader must refer to: [32, 43, 45, 51, 53].

1.3.3 Use of Evolutionary Approximation

In this case, two main approaches are considered: clustering and fitness inheritance. Next, we will briefly discuss each of them.

Clustering is a term used to refer to the unsupervised classification of paterns into groups (which are called *clusters*). The idea is to partition data into different groups either in a hard way (i.e., into well-defined groups) or in a fuzzy way (i.e., using a certain degree of membership to each of the groups) [27].

Although clustering is normally not used as a specific technique to reduce objective function evaluations, this sort of technique is normally adopted in combination with surrogates in order to reduce the size of the training data set. This is an important task, since the use of very large training data sets makes prohibitive the cost of a surrogate method. Clustering is normally adopted in this context to split the data set into several small groups, and then an independent local model is built from each of them.

An example of the use of clustering is the work of Langer et al. [38], in which an integrated approach that adopts computer aided design modeling is combined with a MOEA for solving structural shape and topology optimization problems. The authors were interested in optimizing an instrument panel of a satellite, considering two objectives: (1) minimize the instrument panel mass, and (2) maximize the first eigenfrecuency. The authors solved the optimization problem for three shape and topology optimization cases: (a) a panel without instruments, (b) a panel with instruments at fixed positions, and (c) a panel with instrumental placing. They adopted polynomial based response surface methods in order to reduce the computational cost, and multiple local approximation models were constructed using a clustering technique. The use of parallel techniques was also required in this case (a cluster with 32 processors was adopted by the authors).

Fitness inheritance was originally introduced by Smith et al. [71], with the motivation of reducing the total number of fitness function evaluations performed by an evolutionary algorithm. The idea is that, when assigning fitness to an individual, some times we evaluate the objective function as usual, but the rest of the time, we assign a fitness value equal to the average of the fitness values of its parents. This saves one fitness function evaluation, and is based on the assumption of similarity of an offspring to its parents.

Evidently, fitness inheritance cannot be applied all the time, since it is required to have information from true fitness function evaluations in order to guide the search in a proper way. The percentage of time in which fitness inheritance is applied is called *inheritance proportion*. Clearly, this proportion should be less than one in order to avoid premature convergence [4].

A theoretical model of fitness inheritance was presented by Sastry et al. [69]. Such model was used to obtain the convergence time, the optimal population size and the optimal inheritance proportion (the authors found that values between 0.54 and 0.558 worked best for the inheritance proportion in problems of moderate and large size).

The work of Sastry et al. [69] was extended to the multi-objective case by Chen et al. [4]. In this case, the authors used fitness sharing to maintain diversity in the population with the aim of covering a larger extension of the Pareto front. The problem they solved was a bi-objective extension of the OneMax problem originally solved by Sastry et al. [69] in their study. The authors also presented a generalization (for the multi-objective case) of the theoretical work reported by Sastry et al. [69] regarding convergence time, optimal population sizing and optimal inheritance proportion. The experiments reported by the authors showed that savings of up to 40 % of the total number of evaluations could be achieved when using fitness inheritnance alone. When combining fitness inheritance with fitness sharing, savings of up to 25 % were obtained.

Reyes-Sierra and Coello Coello proposed the use of dynamic rules to assign the inheritance proportion in a multi-objective particle swarm optimizer [55]. Such rules produced savings that were from 19 up to 78 % of the total number of evaluations. However, as expected, the greater the savings in the number of evaluations, the greater was the degradation in the quality of the results. Nevertheless, the authors showed it was possible to obtain savings of up to 49 % without having a significant loss in the quality of the results. The authors adopted the Zitzler-Deb-Thiele (ZDT) test problems in their experiments [89].

It is worth mentioning that some researchers have considered fitness inheritance to be an inappropriate mechanism in complex or real-world problems (see for example [20], in which the authors concluded that fitness inheritance was not useful when the shape of the Pareto front is nonconvex or discontinuous). Such conclusions are valid for the proposal reported in [20]. However, in [56] it is shown that these limitations of fitness inheritance can be overcome, so that this approach can be applied to Pareto fronts having any kind of shape.

For more information on this topic, the interested reader must refer to: [19, 23, 37, 52, 56].

1.4 Other Approaches

There are some other ideas that can be used to tackle problems with computationally expensive objective functions, and which do not fall into any of the categories analyzed in the previous section. Here, we will focus on three types of approaches:

1. Cultural algorithms
2. Use of very small population sizes
3. Use of efficient search techniques

Next, we will briefly discuss each of these three types of approaches.

1.4.1 Cultural Algorithms

Cultural algorithms were originally proposed by Robert Reynolds in the mid-1990s [57, 60]. The core idea behind cultural algorithms is to incorporate domain knowledge extracted during the search to an evolutionary algorithm. Cultural algorithms use, in addition to the population space commonly adopted in evolutionary algorithms, a belief space, which encodes the knowledge obtained from the search points that have been evaluated so far. The belief space is used to influence the evolutionary operators, with the aim of guiding the search in a very efficient way.

At each generation, a cultural algorithm selects some individuals from the population, in order to extract information from them. Such information will then be used to speed up the search. Evidently, the belief space requires some sort of scheme to represent the knowledge extracted during the evolutionary process and this representation is normally specific for each particular problem (or class of problems). It is also necessary to design mechanisms that allow to use this extracted knowledge to influence the way in which the evolutionary operators explore and exploit the search space.

Although cultural algorithms have been adopted for single-objective optimization by several authors (see for example [7, 28, 35, 58, 59]), their use in multi-objective optimization has been very limited until now.

The first proposal to design a cultural algorithm for solving multi-objective optimization problems is the framework described in [12], which uses Pareto ranking, and an approximation of the dimensions of the Pareto front in the belief space. In this proposal, the belief space works as a guide for the individuals to reach regions where nondominated solutions have been found. The belief space includes also a mechanism to obtain a good distribution of the resulting points along the Pareto front (i.e., a density estimator).

The earliest attempt to solve multi-objective optimization problems using cultural algorithms was based on the use of the ε-constraint method [36], since this sort of approach uses a single-objective optimizer rather than a MOEA (the cultural algorithm with differential evolution proposed in [35] was adopted for this sake). This approach turned out to be computational expensive, due to the high number of objective function evaluations required to generate a good approximation of the Pareto front. However, the authors showed that if the aim was to solve very difficult multi-objective optimization problems, then this additional computational cost was worth it. This was illustrated by solving several problems from the Deb-Thiele-Laumanns-Zitzler (DTLZ) [18] and the Walking-Fish-Group (WFG) [25, 26] test suites.

More recently, Best and his collaborators [1, 2] proposed a more general framework for using cultural algorithms with any sort of MOEA. This approach is interesting and incorporates several sources of knowledge, but it did not show a significant reduction of objective function evaluations, which is one of the main motivations for using cultural algorithms. Additionally, the results presented by the authors are not competitive with respect to those obtained by traditional MOEAs using the same number of objective function evaluations, which suggests that it is still required to conduct more research in this area. In fact, the incorporation of knowledge into MOEAs (using any sort of scheme), with the aim of making them more efficient is indeed a very promising research area [37].

1.4.2 Use of Very Small Population Sizes

The use of small population sizes is unusual in the evolutionary algorithms literature in general, mainly because of the evident loss of diversity that is associated to small population sizes, and which normally leads to premature convergence. However, in the genetic algorithms literature, it is known that the use of very small population sizes is possible, if an appropriate reinitialization process is adopted (such approaches are called micro-genetic algorithms (micro-GAs) [13, 14, 34] and they use populations with no more than five individuals).

Krishnakumar [34] proposed the first implementation of a micro-GA. The first micro-GA for multi-objective optimization was introduced in [13, 14]. This approach uses a population size of four individuals, and three forms of elitism: (1) an external archive that adopts the adaptive grid from the Pareto Archived Evolution Strategy (PAES) [33], (2) a population memory, in which randomly generated individuals are replaced by evolved individuals, and (3) a mechanism that retains the two best solutions generated by each run of the micro-GA. The main advantage of this approach is its efficiency (its authors showed that their approach was up to an order of magnitude faster than the NSGA-II [17]). This is the reason why this approach has been used in computationally expensive real-world applications (see for example [8, 9]).

In a further paper, Coello Coello and Pulido introduced the micro-GA^2 [78], which is a fully self-adaptive MOEA that adopts a parallel strategy to adapt the crossover operator and the type of encoding (binary or real numbers) to be used. This approach can even stop automatically (it uses a mechanism based on a performance indicator to decide when to stop the search).

Over the years, other authors have adopted micro-genetic algorithms for solving a variety of problems (see for example [5, 30, 46, 47, 61, 72, 76, 77]). Additionally, the use of very small population sizes has also been attempted with other bio-inspired metaheuristics, such as particle swarm optimization (see [22]).

1.4.3 Use of Efficient Search Techniques

During the last few years, some researchers have proposed schemes that allow a more efficient exploration of the search space through the use of aggressive search engines that produce a few points from the Pareto front and then adopt a local search engine to reconstruct the rest of the front. One example of this sort of hybrid MOEA is DEMORS (differential evolution (DE) for multi-objective optimization with local search based on rough set theory) [64]. This approach operates in two phases. In the first one, a DE-based MOEA produces a rough approximation of the Pareto front using a relatively low number of objective function evaluations (65 % of the total number of objective function evaluations adopted by DEMORS are spent in the first phase). In the second phase, the remainder 35 % of objective function evaluations still available, are spent on the use of a local search procedure based on rough set theory [49, 50], whose task is to reconstruct the missing parts of the Pareto front. DEMORS was validated using several standard test problems taken from the specialized literature, as well as in a real-world problem having 8 objective functions and 160 decision variables in which it was able to outperform NSGA-II.

The same authors experimented with other (similar) hybrids in which DE was replaced by particle swarm optimization [66, 67] or rough sets were replaced by scatter search [65]. All these approaches were found to be very efficient multi-objective optimizers, and seem particularly suitable for real-world applications in which the use of surrogates is not appropriate.

In further related work [63], the same authors compared different surrogate methods (namely, artificial neural networks, radial basis functions and support vector machines) coupled to a MOEA and combined the best performer of them (support vector machines) with rough sets. This sort of scheme was proposed as an alternative for dealing with multi-objective problems that are very expensive (computationally speaking).

1.5 Future Research Paths

There are a number of possible research paths in this area that are worth exploring:

- **Parallel Approaches**: Although parallel MOEAs have been used for several years [11, 48], most of the papers published in that area focus on discussing applications and normally, such papers put little emphasis on the development of innovative algorithmic designs. Nowadays, the use of grid computing and Graphics Processing Units (GPUs) opens new and promising venues for future research in this area (see for example [21, 70, 73–75, 80, 88]), particularly regarding the solution of real-world problems having computationally expensive objective functions. The incorporation of surrogate models into parallel MOEAs is another interesting topic that deserves more research and that has been only scarcely explored in the specialized literature until now (see for example [54]).
- **Hybridization**: Coupling gradient-based or direct search methods to MOEAs is another alternative way for dealing with computationally expensive problems. In recent years, several promising hybrids of this sort have been proposed (see for example [39, 40, 44, 83–85]). These approaches can also be combined with surrogates for further efficiency (see for example [86, 87]). However, the use of such hybrid approaches in real-world applications is still rare (see for example [8]). Nevertheless, this situation is expected to change as more research results in this area become available.
- **Sampling techniques**: Surrogate methods heavily rely on the sample and updating technique adopted. In many real-world applications that use surroages, latin hypercubes have been adopted for the initial sampling, with the aim of covering as much as possible of the design (i.e., decision variable) space. At later stages of the search, it may be more relevant to explore the neighborhood of a good solution (see for example [8]). However, sampling is also relevant in other approaches, such as when using small population sizes or when hybridizing a MOEA with a local search engine. Nevertheless, the impact of the sampling technique in the performance of such approaches has not been properly addressed so far, to the author's best knowledge.

1.6 Conclusions

This chapter has provided a quick overview of the most relevant research tools that are currently available to tackle computationally expensive problems using multi-objective evolutionary algorithms. Breadth has been emphasized over depth in the discussions provided herein. However, several additional references have been provided for those interested in getting an in-depth knowledge about any of the topics that have been addressed in this chapter.

One aspect that is worth mentioning is that the presence of computationally expensive objective functions is clearly not the only relevant aspect when solving real-world problems. Other issues such as scalability (in decision variable space or in objective function space, or in both), uncertainty and incorporation of user's preferences, just to name a few, have not been addressed here, mainly because of obvious space limitations. Readers interested in information about these and other relevant topics are invited to visit the EMOO repository [10], which is available at: http://delta.cs.cinvestav.mx/~ccoello/EMOO/EMOObib.html.

References

1. Best C (2009) Multi-objective cultural algorithms. Master's thesis. Wayne State University, Detroit, Michigan, USA
2. Best C, Che X, Reynolds RG, Liu D (2010) Multi-objective cultural algorithms. In: 2010 IEEE congress on evolutionary computation (CEC'2010), Barcelona, Spain, 18–23 July 2010, IEEE Press, pp 3330–3338
3. Nicola B, Naujoks B, Emmerich M (2007) SMS-EMOA: multiobjective selection based on dominated hypervolume. Eur J Oper Res 181(3):1653–1669
4. Chen J-H, Goldberg DE, Ho S-Y, Sastry K (2002) Fitness inheritance in multi-objective optimization. In: Langdon WB, Cantú-Paz E, Mathias K, Roy R, Davis D, Poli R, Balakrishnan K, Honavar V, Rudolph G, Wegener J, Bull L, Potter MA, Schultz AC, Miller JF, Burke E, Jonoska N (eds) Proceedings of the genetic and evolutionary computation conference (GECCO'2002), Morgan Kaufmann Publishers, San Francisco, California, July 2002, pp 319–326
5. Chen Y, Ma Y, Lu Z, Qiu L, He J (2011) Terahertz spectroscopic uncertainty analysis for explosive mixture components determination using multi-objective micro-genetic algorithm. Adv Eng Softw 42(9):649–659
6. Chiba K, Obayashi S, Nakahashi K, Morino H (2005) High-fidelity multidisciplinary design optimization of wing shape for regional jet aircraft. In: Coello Coello CA, Aguirre AH, Zitzler E (eds) Evolutionary multi-criterion optimization. Third international conference, EMO 2005, Guanajuato, México, March 2005, Lecture notes in computer science. Springer, Heidelberg, pp 621–635
7. Chung C-J, Reynolds RG (1998) CAEP: an evolution-based tool for real-valued function optimization using cultural algorithms. J Artif Intell Tools 7(3):239–292
8. Chung HS (2004) Multidisciplinary design optimization of supersonic business jets using approximation model-based genetic algorithms. PhD thesis, Department of Aeronautics and Astronautics, Stanford University, California, USA, March 2004
9. Chung H-S, Alonso JJ (2004) Multiobjective optimization using approximation model-based genetic algorithms. In: Proceedings of the 10th AIAA/ISSMO symposium on multidisciplinary analysis and optimization, Albany, New York, USA, September 2004. Paper AIAA-2004-4325
10. Coello Coello CA (2006) The EMOO repository: a resource for doing research in evolutionary multiobjective optimization. IEEE Comput Intell Mag 1(1):37–45
11. Coello Coello CA, Lamont GB, Van Veldhuizen DA (2007) Evolutionary algorithms for solving multi-objective problems, 2nd edn. Springer, New York. ISBN 978-0-387-33254-3
12. Coello Coello CA, Becerra RL (2003) Evolutionary multiobjective optimization using a cultural algorithm. In: 2003 IEEE swarm intelligence symposium proceedings, Indianapolis, Indiana, USA, April 2003, IEEE Service Center, pp 6–13
13. Coello Coello CA, Pulido GT (2001) A micro-genetic algorithm for multiobjective optimization. In: Zitzler E, Deb K, Thiele L, Coello Coello CA, Corne D (eds) First international conference on evolutionary multi-criterion optimization. Lecture notes in computer science. Springer, Heidulberg, pp 126–140
14. Coello Coello CA, Pulido GT (2001) Multiobjective optimization using a micro-genetic algorithm. In: Spector L, Goodman ED, Wu A, Langdon WB, Voigt H-M, Gen M, Sen S, Dorigo M,

Pezeshk S, Garzon MH, Burke E (eds) Proceedings of the genetic and evolutionary computation conference (GECCO'2001). Morgan Kaufmann, San Francisco, pp 274–282
15. Deb K (2001) Multi-objective optimization using evolutionary algorithms. Wiley, Chichester ISBN 0-471-87339-X
16. Deb K, Goldberg DE (1989) An investigation of niche and species formation in genetic function optimization. In: Schaffer JD (ed) Proceedings of the third international conference on genetic algorithms, San Mateo, California, June 1989, George Mason University. Morgan Kaufmann, Burlington, pp 42–50
17. Deb K, Pratap A, Agarwal S, Meyarivan T (2002) A fast and elitist multiobjective genetic algorithm: NSGA-II. IEEE Trans Evol Comput 6(2):182–197
18. Deb K, Thiele L, Laumanns M, Zitzler E (2005) Scalable test problems for evolutionary multiobjective optimization. In: Abraham A, Jain L, Goldberg R (eds) Evolutionary multiobjective optimization. Theoretical advances and applications. Springer, Heidelberg, pp 105–145
19. Ducheyne EI, De Baets B, De Wulf RR (2008) Fitness inheritance in multiple objective evolutionary algorithms: a test bench and real-world evaluation. Appl Soft Comput 8(1):337–349
20. Ducheyne EI, Baets BD, De Wulf R (2003) Is fitness inheritance useful for real-world applications? In: Fonseca CM, Fleming PJ, Zitzler E, Deb K, Thiele L (eds) Evolutionary multicriterion optimization. Second international conference, EMO 2003, Faro, Portugal, April 2003. Lecture notes in computer science. Springer, Heidelberg, pp 31–42
21. Grzegorz E, Wojciech K, Brdys MA (2008) Grid implementation of a parallel multiobjective genetic algorithm for optimized allocation of chlorination stations in drinking water distribution systems: Chojnice case study. IEEE Trans Syst Man Cybern Part C Appl Rev 38(4):497–509 (July 2008)
22. Cabrera JCF, Coello Coello CA (2010) Micro-MOPSO: a multi-objective particle swarm optimizer that uses a very small population size. In: Nedjah N, dos Santos Coelho L, de Macedo de Mourelle L (eds) Multi-objective swarm intelligent systems. Theory & experiences. Studies in computational intelligence. Springer, Berlin, pp 83–104 ISBN 978-3-642-05164-7
23. Giannakoglou KC, Kampolis IC (2010) Multilevel optimization algorithms based on metamodel- and fitness inheritance-assisted evolutionary algorithms. In: Tenne Y, Goh C-K (eds) Computational intelligence in expensive optimization problems. Springer, Berlin, pp 61–84. ISBN 978-3-642-10700-9
24. Gonzalez LF, Périaux J, Srinivas K, Whitney EJ (2006) A generic framework for the design optimisation of multidisciplinary uav intelligent systems using evolutionary computing. In: AIAA paper 2006-1475, 44th AIAA aerospace science meeting and exhibit, Reno, Nevada, 9–12 January 2006
25. Simon H, Barone L, While L, Hingston P (2005) A scalable multi-objective test problem toolkit. In: Coello Coello CA, Aguirre AH, Zitzler E (eds) Evolutionary multi-criterion optimization. Third international conference,EMO 2005, Guanajuato, México, Mar 2005, Lecture notes in computer science. Springer, Heidelberg, pp 280–295
26. Huband S, Hingston P, Barone L, While L (2006) A review of multiobjective test problems and a scalable test problem toolkit. IEEE Trans Evol Comput 10(5):477–506
27. Jain AK, Murty MN, Flynn PJ (1999) Data clustering: a review. ACM Comput Surv 31(3):264–323
28. Jin X, Reynolds RG (1999) Using knowledge-based evolutionary computation to solve nonlinear constraint optimization problems: a cultural algorithm approach. In: 1999 congress on evolutionary computation, Washington, DC, July 1999, IEEE Service Center, pp 1672–1678
29. Jin Y (2005) A comprehensive survey of fitness approximation in evolutionary computation. Soft Comput 9(1):3–12
30. Kim Y, Gotoh K, Toyosada M, Park J (2002) Micro-genetic algorithms (μGAs) for hard combinatorial optimisation problems. In: The 12th international offshore and polar engineering conference 2002 (ISOPE 2002), Kitakyushu, Japan, 26–31 May 2002. International society of offshore and polar engineers, pp 230–236
31. Joshua K, David C (2003) Properties of an adaptive archiving algorithm for storing nondominated vectors. IEEE Trans Evol Comput 7(2):100–116

32. Knowles J, Nakayama H (2008) Meta-modeling in multiobjective optimization. In: Branke J, Deb K, Miettinen K, Slowinski R (eds) Multiobjective optimization. Interactive and evolutionary approaches. Lecture notes in computer science. Springer, Berlin, pp 245–284
33. Knowles JD, Corne DW (2000) Approximating the nondominated front using the pareto archived evolution strategy. Evol Comput 8(2):149–172
34. Krishnakumar K (1989) Micro-genetic algorithms for stationary and non-stationary function optimization. SPIE Proc Intell Control Adapt Syst 1196:289–296
35. Becerra RL, Coello Coello CA (2005) Optimization with constraints using a cultured differential evolution approach. In: Beyer H-G et al (eds) Genetic and evolutionary computation conference (GECCO'2005). ACM, Washington, DC, pp 27–34. ISBN 1-59593-010-8
36. Becerra RL, Coello Coello CA (2006) Solving hard multiobjective optimization problems using ε-constraint with cultured differential evolution. In: Runarsson TP, Beyer H-G, Edmund B, Merelo-Guervós JJ, Whitley LD, Yao X (eds) Parallel problem solving from nature—PPSN IX, 9th international conference, Reykjavik, Iceland, September 2006. Lecture notes in computer science. Springer, Heidelberg, pp 543–552
37. Landa-Becerra R, Santana-Quintero LV, Coello Coello CA (2008) Knowledge incorporation in multi-objective evolutionary algorithms. In: Ghosh A, Dehuri S, Ghosh S (eds) Multi-objective evolutionary algorithms for knowledge discovery from data bases. Springer, Berlin, pp 23–46
38. Langer H, Pühlhofer T, Baier H (2004) A multi-objective evolutionary algorithm with integrated response surface functionalities for configuration optimization with discrete variables. In: AIAA paper 2004–4326, 10th AIAA/ISSMO symposium on multidisciplinary analysis and optimization conference, Albany, New York, 30 August–1 September 2004
39. Lara A, Sanchez G, Coello Coello CA, Schütze O (2010) Hcs: a new local search strategy for memetic multi-objective evolutionary algorithms. IEEE Trans Evol Comput 14(1):112–112
40. Lara A, Schütze O, Coello Coello CA (2013) On gradient-based local search to hybridize multi-objective evolutionary algorithms. In: Tantar E, Tantar A-A, Bouvry P, Moral PD, Legrand P, Coello Coello CA, Schütze O (eds) EVOLVE -A bridge between probability, set oriented numerics and evolutionary computation, chapter 9. Studies in computational intelligence. Springer, Heidelberg, pp 305–332. ISBN 978-3-642-32725-4
41. Lee DS, Gonzalez LF, Srinivas K, Periaux J (2007) Multi-objective robust design optimisation using hierarchical asynchronous parallel evolutionary algorithms. In: AIAA paper 2007-1169, 45th AIAA aerospace science meeting and exhibit, Reno, Nevada, 8–11 January 2007
42. Lee DS, Gonzalez LF, Periaux J, Srinivas K (2008) Robust design optimisation using multi-objective evolutionary algorithms. Comput Fluids 37:565–583
43. Lim D, Jim Y, Ong Y-S, Bernhard S (2010) Generalizing surrogate-assisted evolutionary computation. IEEE Trans Evol Comput 14(3):329–355
44. López AL, Coello Coello CA, Schuetze O (2010) A painless gradient-assisted multi-objective memetic mechanism for solving continuous bi-objective optimization problems. In: 2010 IEEE Congress on evolutionary computation (CEC'2010), Barcelona, Spain, 18–23 July 2010, IEEE Press, pp 577–584
45. Loshchilov I, Schoenauer M, Sebag M (2010) Dominance-based pareto-surrogate for multi-objective optimization. In: Deb K, Bhattacharya A, Chakraborti N, Chakroborty P, Das S, Dutta J, Gupta SK, Jain A, Aggarwal V, Branke J, Louis SJ, Tan KC (eds) Simulated evolution and learning, 8th international conference, SEAL 2010, Kanpur, India, 1–4 December 2010. Lecture notes in computer science, vol. 6457. Springer, Heidelberg, pp 230–239
46. Mendoza JE, López ME, Coello Coello CA, López EA (2009) Microgenetic multiobjective reconfiguration algorithm considering power losses and reliability indices for medium voltage distribution network. IET Gener Transm Distrib 3(9):825–840
47. Mendoza J, Morales D, López R, López E, Vannier J-C, Coello Coello CA (2007) Multi-objective location of automatic voltage regulators in a radial distribution network using a micro genetic algorithm. IEEE Trans Power Syst 22(1):404–411
48. Nebro AJ, Luna F, Talbi E-G, Alba E (2005) Parallel multiobjective optimization. In: Alba E (ed) Parallel metaheuristics. Wiley, New Jersey, pp 371–394. ISBN 13-978-0-471-67806-9

49. Pawlak Z (1982) Rough sets. Int J Comput Inform Sci 11(1):341–356
50. Pawlak Z (1991) Rough sets: theoretical aspects of reasoning about data. Kluwer Academic, Dordrecht. ISBN 0-471-87339-X
51. Pilát M, Neruda R (2012) An evolutionary strategy for surrogate-based multiobjective optimization. In: 2012 IEEE congress on evolutionary computation (CEC'2012), Brisbane, Australia, 10–15 June 2012, IEEE Press, pp 866–872
52. Pilato C, Palermo G, Tumeo A, Ferrandi F, Sciuto D, Lanzi PL (2007) Fitness inheritance in evolutionary and multi-objective high-level synthesis. In: 2007 IEEE congress on evolutionary computation (CEC'2007), Singapore, September 2007, IEEE Press, pp 3459–3466
53. Ray T, Isaacs A, Smith W (2009) Surrogate assisted evolutionary algorithm for multi-objective optimization. In: Pandu RG (ed) Multi-objective optimization techniques and applications in chemical engineering. World Scientific, Singapore, pp 131–152
54. Ray T, Smith W (2006) A surrogate assisted parallel multiobjective evolutionary algorithm for robust engineering design. Eng Optim 38(8):997–1011
55. Reyes-Sierra M, Coello Coello CA (2006) Dynamic fitness inheritance proportion for multi-objective particle swarm optimization. In: Keijzer M et al (ed) 2006 Genetic and evolutionary computation conference (GECCO'2006), vol. 1, Seattle, Washington, July 2006. ACM Press, pp 89–90. ISBN 1-59593-186-4
56. Sierra MMR (2006) Use of coevolution and fitness inheritance for multiobjective particle swarm optimization. PhD thesis, Computer science section, Department of Electrical Engineering, CINVESTAV-IPN, Mexico, August 2006
57. Reynolds RG (1994) An introduction to cultural algorithms. In: Sebald AV, Fogel LJ (eds) Proceedings of the third annual conference on evolutionary programming. World Scientific, New Jersey, pp 131–139
58. Reynolds RG, Chung C-J (1997) A cultural algorithm framework to evolve multi-agent cooperation with evolutionary programming. In: Ep '97: Proceedings of the 6th international conference on evolutionary programming VI. Springer, Heidelberg, pp 323–334
59. Reynolds RG, Chung C-J (1997) Knowledge-based self-adaptation in evolutionary programming using cultural algorithms. In: Proceedings of 1997 IEEE international conference on evolutionary computation (ICEC 97), pp 71–76, 1997
60. Reynolds RG, Michalewicz Z, Cavaretta M (1995) Using cultural algorithms for constraint handling in GENOCOP. In: McDonnell JR, Reynolds RG, Fogel DB (eds) Proceedings of the fourth annual conference on evolutionary programming. MIT Press, Cambridge, pp 298–305
61. Ribas PC, Yamamoto L, Polli HL, Arruda LVR, Neves-Jr F (2013) A micro-genetic algorithm for multi-objective scheduling of a real world pipeline network. Eng Appl Artif Intell 26(1):302–313
62. Santana-Quintero LV, Montano AA, Coello Coello CA (2010) A review of techniques for handling expensive functions in evolutionary multi-objective optimization. In: Tenne Y, Goh C-K (eds) Computational intelligence in expensive optimization problems. Springer, Berlin, pp 29–59. ISBN 978-3-642-10700-9
63. Santana-Quintero LV, Coello Coello CA , Hernández-Díaz AG (2008) Hybridizing surrogate techniques, rough sets and evolutionary algorithms to efficiently solve multi-objective optimization problems. In: 2008 Genetic and evolutionary computation conference (GECCO'2008), Atlanta, July 2008. ACM Press, pp 763–764. ISBN 978-1-60558-131-6
64. Santana-Quintero LV, Hernández-Díaz AG, Molina J, Coello Coello CA, Caballero R (2010) DEMORS: a hybrid multi-objective optimization algorithm using differential evolution and rough sets for constrained problems. Comput Oper Res 37(3):470–480
65. Santana-Quintero LV, Ramírez N, Coello Coello C (2006) Multi-objective particle swarm optimizer hybridized with scatter search. In: Gelbukh A, Reyes-Garcia V (eds) MICAI 2006: advances in artificial intelligence, 5th mexican international conference on artificial intelligence, November 2006. Lecture notes in artificial intelligence, vol 4293. Springer, Mexico, pp 294–304
66. Santana-Quintero LV, Ramírez-Santiago N, Coello Coello CA (2008) Towards a more efficient multi-objective particle swarm optimizer. In: Bui LT, Alam S (eds) Multi-objective optimization

in computational intelligence: theory and practice. Information Science Reference, Hershey, pp 76–105. ISBN 978-1-59904-498-9
67. Santana-Quintero LV, Ramírez-Santiago N, Coello Coello CA, Luque JM, Hernández-Díaz AG (2006) A new proposal for multiobjective optimization using particle swarm optimization and rough sets theory. In: Runarsson TP, Beyer H-G, Burke E, Merelo-Guervós JJ, Whitley LD, Yao X (eds) Parallel problem solving from nature-PPSN IX, 9th international conference, Reykjavik, Iceland, September 2006. Lecture notes in computer science, vol 4193. Springer, Heidelberg, pp 483–492
68. Sasaki D, Obayashi S, Nakahashi K (2002) Navier-stokes optimization of supersonic wings with four objectives using evolutionary algorithm. J Aircr 39(4):621–629
69. Sastry K, Goldberg DE, Pelikan M (2001) Don't evaluate, inherit. Proceedings of genetic and evolutionary computation conference. Morgan Kaufmann, Burlington, pp 551–558
70. Sharma D, Collet P (2010) GPGPU-compatible archive based stochastic ranking evolutionary algorithm (G-ASREA) for multi-objective optimization. In: Schaefer R, Cotta C, Kołodziej J, Rudolph G (eds) Parallel problem solving from nature-PPSN XI, 11th International conference, proceedings Part II, September 2010. Lecture Notes in Computer Science, vol 6239. Springer, Poland, pp 111–120
71. Smith RE, Dike BA, Stegmann SA (1995) Fitness inheritance in genetic algorithms. In: SAC '95: proceedings of the 1995 ACM symposium on applied computing, New York, NY, 1995, ACM Press, pp 345–350
72. Szollos A, Smid M, Hajek J (2009) Aerodynamic optimization via multi-objective micro-genetic algorithm with range adaptation, knowledge-based reinitialization, crowding and epsilon-dominance. Adv Eng Softw 40(6):419–430
73. Tagawa K, Shimizu H, Nakamura H (2011) Indicator-based differential evolution using exclusive hypervolume approximation and parallelization for multi-core processors. In: 2011 Genetic and evolutionary computation conference (GECCO'2011), Dublin, Ireland, 12–16 July 2011. ACM Press, pp 657–664
74. Talbi E-G, Cahon S, Melab N (2007) Designing cellular networks using a parallel hybrid metaheuristic on the computational grid. Comput Commun 30(4):498–713
75. Khaled Ahsan Talukder AKM, Kirley M, Buyya R (2009) Multiobjective differential evolution for scheduling workflow applications on global grids. Concurrency Comput-Pract Exp 21(13):1742–1756
76. Tiwari S, Fadel G, Deb K (2011) AMGA2: improving the performance of the archive-based micro-genetic algorithm for multi-objective optimization. Eng Optim 43(4):377–401
77. Tiwari S, Koch P, Fadel G, Deb K (2008) AMGA: An archive-based micro genetic algorithm for multi-objective optimization. In: 2008 Genetic and evolutionary computation conference (GECCO'2008), Atlanta, July 2008. ACM Press, pp 729–736. ISBN 978-1-60558-131-6
78. Pulido GT, Coello Coello CA (2003) The micro genetic algorithm 2: towards online adaptation in evolutionary multiobjective optimization. In: Fonseca CM, Fleming PJ, Zitzler E, Deb K, Thiele L (eds) Evolutionary multi-criterion optimization. Second international conference, EMO 2003, April 2003. Lecture notes in computer science, vol 2632. Springer, Portugal, pp 252–266
79. Pulido GT, Coello Coello CA (2004) Using clustering techniques to improve the performance of a particle swarm optimizer. In: Deb K et al (eds) Genetic and evolutionary computation-GECCO 2004. Proceedings of the genetic and evolutionary computation conference, part I, Seattle, Washington, June 2004. Lecture notes in computer science, vol 3102. Springer, Heidelberg, pp 225–237
80. Van Luong T, Melab N, Talbi E-G (2011) GPU-based approaches for multiobjective local search algorithms. A case study: the flowshop scheduling problem. In: Merz P, Hao J-K (eds) Evolutionary computation in combinatorial optimization, 11th European conference, EvoCOP 2011, Torino, Italy, 27–29 April 2011. Lecture notes in computer science, Vol. 6622. Springer, Heidelberg, pp 155–166
81. Voutchkov I, Keane AJ, Fox R (2006) Robust structural design of a simplified jet engine model, using multiobjective optimization. In: AIAA Paper 2006–7003, Portsmouth, Virginia, 6–8 September 2006

82. Wang Y-N, Wu L-H, Yuan X-F (2010) Multi-objective self-adaptive differential evolution with elitist archive and crowding entropy-based diversity measure. Soft Comput 14(3):193–209
83. Martínez SZ, Coello Coello CA (2008) A proposal to hybridize multi-objective evolutionary algorithms with non-gradient mathematical programming techniques. In: Rudolph G, Jansen T, Lucas S, Poloni C, Beume N (editors) Parallel problem solving from nature-PPSN X, Dortmund, Germany, September 2008. Lecture notes in computer science, Vol. 5199. Springer, Heidelberg, pp 837–846
84. Martínez SZ, Coello Coello CA (2012) A direct local search mechanism for decomposition-based multi-objective evolutionary algorithms. In: 2012 IEEE congress on evolutionary computation (CEC'2012), Brisbane, Australia, 10–15 June 2012. IEEE Press, pp 3431–3438
85. Martínez SZ, Coello Coello CA (2013) A hybridization of MOEA/D with the nonlinear simplex search algorithm. In: Proceedings of the 2013 IEEE symposium on computational intelligence in multicriteria decision making (MCDM'2013), Singapore, 16–19 April 2013. IEEE Press, pp 48–55
86. Martínez SZ, Coello Coello CA (2013) Combining surrogate models and local search for dealing with expensive multi-objective optimization problems. In: 2013 IEEE congress on evolutionary computation (CEC'2013), Cancún, México, 20–23 June 2013. IEEE Press, pp 2572–2579
87. Martínez SZ, Coello Coello CA (2013) MOEA/D assisted by RBF networks for expensive multi-objective optimization problems. In: 2013 Genetic and evolutionary computation conference (GECCO'2013), New York, July 2013. ACM Press, pp 1405–1412. ISBN 978-1-4503-1963-8
88. Zhu W, Yaseen A, Li Y (2011) DEMCMC-GPU: an efficient multi-objective optimization method with gpu acceleration on the fermi architecture. New Generation Comput 29(2):163–184
89. Zitzler E, Deb K, Thiele L (2000) Comparison of multiobjective evolutionary algorithms: empirical results. Evol Comput 8(2):173–195

Chapter 2
Hybrid Optimization Algorithms and Hybrid Response Surfaces

George S. Dulikravich and Marcelo J. Colaço

Abstract In this paper we will present some hybrid methodologies applied to optimization of complex systems. The paper is divided in two parts. The first part presents several automatic switching concepts among constituent optimizers in hybrid optimization, where different heuristic and deterministic techniques are combined to speed up the optimization task. In the second part, several high dimensional response surface generation algorithms are presented, where some very basic hybridization concepts are introduced.

Keywords Multiobjective optimization · Response surfaces · Hybrid optimizers

2.1 Introduction

Design of complex nonlinear engineering systems usually requires a large computational effort in the case of simulation, or a large amount of human and experimental resources in the case of experiments. Multi-dimensional topology of the objective function space of such problems has multiple local minima and large domains of possible variations of the design variables search space. A typical approach to finding the global minimum is to start with a large search space utilizing an entire population of initial guesses and advancing them simultaneously using any of the evolutionary optimization algorithms. Once the search space has narrowed sufficiently, the search process is switched to a fast and accurate gradient-based search algorithm to converge on the minimum. However, this simplistic semi-manual approach to sequential

G.S. Dulikravich (✉)
Multidisciplinary, Inverse Design, Robust Optimization and Control (MAIDROC) Laboratory, Department of Mechanical and Materials Engineering, Florida International University, EC 3462, 10555 West Flagler St, Miami, FL 33174, USA
e-mail: dulikrav@fiu.edu

M.J. Colaço
Department of Mechanical Engineering, Biofuels and Energy Efficiency Program - Brazilian National Agency of Oil, Gas and Biofuels, Federal University of Rio de Janeiro, Cx. Postal 68503, Rio de Janeiro, RJ 21941-972, Brazil
e-mail: colaco@ufrj.br

D. Greiner et al. (eds.), *Advances in Evolutionary and Deterministic Methods for Design, Optimization and Control in Engineering and Sciences*, Computational Methods in Applied Sciences 36, DOI 10.1007/978-3-319-11541-2_2

hybrid optimization is not reliable since it utilizes only one evolutionary optimizer and one gradient-based optimizer, each of which has its own intrinsic deficiencies. A more robust and faster hybrid optimization approach utilizes a collection of several evolutionary optimizers and several gradient-based optimizers and automatically switches among them. This chapter will focus on these types of hybrid optimizers.

When systems having large number of design variables, objective functions and constraints need to be optimized, this implies the evaluation of thousands and even millions of candidate solutions, which can make this task impossible from a very practical point of view, especially if each such high fidelity evaluation of the objective function is time-consuming or expensive. Thus, it is important to develop surrogate models, also called metamodels, which approximate the response of the original problem, but using a much simpler mathematical formulation. The objective of this chapter is to present several common response surface models existing in the literature, and some hybridization among them. Some hybrid optimizers are also presented, based on heuristic and deterministic methods, which take advantage of these hybrid response surface models to improve the efficiency of the optimization task of complex engineering systems.

2.2 Hybrid Optimization Algorithm Concepts

It is well known that each optimization algorithm provides a unique approach to optimization with varying degrees of convergence, reliability, and robustness at different stages during the iterative process. The "no free lunch theorem" states [1] that no individual optimization algorithm is better than all the other optimization algorithms for all classes of optimization problems (Fig. 2.1). A natural response to this problem is to use hybrid optimization algorithms that combine individual constituent optimization algorithms in a sequential or parallel manner so that resulting software can utilize the specific advantages of each constituent algorithm. That is, a variety of individual constituent optimization algorithms that rely on different principles of operation are combined in a hybrid optimization algorithm as subroutines where a set of specified heuristic measures of the iterative convergence process is used to perform automatic switching among the constituent algorithms. This allows for automatic use of the most appropriate constituent optimization algorithm at each step of the global iterative optimization process. The automatic back-and-forth switching [2] among the constituent optimization algorithms can also be viewed as an automatic backup strategy so that, if one optimization algorithm fails, another optimization algorithm can automatically take over.

The key to the success of this hybrid optimization concept is the automatic switching strategy [2, 3] among the constituent optimization algorithms. One of the early single-objective hybrid optimization algorithms [4, 5] had three gradient-based (Davidon-Fletcher-Powell algorithm, Sequential Quadratic Programming and quasi-Newton algorithm of Pshenichny-Danilin) and three non-gradient-based (Genetic Algorithm, Nelder-Mead simplex algorithm, and Differential Evolution

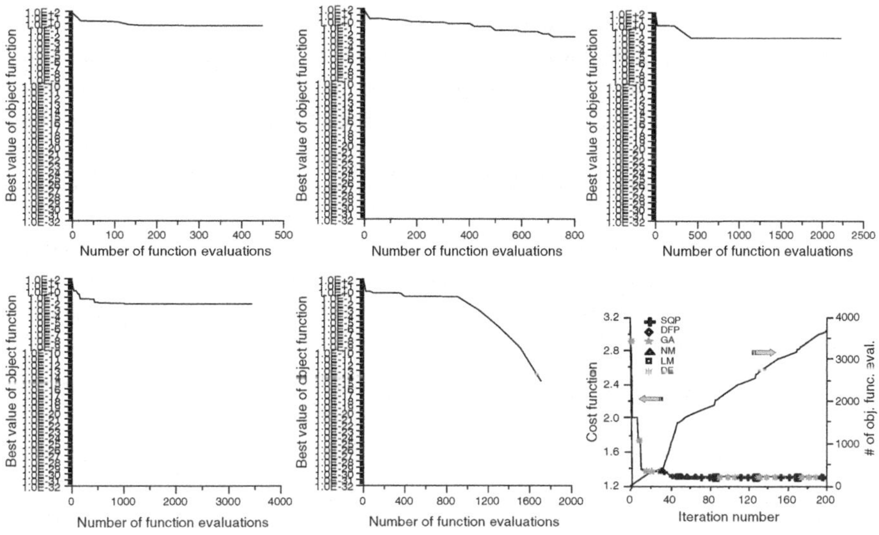

Fig. 2.1 Convergence histories for Griewank's test function no. 8 using (from *left* to *right*) individual: BFGS algorithm, differential evolution, simulated annealing, particle swarm, and our hybrid optimizer [27]. The last figure illustrates the convergence history of one hybrid optimizer using automatic switching process [5]

algorithm) constituent optimization algorithms that were automatically switching back-and-forth each time when a particular heuristic prescribed convergence measure was reached [5].

This hybrid single-objective optimizer only restarts with a single design (the "best" from the "previous" iteration). In other words, when switching from one of the population-based constituent optimizers to a gradient-based constituent optimizer, only the best design from that population, and not the entire population, is used as the initial guess for the gradient-based constituent algorithm.

For population-based constituent optimizers used in this hybrid optimizer, the population matrix was updated every iteration with new designs and ranked according to the value of the objective function. The optimization problem was completed when: (1) the maximum number of iterations or objective function evaluations was exceeded, or (2) the best design in the population was equivalent to a target design, or (3) the optimization program tried all four algorithms, but failed to produce a decrease in the objective function.

Another hybrid single-objective optimization algorithm was developed by combining three of the fastest gradient-based and evolutionary optimization algorithms [5], namely: the Broyden-Fletcher-Goldfarb-Shanno quasi-Newton algorithm, the Differential Evolution algorithm, and the Particle Swarm algorithm. It was found that the most often automatically used constituent optimization module is the Particle Swarm algorithm. When a certain percentage of the particles find a minimum, the algorithm switches automatically to the Differential Evolution algorithm and the particles are forced to breed. If there is an improvement in the objective function, the algorithm returns to the Particle Swarm method, meaning that some other region is more likely to have a global minimum. If there is no improvement in the objective function value, this can indicate that this region already contains the global value expected and the algorithm automatically switches to the Broyden-Fletcher-Goldfarb-Shanno algorithm in order to quickly and accurately find the location of the minimum.

One of the most recent switching algorithms is the Search Vector-based Hybrid (SVH) [2] which automatically changes search directions during the optimization process. During each iteration, the SVH will generate the SVs based on a predetermined formula or quality. Some examples of the SVs include:

1. Global Best vector (GB) which is the fittest design vector currently in the population.
2. Population Weighted Average vector (PWA): The population is ranked from best to worst, with the best receiving a rank equal to the population size, and the worst having a rank of one. The ranks are then used as weights, and the standard weighted arithmetic mean procedure is used to create this SV.

After the SVs have been evaluated, the fittest SV is selected as the SV for that iteration. Once the SV has been selected, the constituent algorithm selection process begins. First, each constituent algorithm is executed so that it generates a temporary small population. This temporary population will not be evaluated. Instead, it will be used as an indication of the behavior of the constituent algorithm for a given topology. For example, suppose the SVH has two constituent algorithms called CA1 and CA2 where CA1 will use the current population to generate a temporary population which will be situated in one part of the space of design variables, while CA2 will create a temporary population shifted to another part of the space of design variables. In order to select the most appropriate constituent optimization algorithm, the Euclidean distance between the endpoint of each centroid vector of the temporary populations and the selected SV is calculated and stored. Then, each centroid is evaluated. The constituent algorithms are then ranked using the Pareto dominance scheme based on two objectives: (1) minimize distance between the centroid and the SV, and (2) minimum objective function value of the centroid. The constituent algorithm to be used is randomly selected from the Pareto front. In order for the centroids of the temporary populations for the constituent algorithms to be statistically meaningful, the constituent algorithms are executed $10\times$ each iteration. Once a constituent algorithm has been selected, it is then executed one last time. This time, the population is permanently changed, and the objective function for each design vector is evaluated. This completes one full iteration of the SVH. This strategy [2] differs from any other known work in that it uses a collection of different search directions, each with its own unique formulation, and chooses among them. The method presented by Ahrari et al. [6], like most other hybrid algorithms, generates the search direction and keeps it fixed throughout the entire optimization process.

In multi-objective optimization it would be onerous to use a single value and compare the quality of one Pareto approximation to another [7]. Instead, multiple attributes of two successive Pareto approximations should be considered to determine if a multi-objective optimizer is converging on non-dominated set. Since "no free lunch theorem" applies to multi-objective optimizers as well as single-objective optimizers, hybrid optimization is highly advisable in the multi-objective optimization problems. Such hybridization can be implemented by using a suite of multi-objective optimization algorithms in the form of either High-level Relay Hybrid (HRH) algorithms where each of the constitutive algorithms run on its own in a sequential non-parallelized scheme, or as High-level Teamwork Hybrid (HTH) metaheuristic algorithms [8] where constitutive optimization algorithms run in parallel and contribute a portion of each new generation's population. The portion that each search contributes to the new generation is dependent on the success of the algorithm to provide past useful solutions to the search. One such HTH algorithm is AMALGAM [9], which utilizes NSGA-II [7] and outperforms NSGA-II.

A robust and accurate HRH type concept is Multi-Objective Hybrid Optimizer (MOHO) [10] which currently uses three multi-objective optimization algorithms: Strength Pareto Evolutionary Algorithm (SPEA-2), a multi-objective implementation of the single objective Particle Swarm, and a Non-Sorting Differential Evolution (NSDE) algorithm which is a low level hybrid metaheuristic search combining NSGA-II [7] and Differential Evolution. MOHO starts by creating the population that will be used for the optimization run. The population contains the decision vector and the objective vector for all population points and stores the Pareto approximation and clustering routine. Clustering is performed by the population on the object vectors of the Pareto approximation. The decision vectors of all population points are evenly distributed over the decision space using Sobol's pseudo random sequence generator [11]. The software then passes the population from optimization routine to optimization routine as the switching criteria dictates. The constitutive algorithm that is selected at each generation makes a new generation using any or all of the information provided to it: the last generation's population and the latest non-dominated set. Then MOHO combines the new generation and the latest non-dominated set to create a new non-dominated set. The switching algorithm compares the non-dominated set from the current generation to the non-dominated set of the previous generation. The comparison process consists of looking at five desired improvements to the Pareto approximation [10]. The improvements are actually gains in five performance criteria (quality factors). If the particular search algorithm can achieve at least two of any of the five specified improvements [10], this algorithm is allowed to create the next generation. The five criteria (aspects) are:

1. The new population changes the number of points in the Pareto set. When this happens, either points are being added to the approximation, or, more importantly, a new point is found that causes points to be deleted from the Pareto set.
2. The new population has at least one point that dominates a point, or points, in the current Pareto approximation. This means that the Pareto approximation is being improved.
3. The hyper volume of the dominated space changes. When the optimization software starts, it picks a worst case objective vector from its initial population guess.

At each constituent optimization algorithm iteration, hyper cubes are created, with one vertex of the diagonal being a point on Pareto approximation and the other vertex of the diagonal being the worst case objective vector. Then the volume of the union of all the hyper cubes is calculated. The union is defined as the Boolean union of the cubes; in the same sense as this operation is performed in Constructive Solid Geometry for CAD applications. When this occurs, the Pareto approximation is changing geometry.

4. The new population generation causes the average distance of Pareto approximation from the objective space origin to change. This also denotes a change on the Pareto approximation geometry. This is a backup to criterion (3) where two approximations may have the same volume, but different average distances.
5. The new population causes the maximum spread of the Pareto approximation to increase. The formula for calculating the spread developed by Zitzler is shown by Deb [7].

At the end of each iteration, the population of design vectors assigns itself a grade point for each of the above criteria that its new generation meets. If the new population earns a grade of 2 or more, the current optimization routine is allowed to continue running. When the grade falls below 2, the software switches to the next optimizer in its repository. If the grade is 0 or 1, the reason to switch to another constituent optimizer is because the currently used constituent algorithm is not contributing to improving the Pareto approximation. As an example, the population gives itself a grade of 1 because it meets criterion 4. This change could be caused just by clustering of the previous Pareto approximation and the new population. While this type of change in the Pareto approximation has its uses, it has been found that the multi-objective routines used here can cause these kind of changes to the Pareto approximation *ad infinitum*, when the Pareto approximation is very near the actual non-dominated set of the objective space. It has been found that by enforcing at least 2 of the criteria, these situations are avoided.

The other limiting factor on how many consecutive iterations a given optimization routine can run is the sub-iteration limit. Although a routine may be able to score a grade of at least 2 indefinitely, for each new generation, there may be an optimization method available that can do a better job. For this reason, each constituent optimizer is limited to a user defined maximum sub-iteration limit. This limit gives all the constituent optimizers a chance to run.

2.3 Hybrid Response Surface Generation Concepts

Optimization of systems with a large number of design variables where the objective function is given in a pointwise fashion requires creation of a hypersurface that fits through the given values of the objective function in a multi-dimensional space where dimensionality corresponds to the number of the design variables. It is well known that the locations of the training points are crucial for the proper construction of the response surface. When we are given a freedom to choose the locations of the support points of a multi-dimensional response surface, a typical approach is to use *Design of Experiments* (DoE) for this purpose. For high dimensional problems, strategies such as Latin Hypercube Sampling [11], Sobol quasi-random sequences of numbers [12], and a variety of *de facto* random number generators are most often used. However, when we do not have freedom to choose the number and the locations of the support points for generation of a response surface, all existing methods for generating response surfaces have serious problems with accuracy and robustness. This is mainly because arbitrary data sets provide inadequate uniformity of coverage of space of the design variables and clustering of the support points that leads to spurious oscillations of the response surfaces.

2.3.1 Polynomial Regression

The use of polynomial regression is one of the earliest attempts to generate response surface models [13, 14]. The idea is to approximate an unknown function $f(\mathbf{x})$ by an approximation $s(\mathbf{x})$ in the following general form

$$\begin{aligned} f(\mathbf{x}) \approx s(\mathbf{x}) = a_0 + \sum_{i=1}^{n} a_i x_i + \sum_{i=1}^{n} \sum_{j \le i}^{n} a_{ij} x_i x_j \\ + \sum_{i=1}^{n} \sum_{j \le i}^{n} \sum_{k \le j}^{n} a_{ijk} x_i x_j x_k + \cdots \end{aligned} \tag{2.1}$$

where n is the number of dimensions of the problem. Notice that we can write Eq. (2.1) as

$$\mathbf{f} = \mathbf{X}\mathbf{a} + \varepsilon \tag{2.2}$$

where ε is an approximation error with zero mean and variance, σ^2. If the functions are given at certain known locations, the unknowns $\mathbf{a}$ can be found by least squares

$$\mathbf{a} = \left(\mathbf{X}^T\mathbf{X}\right)^{-1} \mathbf{X}^T\mathbf{f} \tag{2.3}$$

It is well known that the locations of the training points are crucial for the proper construction of the response surface. Such choice is known in the literature as *Design of Experiments* (DoE). For low dimension problems, the classical method of choice was the factorial design [13], which is not practical for high dimensional problems. Other strategies for high dimensions include the Latin Hypercube Sampling [11], Sobol quasi-random numbers [12], etc.

Although the polynomial regression technique seems very attractive in view of its simplicity, it is not practical when the number of dimensions of the problem becomes very high. In this case, there are other relatively recent techniques more appropriate.

Among those techniques, the most used are Kriging [15], Radial Basis Functions [16], and Neural Networks [17], among others.

2.3.2 Self Organizing Algorithms [19, 20]

Self-organizing algorithms come from the field of cybernetics [18]. The idea is that the program "learns" the black box model as it is trying to mimic and makes the response surface as complex as is required. By allowing the program to gradually complicate the final model, the construction and evaluation time of the surrogate model is automatically optimized for a given task. The black box model is the test function being used to evaluate the RSM method.

The self-organizing algorithm presented here [19], is the multilayer algorithm where the design variables are permutated, in pairs, to form nodes. At each node a least squares regression is performed using the two variables input to the nodes. These are variable vectors that are the size of the sample population. Thus, the output of the node is a vector of the predicted values from the regression. The polynomial used for the regression is a first order or second order polynomial. For instance, a second order basis polynomial would be:

$$\begin{aligned} y_{i,j}^{k,n} &= a_0^{k,n} + a_1{}^{k,n}x_i^{k,n} + a_2{}^{k,n}x_j^{k,n} + a_3{}^{k,n}x_i^{k,n}x_j^{k,n} \\ &+ a_4{}^{k,n}x_i^{k,n}x_i^{k,n} + a_5{}^{k,n}x_j^{k,n}x_j^{k,n} \quad i \neq j \end{aligned} \tag{2.4}$$

where $i, j = 1, 2, \ldots,$ number of inputs to a given layer, k is the current layer, and n is the node number at current layer. Equation (2.4) would then be a row in the Vander Monde matrix for a regression using a second order basis polynomial of two variables. The output of the node is the vector of predicted y values for the given input. The output of a node in layer $k-1$ becomes the input (provides an x_i vector) for layer k.

The notation in Eq. (2.4) is designed to inform the reader that the functions and polynomial coefficients pertain to a particular layer and particular node in the layer. The notation should also give the reader a feel for the computational resources needed to create and maintain a multilayer self-organizing model.

Figure 2.2 shows a possible multilayer surrogate model for a three variable engineering model. In the bottom layer (the zero layer) actual design variables are the nodes in the layer. These become the x inputs to layer 1. The nodes for layer 1 are created by permuting the input variables and performing a least squares fit using Eq. (2.4), and the actual responses from the actual function (that is, objective function, engineering simulation, etc.). Once layer 1 is made, layer 2 is created, but now the nodes of layer 1 provide the x's to make the new nodes using Eq. (2.4). When layer 3 is to be created, the 3rd node of layer 2 is not included. For now, we will just say that the results of that regression were not good enough to be used to make the 3rd layer. Since only two nodes from layer 2 were used to make layer 3, only one node can be created in layer 3 and the model making process ends there.

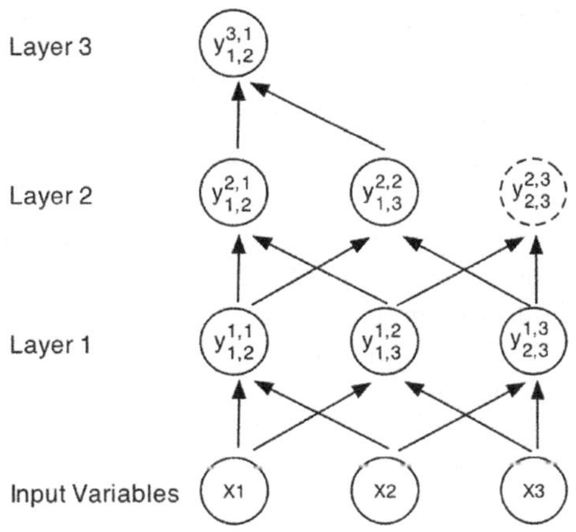

Fig. 2.2 Sample multilayer model for self-organizing response surface creation [19]

The third node in layer 2 is not used to make nodes in layer 3 in Fig. 2.2. A selection criterion is used to determine if the information in a node gets passed on to the next layer. Madalaand and Ivakhnenko [18] suggest that using the following equation is an appropriate means for checking the quality of a node and can be a selection criterion

$$\Delta^2(B) = \frac{\sum_{p \in N_B} (y - \widehat{y})_p^2}{\sum_{p \in N_B} (y_p - \bar{y})^2} \tag{2.5}$$

where y_p are the desired values, $\widehat{y}$ are the predicted values and $\bar{y}$ are the mean of the desired values. In the multilayer algorithm a threshold is set for the maximum acceptable value of Eq. (2.5). Nodes that are within the threshold are passed on to the next layer. For each new layer the threshold is made smaller. This serves to minimize the amount of nodes in each layer to only the information that is needed to improve the network. This trimming of nodes is crucial to keeping the method compact and efficient [19]. The reader is urged to follow the example in Fig. 2.2, but first setting the number of input variables to 4. The rate of the growth of layers can be very large.

The building of the multilayer network can be terminated in two ways (in practice): A) Build a predetermined number of layers and chose the node in the last layer with the best value of Eq. (2.5) to be the model output. B) Build layers until all nodes are unable to meet the threshold value, chose the best-valued node as the output of the model. Once the output node is chosen, the polynomial coefficients pertaining to all the nodes used to create the output node are stored for evaluation of the model (extraction of a predicted value).

2.3.3 Kriging

The Kriging technique was named after the initial work of the South African mining engineer Krige [20]. Details of this technique can be found in the classical papers of Sacks et al. [21, 22] and Matheron [15]. It is worth mentioning also the excellent work of Jones et al. [23] where they proposed an efficient method to global optimization. Later, Huang et al. [24] extended Jones et al. work to deal with model uncertainties. Following procedure described in the works of Jones et al. [23] and Sacks et al. [21, 22], the Kriging method starts by constructing a stochastic process model for the function as

$$f\left(\mathbf{x}^i\right) = \sum_{j=1}^{m} a_j g_j\left(\mathbf{x}^i\right) + \varepsilon^i \tag{2.6}$$

where the superscript i is used to denote the ith point in the space of design variables $\mathbf{x}$. In Eq. (2.6), g_j is a set of m linear or non-linear functions and ε is an approximation error with zero mean and variance, σ^2. In the classical Kriging model, the approximation error is supposed to be function of the design variables, such that $\varepsilon^i = \varepsilon(\mathbf{x}^i)$. Some recent works [24] also include a measurement error in Eq. (2.6), but this will not be discussed here. A usual hypothesis in this model is that if two points $\mathbf{x}^i$ and $\mathbf{x}^j$ are close, then their approximation errors $\varepsilon(\mathbf{x}^i)$ and $\varepsilon(\mathbf{x}^j)$ are also close, meaning that $\varepsilon(\mathbf{x}^i)$ and $\varepsilon(\mathbf{x}^j)$ are correlated. The correlation function between those two errors can be given as a function of the weighted distance between then [21, 22].

$$d\left(\mathbf{x}^i, \mathbf{x}^j\right) = \sum_{k=1}^{n} \theta_k \left|x_k^i - x_k^j\right|^{P_k} \qquad \theta_k \geq 0,\ P_k \in [1, 2] \tag{2.7}$$

where n is the number of dimensions of the problem. The correlation is given as

$$\mathrm{Corr}\left[\varepsilon\left(\mathbf{x}^i\right), \varepsilon\left(\mathbf{x}^j\right)\right] = \exp\left[-d\left(\mathbf{x}^i, \mathbf{x}^j\right)\right] \tag{2.8}$$

According to Jones et al. [23], such model is so powerful that we can rewrite Eq. (2.6) in terms of the mean of the stochastic process μ, such that

$$f\left(\mathbf{x}^i\right) = \mu + \varepsilon\left(\mathbf{x}^i\right) \tag{2.9}$$

Following Jones et al. [23] we can then obtain the approximate function for a new point $\mathbf{x}^*$ as

$$f\left(\mathbf{x}^*\right) = \hat{\mu} + \mathbf{r}^T\mathbf{R}^{-1}\left(\mathbf{f} - \mathbf{1}\hat{\mu}\right) \tag{2.10}$$

where $\mathbf{1}$ is vector of ones, $R_{i,j} = \mathrm{Corr}[\varepsilon(\mathbf{x}^i), \varepsilon(\mathbf{x}^j)]$, $r_i = \mathrm{Corr}[\varepsilon(\mathbf{x}^*), \varepsilon(\mathbf{x}^i)]$, $\mathbf{f}$ is the vector of functions at the known locations, and

$$\hat{\mu} = \frac{\mathbf{1}\mathbf{R}^{-1}\mathbf{f}}{\mathbf{1}^T\mathbf{R}^{-1}\mathbf{1}} \tag{2.11}$$

Kriging also predicts the mean squared error of the estimates [21–23] and this has been used as a predictor to locations where to add points in the response surface model. Locations of the domain where the mean squared error of the estimates are large, usually require the addition of extra points to increase the local accuracy.

2.3.4 Radial Basis Functions

The use of Radial Basis Functions (RBFs), initially proposed in the work of Hardy [25] on multivariate approximation, is now becoming an established approach. RBFs may be classified into two main groups:

- The globally supported ones namely the multiquadrics (MQ, $\sqrt{(x-x_j)^2+c_j^2}$, where c_j is a shape parameter), the inverse multiquadrics, thin plate splines, Gaussians, etc;
- The compactly supported ones such as the Wendland [26] family (for example, $(1-r)_+^n + p(r)$ where $p(r)$ is a polynomial and $(1-r)_+^n$ is 0 for r greater than the support).

Let us suppose that we have a function of L variables x_i, $i = 1, \ldots, L$. One possible RBF approximation [27] can be written as

$$f(\mathbf{x}) \approx s(\mathbf{x}) = \sum_{j=1}^{N} \alpha_j \phi\left(\left|\mathbf{x} - \mathbf{x}_j\right|\right) + \sum_{k=1}^{M}\sum_{i=1}^{L} \beta_{i,k} p_k(x_i) + \beta_0 \tag{2.12}$$

where $\mathbf{x} = \{x_1, \ldots, x_i, \ldots, x_L\}$ and $f(\mathbf{x})$ is known for a series of points $\mathbf{x}$. Here, $p_k(x_i)$ is one of the M terms of a given basis of polynomials [28]. This approximation is solved for the α_j and $\beta_{i,k}$ unknowns from the system of N linear equations, subject to the conditions (for the sake of uniqueness)

$$\begin{gathered} \sum_{j=1}^{N} \alpha_j p_k(x_i) = 0 \\ \vdots \\ \sum_{j=1}^{N} \alpha_j p_k(x_L) = 0 \end{gathered} \tag{2.13}$$

$$\sum_{j=1}^{N} \alpha_j = 0 \tag{2.14}$$

One of the possible RBFs are the multiquadrics radial functions

$$\phi\left(\left|\mathbf{x}_i-\mathbf{x}_j\right|\right)=\sqrt{\left(\mathbf{x}_i-\mathbf{x}_j\right)^2+c_j^2} \tag{2.15}$$

where the shape parameter c_j must be adjusted. According to Baxter [29], usually large values of c_j provide the best approximations.

2.3.5 Wavelet Based Neural Networks [31, 32]

Wavelets occur in the family of functions generated from a mother wavelet $\psi(x)$. Each wavelet in it is defined by dilatation vector, a_i, which controls the scaling, and translation vector, t_i, which controls the position. Given a training set, the overall response of a WNN can be arithmetically written as

$$s\left(\mathbf{x}\right)=W_0+\sum_{i=1}^{N_p} W_i\psi_i\left(\frac{\bar{x}-\bar{t}}{\bar{a}}\right) \tag{2.16}$$

where N_P is the number of wavelet nodes in the hidden layer and w_i is the synaptic weight for each hidden node in the WNN. The dilatation and translation vectors have size equal to the number of variables in the estimated function. Such a network can be used to approximate any function

$$f\left(\mathbf{x}\right)=s\left(\mathbf{x}\right)+\varepsilon \tag{2.17}$$

where s is a regression function and the error term ε is a zero-mean random variable of disturbance.

One of the well known approaches for constructing WNN [30] requires the generation of a wavelet library, W. This library is composed of discretely dilatated and translated versions of mother wavelet function, $\psi\left(x\right)$. The next step is selecting the best wavelets based on the training data from this library to build the regression. This approach for building WNN becomes prohibitively computationally expensive when the estimated function has a large number of variables. This is due to exponential increase of the size of the wavelet library W with the dimension of the estimated function. Searching such a huge library one-by-one is computationally redundant. Therefore, a stochastic approach should be used for searching the best wavelets for the WNN hidden nodes [31].

2.4 Hybrid Methods for Response Surfaces

In this section we will present some hybrid response surface methods. The accuracy of these methods, along with their comparison against other strategies, will be presented in the next section.

2.4.1 Fittest Polynomial Radial Basis Function (FP-RBF) [28]

The FP-RBF hybrid method [27] consists of choosing the best possible combination of RBF, polynomial order, variable scaling, and shape parameter for a given problem.

In the work of Colaço et al. [27], the polynomial part of Eq. (2.12) was taken as

$$p_k(x_i) = x_i^k \tag{2.18}$$

and the radial basis functions were selected among the following

$$\text{Multiquadrics:}\quad \phi\left(|x_i - x_j|\right) = \sqrt{\left(x_i - x_j\right)^2 + c_j^2} \tag{2.19}$$

$$\text{Gaussian:}\quad \phi\left(|x_i - x_j|\right) = \exp\left[-c_j^2\left(x_i - x_j\right)^2\right] \tag{2.20}$$

$$\text{Squared multiquadrics:}\quad \phi\left(|x_i - x_j|\right) = \left(x_i - x_j\right)^2 + c_j^2 \tag{2.21}$$

$$\text{Cubical multiquadrics:}\quad \phi\left(|x_i - x_j|\right) = \left[\sqrt{\left(x_i - x_j\right)^2 + c_j^2}\right]^3 \tag{2.22}$$

Some tests were made using the cross-product polynomials ($x_i x_j x_k \ldots$), but the improvements on the results were found out to be irrelevant [27]. Also, other types of RBFs were previously considered by the authors [27], but no improvement in the accuracy of the interpolation was observed.

Therefore, a polynomial of order M is added to the radial basis function. After inspecting Eqs. (2.12)–(2.14), (2.18), one can easily check that the final linear system has [(N+M*L)+1] equations that can be solved by any traditional technique.

In the technique presented by Colaco et al. [27], initially all variables have to be normalized. Then, the initial guess for the shape parameter c is set as the minimum distance between two points in the training set of variables. Shape parameter, c, is then increased until the best solution is obtained. Also, different scaling of the variables are tried to give the best fit for the function. The choice of which polynomial order, which shape parameter and scaling of the variables, and which RBF are the best for fitting a specific data set was made based on a cross-validation procedure. Let us suppose that we have PTR training points, which are the locations in the multidimensional space where the values of the function are known. Such set of training points is equally subdivided into two subsets of points, named PTR1 and PTR2. The Eqs. (2.12)–(2.14) are solved for a polynomial of order zero and for one of the RBF expression given by Eqs. (2.19)–(2.22) using the subset PTR1. Then, the value of the interpolated function is checked against the known values of the function that are in the subset PTR2. The error is recorded as

$$RMS_{PTR1,M=0,RBF1} = \sum_{i=1}^{P_{TR2}} \left[s(\mathbf{x}_i) - f(\mathbf{x}_i)\right]^2 \tag{2.23}$$

Then, the same procedure is repeated by using the subset PTR2 to solve the equations and the subset PTR1 to calculate the error as

$$RMS_{PTR2,M=0,RBF1} = \sum_{i=1}^{P_{TR1}} \left[s(\mathbf{x}_i) - f(\mathbf{x}_i) \right]^2 \tag{2.24}$$

Finally, the total error for the polynomial of order zero using one of the RBF expressions given by Eqs. (2.19)–(2.22) is obtained as

$$RMS_{M=0,RBF1} = \sqrt{RMS_{PTR1,m=0,RBF1} + RMS_{PTR2,m=0,RBF1}} \tag{2.25}$$

This procedure is repeated for all polynomial orders, up to $M = 10$ and for each one of the RBF expressions given by Eqs. (2.19)–(2.22). The best combination is the one that returns the lowest value of the RMS error. Although this cross-validation procedure is quite simple, it worked very well for practical test cases including optimization of chemical compositions of alloys [32, 33] ,maghnetohydrodynamic flow in cavities [34], energy/exergy optimization [35, 36] and Bayesian inverse problems in heat transfer [37].

2.4.2 Kriging Approximation with Fittest Polynomial Radial Basis Function (KRG-FP-RBF)

A new method is proposed in this paper, which combines the very high accuracy of the FP-RBF approximation with the stochastic appealing of the Kriging method. The idea is conceptually simple, although the computational implementation requires some effort. Results presented in this paper are still preliminary and need further investigation.

Referring to Jones et al. [23], the Kriging approximation given by Eq. (2.6) is so powerful that the base function $g(\mathbf{x})$ can be written as the mean of the stochastic process, μ. However, we propose to extend even more such accuracy by using the FP-RBF approximation as a base function for the Kriging process. In other words, once the FP-RBF method has been adjusted and fitted to a particular set of data, we use this approximation as the $g(\mathbf{x})$ function in the Kriging model given by Eq. (2.6). Inserting the FP-RBF definition given by Eq. (2.12) into the Kriging model of Eq. (2.6), we have the following stochastic process model for the function as

$$f\left(\mathbf{x}^i\right) = a\, s\left(\mathbf{x}^i\right) + \varepsilon^i \tag{2.26}$$

Notice that $s(\mathbf{x})$ has to be built using the FP-RBF model explained earlier. By doing this, we are assuming that ε is the FP-RBF approximation error, which has (by hypothesis) zero mean and variance σ^2. Once the approximation was built by the FP-RBF method, the Kriging approximation is used to model such error in a stochastic way. Then, following Sacks et al. [21, 22] derivations, adapted to our nomenclature, we can obtain the best linear unbiased predictor at a new point $\mathbf{x}^*$ as (mathematical details are omitted for lack of space, but the interested readers are urged to read publications by Sacks et al. [21, 22] for all details of this derivation)

$$\tilde{s}\left(\mathbf{x}^*\right) = a\, s\left(\mathbf{x}^*\right) + \mathbf{r}^T\left(\mathbf{x}^*\right)\mathbf{R}^{-1}\left[\mathbf{F} - a\,\mathbf{S}\right] \tag{2.27}$$

where $\mathbf{F}$ is the vector of the exact function evaluated at the training points locations and $\mathbf{S}$ is vector of FP-RBF approximation at these same locations. Notice that if $a = 1$, and $\mathbf{F}=\mathbf{S}$, then the Kriging model reduces to the FP-RBF model. Thus, the second term gives some measure of the error between the real data and the FP RBF approximation. In this equation, $\mathbf{R}$ is the full correlation matrix between the training points, given by Eq. (2.8) and $\mathbf{r}$ is the correlation vector between the evaluation point $\mathbf{x}^*$ and the training points, also given by Eq. (2.8). The parameters θ and P, appearing in Eq. (2.7) can be obtained by minimizing $[(\det\mathbf{R})^{1/m}\sigma^2]$ where m is the number of training points [21, 22]. In this paper, such minimization was performed by the Particle Swarm method. The other parameters appearing in these equations are given as [21, 22]

$$\sigma^2 = \frac{1}{m}\left(\mathbf{F} - a\,\mathbf{S}\right)^T\mathbf{R}^{-1}\left(\mathbf{F} - a\,\mathbf{S}\right) \tag{2.28}$$

$$a = \left(\mathbf{S}^T\mathbf{R}^{-1}\mathbf{S}\right)^{-1}\mathbf{S}^T\mathbf{R}^{-1}\mathbf{F} \tag{2.29}$$

where Eq. (2.28) gives the maximum likelihood estimation of the variance and Eq. (2.29) is the generalized least-squares estimate [21, 22] of a. Thus, this procedure uses the Kriging method to model the approximation error of the FP-RBF approximation.

2.4.3 Hybrid Self Organizing Model With RBF [20]

The best known application of self-organizing method is in the commercial software IOSO [38] which uses quadratic local fitting polynomials. A more general idea is to use the self-organizing method given by Eq. (2.4) to choose the best local fitting functions (linear, quadratic, cubic or quadratic) to generate a response surface thus capturing the major topology of the response multi-dimensional hyper-surface. However, this metamodel does not force the hyper-surface to pass exactly through the provided support points. The difference between the actual values of the objective function at the support points and the fitted values at the support points represents a much less challenging topology which it then fitted using RBF method [19]. Such a combined response surface fitting method is much more robust and accurate (Figs. 2.3 and 2.4) than either of the separate methods used in this hybrid [19]. This hybrid

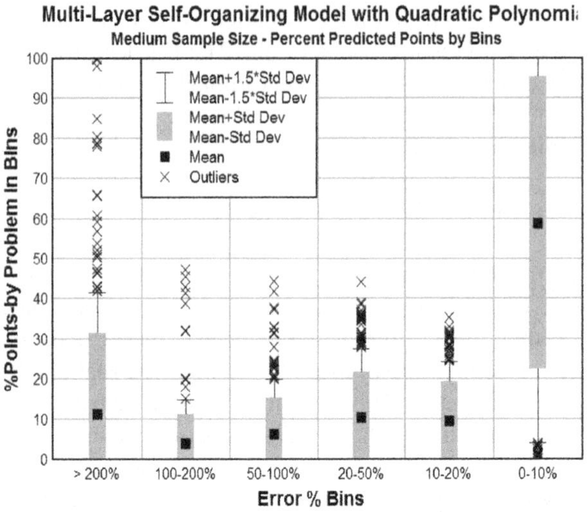

Fig. 2.3 Self-organizing method using quadratic local polynomials applied to a medium size data set; model accuracy [19]

method is simple to implement, although computationally costly, where computing cost is mainly due to the self-organizing method. This suggests a need for research into reducing the computing cost of the self-organizing method by using more efficient pruning algorithms to eliminate those branches in the genetic tree that are terminating.

2.4.4 Genetic Algorithm Based Wavelet Neural Network (HYBWNN) [31, 32]

Another hybrid method used for fitting high dimensional functions is the Genetic Algorithm Wavelet Based Neural Network (WNN) model presented by Sahoo and Dulikravich [31] with 5 neural subnets. Typically, the mother wavelet used in the WNN is Mexican Hat wavelet given by

$$\psi(x) = \left(\frac{2}{\sqrt{3}}\pi^{-1/4}\right)\left(1 - x^2\right)\exp\left(\frac{-x^2}{2}\right) \tag{2.30}$$

Gaussian wavelets were also used along with this mother wavelet to construct the WNN [31]. For each node of the WNN, genetic algorithm was used to search the best Mexican Hat wavelet and the best Gaussian wavelet. The one having a lower norm of residue after performing multiple linear regression was selected and used in the WNN architecture. The concept of binary genetic algorithm was used to

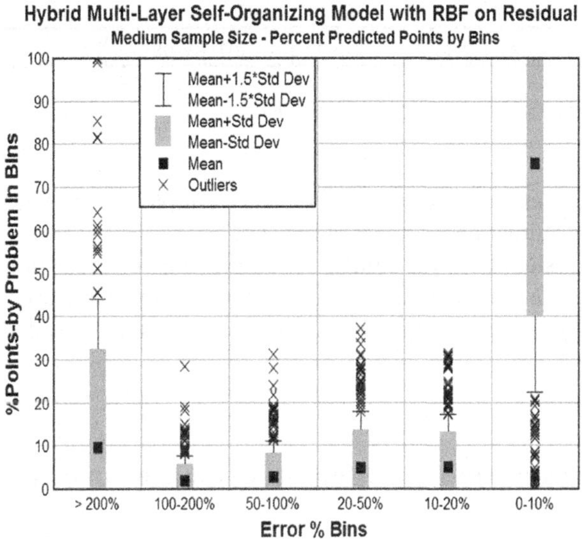

Fig. 2.4 Self-organizing method followed by RBF used on residuals applied to a medium size data set; model accuracy [19]

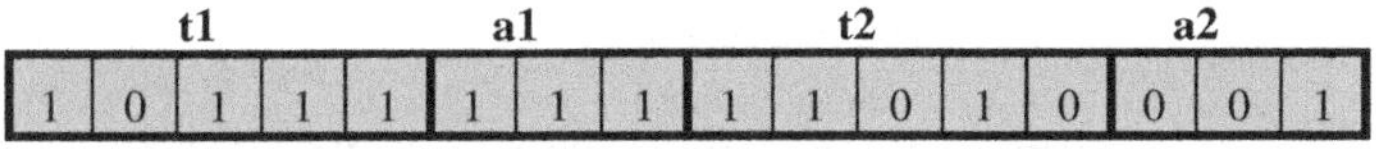

Fig. 2.5 Binary string representation of a 2-D wavelet

search for the wavelets required for the hidden nodes in the WNN. The dilatation and translation factors (binary representation) in Eq. (2.16) for each dimension of the wavelet were concatenated to form the chromosomes in the GA population. A typical representation of a wavelet in 2-D function estimation is shown in Fig. 2.5.

It has two dilatation factors specifying the scaling and two translation factors specifying the position of the wavelet in each dimension. The variables space is normalized so the translation factors can vary within $[-1, 1]$ and the dilatation factors can vary within $[0.1, 0.8]$. The fitness for selecting the wavelet was defined as the norm of the residue obtained by doing multiple linear regression of the values given by the wavelet transform of the training data versus the real function values. The GA was run for a sufficient number of generations to select a wavelet. Subsequent wavelets were searched by the GA based on the residue obtained in former step set as target values. This approach was unable to search for proper wavelets when the number of variables in the estimated function went beyond ten. The chromosome length for such functions was huge and the binary GA became inefficient. Therefore, a real numbers GA search was proposed where the wavelet is represented as a string of real number instead of a binary string. The range for searching for the values of dilation factors was relaxed to $[0.005, 5.00]$. This gave more flexibility to the GA for

a1	t1	a2	t2	a3	t3	a4	t4
2.8424	0.25483	2.4672	0.000922	1.7407	0.98521	2.5848	-0.61214

Fig. 2.6 Real string representation of a 4-D wavelet

searching appropriate wavelets [31]. A typical example of a wavelet representation in 4-D function estimation is shown in Fig. 2.6.

The fitness assignment was similar to the previous method. In addition, whole arithmetic crossover and floating point mutation operators were used. Separate GAs were run serially [36] for finding the activation function in each node of the WNN architecture.

2.5 Comparison Among Different Response Surface Algorithms

Performance of different hybrid response surface algorithms was evaluated on data sets containing either scarce (3*L*), small (10*L*), medium (30*L*) or large (50*L*) number of points, where *L* designates the dimensionality of the problem [39, 40].

2.5.1 Fittest Polynomial RBF Versus Hybrid Wavelet Neural Network [42]

In order to compare the accuracy of the FP-RBF [27] model against the Hybrid Wavelet Neural Network (WNN) [31], 13 test cases were used, representing linear and non-linear problems with up to 16 variables. These test cases, defined as problems 1–13 were selected by Jin et al. [41] in a comparative study among different kinds of meta-models. Such problems were selected from a collection of 395 problems (actually 296 test cases), proposed by Hock and Schittkowski [42] and Schittkowski [43]. For the other comparison presented in this paper, all 296 test cases will be presented. The reason is the very high computational cost associated with the WNN method that restricted us to these 13 test cases initially. Also, for these test cases, the polynomial degree of the FP-RBF model was fixed in a pre-specified value, and the shape parameter was set to 1/*N*, where *N* is the number of training points. For the other sections of this paper, those two parameters were allowed to vary according to the cross-validation procedure defined previously.

The first 12 problems do not have random errors added to the original function, while the problem no. 13 has a noise added with the following form

$$\varepsilon\left(x_1, x_2\right) = \sigma r \tag{2.31}$$

where σ is the standard deviation and r is a random number with Gaussian distribution and zero mean.

For accuracy, the goodness of fit obtained from "training" data is not sufficient to assess the accuracy of newly predicted points. For this reason, additional confirmation samples are used to verify the accuracy of the metamodels. To provide a more complete picture of metamodel accuracy, two different metrics are used: R Square, and Relative Average Absolute Error (RAAE) [41].

(a) R Square (R2)

$$R^2 = 1 - \frac{\sum_{i=1}^{n} \left(y_i - \hat{y}_i\right)^2}{\sum_{i=1}^{n} \left(y_i - \bar{y}\right)^2} = 1 - \frac{MSE}{\text{variance}} \tag{2.32}$$

where $\hat{y}_i$ is the corresponding predicted value for the observed value y_i; $\bar{y}$ is the mean of the observed values. While *MSE* (Mean Square Error) represents the departure of the metamodel from the real simulation model, the variance captures how irregular the problem is. The larger the value of R2, the more accurate the metamodel.

(b) Relative Average Absolute Error (RAAE)

$$RAAE = \frac{\sum_{i=1}^{n} \left|y_i - \hat{y}_i\right|}{n^* STD} \tag{2.33}$$

where *STD* stands for standard deviation. The smaller the value of RAAE, the more accurate the metamodel.

The FP-RBF model presented here was compared against the WNN method for the 13 selected analytical test cases. In order to check the accuracy of the metamodel when different samples were employed, three different sets of training points were used, as suggested by Jin et al. [41]. Table 2.1 gives the number of training points, testing points, minimum and maximum value of each test function, as well as the standard deviation and average value of each test function.

Initially, the results obtained with the FP-RBF model, with a polynomial of order 10 using a *large* number of training points and the results obtained with a polynomial of order 1 for small and scarce sets of training points were compared with the results obtained by using WNN method [44]. Only problems no. 1–5 were tested for a scarce set of training points, as suggested by Jin et al. [41].

Figure 2.7 demonstrates that when considering R2 metric, for large and small sets of training points, the RBF was better than the WNN, while for a scarce number of training points, the WNN was more accurate. On the contrary, it appears that when considering RAAE metric, for large and small sets of training points, the WNN was better than the RBF, while for a scarce number of training points, the RBF was more accurate.

Table 2.1 Parameters for the 13 analytical test functions used by Jin et al. [41]

		Number of training and testing points				Non Linearity	Minimum value of f	Maximum value of f	Standard deviation of f	Average value of f
		Training			Testing					
PB #	Vars	Scarce	Small	Large						
PB1	10	30	100	198	1,000	High	−21.01	29.25	8.92	8.60
PB2	10	30	100	198		Low	−1717.43	−539.36	187.40	−1146.82
PB3	10	30	100	198		High	−1327241.16	−1.14	224561.73	−213124.84
PB4	10	30	100	198		Low	258.78	4779.01	1002.27	2185.25
PB5	16	48	160	459		Low	7136.99	195608.81	27294.72	69060.73
PB6	2	N/A	9	100		High	98.94	187.53	17.75	127.73
PB7	2		9	100		High	−0.26	0.26	0.15	0.00
PB8	2		9	100		Low	−1.51	8.91	2.08	1.66
PB9	3		27	125		High	2.15	146369.38	21527.31	10842.11
PB10	3		27	125		High	3.42	32.84	7.70	29.13
PB11	3		27	125		Low	40829.97	−40749.42	21.60	−40790.41
PB12	2		9	100		Low	−12.72	13.75	5.74	0.52
PB13	2		9	100		Low	−54.77	77.80	30.94	12.60

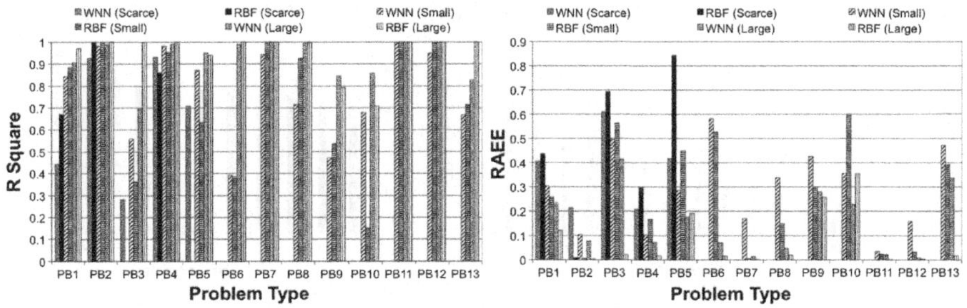

Fig. 2.7 R2 and RAAE metrics for WNN and FP-RBF

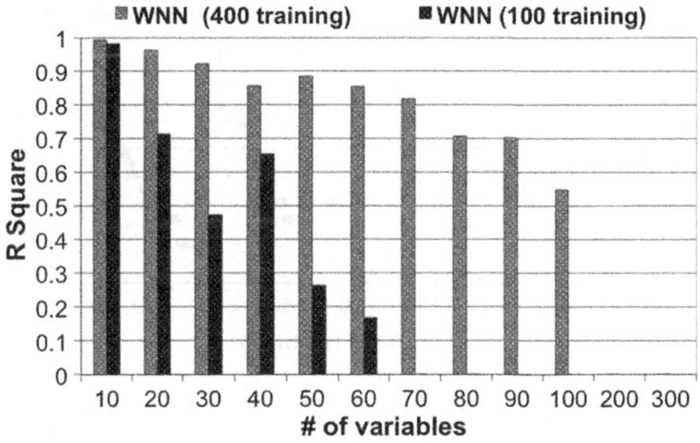

Fig. 2.8 R2 results with WNN for a large number of variables (WNN with five subnets)

However, one of the major problems with WNN is its rapidly decreasing accuracy with increasing dimensionality of the problem. Figure 2.8 demonstrates the results for test problem no. 2 when using the WNN. One can see that the accuracy, given by the R2 metric, decreases rapidly when using 100 training points. Also, for 400 training points, the R2 goes to a negative value when using more than 100 variables. Colaço et al. [39] demonstrated that the RBF model was able to maintain a very high accuracy even when the number of variables increased to 500.

Figure 2.9a shows the computational time required to run this test case using the FP-RBF model. The code was written in Fortran 90 and the CPU was an Intel T2300 1.66 Ghz (Centrino Duo) with 1 Gb RAM. Figure 2.9b shows the computational time required by the WNN where one can notice the extremely high computational cost. The code for the WNN was written in Matlab 7.0.4 and the CPU was an Intel T2300 1.66 Ghz (Centrino Duo) with 1 Gb RAM. Some improvement in the performance could be expected by converting this code to Fortran90 or C++. However, the computational cost for the WNN for a problem with 300 variables and 400 training points, even with different programming languages (Matlab and Fortran90) was approximately 6,000 times greater than for the RBF.

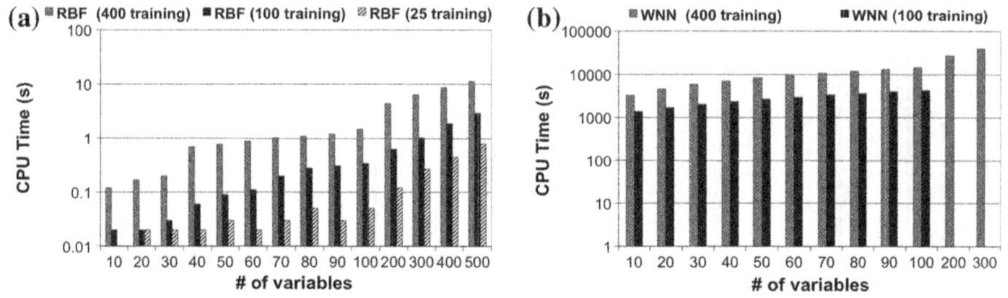

Fig. 2.9 Computing time for a large number of variables: (**a**) RBF with M = 1, (**b**) WNN with five subnets

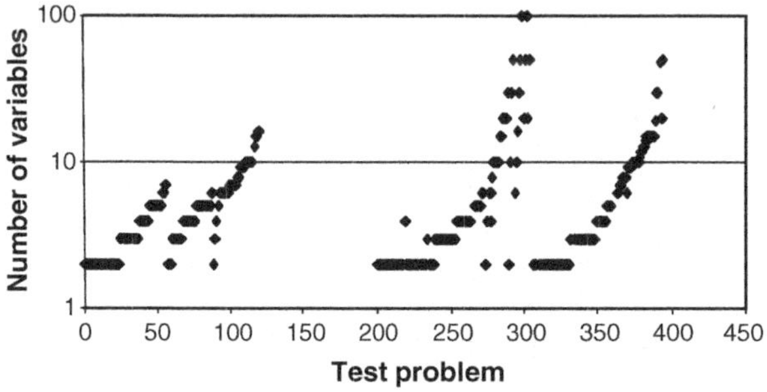

Fig. 2.10 Number of variables for each problem considered in the Schittkowski suite of analytical test cases [42, 43]

Table 2.2 Number of training and testing points

	Number of training points (L)	Number of testing points (L)
Scarce set	3	300
Small set	10	1,000

2.5.2 Fittest Polynomial RBF Versus Kriging

In this section we will compare the accuracy and computing time requirement of the FP-RBF model against the one given by the Kriging model proposed by Jones et al. [23]. From now on, 296 test cases will be used, representing linear and non-linear analytical problems with up to 100 variables. These test problems were selected from a collection of 395 (actually 296) test cases proposed by Hock and Schittkowski [42] and Schittkowski [43]. Figure 2.10 shows the number of variables of each test case analyzed. To verify the accuracy of the interpolation over different number of training points, two sets were defined. Also, the number of testing points varied, according to the number of training points. Table 2.2 presents these two sets, based on the number of dimensions (variables) L of the problem.

Initially, Fig. 2.11 presents the values of R2 and RAAE metrices for the FP-RBF model, considering a scarce set of data. As one can see, most of the test cases

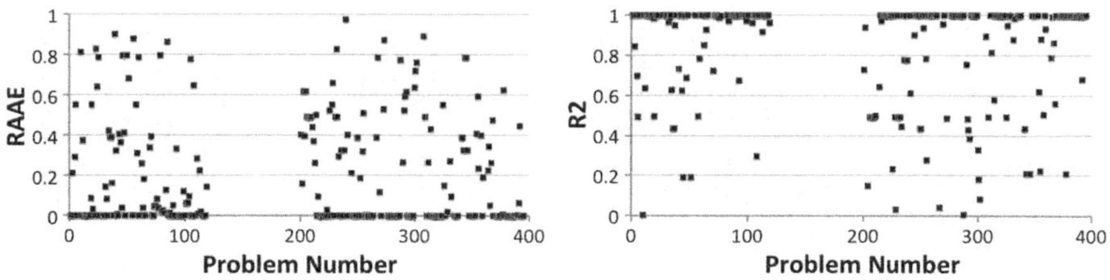

Fig. 2.11 Metrics for the FP-RBF method (scarce set of data)

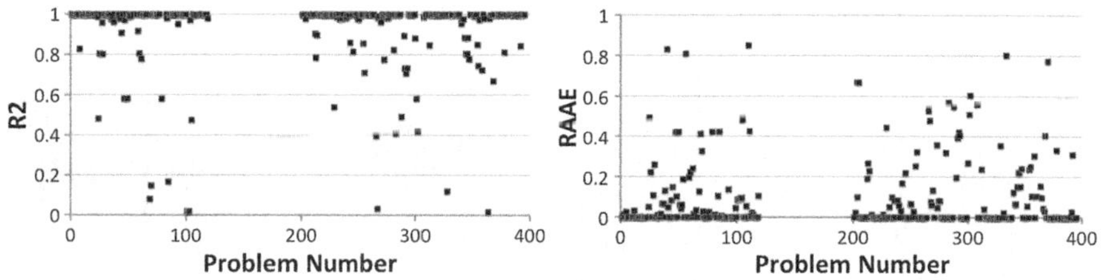

Fig. 2.12 Metrics for the FP-RBF method (*small* set of data)

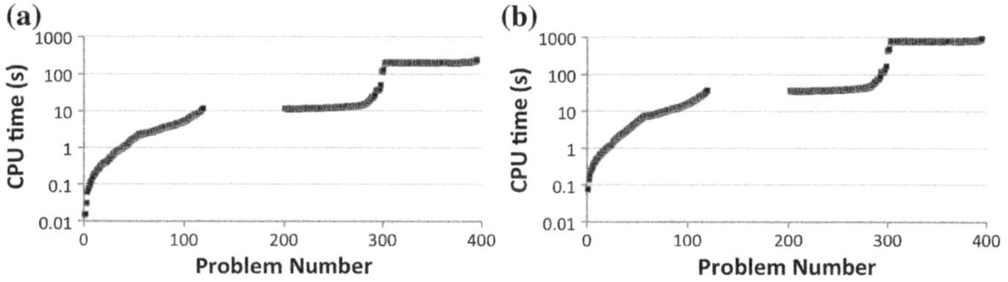

Fig. 2.13 CPU time for the FP-RBF method. **a** scarce set of data, **b** small set of data

have high values of R2 and low values of RAAE, indicating a good approximation, even for such small number of training points. When the number of training points is increased from scarce to small, results look better, as one can notice from the analysis of Fig. 2.12 for the R2 and RAAE metrics.

Figure 2.13 shows the CPU time for FP-RBF method. All test cases ran on an Intel i7 2GHz with 4 Gb of RAM (Windows XP emulated under Mac OS X 10.8.4) and codes were written in Fortran 90. Besides running on different processors, CPU times in these test cases are a little higher than in previous one, since now we are also optimizing the shape parameter, the RBF polynomial degree and the scaling of the variables, as discussed before. Notice that computing time increased only slightly by going from a scarce data set to a small data set.

Figure 2.14 shows difference of R2 and RAAE metrics between the Kriging method and the FP-RBF model for 296 test-cases studied in this section, using the scarce set of data. Since higher R2 and lower RAAE values indicate a good accuracy, in these graphics, negative values of Delta R2 and positive values of RAAE indicate

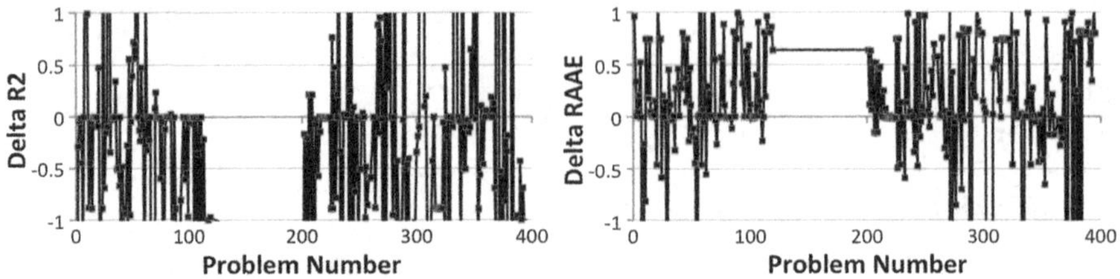

Fig. 2.14 Difference of the R2 and RAAE metrices between the Kriging model proposed by Jones et al. [23] and FP-RBF methods for the scarce set of data (negative values of Delta R2 and positive values of Delta RAAE indicates superiority of the FP-RBF method)

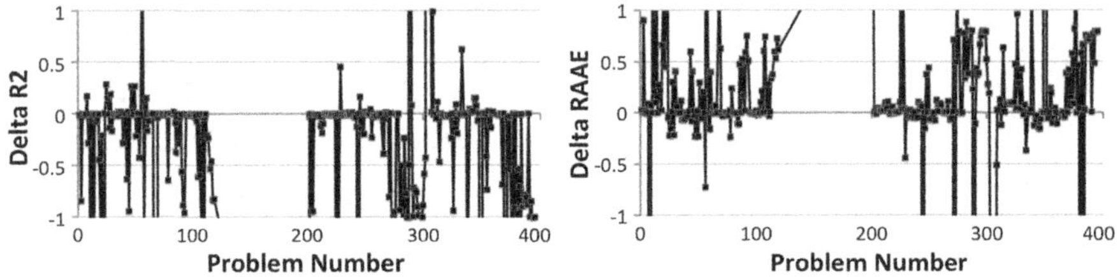

Fig. 2.15 Difference of the R2 and RAAE metrics between the Kriging model proposed by Jones et al. [23] and FP-RBF methods for the small set of data (negative values of Delta R2 and positive values of Delta RAAE indicates superiority of the FP-RBF method)

the superiority of the FP-RBF method. As a general trend, the FP-RBF method performs better than the original Kriging method, given by Eq. (2.9), for most of the test cases, although there are some functions were Kriging has a better accuracy. For the small set of data, such comparison is presented in Fig. 2.15, where one can notice the superiority of the FP-RBF method over Kriging.

Figure 2.16 shows the CPU time ratio between Kriging and FP-RBF for scarce and small sets of data. In this figure, values greater than one indicate how many times the Kriging is slower than the FP-RBF. As a general trend, for the scarce set of data, Kriging is one order of magnitude slower than the FP-RBF, whereas for the small set of data it is two orders of magnitude slower. Two factors contribute for the high computing cost of Kriging: (i) the need to invert the covariance matrix **R** in Eqs. (2.27)–(2.29) and the minimization of $[(\det\mathbf{R})^{1/m}\sigma^2]$ by the Particle Swarm method. We intend to investigate ways to reduce this computational cost, since Kriging seems to have some advantage over FP-RBF model when applied to certain functions, as shown above. It is also worth noting that the CPU time ratio is almost constant over all test problems. In fact, going from test function number 200–295 the CPU time ratio decreases, when the number of dimensions of the problems varies considerably, as one can check from Figs. 2.10 and 2.16.

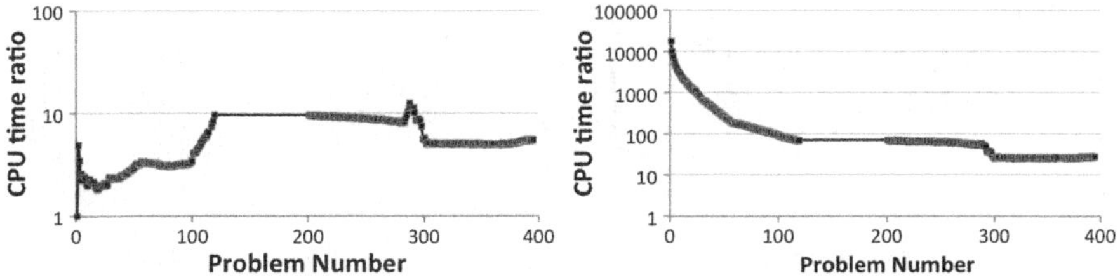

Fig. 2.16 CPU time ratio between the Kriging model proposed by Jones et al. [23] and FP-RBF methods for the scarce and small sets of data (values higher than one indicate the Kriging method is more expensive than FP-RBF)

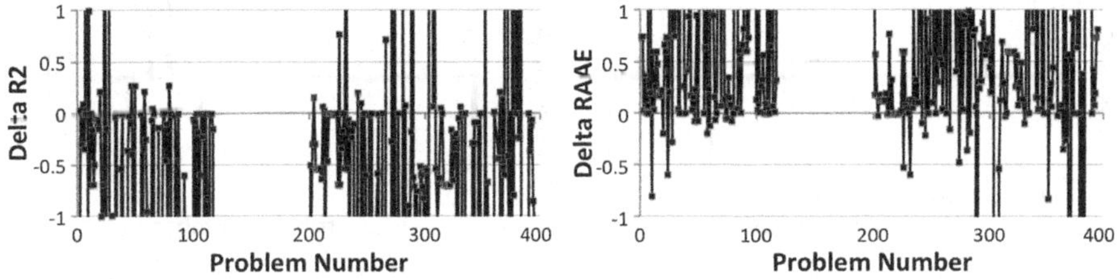

Fig. 2.17 Difference of the R2 and RAAE metrics between the HYBSORSM and FP-RBF methods for the scarce sets of data (negative values of Delta R2 and positive values of Delta RAAE indicates superiority of the FP-RBF method)

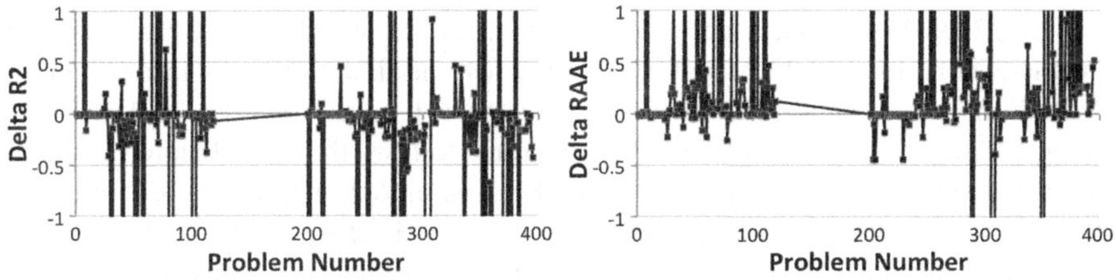

Fig. 2.18 Difference of the R2 and RAAE metrics between the HYBSORSM and FP-RBF methods for the small sets of data (negative values of Delta R2 and positive values of Delta RAAE indicates superiority of the FP-RBF method)

2.5.3 Fittest Polynomial RBF Versus Hybrid Self Organizing Response Surface Method—HYBSORSM

Results in Fig. 2.17 indicate better accuracy of the FP-RBF method over HYBSORM. From the analysis of this figure it is clear the superiority of FP-RBF model. Comparing Fig. 2.17 with Fig. 2.14, it is evident that for this set of data, the Kriging model is also superior to HYBSORSM. However, when the small set of data is used (see Fig. 2.18) the HYBSORM method improves its performance, but is still outperformed by the FP-RSM method. Comparing now Figs. 2.15 and 2.18, the HYBSORM has a better performance than the Kriging model for the small sets of data.

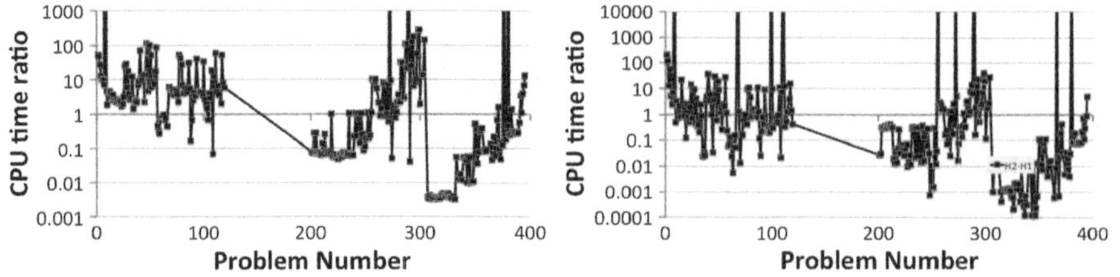

Fig. 2.19 CPU time ratio between the HYBSORSM and FP-RBF methods for the scarce and small sets of data (values higher than one indicate the method is more expensive than FP-RBF)

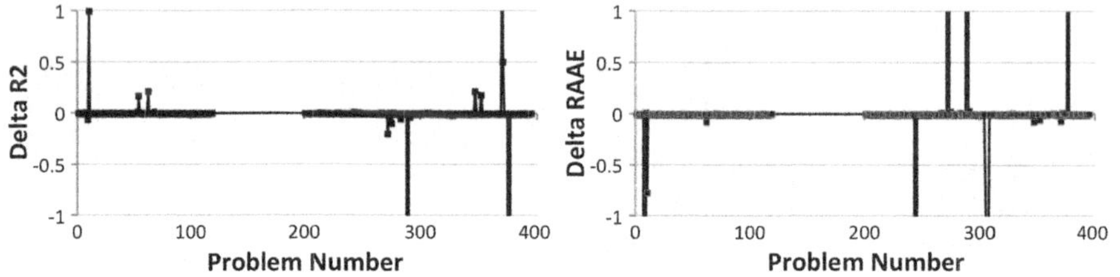

Fig. 2.20 Difference of the R2 and RAAE metrics between the KRG-FP-RBF and FP-RBF methods for the scarce sets of data (negative values of Delta R2 and positive values of Delta RAAE indicates superiority of the FP-RBF method)

Figure 2.19 shows some interesting results regarding the CPU time ratio between HYBSORM and FP-RSM model. In some cases, the first method is faster then the FP-RBF model, whereas it is slower for other ones. Comparing Figs. 2.10 and 2.19, it is interesting to notice that the CPU time ratio follows the same behaviour as the number of variables. Thus, opposite to the Kriging model, where the CPU time was almost constant with the number of variables, the HYBSORSM method requires more computational effort for problems where the number of dimensions is high.

2.5.4 Fittest Polynomial RBF Versus Kriging Approximation with Fittest Polynomial Radial Basis Function—KRG-FP-RBF

This section compares the results of the FP-RBF method with the ones obtained by the combined (hybrid) KRG-FP-RBF method. Figures 2.20 and 2.21 show Delta R2 and Delta RAAE for the scarce and small sets of data. In general, the hybrid KRG-FP-RBF method does not modify the accuracy of the FP-RBF method, except in a few cases. Although for some cases the performance decreases, for most of the cases where the hybrid KRG-FP-RBF method changes the FP-RBF performance, it improves the solution. These results are still in a very early stage of development and we believe this approach might improve the overall performance of the FP-RBF method if some better strategy to minimize $[(\det\mathbf{R})^{1/m}\sigma^2]$ is used.

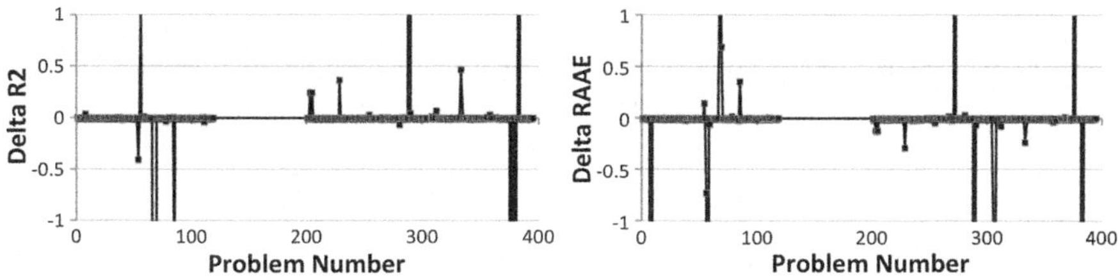

Fig. 2.21 Difference of the R2 and RAAE metrics between the KRG-FP-RBF and FP-RBF methods for the small set of data (negative values of Delta R2 and positive values of Delta RAAE indicates superiority of the FP-RBF method)

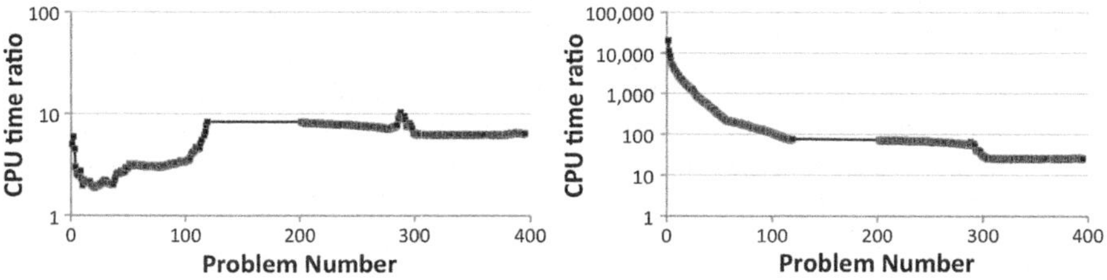

Fig. 2.22 CPU time ratio between the KRG-FP-RBF and FP-RBF methods for the scarce and small sets of data (values higher than one indicate the method is more expensive than FP-RBF)

Finally, Fig. 2.22 shows the CPU time ratio between the KRG-FP-RBF method and the FP-RBF method. This ratio can be reduced if better optimization procedures are used in the Kriging part of the code.

2.6 Conclusions

A number of concepts for constructing hybrid optimization algorithms with focus on automatic switching logic have been described. Also, a number of multi-dimensional response surface fitting algorithms and their hybrids have been described and their performances compared for scarce, small and medium data sets. Fittest polynomial radial basis function (FP-RBF) method appears to offer the best overall performance concerning high accuracy of fitting arbitrary data sets and low computing time requirements. Possible hybridization of Kriging and FP-RBF was also thoroughly tested showing its promises as far as increased robustness of such hybrids, although at significant increase in the computing time.

Acknowledgments This work was partially funded by the US Air Force Office of Scientific Research under grant FA9550-12-1-0440 monitored by Dr. Ali Sayir. The views and conclusions contained herein are those of the authors and should not be interpreted as necessarily representing the official policies or endorsements, either expressed or implied, of the US Air Force Office of Scientific Research or the U.S. Government. The U.S. Government is authorized to reproduce and distribute reprints for government purposes not with standing any copyright notation thereon. Dr. Colaço also acknowledges the financial support provided by CNPq, CAPES, FAPERJ and ANP/PRH37 (http:// www.prh.mecanica.ufrj.br/), Brazilian agencies for fostering science and technology.

References

1. Wolpert DH, Macready WG (1997) No free lunch theorems for optimization. Evol Comput IEEE Trans Evol Comput 1(1):67–82
2. Dulikravich GS, Martin TJ, Colaço MJ, Inclan EJ (2013) Automatic switching algorithms in hybrid single-objective optimizers. FME Trans 41(3):167–179
3. Foster NF, Dulikravich GS (1997) Three-dimensional aerodynamic shape optimization using genetic and gradient search algorithms. AIAA J Spacecr Rockets 34(1):36–42
4. Dulikravich GS, Martin TJ, Dennis BH, Foster NF (1999) Multidisciplinary hybrid constrained GA optimization. In: Miettinen K, Makela MM, Neittaanmaki P, Periaux J (eds) EUROGEN99-evolutionary algorithms in engineering and computer science: recent advances and industrial applications, Jyvaskyla, Wiley, pp 233–259, 30 May–3 June 1999
5. Colaço MJ, Dulikravich GS, Orlande HRB, Martin TJ (2005) Hybrid optimization with automatic switching among optimization algorithms. In: Annicchiarico W, Périaux J, Cerrolaza M, Winter G (eds) Evolutionary algorithms and intelligent tools in engineering optimization, CIMNE, Barcelona, Spain/WITpress, pp 92–118
6. Ahrari A, Shariat-Panahi M, Atai AA (2009) GEM: a novel evolutionary optimization method with improved neighborhood search, Appl Math Comput 210(2):376–386
7. Deb K (2009) Multi-objective optimization using evolutionary algorithms. Wiley, Chichester
8. Talbi E-G (2002) A taxonomy on hybrid metaheuristics. J Heuristics 8:541–562
9. Vrugt JA, Robinson BA (2007) Improved evolutionary optimization from genetically adaptive multimethod search. Proc Nat Acad Sci (US) 104(3):708–711
10. Moral RJ, Dulikravich GS (2008) Multi-objective hybrid evolutionary optimization with automatic switching among constituent algorithms. AIAA J 46(3):673–700
11. McKay M, Conover W, Beckman RA (1979) A comparison of three methods for selecting values of input variables in the analysis of output from a computer code. Technometrics 21:239–245
12. Sobol IM (1967) Distribution of points in a cube and approximate evaluation of integrals. USSR Comput Math Math Phys 7:86–112
13. Box GEP, Draper NR (1987) Empirical model-building and response surfaces,vol 157, Wiley Series in Probability and Statistics
14. Meyers RH, Montgomery DC, Anderson-Cook CM (2009) Response surface methodology: process and product optimization using designed experiments, vol 705, Wiley Series in Probability and Statistics
15. Matheron G (1963) Principles of geostatistics. Econ Geol 58:1246–1266
16. Hardy RL (1971) Multiquadric equations of topography and other irregular surfaces. J Geophys Res 176:1905–1915
17. Haykin S (2008) Neural networks and learning machines, Prentice Hall, Englewood Cliffs
18. Madalaand HR, Ivakhnenko AG (1994) Inductive learning algorithms for complex systems modeling. CRC Press, Boca Raton
19. Moral RJ, Dulikravich GS (2008) A hybrid self-organizing response surface methodology, paper AIAA-2008-5891. In: 12th AIAA/ISSMO multi-disciplinary analysis and optimization conference, Victoria, 10–12 September 2008
20. Krige DG (1951) A statistical approach to some basic mine valuation problems on the witwatersrand. J Chem Metal Mining Soc South Africa 52(6):119–139
21. Sacks J, Schiller SB, Welch WJ (1989) Design for computer experiments. Technometrics 31(1):41–47

22. Sacks J, Welch WJ, Mitchell TJ, Wynn HP (1989) Design and analysis of computer experiments. Stat Sci 4(4):409–435
23. Jones DR, Schonlau M, Welch WJ (1998) Efficient global optimization of expensive black-box functions. J Glob Optim 13:455–492
24. Huang D, Allen TT, Notz WI, Zeng N (2006) Global optimization of stochastic black-box systems via sequential kriging meta-models. J Glob Optim 34:441–466
25. Hardy RL (1971) Multiquadric equations of topography and other irregular surfaces. J Geophys Res 176:1905–1915
26. Wendland H (1998) Error estimates for interpolation by compactly supported radial basis functions of minimal degree. J Approx Theory 93:258–272
27. Colaço MJ, Dulikravich GS, Sahoo D (2008) A response surface method based hybrid optimizer. Inverse Prob Sci Eng 16:717–742
28. Buhmann MD (2003) Radial basis functions on grids and beyond. In: International workshop on meshfree methods, Lisbon, Portugal
29. Baxter BJC (1992) The interpolation theory of radial basis functions, PhD thesis, Cambridge
30. Zhang Q (1997) Using wavelet network in nonparametric estimation. IEEE Trans Neural Netw 8:227–236
31. Sahoo D, Dulikravich GS (2006) Evolutionary wavelet neural network for large scale function estimation in optimization, In: 11th AIAA/ISSMO multidisciplinary analysis and optimization conference, Portsmouth, 6–8 September 2006
32. Dulikravich GS, Egorov IN, Colaço MJ (2008) Optimizing chemistry of bulk metallic glasses for improved thermal stability. Model Simul Mater Sci Eng 16:075010
33. Bhargava S, Dulikravich GS, Murty GS, Agarwal A, Colaço MJ (2011) Stress corrosion cracking resistant aluminum alloys: optimizing concentrations of alloying elements and tempering. Mat Manuf Process 26:363–374
34. Colaço MJ, Dulikravich GS, Orlande HRB (2009) Magnetohydrodynamic simulations using radial basis functions. Int J Heat Mass Trans 52:5932–5939
35. Colaço MJ, Teixeira CV, Dutra LM (2010) Thermodynamic simulation and optimization of diesel engines operating with diesel and biodiesel blends using experimental data. Inverse Prob Sci Eng 18:787–812
36. Pires TS, Cruz MEC, Colaço MJ (2013) Response surface method applied to the thermoeconomic optimization of a complex cogeneration system modeled in a process simulator. Energy 52:44–54
37. Orlande HRB, Colaço MJ, Dulikravich GS (2008) Approximation of the likelihood function in the bayesian technique for the solution of inverse problems. Inverse Prob Sci Eng 16:677–692
38. IOSO NM (2003) Version 1.0, user's guide, IOSO Technology Center, Moscow, Russia
39. Colaço MJ, Silva WB, Magalhães AC, Dulikravich GS (2008) Response surface methods applied to scarce and small sets of training points A comparative study. In: EngOpt 2008 - international conference on engineering optimization, Rio de Janeiro, Brazil
40. Colaço MJ, Dulikravich GS (2008) A hybrid RBF based methods for highly multidimensional response surfaces using scarce data sets. In: 12th AIAA/ISSMO multidisciplinary analysis and optimization conference, Victoria, British Columbia
41. Jin R, Chen W, Simpson TW (2000) comparative studies of metamodeling techniques under multiple modeling criteria. In: 8th AIAA/USAF/NASA/ISSMO multidisciplinary analysis & optimization symposium, AIAA 2000–4801, Long Beach, 6–8 September 2000
42. Hock W, Schittkowski K (1981) Test examples for nonlinear programming codes. Lecture notes in economics and mathematical systems, vol 187. Springer, Berlin
43. Schittkowski K (1987) More test examples for nonlinear programming. Lecture notes in economics and mathematical systems, vol 282. Springer, Berlin
44. Colaço MJ, Dulikravich GS, Sahoo D (2007) A comparison of two methods for fitting high dimensional response surfaces. In: Inverse problems, design and optimization symposium, Miami, 16–18 April 2007

Chapter 3
A Genetic Algorithm for a Sensor Device Location Problem

Egidio D'Amato, Elia Daniele and Lina Mallozzi

Abstract In this paper we present a noncooperative game theoretical model for the well known problem of experimental design. A virtual player decides the design variables of an experiment and all the players solve a Nash equilibrium problem by optimizing suitable payoff functions. We consider the case where the design variables are the coordinates of n points in a region of the plane and we look for the optimal configuration of the points under some constraints. Arising from a concrete situation, concerning the ARGO-YBJ experiments, the goal is to find the optimal configuration of the detector, consisting of a single layer of resistive plate counters. Theoretical and computational results are presented for this location problem.

Keywords Facility location · Nash equilibrium · Constrained optimization

3.1 Introduction

The scope of the ARGO-YBJ project is to study cosmic gamma radiation, identifying transient emissions and performing a systematic search of steady sources [1]. The detection of very small size air showers (at low energy $< TeV$), is needed to reach this scope, because standard ones would sample only a small percentage of the shower particles. The achieving of the objective is committed to a new instrument located in Yangbajing Laboratory (Tibet, China), at a very high altitude (4,000 m a.s.l.) in order to approach the maximum size development of low energy showers. This detector

E. D'Amato
Dipartimento di Ingegneria Industriale e dell'Informazione, Seconda Università degli Studi di Napoli, Via Roma 29, 80039 Aversa, Italy
e-mail: egidio.damato@unina2.it

E. Daniele
Fraunhofer Institut für Windenergie und Energiesystemtechnik - IWES, Ammerländer Heerstraße 136, 26129 Oldenburg, Germany
e-mail: elia.daniele@iwes.fraunhofer.de

L. Mallozzi (✉)
Department of Mathematics and Applications "R.Caccioppoli", University of Naples "Federico II", Via Claudio 21, 80125 Naples, Italy
e-mail: mallozzi@unina.it

D. Greiner et al. (eds.), *Advances in Evolutionary and Deterministic Methods for Design, Optimization and Control in Engineering and Sciences*, Computational Methods in Applied Sciences 36, DOI 10.1007/978-3-319-11541-2_3

uses a full coverage layer of Resistive Plate Counters (RPCs) that can provide a high granularity sampling of particle showers. It covers an area of about 6,700 m^2 and allows a detailed space-time picture of the shower front.

This work is related to the optimization of RPCs location on the layer, to capture a uniform cosmic source distribution, constrained by a limited number of receivers due to a budget limitation.

Considering the capture surface of a single receiver shaped as a circular area, the problem has many points in common with a classic sphere packing problem [5, 13, 18]. The problem of packing circles in different geometrical shapes in the plane has always attracted researchers for the large amount of fields on which it can be applied. In the last decades many results, mainly for small packings, were obtained. The increasing performance of computing systems and the development of new optimization algorithms for large problems have recently brought to the forefront this kind of problems. Usually the circle packing problem can be stated as that of spreading points and it's needed to find a configuration of points in the given region such that the minimum mutual distance between the points is as large as possible. The packing problem is dual to the covering one, in which the optimal location of points is needed to cover as much as possible the area of interest. Typical solutions can be found in several fields and are addressed using several algorithmic optimization procedures [3, 11, 14, 16, 19].

In this paper we are interested in finding the optimal location of a limited number of receivers to maximize the total detection area. This experimental design problem can be faced as a Nash equilibrium problem as stated in Game Theory: the choice of the variables in n experiments is made by n players, each of them has to decide his location far as possible from the opponents and also from the border of the region. On this model, it is possible to compute the equilibria by using a numerical procedure based on a genetic algorithm [4, 6, 10, 15, 17, 20]. In Sect. 3.2 the constrained location problem is introduced and the procedure to solve it by a Nash game is shown; in Sect. 3.3 the Nash genetic algorithm for the facility location game is presented with several test cases. In Sect. 3.4 concluding remarks and some further developments are discussed.

3.2 Constrained Location Problem

3.2.1 Preliminaries

Let us consider an n-player normal form game Γ ($n \in \mathscr{N}$, where $\mathscr{N}$ is the set of natural numbers), that consists of a tuple

$$\Gamma = \langle N; X_1, \ldots, X_n; f_1, \ldots, f_n \rangle$$

where $N = \{1, 2, \ldots, n\}$ is the finite player set, for each $i \in N$ the set of player i's strategies is X_i (i.e. the set of player i's admissible choices) and $f_i : X_1 \times \cdots \times X_n \to \mathscr{R}$ is player i's payoff function ($\mathscr{R}$ is the set of real numbers). We suppose here that players are cost minimizing, so that player i has a cost $f_i(x_1, x_2, \ldots, x_n)$ when player 1 chooses $x_1 \in X_1$, player 2 chooses $x_2 \in X_2$, ..., player n chooses $x_n \in X_n$.

We define $X = X_1 \times \cdots \times X_n$ and for $i \in N$: $X_{-i} = \Pi_{j \in N \setminus \{i\}} X_j$. Let $\mathbf{x} = (x_1, x_2, \ldots, x_n) \in X$ and $i \in N$. Sometimes we denote $\mathbf{x} = (x_i, \mathbf{x}_{-i})$, where $\mathbf{x}_{-i} = (x_1, \ldots, x_{i-1}, x_{i+1}, \ldots, x_n)$.

A *Nash equilibrium* [2, 12] for Γ is a strategy profile $\hat{\mathbf{x}} = (\hat{x}_1, \hat{x}_2, \ldots, \hat{x}_n) \in X$ such that for any $i \in N$ and for any $x_i \in X_i$ we have that

$$f_i(\hat{\mathbf{x}}) \leq f_i(x_i, \hat{\mathbf{x}}_{-i}).$$

Such a solution is self-enforcing in the sense that once the players are playing such a solution, it is in every player's best interest to remain in his strategy. We denote by $NE(\Gamma)$ the set of the Nash equilibrium strategy profiles.

Any $\hat{\mathbf{x}} = (\hat{x}_1, \ldots, \hat{x}_n) \in NE(\Gamma)$ is a vector such that for any $i \in N$, $\hat{x}_i$ is solution to the optimization problem

$$\min_{x_i \in X_i} f_i(x_i, \hat{\mathbf{x}}_{-i}).$$

3.2.2 The Facility Location Game

We consider the unit square $\Omega = [0, 1]^2$: the problem is to decide for two variables x and y the values of n available experiments ($n \in \mathcal{N}$ given).

Problem 1 Experimental Design (ED)

The problem is to sattle n points $P_1, P_2, \ldots, P_n$ in the square Ω in such a way that they are far as possible from the rest and from the boundary of the square.

This implies to maximize the *dispersion* of the points in the interior and the distance from the boundary of Ω as in experimental design ([9]). Various concrete situations satisfy these requirement, for example the location of sensor device to capture cosmic rays in a region that will be discussed in the next section.

There is a competition between the points in the square, because the dispersion depends on the mutual position of all the points, also with respect to the boundary of Ω, so we use a game theoretical model and assign each point to a virtual player, whose decision variables are the coordinates and whose payoff function translates the dispersion in terms of distances.

As it happens in applications, forbidden places may be present inside the square. We consider the location problem in the constrained case depending on the admissible subregion of Ω, say $\Omega_c \subset \Omega$.

In the constrained case we define the following n-player normal form game $\Gamma_c = \langle N; \Omega_c, \ldots, \Omega_c; f_1, \ldots, f_n \rangle$ where each player in $N = \{1, 2, \ldots, n\}$, for each $i \in N$, minimizes the cost $f_i : A_c \to \mathcal{R}$ defined by

$$f_i(P_1, \ldots, P_n) = \sum_{1 \leq j \leq n, j \neq i} \frac{1}{d(P_i, P_j)} + \frac{1}{\sqrt{2d(P_i, \partial\Omega)}}$$

being $A_c = \{(P_1, \ldots, P_n) \in \Omega_c^n : P_i \in (]0, 1[)^2, P_i \neq P_j \; \forall i, j = 1, \ldots, n, j \neq i\}$ and d is the Euclidean metric in $\mathscr{R}^2$. The first $n - 1$ terms in the definition of f_i give the distance between the point P_i and the rest of the points, the last term an decreasing function of the distance of P_i from the boundary of the square.

Definition 1 Any $(\hat{x}_1, \hat{y}_1, \ldots, \hat{x}_n, \hat{y}_n) \in A_c$ that is a Nash equilibrium solution of the game Γ_c is an optimal solution of the problem (*ED*). For any $i \in N$, $(\hat{x}_i, \hat{y}_i)$ is solution to the optimization problem

$$\min_{(x_i, y_i) \in \Omega_c} f_i(\hat{x}_1, \hat{y}_1, \ldots, \hat{x}_{i-1}, \hat{y}_{i-1}, x_i, y_i, \hat{x}_{i+1}, \hat{y}_{i+1}, \ldots, \hat{x}_n, \hat{y}_n)$$

with $(x_1, y_1, \ldots, x_n, y_n) \in A_c$.

A very common situation is to consider $\Omega_c = \Omega \backslash T$ with T a closed subset of Ω (a triangle, a circle, etc.) that corresponds to a facility location problem with an obstacle (a lake, a mountain, etc.). Other concrete cases for the admissible region Ω_c can be considered: in the following Section we will examine the location problem when the admissible region is given by a set of segments.

3.2.3 Location of Sensor Devices on a Grid

Given the set $\{h_1, \ldots, h_k\}$ ($h_i \in]0, 1[$, i = 1, ..., k) we consider the set of possible location of n sensor devices able to capture cosmic particles

$$\Omega_c = \{[0, 1] \times \{h_1\}, \ldots, [0, 1] \times \{h_k\}\}.$$

We are obliged to locate the sensors on the given k segments in the square: for example because of electricity constraints.

In terms of coordinates, if $P_i = (x_i, y_i)$, $i \in N$ the distance of a point $P = (x, y)$ from the set $\partial\Omega$, the boundary of Ω, is

$$d(P, \partial\Omega) = \min_{Q \in \partial\Omega} d(P, Q) = \min\{x, y, 1 - x, 1 - y\}$$

and we have for $(x_1, y_1, \ldots, x_n, y_n) \in A_c$

$$f_i(x_1, y_1, \ldots, x_n, y_n) = \sum_{1 \leq j \leq n, j \neq i} \frac{1}{\sqrt{(x_i - x_j)^2 + (y_i - y_j)^2}} + \frac{1}{\sqrt{2} \min\{x_i, y_i, 1 - x_i, 1 - y_i\}}$$

for $(x_1, y_1, \ldots, x_n, y_n) \in \Omega_c \cap A_c$.

The optimal location of the sensors will be the Nash equilibrium solutions of the game $\Gamma_c = \langle N; \Omega_c, \ldots, \Omega_c; f_1, \ldots, f_n \rangle$, where each player in $N = \{1, 2, \ldots, n\}$, for each $i \in N$, minimizes the cost $f_i : A_c \to \mathscr{R}$ for $(x_1, y_1, \ldots, x_n, y_n) \in \Omega_c \cap A_c$.

3.3 Nash Genetic Algorithm for the Location Problem

3.3.1 Genetic Algorithm

Let $X_1, X_2, \ldots, X_n$ be compact subsets of an Euclidean spaces, denoted as search space. Let $f_1, f_2, \ldots, f_n$ be real valued functions, defined on $X_1 \times X_2 \times \cdots \times X_n$, representing the objective functions to be maximized.

Let $s = x_1, x_2, \ldots, x_n$ be the individual (or chromosome) representing a feasible solution in the search space. A finite set of individuals make up a population. It can be viewed as a sampling of the problem domain that generation by generation maps zones with an higher probability of presence of the optimum ([10]).

A typical genetic algorithm consists of several steps.

- Population initialization: at the first step, a random population is set to map the search domain.
- Selection: on the sorted population, a probabilistic based selection of parents is made to permit coupling of best individuals without wasting worst chromosomes that may be useful to move towards unexplored zones of search space.
- Crossover: on selected parents, a crossover operator is applied to create two new individuals. This operator may be applied in several forms.
- Mutation: to avoid premature stagnation of the algorithm a mutation operator is used, randomly changing a bit of the just created chromosomes.
- Fitness computation: objective function and constraints must be evaluated to sort individuals in the population.
- Termination criterion: usually two criteria are defined in a GA, one on the maximum number of total generations and one on the maximum number of total generations without improvements on the best chromosome.

3.3.2 Nash Equilibrium Game

According to the definition of Nash equilibrium presented in 3.2.3, the algorithm for a n players Nash equilibrium game is presented [6–8, 15].

The algorithm is based on the *Nash adjustment process* [12], where players take turns setting their outputs, and each player's chosen output is a best response to the output that his opponent chose the period before. If the process does converge, the solution is an optimal location of the n sensor devices.

Let $\underline{x} = \underline{x}_1, \ldots, \underline{x}_n$ be a feasible solution for the n player Nash problem. Then x_i denotes the subset of variables handled by player i, belonging to a metric space X_i, and optimized by an objective function called f_i. Player i search the optimal solution with respect to his objective function by modifying $\underline{x}_i$.

At each step k of the optimization algorithm, player i optimizes x_i^k using $x_{(-i)}^{k-1} = \underline{x}_1^{k-1}, \ldots, \underline{x}_{i-1}^{k-1}, \underline{x}_{i+1}^{k-1}, \ldots \underline{x}_n^{k-1}$.

The first step of the algorithm consists of creating n different populations, one for each player. Player i's optimization task is performed by population i. Let $\underline{x}_i^{k-1}$ be the best value found by player i at era $k-1$. At era k, player i optimizes $\underline{x}_i^k$ using $\underline{x}_{-i}^{k-1}$ in order to evaluate the chromosome. At the end of kth era optimization procedure players $-i$ communicate their own best value $\underline{x}_{-i}^k$ to player i who will use it at era $k+1$ to generate their entire chromosome, using only $\underline{x}_i^k$ for common GAs crossover and mutation procedures. A Nash equilibrium is reached when no player can further improve his objective function, or a generation number limit is reached.

3.3.3 Test Cases

In this section, numerical results for the constrained location model are shown. They have been obtained using the Nash Genetic Algorithm presented above, with parameters summarized in Table 3.1.

First results are relative to the grid constrained case, in which RPCs can be located only at defined values of the second coordinate $h_1, \ldots, h_k$. In this case, the genetic algorithm is modified to handle a discrete variable $y \in Y$, where $Y = \{h_1, \ldots, h_k\}$ is the set of feasible bands.

In Figs. 3.1, 3.2 and 3.3 the comparison for uncostrained and constrained cases are shown, changing the number of rows on which the RPCs are contrained case by case, depending on the results of the unconstrained cases. The optimal location points are denoted by blue circles in the unconstrained case, and by red squares in the constrained case.

Table 3.1 Genetic algorithms characteristics

Parameter	Value or type
Chromosome	Binary string
Crossover	Multi-cut
Mutation probability	0.01
Population size	100
Mating-pool	50

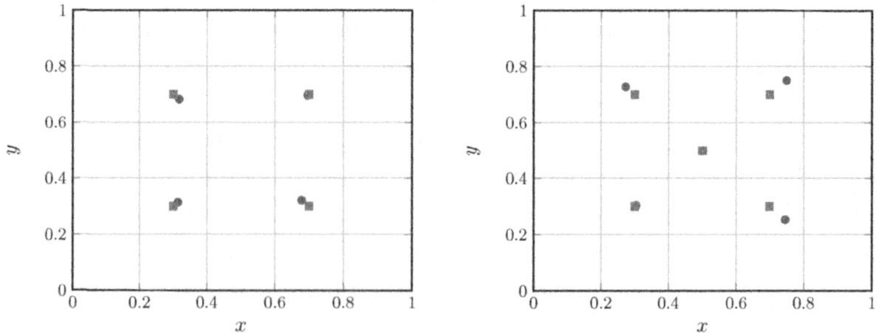

Fig. 3.1 Cases $n = 4$ and $Y = \{0.3, 0.7\}$; $n = 5$ and $Y = \{0.3, 0.5, 0.7\}$

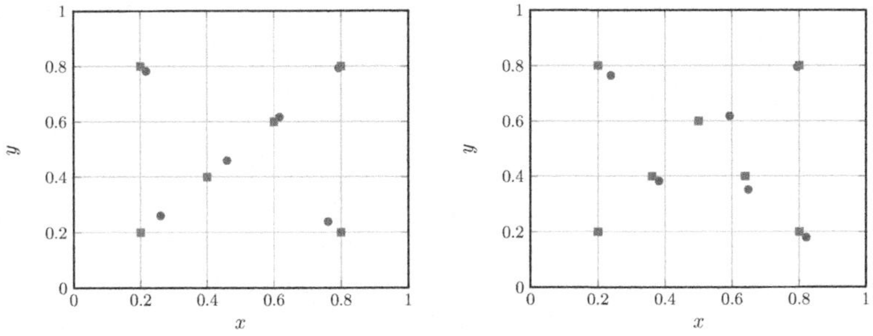

Fig. 3.2 Cases $n = 6, 7$ and $Y = \{0.2, 0.4, 0.6, 0.8\}$

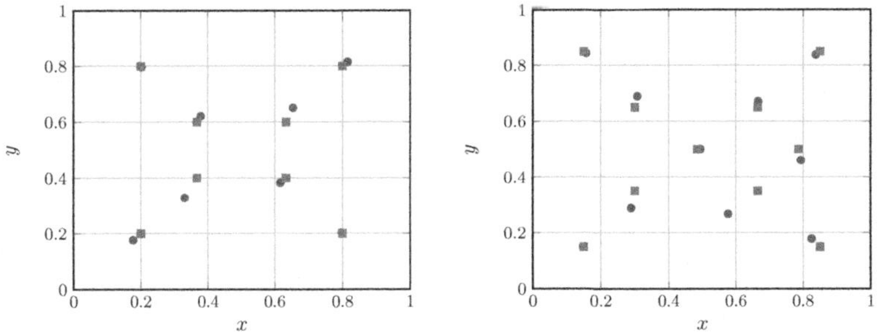

Fig. 3.3 Cases $n = 8$ and $Y = \{0.2, 0.4, 0.6, 0.8\}$; $n = 10$ and $Y = \{0.15, 0.35, 0.5, 0.65, 0.85\}$

3.4 Conclusions

In this paper the problem of locating a given number of sensor devices has been solved by means of a facility location problem whose solutions are the Nash equilibrium profiles of a suitable normal form game. The objective functions are given according to physical requirements. For such a problem a numerical procedure based on a genetic type of algorithm has been used to compute the final configurations. We considered the special case where the admissible region is made by a set of parallel segments, due to operative constraints (for example, electricity lines).

Other possible cases could be examined, for example the case where in the admissible region a convex obstacle is present. In this case the optimal location of the sensors will be the Nash equilibrium solutions of the game $\Gamma_c = \langle N; \Omega_c, \ldots, \Omega_c; f_1, \ldots, f_n \rangle$, where each player in $N = \{1, 2, \ldots, n\}$, for each $i \in N$, minimizes the cost $f_i : A_c \to \mathscr{R}$ for $(x_1, y_1, \ldots, x_n, y_n) \in \Omega_c \cap A_c$ and $\Omega_c = \Omega \setminus T$ with T a closed subset of Ω (a triangle, a circle, etc.). In the numerical procedure the objective functions can been modified to handle obstacles as penalty functions applied to the principal objective. In particular, f_i the objective function relative to the ith player, it is penalized by:

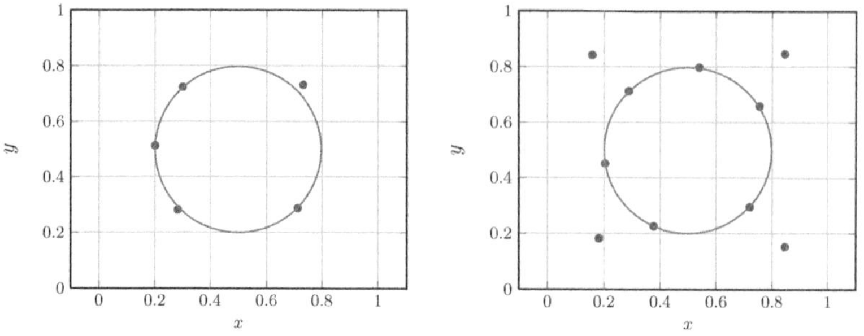

Fig. 3.4 Cases for $n = 5, 10$ with *circle shaped* obstacle

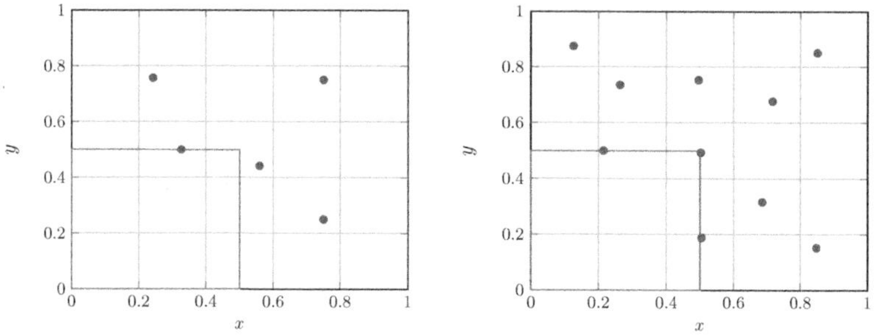

Fig. 3.5 Cases for $n = 5, 10$ with *box shaped* obstacle $T = [0, 0.5]^2$

$$f_i = f_i / f_{pen}$$

where $f_{pen} \in [0, 1]$ is a suitable penalty function.

For example, for a circular obstacle $f_{pen} = d(x, y)/r_c$, where $d(x, y)$ is the minimum distance between the sensor (x, y) and the center of the circular obstacle, r_c is the radius of the circle. Two test cases are shown in Fig. 3.4 with T given by the circle centered at $(0.5, 0.5)$ with radius 0.25.

In other cases, for example if we have a rectangular obstacle, a constant penalty $(f_{pen} = 0.1)$ can be applied for each sensor located in the unfeasible region. Two test cases are shown in Fig. 3.5.

A more systematic study of the constrained case from a theoretical as well as from a numerical point of view will be the object of future research.

Acknowledgments This work has been partially supported by F.A.R.O. 2012 "Metodi Matematici per la modellizzazione di fenomeni naturali", University of Naples Federico II, Italy.

References

1. Bacci C, Bao KZ, Barone F, Bartoli B, Bernardini P, Buonomo R, Bussino S, Calloni E, Cao BY, Cardarelli R et al (2002) High altitude test of RPCs for the Argo YBJ experiment. In: Nuclear instruments and methods in physics research section a: accelerators, spectrometers, detectors and associated equipment, vol 443(2), Elsevier, Amsterdam, pp 342–350
2. Başar T, Olsder GJ (1999) Dynamic noncooperative game theory, Reprint of the second (1995) edition, Classics in applied mathematics, Society for industrial and applied mathematics (SIAM), Philadelphia
3. Benabbou A, Borouchaki H, Laug P, Lu J (2008) Sphere packing and applications to granular structure modeling. In: Garimella RV (ed) Proceedings of the 17th international meshing roundtable, Springer, 12–15 October 2008
4. Clarich A, Periaux J, Poloni C (2003) Combining game strategies and evolutionary algorithms for CAD parametrization and multi-point optimization of complex aeronautic systems. In: EUROGEN 2003, Barcelona
5. Conway JH, Sloane NJA (1998) Sphere packings, lattices and groups. Springer, New York
6. D'Amato E, Daniele E, Mallozzi L, Petrone G (2012) Equilibrium strategies via GA to stackelberg games under multiple follower best reply. Int J Intell Syst 27:74–85
7. D'Amato E, Daniele E, Mallozzi L, Petrone G, Tancredi S (2012) A hierarchical multi-modal hybrid Stackelberg-Nash GA for a leader with multiple followers game. In: Sorokin A, Murphey R, Thai MT, Pardalos PM (ed) Dynamics of information systems: foundations mathematical, Springer proceedings in mathematics & statistics. Springer, New York, pp 267–280
8. D'Amato E, Daniele E, Mallozzi L, Petrone G (2012) Waiting time costs in a bilevel location-allocation problem. Contrib Game Theor Manage 5:178–186
9. Dean AM, Voss D (2000) Design and analysis of experiments series, Springer texts in statistics. Springer, New York
10. Deb K (2001) Multi-objective optimization using evolutionary algorithms. Wiley, Chichester
11. Donev A, Torquato S, Stillinger FH, Connelly R (2004) A linear programming algorithm to test for jamming in hard-sphere packings. J Comput Phys 197:139–166
12. Fudenberg D, Tirole J (1993) Game theory. The MIT Press, Cambridge
13. Hales TC (1992) The sphere packing problem. J Comput Appl Math 42:41–76
14. Mallozzi L (2007) Noncooperative facility location games. Oper Res Lett 35:151–154
15. Mallozzi L, D'Amato E, Daniele E, Petrone G (2011) N leader—M follower coalition games with genetic algorithms and applications. In: Poloni C, Quagliarella D, Periaux J, Gauger N, Giannakoglou K (eds) Evolutionary and deterministic methods for design, optimization and control. CIRA, Capua
16. Nurmela KJ (1997) Stochastic optimization methods in sphere packing and covering problems in discrete geometry and coding theory, PhD Thesis, Helsinki University of Technology, printed by Picaset Oy
17. Periaux J, Chen HQ, Mantel B, Sefrioui M, Sui HT (2001) Combining game theory and genetic algorithms with application to DDM-nozzle optimization problems. Finite Elem Anal Des 37:417–429
18. Sloane NJA (1998) The sphere packing problem, Shannon lecture. AT & T Shannon Lab, Florham Park, pp 07932–0971
19. Sutou A, Dai Y (2002) Global optimization approach to unequal sphere packing problems in 3D. J Optim Theor Appl 114(3):671–694
20. Wang JF, Periaux J (2001) Multi-point optimization using GAS and Nash/Stackelberg games for high lift multi-airfoil design in aerodynamics. In: Proceedings of the 2001 congress on evolutionary computation CEC2001, May 2001, pp 552–559

Chapter 4
The Role of Artificial Neural Networks in Evolutionary Optimisation: A Review

M. Maarouf, A. Sosa, B. Galván, D. Greiner, G. Winter, M. Mendez and R. Aguasca

Abstract This paper reviews the combination of Artificial Neural Networks (ANN) and Evolutionary Optimisation (EO) to solve challenging problems for the academia and the industry. Both methodologies has been mixed in several ways in the last decade with more or less degree of success, but most of the contributions can be classified into the two following groups: the use of EO techniques for optimizing the learning of ANN (EOANN) and the developing of ANNs to increase the efficiency of EO processes (ANNEO). The number of contributions shows that the combination of both methodologies is nowadays a mature field but some new trends and the advances in computer science permits to affirm that there is still room for noticeable improvements.

Keywords Artificial neural networks · Evolutionary optimisation · Evolutionary algorithm

4.1 Introduction

Artificial Neural Network (ANN) and Evolutionary Algorithm (EA) are relatively young research areas that were subject to a steadily growing interest nowadays; represent two evolving technologies that are inspired by biological information science.

ANN is derived from brain theory to simulate learning behavior of an individual, which is, used for approximation and generalization, while EA is developed from the evolutionary theory raised by Darwin to evolve the whole population for better fitness. Evolutionary Algorithm is actually used as an optimisation algorithm, and not a learning algorithm.

In particular, EO has been used to search for the design and structure of the network and to select the most relevant features of the training data. It is well known that to solve nonlinearly separable problems, the network must have at least one

M. Maarouf (✉) · A. Sosa · B. Galván · D. Greiner · G. Winter · M. Mendez · R. Aguasca
División de Computación Evolutiva y Aplicaciones (CEANI), Instituto Universitario de Sistemas Inteligentes y Aplicaciones Numéricas en Ingeniería (SIANI), Universidad de Las Palmas de Gran Canaria, Islas Canarias, Spain
e-mail: maaroufmusta@gmail.com

D. Greiner et al. (eds.), *Advances in Evolutionary and Deterministic Methods for Design, Optimization and Control in Engineering and Sciences*, Computational Methods in Applied Sciences 36, DOI 10.1007/978-3-319-11541-2_4

hidden layer; but determining the number and the size of the hidden layers is mostly a matter of trial and error. EOs has been used to search for these parameters, to generate the appropriate network to solve specific problems.

On the other hand, ANN has yield many benefits solving a lot of problems in fields as diverse as biology, physics, computer science and engineering. In many applications, the real-time solutions of optimisation problems are widely required. However, traditional algorithms may not be efficient since the computing time required for a solution is greatly dependent on the structure of the problems and there dimension.

A promising approach to solving such problems in real time is to employ artificial neural networks based on circuit implementation [133]. ANNs possess many desirable properties such as real-time information processing. Therefore, neural networks for optimisation, control, and signal processing received tremendous interests. The theory, methodology, and applications of ANNs have been widely investigated.

The present work focuses on Evolutionary Algorithms (EO), Artificial Neural Networks (ANN) and their joint applications as a powerful tool to solve challenging problems for the academia and the industry. In this section the essentials and importance of both methodologies are presented.

4.1.1 Evolutionary Algorithms

Since 1960s, there has been an increasing interest in emulating evolutionary process of living beings to solve hard optimisation problems [34, 101]. Simulating these features of living beings, yields stochastic optimisation procedures called Evolutionary Optimisation (EO). EO belongs to global search meta-heuristics methods since, by its own nature, explores the whole decisional space for global optima.

Evolutionary algorithms (EAs) are a class of stochastic and probabilistic optimisation methods that are inspired by some presumed principles of evolution; attempt to emulate the biological process of evolution, incorporating concepts of selection, reproduction, and mutation.These techniques, inspired in Darwinian evolution postulates, consider a population of individuals on which selection and diversity generation procedures are performed, guaranteeing better fitted individuals to survive through successive iterations [12, 46, 47]. Each individual is a potential solution of the optimisation problem, so it belongs to decision space. Every iteration (generation), individual features are combined by means of recombination operators such selection, crossover and mutation, driving solutions to global optima. By mimicking this process, EAs are able to evolve solutions to real world problems, if they have been suitably encoded.

Evolutionary Optimisation has demonstrated to be effective in engineering and science optimisation problems in several fields such as: Aerospatiale applications [15], energy [9], transport planning, RAMS [87, 96, 97, 111], task scheduling and so on.

The applications mentioned above usually results in high dimensional search spaces, highly non-linear optimisation problems, non-convex optimisation, highly constrained problems, uncertainty effects and/or multicriteria paradigm. These numerical difficulties are commonly tackled by these meta-heuristics, which usually outperforms traditional optimisation strategies with lower probability of being stacked in local optima, and being able to yield a Pareto set in a single run of the algorithm (multi-objective approach). Because of this, Evolutionary Optimisation has been an important R&D matter in the last decade.

Among the variety of EO meta-heuristics, the most relevant nowadays are: Genetic Algorithm (GA), Evolutionary Strategy (ES), Evolutionary Programming (EP), Collective Intelligence, Memetic Algorithms, and Differential Evolution.

In recent years, there has been an increase in the use of evolutionary approaches in the training and optimisation of artificial neural networks(ANNs). Different works are presented in Sect. 4.2.1.

4.1.2 Artificial Neural Networks ANN

Artificial neural networks (ANNs) are biologically inspired computer programs, inspired from the morphological and biophysical properties of neurons in the brain. ANNs are designed to simulate the way the human brain processes information. Neural networks are similar to the human brain in the following two ways:

1. A neural network acquires knowledge by learning.
2. The knowledge of a neural network is stored in the connections between neurons known as synaptic weights.

[98] were the first patterned biological neurons from the binary automata. A second generation of neurons integrates a nonlinear activation function that has allowed growing up the interest in ANNs [69], allowing to solve nonlinear problems.

The power and utility of artificial neural networks have been shown in several applications including speech synthesis, diagnostic problems, medicine, business and finance, control, robotics, signal processing, computer vision and many other industrial problems that are included in the category of pattern recognition. But knowing that there exists a suitable network for a specific problem is interesting, finding it proved to be difficult. Although there exist some algorithms to set the weights by learning from data training given a fixed topology, even if get stuck in local minima. To lead to good results, they strongly depend on problem specific parameter settings and on the topology of the network.

Training procedures of neural networks are optimisation algorithms aim to minimize the global error output respect to connection weights under conditions of noise, qualitative uncertainty, nonlinearity, etc. Therefore training procedures of neural networks provide a common approach to the optimisation task in process optimisation and control applications [149]. However, this is an important issue because there are strong biological and engineering evidences to support that an ANN is determined by its architecture. Using EOs as a procedure to assist neural network design and training seems to be a straightforward idea. Branke in [23] explain how EO improve design and training ANNs.

Today, significant progress has been made in the field of neural networks, enough to attract a lot of attention and research funding. Research in the field is advancing on many fronts. New neural concepts are emerging and applications to complex problems developing. Clearly, today is a period of transition for neural network technology.

ANNs are inherently parallel architectures which can be implemented in software and hardware. One important implementation issue is the size of the neural network and its weight adaptation. This makes the hardware implementation complex and software learning slower.

ANNs has two distinct steps [172];

1. Choosing proper network architecture.
2. Adjusting the parameters of a network so as to minimize certain fit criterion.

Even if, the most of problems treated in real study are complex, so the unknown architecture of the ANN is set arbitrarily or by trial and error [121], and small networks cannot achieve the solution in much iteration, but if the network is too large, it leads to overfitting and a bed generalization, and the majority of neural networks suffers of premature convergence and low global convergence speed etc. In order to overcome these limitations, some improvements were made for EOs in the last decade on ANNs.

4.2 Different Use of ANNEO and EOANN

In the last decade, there has been a great interest combining learning and evolutionary techniques in computing science to solve complex problems for different fields. Different works presented how ANNs are an optimisation tools [33, 120, 131, 136, 173, 175]. In this paper we just limit this review to several ways in which EOs and ANNs may be combined.

The next section presents different approaches of combining EOs and ANNs.

4.2.1 The Use of EOs in ANNs: EOANN

In recent years, evolutionary algorithms (EAs) have been applied to the ANN's optimisation. The first applications of EAs to ANN parameter learning date back to the late 80s in the fields of Genetic Algorithms (GAs) [45, 47] and Evolutionary Programming (EP) [47, 103]. EAs are powerful search algorithms based on the mechanism of natural selection. Since the 90s, EAs have been successfully used for optimizing the design and the parameters of ANNs [132, 150]. This special class of ANNs in which evolution is another fundamental form of adaptation in addition to learning creating an EOANN [40, 167]. EOANN are used to find the best data subset which optimizes the ANN training for a specific problem.

The result of EOANN is an ANN with optimal performance to estimate the value of one or more variables and the estimation error is strongly related to the quality of the training set in terms of size and treatment of the possible outputs. Another approach is the evolutionary weight training process, to optimizing connection weights, learning rules, and optimizing the network architecture by identifying how many inputs neurons, hidden neurons, hidden layers we have to use, to get a good performance [18, 39, 88, 92, 128]. Considerable research has been conducted on the evolution of topological structures of networks using evolutionary algorithms [3, 6, 13, 14, 20, 73, 84].

In the other hand, an essential issue is to improve generalization of the neural network training models. Early stopping, weight decay and curvature-driven smoothing and other techniques are been used to resolve this problem, another approach is including an additional term in the cost function of learning algorithms, which penalizes overly high model complexity. Regularization of neural training was treated with Eos in [1, 75].

In general, constructing neural network consists of two major steps, design and training component networks, combining of the component networks predictions to produce the neural networks solutions. ANN training method has some limitations associated with overfitting, local optimum problems and slow convergence rate. In order to overcome the limitations, some scientist proposed particle evolutionary algorithm to train ANN.

The research use EOs to evolve and design the structure architecture or the selection of the training algorithms and optimisation of its synaptic weight initialization, thresholds, training ratio, momentum factor, etc., of neural network roundly. The scientist object is to accelerate the convergence speed of network and optimize the result in case of trapping into local optimal value, and a better searching space is found out in the solution space.

The initial set of weights to be used in learning of ANN has a strong influence in the learning speed and in the quality of the solution obtained after training. An inadequate initial choice of the weight values may cause the training process to get stuck in a poor local minimum or more time to converge. Inappropriate topology selection and learning algorithm are frequently bed; there is little reason to expect that one can find a uniformly best algorithm for selecting the weights in an ANN [94]. This kind of topology was chosen for the following main reasons:

1. Additional links that skip adjacent layers allow the genetic algorithm to remove a whole layer of nodes while still keeping the network functional.
2. Some additional shortcut connections, e.g., those between the network input and output, may ease training and therefore the whole genetic topology optimisation may become faster.

There are different ways to evaluate weights of the component networks. For example Jimenez [74] use weights determined by confidence of the component networks. Zhou [184] utilize the genetic algorithm to find proper weights for each member of an ensemble. In [143], present and define many operators and crossover applied to weights of an ANN. The importance of a good choice for the initial set of weights is stressed by Kolen and Pollak [83].

ANNs strongly depend on the network topology, the neurons activation function, the learning rule, etc. optimisation for these factors are usually unknown a priori because they depend mainly on the particular training set to be considered and on the nature of the solution [137]. So, for practical purposes, the learning rule should be based on optimisation techniques that employ local search to find the optimal solution [124, 126].

Evolutionary approaches have been shown to be very effective as an optimisation technique, their efficiency could he exploited in training and constructing neural networks, there architecture/design and learning, they can evolve towards the optimal architecture without outside interference, thus eliminating the tedious trial and error work of manually finding an optimal network, adapting the connection weights and learning algorithms according to the problem environment. Many considerable efforts at obtaining optimal ANNs based on EAs have been reported in the literature [4–7, 10, 13, 24–28, 31, 52, 53, 58, 63–67, 82, 99, 117, 122, 132, 139, 144, 166, 168–171].

The EAs are using diver methods to encode ANNs for the purpose of training and design. The common approach is to encode the ANN weights into genes that are then concatenated to build the genotype. Encoding methods can be divided in three main groups according to the process of creating the network from the encoded genome: direct, parametric and indirect encoding. A real coded genetic algorithm is used to optimize the mean square of the error produced by training a neural network established by Aljahdali in [11]. Benaddy et al. [17] present a real coded genetic algorithm that uses the appropriate operators type to train feed-forward neural network. Larranaga et al. [86] describes various methods used to encode artificial neural networks to chromosomes to be used in evolutionary computation.

Another important point to note is the use of EOs to extract rules from neural networks trained. Rule extraction from neural networks is attracting wide attention because of its computational simplicity and ability to generalize [49–51, 141].

4.2.2 The Use of ANNs in EO: ANNEO

In real world applications, sometimes it is not easy to obtain the objective value. Therefore we needs complicated analysis or time consuming simulation to evaluate the performance of design variables [147]. As ANNs represent a nonlinear robust modeling technique which are developed, or trained, based on analytical or simulated results of a subset of possible solutions [35], give to ANNs an important role in solving problems with extremely difficult or unknown analytical solution. ANNs can be used with a huge reduction of cost in terms of objective function evaluations [22]. One of the first pioneers in ANNEO were Hopfield and Tank who presented an ANN for solving combinatorial problems, that was mapped into a closed-loop circuit [68, 133], named Hopfield Neural Network (HNN). HNN is a continuously operating model very close to analog circuit implementation. Since 1985 a wide variety of ANNs have been designed for improving the performance of HNN.

In ANNEO fast objective functions evaluations are performed using pre-trained ANN. Normally in iterative process subsets of objective function values obtained using exact procedures are used in an embedded EOANN algorithm, and some of the new objective function evaluations are performed using the ANN. The result is an EO which evolves faster than conventional ones but special care must be paid to the selection of appropriate training subsets and the number of objective functions evaluated using ANN in order to avoid convergence problems. Acceleration of the convergence speed is done in [176] as an ANN model trained to approximate the fitness function according to an adaptive scheme for increasing the number of network fitness calculation.

EOs usually needs a large number of fitness evaluations before a satisfying result can be obtained. And as an explicit fitness function does not exist, or his evaluation is computationally very expensive. It is necessary to estimate the fitness function by constructing an approximate model or presenting an interpolation of the true fitness function via some interpolation technique as ANNs [22, 55] employ a Feedforward neural networks for fitness estimation, reduce the number of expensive fitness function evaluations in evolutionary optimisation. The idea of the implementation of an ANN, that approximates the fitness function, comes from the universal approximation capability of multi-layer neural networks [69]. An artificial neural networks model is used in order to reduce the number of time-consuming fitness evaluations [55].

Real-time solutions to resolve problems are often needed in engineering applications. Solve many problem of optimisation in real time usually contain time-varying parameters, such in signal processing, robotic, time series, etc., and we have to reduce and optimize the performance. The numbers of decision variables and constraints are usually very large and large-scale optimisation problems are even more challenging when they have to be solved in real time to optimize the performance of dynamical systems. For such applications, Conventional numerical methods may not be effective at all due to the problem dimensionality and stringent requirement on computational time [16]. The employment of ANNs techniques as Recurrent Neural Networks (RNN) [21, 33, 48, 79, 89, 91, 113, 115, 127, 133, 145, 146, 160, 165, 177] (Papers below proposed neural networks guaranteed to be globally convergent in finite time to the optimal solutions), Fuzzy Neural Network (FNN) are a promising approach to resolve this inefficiency [123].

Application of ANNs algorithms receive increase interests for optimisation as we see in [21, 32, 33, 37, 79, 93, 105–107, 118, 133, 134, 151, 154, 156, 157, 182], using gradient and projection methods [91, 104], Bouzerdoum and Pattison [21] presented a neural network for solving quadratic optimisation problems with bounded variables only, which constitutes a generalization of the network described by Sudharsanan [127]. Rodrguez-Vzquez et al. [118] proposed a class of neural networks for solving optimisation problems, in which their design does not require the calculation of a penalty parameter. To avoid using finite penalty parameters, many other studies have been done in [21, 56, 70, 78, 118, 134, 148, 153, 156, 157, 159, 182, 183]. In [151–154], the authors presented several neural networks for solving linear and quadric programming problems with no unique solutions, which are proved to be globally convergent to exact solutions, and in which there is no variable parameter to tune.

Romero [119] approached optimisation problems with a multilayer neural network. Da Silva [36] coupled fuzzy logic with Hopfield Neural Networks to solve linear and non-linear optimisation problems. Case studies with convex and non-convex optimisation problems are studied in illustrate the approach with a trained Multilayer neural networks [143]. Xia et al. [155, 158, 160] proposed a general projection neural network, that includes projection neural network, the primal-dual neural network, and/or the dual neural network, as special cases, for solving a wider class of variational inequalities and related optimisation problems.

A reliability network reflect a network optimized whose topology is optimist, at minimal cost, under the constraint that every pair of neurons can communicate with each other. Abo El fotoh et al. [2] presents an ANN for solving this problem o reliability, by constructing an energy function whose minimization process drives the neural network into one of its stable states.

An ANN aided with other algorithms as simulated-annealing (SA) algorithm, can be usefully used to resolve optimisation problems as they do in [85]. Chaotic artificial neural are studied and established as optimisation model in [8, 85]. Another point to be mentioned is the use of ANNs to re-optimisation online as it presented in [138] .

4.3 Some Applications Using ANNEO and EOANN

Many academic papers show the applicability of EOANN to optimize different parameters of ANN, to improve their training and their stability [13, 20, 43, 71–73, 76, 84] and the papers cited below in Sect. 4.2.2. Other EOANN applications were performed in several fields such as Financial Engineering [41, 59], grammatical inference [19], Chemical Reaction [178], Hydrochemistry [142], Time series prediction [77, 81, 180], Classification Process [29, 30, 44, 80, 90, 95, 109, 112, 135], Medicine [54], Diagnosis problems [17, 72, 129, 130], Diverse Engineering Applications [57, 100, 125, 135, 174], Robotic [161], Monitoring [38], Traffic Prediction [114], Control Systems [181], Neutron spectrometry and Dosimetry research areas [108], Multi-agent systems [81], Regression problems [61], Chaos dynamics Problem [179, 180], Reliability [62], etc.

Also, the role of ANNs in EO is presented in many academic papers as in [22, 147, 176], and in many fields like Reliability Systems [35], Electromagnetics optimisation [22], pressure vessel design problem [147], aerodynamic design optimisation [73], Electric-Field optimisation [85, 116, 164], Cybernetics [162], Design optimisation [102, 163], Diagnosis [60], Power Technology [110, 140], Resource Management optimisation [42], etc.

4.4 Conclusions

We can deduce that if the purpose of using ANNs is to find the best network configuration for solving particular problems, is has been possible employing EOs, as it have mentioned above in several works. EOs provides good approximation to get success and speed of training of neural network based on the initial parameter settings, such as architecture, initial weights, learning rates, and others.

In the other hand, ANNs based objective function allows the fitness to be evaluated in a small fraction of the time taken to perform first principles analysis and permits the EAs to complete in a reasonably small amount of time.

Due to the last advances in both methodologies, there are several chances for future develops of joint procedures especially when complex industrial applications are addressed. Anyway the use of big computing facilities will continue being still necessary for that applications.

References

1. Abbass HA (2003) Speeding up back-propagation using multiobjective evolutionary algorithms. Neural Comput 5:2705–2726
2. Abo El Fotoh HMF, Al-Sumait LS (2001) A neural approach to topological optimization of communication networks, with reliability constraints. IEEE Trans Reliab 50(4):397–408
3. Abraham A, Nath B (2001) ALEC—an adaptive learning framework for optimizing artificial neural networks. In: Alexandrov VN et al (eds) Computational science, Springer, Germany, San Francisco, 171–180
4. Abraham A (2002) Optimization of evolutionary neural networks using hybrid learning algorithms. Int Symp Neural Netw 3:2797–2802. doi:10.1109/IJCNN.2002.1007591
5. Abraham A (2004) Meta learning evolutionary artificial neural networks. Neurocomputing 56:138
6. Aguilar J, Colmenares A (1997) Recognition algorithm using evolutionary learning on the random neural networks. International symposium on neural networks. doi:10.1109/ICNN.1997.616168
7. Ajith A (2004) Meta-learning evolutionary artificial neural networks. Neurocomputing 56:1–38. doi:10.1016/S0925-2312(03)00369-2
8. Akkar HAR (2010) Optimization of artificial neural networks by using swarm intelligent
9. Alarcon-Rodriguez A, Ault G, Galloway S (2010) Multi-objective planning of distributed energy resources: a review of the state-of-the-art. Renew Sustain Energy Rev 14(5):1353–1366
10. Alexandridis A (2012) An evolutionary-based approach in RBF neural network training. IEEE workshop on evolving and adaptive intelligent systems. doi:10.1109/EAIS.2012.6232817
11. Aljahdali S, Buragga KA (2007) Evolutionary neural network prediction for software reliability modeling. The 16th international conference on software engineering and data engineering
12. Alonso S (2006) Propuesta de un Algoritmo Flexible de Optimización Global. Ph.D. thesis, Departamento de Informática y Sistemas. Universidad de Las Palmas de Gran Canaria
13. Angeline P, Saunders G, Pollack J (1994) An evolutionary algorithm that constructs recurrent neural networks. IEEE Trans Neural Netw 5(1):54–65
14. Arena P, Caponetto R, Fortuna L, Xibilia MG (1992) Genetic algorithms to select optimal neural network topology. In: Proceedings of the 35th midwest symposium on circuits and systems, vol 2, pp 1381–1383. doi:10.1109/MWSCAS.1992.271082

15. Arias-Montano A, Coello CA, Mezura-Montes E (2012) Multiobjective evolutionary algorithms in aeronautical and aerospace engineering. IEEE Trans Evol Comput 16(5):662–694
16. Bazaraa MS, Sherali HD, Shetty CM (1993) Nonlinear programming: theory and algorithms, 2nd edn. Wiley, New York
17. Benaddy M, Wakrim M, Aljahdali S (2009) Evolutionary neural network prediction for cumulative failure modeling. ACS/IEEE international conference on computer systems and applications. doi:10.1109/AICCSA.2009.5069322
18. Bevilacqua V, Mastronardi G, Menolascina F, Pannarale P, Pedone A (2006) A novel multi-objective genetic algorithm approach to artificial neural network topology optimisation: the breast cancer classification problem. IJCNN '06. International joint conference on neural networks, pp 1958–1965
19. Blanco A, Delgado M, Pegalajar M (2000) A genetic algorithm to obtain the optimal recurrent neural network. Int J Approx Reason 23(1):67–83
20. Bornholdt S, Graudenz D (1993) General asymmetric neural networks and structure design by genetic algorithms: a learning rule for temporal patterns. In: International conference on systems, man and cybernetics. Systems engineering in the service of humans, conference proceedings, vol 2, pp 595–600
21. Bouzerdoum A, Pattison TR (1993) Neural network for quadratic optimization with bound constraints. IEEE Trans Neural Netw 4:293–303
22. Bramanti A, Di Barba MF, Savini A (2001) Combining response surfaces and evolutionary strategies for multiobjective pareto-optimization in electromagnetics. Stud Appl Electromag Mech JSAEM 9:231–236
23. Branke J (1995) Evolutionary algorithms for neural network design and training. In: Proceedings of the first nordic workshop on genetic algorithms and its applications, pp 145–163
24. Branke J, Kohlmorgen U, Schmeck H (1995) A distributed genetic algorithm improving the generalization behavior of neural networks. In: Lavrac N et al Proceedings of the European conference on machine learning, pp 107–112
25. Braun H, Weisbrod J (1993) Evolving neural networks for application oriented problems. In: Fogel DB (ed) Proceedings of the second conference on evolutionary programming
26. Braun H (1995) On optimizing large neural networks (multilayer perceptrons) by learning and evolution. In: Proceedings of the third international congress on industrial and applied mathematics, ICIAM
27. Bukhtoyarov VV, Semenkina OE (2010) Comprehensive evolutionary approach for neural network ensemble automatic design. IEEE congress on evolutionary computation p 1–6. doi:10.1109/CEC.2010.5586516
28. Bundzel M, Sincak P (2006) Combining gradient and evolutionary approaches to the artificial neural networks training according to principles of support vector machines. In: International symposium on neural networks, pp 2068–2074. doi:10.1109/IJCNN.2006.246976
29. Cantu-Paz E, Kamath C (2002) Evolving neural networks for the classification of galaxies. Genetic and evolutionary computation conference, pp 1019–1026
30. Castellani M (2006) ANNE—a new algorithm for evolution of artificial neural network classifier systems. doi:10.1109/CEC.2006.1688728
31. Chi-Keong G, Eu-Jin T, Kay CT (2008) Hybrid multiobjective evolutionary design for artificial neural networks. IEEE Trans Neural Netw 19(9):1531–1548. doi:10.1109/TNN.2008.2000444
32. Cichocki A, Unbehauen R (1991) Switched-capacitor artificial neural networks for differential optimization. J Circuit Theory Appl 19:61187
33. Cichocki A, Unbehauen R (1993) Neural networks for optimization and signal processing. Wiley, New York
34. Coello CAC, Lamont GB, Veldhuizen DAV (2006) Evolutionary algorithms for solving multi-objective problems (genetic and evolutionary computation). Springer, New York
35. Coit DW, Smith AE (1996) Solving the redundancy allocation problem using a combined neural network/genetic algorithm approach. Comput Oper Res 23(6):515–526

36. Da Silva IN (1997) A neuro-fuzzy approach to systems optimization and robust estimation. PhD thesis, School of electrical engineering and computer science, State University of Campinas, Brazil
37. Dempsey GL, McVey SE (1993) Circuit implementation of a peak detector neural network. IEEE Trans Circ Syst II 40:585–591
38. Desai CK, Shaikh AA, (2006) Drill wear monitoring using artificial neural network with differential evolution learning. In: IEEE International conference on industrial technology ICIT 2006, pp 2019–2022. doi:10.1109/ICIT.2006.372500
39. Di Muro G, Ferrari S (2008) A constrained-optimization approach to training neural networks for smooth function approximation and system identification. IJCNN 2008. (IEEE world congress on computational intelligence). IEEE international joint conference on neural networks, vol 2353–2359
40. Ding S, Li H, Su C, Yu J, Jin F (2013) Evolutionary artificial neural networks: a review. Artif Intell Rev 39(3):251–260
41. Edwards D, Brown K, Taylor N (2002) An evolutionary method for the design of generic neural networks. In: Proceedings of the 2002 congress on evolutionary computation. CEC '02. vol 2, pp 1769–1774
42. Efstratios FG, Sotiris MG (2007) Solving resource management optimization problems in contact centers with artificial neural networks. Int Conf Tools Artif Intell 2:405–412
43. Farzad F, Hemati S (2003) An algorithm based on evolutionary programming for training artificial neural networks with nonconventional neurons. In: Canadian conference on electrical and computer engineering. vol 3. doi:10.1109/CCECE.2003.1226270
44. Fiszelew A, Britos P, Ochoa A, Merlino H, Fernndez E, Garca-Martnez R (2007) Finding optimal neural network architecture using genetic algorithms. Advances in computer science and engineering research in computing science 27:15–24
45. Fogel DB, Fogel LJ, Porto VW (1990) Evolutionary programming for training neural networks. In: Proceedings of the international joint conference on NNs, San Diego, CA, pp 601–605
46. Fogel D (1997) The advantages of evolutionary computation, bio-computing and emergent computation. In: Lundh D, Olsson B, and Narayanan A (eds) Skve, Sweden, World scientific press, Singapore, pp 1–11
47. Fogel D (1999) Evolutionary computation: towards a new philosophy of machineintelligence. 2nd edn, IEEE press
48. Forti M, Nistri P, Quincampoix M (2004) Generalized neural network for nonsmooth nonlinear programming problems. IEEE Trans Circ Syst I 51(9):1741–1754
49. Fukumi M, Akamatsu N (1996) A method to design a neural pattern recognition system by using a genetic algorithm with partial fitness and a deterministic mutation. In: Proceedings of IEEE international conference on SMC, vol 3, pp 1989–1993
50. Fukumi M, Akamatsu N (1998) Rule extraction form neural networks trained using evolutionary algorithms with deterministic mutation. In: International symposium on neural networks. vol 1. doi:10.1109/IJCNN.1998.682363
51. Fukumi M, Akamatsu N (1999) An evolutionary approach to rule generation from neural networks. IEEE Int Fuzzy Syst Conf Proc 3:1388–1393
52. Funabiki N, Kitamichi J, Nishikawa S (1998) An evolutionary neural network approach for module orientation problems. IEEE Trans Syst Man Cybern—Part B Cybern 28(6):849–855
53. Garcia-Pedrajas N, Hervas-Martinez C, Munoz-Perez J (2003) COVNET: a cooperative coevolutionary model for evolving artificial neural networks. IEEE Trans Neural Netw 14(3):575–596
54. Gorunescu F, Gorunescu M, Gorunescu S (2005) An evolutionary computational approach to probabilistic neural network with application to hepatic cancer diagnosis. In: Proceedings of the 18th IEEE symposium on computer-based medical systems (CBMS05)
55. Grning L, Jin Y, Sendhoff B (2005) Efficient evolutionary optimization using individual-based evolution control and neural networks: a comparative study. The European symposium on artificial neural networks, pp 273–278

56. Guo-Cheng L, Zhi-Ling D (2008) Sub-gradient based projection neural networks for non-differentiable optimization problems. In: Proceedings of the seventh international conference on machine learning and cybernetics, Kunming
57. Gwo-Ching L (2012) Application a novel evolutionary computation algorithm for load forecasting of air conditioning. In: Power and energy engineering conference (APPEEC), 2012 Asia-pacific, pp 1–4. doi:10.1109/APPEEC.2012.6307573
58. Haykin S (1999) Neural network, 2nd edn., A comprehensive foundationPretince Hall, New Jersey
59. Hayward S (2004) Evolutionary artificial neural network optimisation in financial engineering. In: Fourth international conference on hybrid intelligent systems. HIS '04, pp 210–215
60. He Y, Sun Y (2001) Neural network-based L1-norm optimisation approach for fault diagnosis of nonlinear circuits with tolerance. In: Lee proceedings-circuits devices and systems, pp 223–228. doi:10.1049/ip-cds:20010418
61. Hieu TH, Yonggwan W (2008) Evolutionary algorithm for training compact single hidden layer feedforward neural networks. In: International symposium on neural networks, pp 3028–3033: doi:10.1109/IJCNN.2008.4634225
62. Hochman R, Khoshgoftaar TM, Allen EB, Hudepohl JP (1997) Evolutionary neural networks: a robust approach to software reliability problems. In: International symposium on software reliability engineering. doi:10.1109/ISSRE.1997.630844
63. Holland JH (1975) Adaptation in natural and artificial systems. University of Michigan Press, Ann Arbor
64. Holland JH (1980) Adaptive algorithms for discovering and using general patterns in growing knowledge-based. Intl J Policy Anal Inf Syst 4(3):245–268
65. Holland JH (1986) Escaping brittleness: the possibilities of general purpose learning algorithms applied in parallel rule-based systems. In: Michaiski RS, Carbonell JG, Mitchell TM (eds) Machine learning II, pp 593–623
66. Holland JH, Holyoak KJ, Nisbett RE, Thagard PR (1987) Classifier systems, Q-morphisms, and induction. In: Davis L (ed) Genetic algorithms and simulated annealing, pp 116–128
67. Honavar V, Uhr L (1993) Generative learning structures and processes for generalized connectionist networks. Inf Sci 70:75–108
68. Hopfield JJ, Tank DW (1985) Neural Y computation on decisions optimization problem. Biol Cybem 52:141–152
69. Hornik K, Stinchcombe M, White H (1989) Multilayer feedforward networks are universal approximators. Neural Netw 2(5):359–366
70. Hu X, Jun W (2006) Solving pseudomonotone variational inequalities and pseudoconvex optimization problems using the projection neural network. IEEE Trans Neural Netw 17(6):1487–1499
71. Huning H (2010) Convergence analysis of a segmentation algorithm for the evolutionary training of neural networks. In: IEEE symposium on combinations of evolutionary computation and neural networks. doi:10.1109/ECNN.2000.886222
72. Husken M, Gayko JE, Sendhoff B (2000) Optimization for problem classes—neural networks that learn to learn 98–109
73. Husken M, Jin Y, Sendhoff B (2002) Structure optimization of neural networks for evolutionary design optimization. In: Proceedings of the 2002 GECCO workshop on approximation and learning in evolutionary computation, pp 13–16
74. Jimenez D (1998) Dynamically weighted ensemble neural networks for classification. In: Proceedings of the IJCNN-98, vol 1, Anchorage, AK, IEEE Computer Society Press, Los Alamitos, pp 753–756
75. Jin Y, Okabe T, Sendhoff B (2004) Neural network regularization and ensembling using multi-objective evolutionary algorithms. Congr Evol Comput CEC2004 1:1–8
76. Jinn-Moon Y, Jorng-Tzong H, Cheng-Yen K (1999) Incorporation family competition into gaussian and cauchy mutations to training neural networks using an evolutionary algorithm. IEEE Congress on evolutionary computation. doi:10.1109/CEC.1999.785519

77. Katagiri H, Nishizaki I, Hayashida T, Kadoma T (2010) Multiobjective evolutionary optimization of training and topology of recurrent neural networks for time-series prediction. In: Prooceedings of the international conference on information science and applications (ICISA), p 1–8
78. Kazuyuki M (2001) A new algorithm to design compact two-hidden-layer artificial neural networks. Neural Netw 14(9):1265–1278. doi:10.1016/S0893-6080(01)00075-2
79. Kennedy MP, Chua LO (1988) Neural networks for nonlinear programming. IEEE Trans Circ Syst 35(5):554562
80. Khorani V, Forouzideh N, Nasrabadi AM (2010) Artificial neural network weights optimization using ICA. Comparing performances. In: Proceedings of the IEEE workshop on hybrid intelligent models and applications, GA, ICA-GA and R-ICA-GA. doi:10.1109/HIMA.2011.5953956
81. Kisiel-Dorohinicki M, Klapper-Rybicka M (2000) Evolution of neural networks in a multiagent world
82. Kok JN, Marchiori E, Marchiori M, Rossi C (1996) Evolutionary training of CLP—constrained neural networks. In: Proceedings of the 2nd international conference on practical application of constraint technology, pp 129–142
83. Kolen JF, Pollack JB (1990) Back propagation is sensitive to initial conditions. Technical report TR 90-JK-BPSIC
84. Koza JR, Rice JP (1991) Genetic generation of both the weights and architecture for a neural network. In: Proceedings of the international joint conference on neural networks., IJCNN-91-Seattle, vol 2, pp 397–404
85. Lahiri A, Chakravorti S (2005) A novel approach based on simulated annealing coupled to artificial neural network for 3-D electric-field optimization. IEEE Trans Pow Deliv 20(3):2144–2152
86. Larranaga P, Karshenas A, Bielza C, SantanaR (2013) A review on evolutionary algorithms in bayesian network learning and inference tasks. Inform Sci (2013), http://dx.doi.org/10.1016/j.ins.2012.12.051
87. Levitin G (2007) Computational intelligence in reliability engineering: evolutionary techniques in reliability analysis and optimization. Volume 39 of studies in computational intelligence. Springer
88. Lin T, Ping HC, Hsu TH, Wang LC, Chen C, Chen CF, Wu CS, Liu TC, Lin CL, Lin YR, Chang FC (2011) A systematic approach to the optimization of artificial neural networks. In: IEEE 3rd International conference on communication software and networks (ICCSN), pp 76–79
89. Liu Q, Wang J (2008) A one-layer recurrent neural network with a discontinuous hard-limiting activation function for quadratic programming. IEEE Trans Neural Netw 19(4):558–570
90. Liyanage SR, Xu J-X, Guan C, Ang KK, Zhang CS, Lee TH (2009) Classification of self-paced finger movements with EEG signals using neural network and evolutionary approaches. In: Proceedings of the international conference on control and automation. doi:10.1109/ICCA.2009.5410152
91. Long C, Zeng-Guang H, Yingzi L, Min T, Wenjun CZ, Fang-Xiang W (2011) Recurrent neural network for non-smooth convex optimization problems with application to the identification of genetic regulatory networks. IEEE Trans Neural Netw 22(5):714–726
92. Lu M, Shimizu K (1995) An epsi-approximation approach for global optimization with an application to neural networks. In: Proceedings of the IEEE international conference on neural networks, vol 2, pp 783–788
93. Maa CY, Shanblatt MA (1992) Linear and quadratic programming neural network analysis. IEEE Trans Neural Netw 3:580594
94. Macready WG, Wolpert DH (1997) The no free lunch theorems. IEEE Trans Evol Comput 1(1):67–82
95. Malinak P, Jaksa R (2007) Simultaneous gradient and evolutionary neural network weights adaptation methods. IEEE congress on evolutionary computation, pp 2665–2671. doi:10.1109/CEC.2007.4424807

96. Marseguerra M, Zio E, Martorell S (2006) Basics of genetic algorithms optimization for RAMS applications. Reliab Eng Syst Saf 91(9):977–991
97. Martorell S, Sánchez A, Carlos S, Serradell V (2004) Alternatives and challenges in optimizing industrial safety using genetic algorithms. Reliab Eng Syst Saf 86(1):25–38
98. McCulloch WS, and Pitts WH (1943) A logical calculus of the ideas immanent in nervous activity. Bull Math Biophys 5:115–133
99. Miller GF, Todd PM, Hedge SU (1989) Designing neural networks using genetic algorithms. In: Schaffer JD (ed) Proceedings of the third international conference on genetic algorithms, pp 379–384
100. Misgana KM, John WN (2004) Joint application of artificial neural networks and evolutionary algorithms to watershed management. Water Resour Manage 18(5):459–482. doi:10.1023/B:WARM.0000049140.64059.d1
101. Mitsuo-Gen RC (1997) Genetic algorithms and engineering design. Wiley, New York
102. Mohankumar N, Bhuvan B, Nirmala Devi M, Arumugam S (2008) A modified genetic algorithm for evolution of neural network in designing an evolutionary neuro-hardware. In: Proceedings of the international conference on genetic and evolutionary methods, pp 108–111
103. Montana D, Davis L (1989) Training feedforward neural networks using genetic algorithms. In: Proceedings of the 11th international joint conference on AI, Detroit, pp 762–767
104. More JJ, Toroaldo G (1991) On the solution of large quadratic programming problems with bound constraints. SIAM J Optim 1(1):93–113
105. Niknam A, Hoseini P, Mashoufi B, Khoei A (2013) A novel evolutionary algorithm for block-based neural network training. In: Proceedings of the first iranian conference on pattern recognition and image analysis (PRIA), pp 1–6
106. Nobuo F, Junji K, Seishi N (1998) An evolutionary neural network approach for module orientation problems. IEEE Trans Syst Man Cybern 28(6):849–855. doi:10.1109/3477.735394
107. Nunes De Castro L, Iyoda EM, Von Zuben FF, Gudwin RR (1998) Feedforward neural network initialization: an evolutionary approach. Brazilian symposium on neural networks, pp 43–48. doi:10.1109/SBRN.1998.730992
108. Ortiz-Rodrguez JM, del Rosario BM, Gallego E, Vega-Carrillo HR(2008) Artificial neural networks modeling evolved genetically, a new approach applied in neutron spectrometry and dosimetry research areas. In: Proceedings of the electronics, robotics and automotive mechanics conference
109. Pal S, Vipsita S, Patra PK (2010) Evolutionary approach for approximation of artificial neural network. In: Proceedings of the IEEE international advance computing conference. doi:10.1109/IADCC.2010.5423015
110. Pantic D, Trajkovic T, Milenkovic S, Stojadinovic N (1995) Optimization of power VDMOSFET's process parameters by neural networks. In: Proceedings of the 25th European solid state device research conference, ESSDERC '95. 793, 796, pp 25–27
111. Pattison RJA (1999) Genetic algorithms in optimal safety system design. Proc Inst Mech Eng Part E: J Process Mech Eng 213:187–197
112. Pavlidis NG, Tasoulis DK, Plagianakos VP, Nikiforidis G, Vrahatis MN (2004) Spiking neural network training using evolutionary algorithms. In: Proceedings of the international joint conference on neural networks, pp 2190–2194
113. Prez-Ilzarb MJ (1998) Convergence analysis of a discrete-time recurrent neural network to perform quadratic real optimization with bound constraints. IEEE Trans Neural Netw 9:1344–1351
114. Qiang Z, Pan-chi L (2012) Training and application of process neural network based on quantum-behaved evolutionary algorithm. In: Proceedings of the 2nd international conference on computer science and network technology (ICCSNT), pp 929–934
115. Qingshan L, Jun W (2011) Finite-time convergent recurrent neural network with a hard-limiting activation function for constrained optimization with piecewise-linear objective functions. IEEE Trans Neural Netw 22(4):601–614
116. Rayas-Snchez JE (2004) EM-based optimization of microwave circuits using artificial neural networks: the state-of-the-art. IEEE Trans Microw Theory Tech 52(1):420–435

117. Reddipogu A, Maxwell G, MacLeod C, Simpson M (2002) A novel artificial neural network trained using evolutionary algorithms for reinforcement learning. In: Proceeding of the international conference on neural information processing. doi:10.1109/ICONIP.2002.1199013
118. Rodrguez-Vzquez A, Domnguez-Castro R, Rueda A, Huertas JL, Snchez-Sinencio E (1990) Nonlinear switched-capacitor neural networks for optimization problems. IEEE Trans Circ Syst II 37:384397
119. Romero RAF (1996) Otimizao de Sistemas atravs de Redes Neurais Multi-camadas. XI Congresso Brasileiro de Automtica, pp vol 2. So Paulo, Brasil pp 1585–1590
120. Rossana MS, Cruz HMP, Magalhaes RM (2011) Artificial neural networks and efficient optimization techniques for applications in engineering. Artificial neural networks—methodological advances and biomedical applications, pp 45–68
121. Schaffer JD, Whitley D, Eshelman LJ (1992) Combinations of genetic algorithms and neural networks: a survey of the state of the art. International workshop on combinations of genetic algorithms and neural networks (1992) COGANN-92. 1–37. doi:10.1109/COGANN.1992.273950
122. Sebald AV, Chellapilla K (1998) On making problems evolutionarily friendly part 2: evolving the most convenient representations. The seventh international conference on evolutionary programming, EP98, San Diego, pp 281–290
123. Sheng-Fuu L, Jyun-Wei C (2013) Adaptive group organization cooperative evolutionary algorithm for tsk-type neural fuzzy networks design. Int J Adv Res Artif Intell (IJARAI) 2(3):1–9
124. Shepherd AJ (1997) Second-order methods for neural networks fast and reliable methods for multi-layer perceptrons. Springer
125. Shumeet B (1996) Evolution of an artificial neural network based autonomous land vehicle controller. IEEE Trans Syst Man Cybern 26(3):450–463. doi:10.1109/3477.499795
126. Stepniewski SW, Keane AJ (1996) Topology design of feedforward neural networks by genetic algorithms. Parallel problem solving from nature, pp 771–780. doi:10.1007/3-540-61723-1040
127. Sudharsanan SI, Sundareshan MK (1991) Exponential stability and a systematic synthesis of a neural network for quadratic minimization. Neural Netw 4:599–613
128. Suraweera NP, Ranasinghe DN (2008) A natural algorithmic approach to the structural optimisation of neural networks. In: Proceedings of the 4th international conference on information and automation for sustainability, ICIAFS 2008, pp 150–156
129. Taishan Y, Duwu C, Yongqing T (2007) A new evolutionary neural network algorithm based on improved genetic algorithm and its application in power transformer fault diagnosis. doi:10.1109/BICTA.2007.4806406
130. Tai-shan Y (2010) An improved evolutionary neural network algorithm and its application in fault diagnosis for hydropower units. In: Proceedings of the international conference on intelligent computation technology and automation. doi:10.1109/ICICTA.2010.589
131. Talbot C, Massara R (1993) An application oriented comparison of optimization and neural network based design techniques. In: Proceedings of the 36th midwest symposium on circuits and systems. vol 1, pp 261–264
132. Tan ZH (2004) Hybrid evolutionary approach for designing neural networks for classification. Electr Lett 40(15). doi:10.1049/el:20045250
133. Tank DW, Hopfield JJ (1986) Simple neural optimization network: an AD onverter. Signal decision circuit and a linear programming circuit IEEE trans circuits and systems, CAS-33, 533–541
134. Tao Q, Cao JD, Xue MS, Qiao H (2001) A high performance neural network for solving nonlinear programming problems with hybrid constraints. Phys Lett A 288(2):8894
135. Taylor CM (1997) Selecting neural network topologies: a hybrid approach combining genetic algorithms and neural network. B.S. Computer Science Southwest Missouri State University
136. Tenne Y, Goh CKE (2010) Computational intelligence in optimization, volume adaptation, learning, and optimization, vol 7. Springer
137. Thimm G, Fiesler E (1997) High-order and multilayer perceptron initialization. IEEE Trans Neural Netw 8(2):349–359

138. Tian Y, Zhang J, Morris J (2001) On-line re-optimisation control of a batch polymerisation reactor based on a hybrid recurrent neural network model. In: Proceedings of the american control conference. Arlington, pp 350–355
139. Topchy AP, Lebedko OA (1997) Neural network training by means of cooperative evolutionary search. Nucl Instrum Methods Phys Res Sect A Accel Spectrom Detect Assoc Equip 389(1–2):240–241
140. Trajkovic T, Pantic D (1995) Inverse modeling and optimization of low-voltage power VDMOSFET's technology by neural networks. Semiconductor, international conference. doi:10.1109/SMICND.1995.494868
141. Ueda H, Ishikawa M (1997) Rule extraction from data with continuous valued inputs and and discrete valued outputs using neural networks. Technical report IEICE Japan, NC96-121, pp 63–70
142. Valdes J, Barton A (2007) Multi-objective evolutionary optimization of neural networks for virtual reality visual data mining: application to hydrochemistry. In: Proceedings of the international joint conference on neural networks, 2007. IJCNN 2007, pp 2233–2238
143. Velazco MI, Lyra C (2002) Optimization with neural networks trained by evolutionary algorithms. Int Symp Neural Netw 2:1516–1521. doi:10.1109/IJCNN.2002.1007742
144. Vonk E, Lakhmi CJ, Veelenturf LPJ, Johnson R (1995) Automatic generation of a neural network architecture using evolutionary computation. Electron Technol Dir Year 2000:144–149. doi:10.1109/ETD.1995.403479
145. Wai ST, Jun W (2000) Two recurrent neural networks for local joint torque optimization of kinematically redundant manipulators. IEEE Trans Syst Man Cybern Part B: Cybern 30(1):120–128
146. Wang J (1996) Recurrent neural networks for optimization. In: Chen CH (ed) Fuzzy logic and neural network handbook. New York: McGraw-Hill, 4.14.35
147. Wang L (2005) A hybrid genetic algorithmneural network strategy for simulation optimization. Appl Math Comput 170(2):1329–1343
148. Wei B, Xiaoping X (2009) Subgradient-based neural networks for nonsmooth nonconvex optimization problems. IEEE Trans Neural Netw 20(6):1024–1038
149. Werbos P (1991) An overview of neural networks for control. Control Syst IEEE 11(1):40–41
150. Whitley D (2001) An overview of evolutionary algorithms: practical issues and common pitfalls. Inf Softw Technol 43:817–831
151. Wu X, Xia Y, Li J, Chen WK (1996) A high performance neuralnetwork for solving linear and quadratic programming problems. IEEE Trans Neural Netw 7:643–651
152. Xia SY, Wang J (1995) Neural network for solving linear programming problems with bounded variables. IEEE Trans Neural Netw 6:515–519
153. Xia SY (1996) A new neural network for solving linear and quadratic programming problems. IEEE Trans Neural Netw 7:1544–1547
154. Xia SY (1996) A new neural network for solving linear programming problems and its applications. IEEE Trans Neural Netw 7:525–529
155. Xia SY, Wang J (1998) A general methodology for designing globally convergent optimization neural networks. IEEE Trans Neural Netw 9(6):1331–1343. doi:10.1109/72.728383
156. Xia SY, Leung H, Wang J (2001) A dual neural network for kinematic control of redundant robot manipulators. EEE Trans Syst Man Cybern B 31:147–154
157. Xia SY, Leung H, Wang J (2002) A projection neural network and its application to constrained optimization problems. IEEE Trans Circ Syst II 49:447–458
158. Xia Y, Jun W (2003) A general projection neural network for solving optimization and related problems. In: Proceedings of the international joint conference on neural networks, vol 3, pp 2334–2339. doi:10.1109/IJCNN.2003.1223776
159. Xia Y, Jun W (2004) A general projection neural network for solving monotone variational inequalities and related optimization problems. IEEE Trans Neural Netw 15(2):318–328. doi:10.1109/TNN.2004.824252
160. Xia Y, Wang J (2004) A recurrent neural network for nonlinear convex optimization subject to nonlinear inequality constraints. IEEE Trans Circ Syst I 51(7):1385–1394

161. Xiao S, Yo D, Li Y (2006) Parallel learning evolutionary algorithm based on neural network ensemble. In: Proceedings of the international conference on information acquisition. doi:10.1109/ICIA.2006.305824
162. Xiaolin H, Bo Z (2009) An alternative recurrent neural network for solving variational inequalities and related optimization problems. IEEE Trans Syst Man Cybern Part B: Cybern, 39 (6)
163. Xiyu L, Huichuan D, Mingxi T (2005) Design optimization by functional neural networks. Computer supported cooperative work in design. 824–829. doi:10.1109/CSCWD.2005.194292
164. Xue W (2010) Chaotic artificial neural network in reactive power optimization of distribution network. China international conference on electricity distribution
165. Xue-Bin L, Jun W (2000) A recurrent neural network for nonlinear optimization with a continuously differentiable objective function and bound constraints. IEEE Trans Neural Netw 11 (6)
166. Xuejun C, Jianzhou W, Donghuai S, Jinzhao L (2008) A novel hybrid evolutionary algorithm based on PSO and AFSA for feedforward neural network training. In: Proceedings of the international conference on wireless communications, networking and mobile computing. doi:10.1109/WiCom.2518
167. Yao X (1993) A review of evolutionary artificial neural networks. Int J Intell Syst 4:539–567
168. Yao X (1995) Designing artificial neural networks using co-evolution. In: Proceedings of IEEE singapore international conference on intelligent control and instrumentation, pp 149–154
169. Yao X, Liu Y (1997) A new evolutionary system for evolving artificial neural networks. IEEE Trans Neural Netw 8(3):694–713
170. Yao X, Liu Y (1998) Making use of population information in evolutionary artificial neural networks. IEEE Trans Syst Man Cybern Part B: Cybern 28(3):417–425
171. Yao X, Liu Y (1998) Towards designing artificial neural networks by evolution. Appl Math Comput 91(1):83–90
172. Yao X (1999) Evolving artificial neural networks. Proc IEEE 87(9):14231447
173. Yaochu J, Tatsuya O, Bernhard S (2004) Neural network regularization and ensembling using multi-objective evolutionary algorithms. IEEE congress on evolutionary computation. doi:10.1109/CEC.2004.1330830
174. Yasin ZM, Rahman TKA, Zakaria Z (2013) Quantum-inspired evolutionary programming-artificial neural network for prediction of undervoltage load shedding. In: Proceedings of the 8th IEEE conference on industrial electronics and applications (ICIEA), pp 583–588. doi:10.1109/ICIEA.2013.6566436
175. Yong C, Xia L, Qi H, Chang-hua Z (2008) An artificial neural network based on CIEA. In: Proceedings of the international conference on computational intelligence and security, pp 35–40. doi:10.1109/CIS.2008.178
176. Young-Seok H, Hungu L, MJT (2010) Acceleration of the convergence speed of evolutionary algorithms using multilayer neural networks. Eng Optim 35:91–102
177. Youshen X, Henry L, Jun W (2002) A projection neural network and its application to constrained optimization problems. IEEE Trans Circuits Syst-I: Fundam Theory Appl 49(4):447–458
178. Yu J, Lam A, Li VK (2011) Evolutionary artificial neural network based on chemical reaction optimization. IEEE Congress on evolutionary computation (CEC), pp 2083–2090
179. Yuji S, Shigei N (1996) Evolutionary algorithms that generate recurrent neural networks for learning chaos dynamics. In: Proceedings of the international conference on evolutionary computation. 144–149. doi:10.1109/ICEC.1996.542350
180. Young-Chin L, Yung-Chien L, Kuo-Lan S, Wen-Cheng C (2009) Mixed-integer evolutionary optimization of artificial neural networks. In: Proceedings of the international conference on innovative computing, information and control. doi:10.1109/ICICIC.2009.260

181. Yun L, Alexander H (1996) Artificial evolution of neural networks and its application to feedback control. Artif Intell Eng 10(2):143–152. doi:10.1016/0954-1810(95)00024-0
182. Zhang S, Zhu X, Zou LH (1992) Second-order neural networks for constrained optimization. IEEE Trans Neural Netw 3:1021–1024
183. Zhang BT, Mhlenbein H (1993) Evolving optimal neural networks using genetic algorithms with Occam's razor. Complex Syst 7(3):199–220
184. Zhou ZH, Wu J, Tang W (2002) Ensembling neural networks: many could be better than all. Artif Intell 137(1–2):239–263

Chapter 5
Reliability-Based Design Optimization with the Generalized Inverse Distribution Function

Domenico Quagliarella, Giovanni Petrone and Gianluca Iaccarino

Abstract This paper presents an approach to optimization under uncertainty that is very well and naturally suited to reliability-based design optimization problems and it is a possible alternative to traditional approaches to robust design based on the optimization of statistical moments. The approach shown here is based on the direct use of the generalized inverse distribution function estimated using the empirical cumulative distribution function (ECDF). The optimization approach presented is illustrated with the application to some test functions for both robust optimization and reliability-based design optimization. In the robust optimization test case, the bootstrap statistical technique is used to estimate the error introduced by the usage of the ECDF for quantile estimation.

Keywords Optimization under uncertainty · Reliability based design · Robust design · Generalized inverse distribution function · Bootstrap method

5.1 Introduction

Many industrial optimization processes must take account of the stochastic nature of the system and processes to design or re-design and consider the variability of some of the parameters that describe them. Thus it is necessary to characterize the system that is being studied from various points of view related to the treatment of uncertainty.

D. Quagliarella (✉)
Department of Fluid Mechanics, Italian Aerospace Research Center,
Via Maiorise snc, 81043 Capua, Italy
e-mail: d.quagliarella@cira.it

G. Petrone
Aerospace, Automotive and Turbo CFD Team, ANSYS UK Ltd,
Sheffield Business Park, S6 Europa View S9 1XH, UK
e-mail: giovanni.petrone@ansys.com

G. Iaccarino
Department of Mechanical Engineering, Stanford University,
Stanford, CA 94305, USA
e-mail: jops@stanford.edu

D. Greiner et al. (eds.), *Advances in Evolutionary and Deterministic Methods for Design, Optimization and Control in Engineering and Sciences*, Computational Methods in Applied Sciences 36, DOI 10.1007/978-3-319-11541-2_5

In particular, it is necessary to consider the sensitivity of the system to the uncertain parameters and assess its reliability. Having established the ability to characterize the system from this point of view, it is necessary to build an optimization loop that can improve its reliability, or that is capable of providing a robust optimum, or that could withstand acceptably random perturbations of design parameters or operating conditions. The classical approach to this problem is the so-called "robust design optimization" (RDO), which tries to maximize the performance and simultaneously to minimize the performance sensitivity with respect to random parameters. Instead, the "reliability based design optimization" (RBDO) tries to find the optimum design by explicitly assigning a specific level of risk and a given level of reliability. This is equivalent to assigning a quantile of the function to be optimized as the actual objective function and, for example, to minimize its value. Therefore, if the goal is that a given objective function is less than a specific value $\bar{q}$ in 75 % of possible cases, this will translate into the following constraint on the corresponding quantile: $q^{0.75} \leq \bar{q}$. Alternatively, the problem can be set as the minimization of $q^{0.75}$ and a function is thus obtained which is less than or equal to the value obtained by optimization of the quantile in 75 % of cases. If, instead, the objective is $\min q^1$, then the purpose of the optimization procedure is to protect against the worst case scenario, as it happens when the problem is of minimax type.

In this work an approach to robust and reliable design optimization based on the use of the generalized inverse distribution function is presented. The robust optimization framework is illustrated and the commonly used techniques to face the problem are briefly summarized making reference to the related literature. A very simple evolutionary multi-objective optimization algorithm based on the usage of the inverse cumulative distribution function is illustrated and discussed with the help of some test problems.

5.2 Robust Optimization

Let Z be a metric space and $z \in Z$ the vector of design variables. Let also $X : \Omega \rightarrow \Xi \subseteq \mathbb{R}$ be a real valued random variable defined in a given $(\Omega, \mathscr{F}, P)$ probability space. We want to deal with an optimization problem where an objective is optimized with respect to $z \in Z$ and depends on the realizations x of X. In other terms we have:

$$y(z, X) \colon z \in Z, X \longrightarrow Y(z)$$

with $Y(z)$ a new random variable, e.g. a new mapping of $(\Omega, \mathscr{F}, P)$ into $\mathbb{R}$, that depends on z. Solving an optimization problem involving $Y(z) = y(z, X)$ means that we want to find a value $\bar{z} \in Z$ such that the random variable $Y(\bar{z})$ is optimal. To establish the optimality of a given $Y(\bar{z})$ with respect to all $Y(z)$, $\forall z \in Z$, a ranking criterion must be defined such that for any couple $z_1, z_2 \in Z$ it is possible to state that $Y(z_1)$ is better or worse than $Y(z_2)$ (from now on, $Y(z_1) \preceq Y(z_2)$ will mean that $Y(z_1)$ is better or equivalent to $Y(z_2)$).

Recalling that a random variable is a measurable function, it seems natural to introduce measures that highlight particular features of the function. This leads to the classical and widely used approach of using the statistical moments to define the characteristics of the probability distribution that are to be optimized. More generally, let's consider an operator

$$\Phi_X : Y(z) = y(z, X) \in Z \times (\Omega, \mathcal{F}, P) \longrightarrow \Phi(z) \in V \subseteq \mathbb{R}$$

that translates the functional dependency on the random variable, Y, into a real valued function of z that represents a deterministic attribute of the function, $Y(z)$. This makes possible to formulate the following optimization problem

$$P_\Phi : \min_{z \in Z} \Phi(z)$$

Without loss of generality, it is possible to identify the random variable Y through its distribution function $f_Y(y)$ or its cumulative distribution function $F_Y(y)$. If $\Phi(\cdot)$ is assumed as the expected value of the objective function ($\mathbb{E}$), the classical formulation of first moment optimization is retrieved:

$$P_{\mathbb{E}} : \min_{z \in Z} \int_{\mathbb{R}} y f_Y(y, z) dy$$

that in terms of the CDF becomes:

$$P_{\mathbb{E}} : \min_{z \in Z} \int_{\mathbb{R}} y dF_Y(y, z)$$

It should be noted that here the distribution function depends also on z, that is the vector of the design variables.

For the purposes of the definition of the problem, it is not necessary to know exactly the distribution f_Y (or F_Y). Indeed, it is possible, as will be shown below, to use an estimate of the distribution having the required accuracy. In particular, the Empirical Cumulative Distribution Function (ECDF) will be used in this work as statistical estimator of the CDF.

The first order moment method is also called mean value approach, as the mean is used as objective to reduce the dependency on Y. This method is widely used, mostly because the mean is the faster converging moment and relatively few samples are required to obtain a good estimate. Often, however, the mean alone is not able to capture and represent satisfactorily the uncertainties embedded in a given design optimization problem. To overcome this drawback, a possible approach is the introduction in the objective function of penalization terms that are function of higher order moments. The drawback of this technique is that the ideal weights of the penalization terms are often unknown. Furthermore, in some cases, an excessive number of higher order moments may be required to adequately capture all the significant aspect of the uncertainty embedded into a given problem. Finally, a wrong choice of the penalties may lead to a problem formulation that does not have any feasible solution. Instead of penalization terms, explicit constraints can be introduced in the robust optimization problem, and the same considerations apply for the advantages and the drawbacks of the technique.

Another possibility is the minimax criterion, very popular in statistical decision theory, according to which the worst case due to uncertainty is the objective for optimization. This ensures protection against worst case scenario, but it is often excessively conservative.

The multi-objective approach [7] based on constrained optimization is also widely adopted. Here different statistical moments are used as independent trade-off objectives. The obtained Pareto front allows an a-posteriori choice of the optimal design between a set of equally ranked candidates. In this case a challenge is posed by the increase in the dimensionality of the Pareto front when several statistical moments are used. The research related to the multi-objective method has led to several extensions of the classical Pareto front concept. In [10], for example, the Pareto front exploration in presence of uncertainties is faced introducing the concept of *probabilistic dominance*, which is an extension of the classical *Pareto dominance*. While in [6], a probabilistic ranking and selection mechanism is proposed that introduces the *probability of wrong decision* directly in the formula for rank computation.

An interesting approach, similar in some aspects to the one here described, is found in [5] where a quantile based approach is coupled with the probability of *Pareto nondominance* (already seen in [6]). Here, contrary to the cited work, the optimization technique introduced relies on direct estimation of the quantile function obtained through the Empirical Cumulative Distribution Function.

5.3 The Generalized Inverse Distribution Function Method

In the methodology presented herein, the operator that is used to eliminate the dependence on random variables is the quantile function of the objective function to be optimized, calculated in one or more points of its domain of definition.

Before going into the details of the exposure, the definitions of *Cumulative Distribution Function* (CDF) and *Generalized Inverse Distribution Function* (GIDF) that will be used are reported.

The "cumulative distribution function" (CDF), or just "distribution function", describes the probability that a real-valued random variable Q with a given probability distribution will be found at a value less than or equal to q. Intuitively, it is the "area so far" function of the probability distribution. The CDF is one of the most precise, efficient and compact ways to represent information about uncertainty, and a new CDF based approach to robust optimization is described.

If the CDF is continuous and strictly monotonic then it is invertible, and its inverse, called quantile function or inverse distribution function, returns the value below which random draws from the given distribution would fall, $s \times 100$ percent of the time. That is, it returns the value of q such that

$$F_Q(q) = \Pr(Q \leq q) = s \tag{5.1}$$

Hence $F^{-1}(s),\ s \in [0, 1]$ is the unique real number q such that $F_Q(q) = s$.

Unfortunately, the distribution does not, in general, have an inverse. If the probability distribution is discrete rather than continuous then there may be gaps between values in the domain of its CDF, while, if the CDF is only weakly monotonic, there may be "flat spots" in its range. In general, in these cases, one may define, for $s \in [0, 1]$, the "generalized inverse distribution function" (GIDF)

$$q^s = Q(s) = F_Q^{-1}(s) = \inf\{q \in \mathbb{R} : F(q) \geq s\}$$

that returns the minimum value of s for which the previous probability statement (5.1) holds. The infimum is used because CDFs are, in general, weakly monotonic and right-continuous (see [15]).

Now that the CDF and the GIDF have been introduced, it becomes easy to define, within the framework of multi-objective optimization, a robust optimization problem in terms of an arbitrary number of quantiles to optimize:

$$P_{Q(s_i)}\colon \min_{z \in Z} q^{s_i}(z) = \min_{z \in Z} \inf\left\{q(z) \in \mathbb{R} : F_Q(q(z)) \geq s_i\right\} \quad i = 1, \ldots, n \tag{5.2}$$

where n is the number of objectives chosen. The approach, then, can be further extended by introducing objectives that are arbitrary functions of quantiles.

Of course, the problem now is focused on how to satisfactorily calculate the quantiles required by the method. In this work the *Empirical Cumulative Distribution Function* (ECDF) is used for this purpose. The definition of ECDF, taken from [16], is reported here for the sake of completeness.

Let $X_1, \ldots, X_n$ be random variables with realizations $x_i \in \mathbb{R}$, the empirical distribution function is an indicator function that estimates the true underlying CDF of the points in the sample. It can be defined by using the order statistics $X_{(i)}$ of X_i as:

$$\widehat{F}_n(x, \omega) = \begin{cases} 0 & \text{if } x < x_{(1)}; \\ \frac{1}{n} & \text{if } x_{(1)} \leq x < x_{(2)}, 1 \leq k < 2; \\ \frac{2}{n} & \text{if } x_{(2)} \leq x < x_{(3)}, 2 \leq k < 3; \\ \vdots & \\ \frac{i}{n} & \text{if } x_{(i)} \leq x < x_{(i+1)}, i \leq k < i+1; \\ \vdots & \\ 1 & \text{if } x \geq x_{(n)}; \end{cases}$$

where $x_{(i)}$ is the realization of the random variable $X_{(i)}$ with outcome (elementary event) $\omega \in \Omega$.

From now on, therefore, when the optimization algorithm requires the calculation of the $F_Q(s)$, it will used instead its estimator $\widehat{F}_{Q_n}(s)$, where n indicates the number of samples used to estimate this ECDF.

Note that each indicator function, and hence the ECDF, is itself a random variable. This is a very delicate issue to consider. Indeed, if the EDCF is used to approximate the deterministic operator $Q(s)$, a direct residual influence of the random variables that characterize the system under investigation remains on $P_{Q(s)}$. In other words $Q(s)$ behaves as a random variable, but with the important difference that its variance tends to zero when the ECDF approximates the CDF with increasing precision. It is possible to demonstrate that the estimator $\widehat{F}_{Q_n}(s)$ is consistent, as it converges almost surely to $F_Q(s)$ as $n \to \infty$, for every value of s [11]. Furthermore, for the Glivenko-Cantelli theorem [8], the convergence is also uniform over s. This implies that, if the ECDF is calculated with sufficient accuracy, it can be considered and treated as a deterministic operator. On the other hand, if the number of samples, or the estimation technique of the ECDF, do not allow as such, one can still correlate the variance of the ECDF with the precision of the obtained estimate. Of course, if the ECDF is estimated in a very precise way, it is possible to use an algorithm conceived for deterministic problems, provided that it has a certain resistance to noise. Conversely, if the ECDF is obtained from a coarse sample, its practical use is only possible with optimization algorithms specifically designed for that purpose.

For the same reason, it is often convenient, especially in applications where the ECDF is defined with few samples, to use q^ε instead of q^0, with $\varepsilon > 0$ and small, but such that a not excessive variance of the estimate of q^ε is ensured.

5.4 A Robust Optimization Test Case

The function reported in Table 5.1, taken from [13], is used as a benchmark to test the GIDF based approach to robust optimization. With respect to the function reported in the reference, the following changes have been introduced: the ranges of design and uncertain parameters have been changed as reported in table, and a multiplicative factor equal to $1/n$ has been introduced to make easier the result comparison when the dimension of the parameter space changes. The random variables $\mathbf{u}$ have a uniform distribution function. Table 5.2 reports the solutions to the optimization problems

$$\min_{\mathbf{d} \in D, \mathbf{u} \in U} f(\mathbf{d}, \mathbf{u})$$

$$\min_{\mathbf{d} \in D} \max_{\mathbf{u} \in U} f(\mathbf{d}, \mathbf{u})$$

over the cartesian product of D and U. The first problem represents the best possible solution obtainable if the $\mathbf{u}$ are considered as design parameters varying in

Table 5.1 Benchmark functions table

ID	Function	Ranges	Dimension
MV4	$f = \frac{1}{n}\sum_{i=1}^{n}(2\pi - u_i)\cos(u_i - d_i)$	$\mathbf{u} \in [0, 3]^n$, $\mathbf{d} \in [0, 2\pi]^n$	1, 2 and 6

Table 5.2 Benchmark functions table results

ID	$\min_{\mathbf{d}\in D,\mathbf{u}\in U} f(\mathbf{d}, \mathbf{u})$			$\min_{\mathbf{d}\in D}\max_{\mathbf{u}\in U} f(\mathbf{d}, \mathbf{u})$		
	$\mathbf{d}$	$\mathbf{u}$	f	$\mathbf{d}$	$\mathbf{u}$	f
MV4	$[3.1416]^n$	$[0]^n$	−6.283185...	$[4.6638]^n$	$[0]^n$	−0.305173...

U. The second one, instead, minimizes the maximum possible loss or, alternatively, maximizes the minimum gain, according to the framework of decision theory [12]. These solutions have been obtained analytically and verified by exhaustive search for $n = 1$. It is worth to note that these particular optimal solutions are the same whatever is the dimension of the search space.

The optimization algorithm used here is a simple multi-objective genetic algorithm not specially conceived for optimization under uncertainty. The algorithm is based on the Pareto dominance concept and on local random walk selection [14]. The crossover operator is the classical *one-point* crossover which operates at bit level, while mutation operator works at the level of the design vector parameters (which are real numbers). A parameter, called *mutation rate* controls the operator activation probability for each variable vector element, while a further parameter, called *strength*, is the maximum relative value for the uniform word mutation. The word mutation value is given by strength $\cdot\,(r - 0.5)(u - l)$ where $r \in [0, 1]$ is the uniform random number, u is the upper variable bound and l is the lower variable bound. An elitist strategy was adopted in the optimization runs. It consists in replacing 20 % of the population calculated at each generation with elements taken at random from the current Pareto front. Obviously, the elements of the population are used to update the current Pareto front before the replacement, in order to avoid losing non-dominated population elements.

The multi-objective runs were performed using 100 % crossover activation probability and word mutation with *mutation rate* equal to 50 % and *strength* equal to 0.06. The initial population was obtained using the quasi-random low-discrepancy Sobol sequence [1]. The ECDF used to estimate the CDF was obtained with 2,500 Montecarlo samples in all runs. The population size was set to 4,000 elements for all runs, while the number of generations was set to 10 for $n = 1$, 200 for $n = 2$ and 1,000 for $n = 6$. The problem solved was $\min_{\mathbf{z}\in Z}\left(q^{\varepsilon}, q^{1-\varepsilon}\right)$.

Figure 5.1 reports the Pareto fronts and the deterministic min and min max solutions obtained for the MV4 test case at different values of the design space size n. It can be easily observed that, in the case $n = 1$, the extremes of the front are practically coincident with the deterministic solutions, while, in the case $n = 2$, the

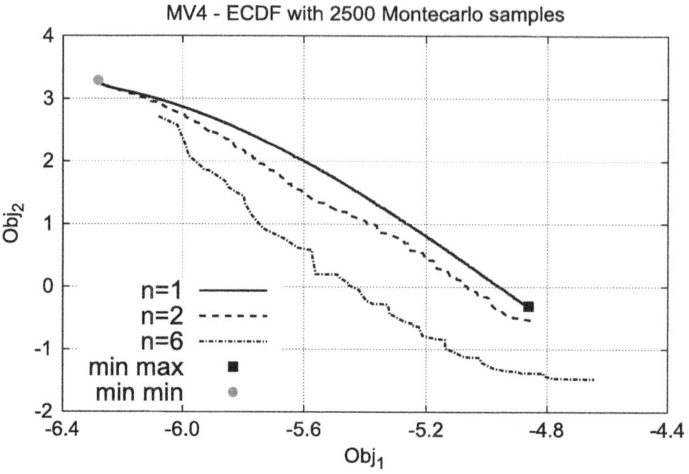

Fig. 5.1 Pareto fronts and deterministic min and min max solutions for the MV4 test case

solution of the Pareto front which minimizes the second objective ($q^{1-\varepsilon}$) underestimates the min max solution. The trend is even more evident in the case $n = 6$, where also the extreme of the front that minimizes the first goal (q^{ε}) overestimates the value obtained from the min problem. This can be explained by the fact that the two deterministic solutions are located in correspondence with the extremes of variation of the random variables of the problem. Therefore, as the number of random variables increases, in accordance with the central limit theorem [9], it becomes less likely that all random variables are located in correspondence of one of their limits of variation. However, as illustrated in Fig. 5.2, when the Pareto front obtained with the sample size m equal to 2,500 is re-evaluated with a larger Montecarlo sample, the curve is a quite acceptable approximation of the Pareto front obtained with m =100,000.

Figures 5.3 and 5.4 show the ECDF corresponding to the extremes of the Pareto front, respectively for the cases $n = 1$ and $n = 6$. It is noted, again in accordance with the central limit theorem, that, in the case $n = 6$, the ECDF curves are very close to those related to a Gaussian distribution.

5.5 Evaluating and Improving the Quantile Estimation

The example in the previous section shows very clearly that the results of the proposed method may depend in an essential way on the quality of estimation of quantiles that is obtained through the ECDF. This leads in a natural way to deal with two issues: how to evaluate the quality of the estimation of the quantiles used in the multi-objective optimization problem, and how to possibly get a better quantile estimate with a given computational effort.

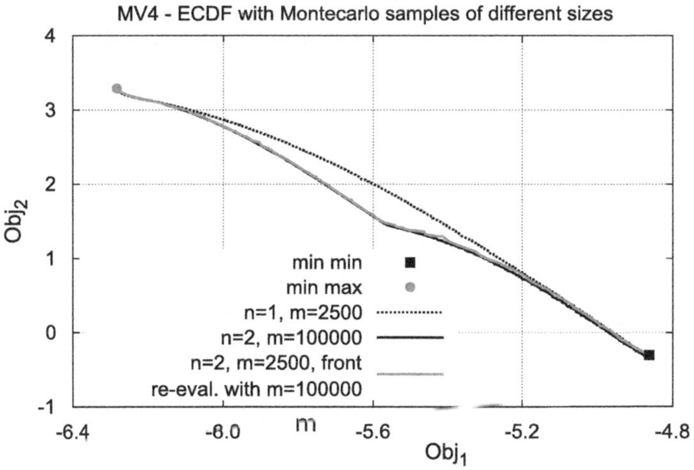

Fig. 5.2 Pareto fronts for the MV4 test case obtained with different sizes for Montecarlo sampling

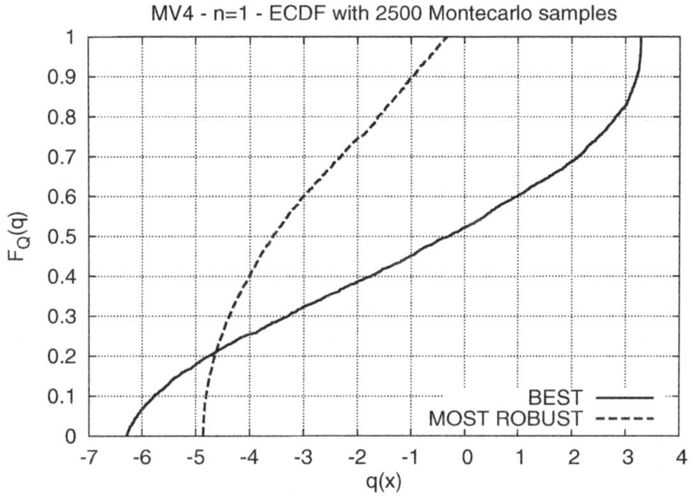

Fig. 5.3 Optimal ECDF curves for the MV4 with $n = 1$

The approach here proposed for assessing the quality of the estimate of the quantile is based on the bootstrap method introduced by Efron in 1977 [3, 4].

This method represents a major step forward in the statistical practice because it allows to accurately assess the variability of any statistical estimator without making any assumption about the type of distribution function involved. Suppose that a statistic $T(x_1, x_2, \ldots, x_n)$ is given, evaluated on a set of data $x_1, x_2, \ldots, x_n$ belonging to an assigned space X. The bootstrap essentially consists in the repeated recalculation of the statistic T employing a tuple of new samples $x_1^*, x_2^*, \ldots, x_n^*$

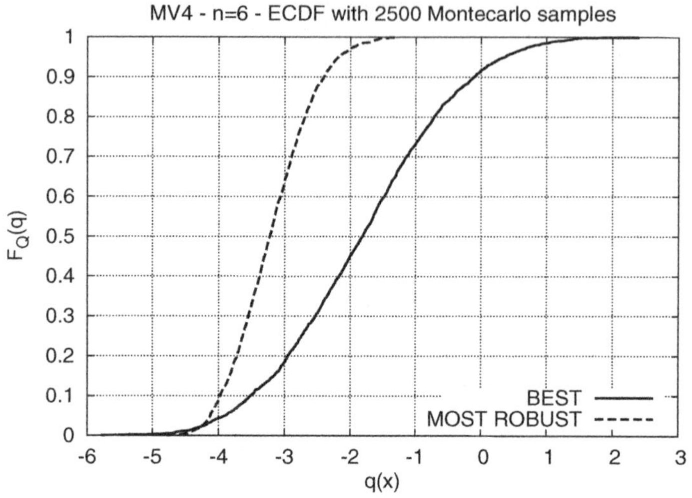

Fig. 5.4 Optimal ECDF curves for the MV4 with $n = 6$

obtained by selecting them from the collection $\{x_1, x_2, \ldots, x_n\}$ by replacement. The repeated calculation of $T\left(x_1^*, x_2^*, \ldots, x_n^*\right)$ gives a set of values that is a good indication of the distribution of T.

Therefore, to calculate the accuracy of a generic quantile q^s, obtained by the estimator $\widehat{F}_{Q_n}(s)$, the bootstrap procedure can be applied to the samples that define the estimator. This allows to calculate the corresponding distribution of q^s for a fixed value of s.

Figure 5.5 reports the ECDF related to the solution labeled as "MOST ROBUST" in Fig. 5.4. The bootstrap was applied to this ECDF repeating the sampling process 2,000 times. The area in gray color represents the superposition of all the curves obtained in this way. From the bootstrap data it is then possible to evaluate the accuracy of a given quantile estimate. According to [3], an accuracy measure for q^s can be obtained considering the central 68 % of bootstrap samples. These values lay between an interval $[q_\ell^s, q_u^s]$ centered on the observed value q^s. Half the length of this interval gives a measure of the accuracy for the observed value that corresponds to the traditional concept of "standard error". Here this value is indicated with $\widehat{SE}$ to distinguish it from the true standard error SE.

Table 5.3 reports the computed accuracy values for the considered quantiles for the above cited "MOST ROBUST" solution obtained from an ECDF with 2,500 Montecarlo samples. The fourth column reports, finally, the maximum estimated error $\widehat{ME}$.

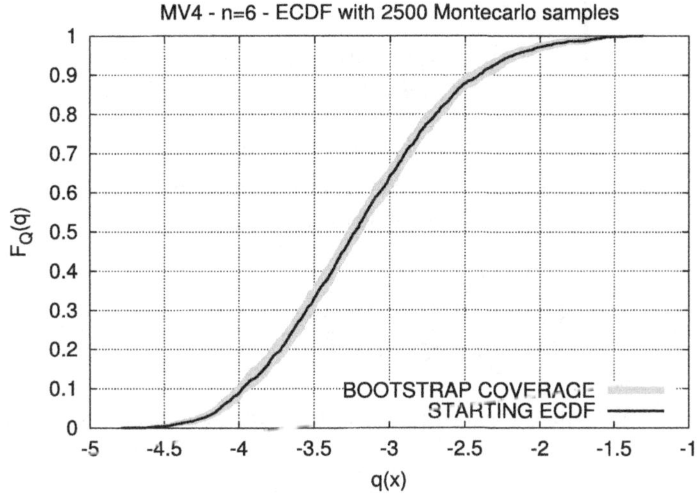

Fig. 5.5 ECDF corresponding to the most robust solution and related bootstrap coverage

Table 5.3 Quantile estimates and related accuracy for MV4 "MOST ROBUST" solution with $n = 6$

s	q^s	$\widehat{SE}$	$\widehat{ME}$
0.001000	−4.630433	±0.090423	±0.117169
0.500000	−3.230388	±0.018834	±0.054983
0.999000	−1.425868	±0.013192	±0.136330

5.6 Single and Multi-objective Reliability Optimization Tests

A very complete review of reliability-based optimization and of the solution techniques based on evolutionary algorithms can be found in [2].

A reliability based single or multi-objective problem can be written as follows:

$$\begin{cases} \min\limits_{\mathbf{z},\mathbf{x}} \mathbf{f}(\mathbf{x}, \mathbf{z}) \\ \text{s. to: } g_i(\mathbf{z}, \mathbf{x}) \leq 0 \quad i = 1, \ldots, n \\ \qquad \mathbf{x} \in \mathbf{X}, \quad \mathbf{z} \in \mathbf{Z} \end{cases} \tag{5.3}$$

with $\mathbf{x}$ representing deterministic and $\mathbf{z}$ stochastic design variables. In order to find a solution to this problem, a reliability measure R is introduced that means that the probability of having an infeasible solution instance will be at most $(1 - R)$, and the problem is reduced to the following one:

$$\begin{cases} \min\limits_{\mathbf{z},\mathbf{x}} \mathbf{f}(\mathbf{x}, \mu_{\mathbf{z}}) \\ \text{s. to: } P_i \left(g_i(\mathbf{z}, \mathbf{x}) \leq 0\right) \geq R \quad i = 1, \ldots, n \\ \qquad \mathbf{x} \in \mathbf{X}, \quad \mathbf{z} \in \mathbf{Z} \end{cases} \tag{5.4}$$

where $\mu_{\mathbf{z}}$ is the mean vector of the uncertain variables $\mathbf{z}$, and $P_i()$ is the probability that the ith constraint is feasible. The constraints in problem (5.4) can be immediately expressed in terms of generalized inverse distribution function:

$$\begin{cases} \min\limits_{\mathbf{z},\mathbf{x}} \mathbf{f}(\mathbf{x}, \mu_{\mathbf{z}}) \\ \text{s. to: } q^R \left(g_i(\mathbf{z}, \mathbf{x})\right) \leq 0 \quad i = 1, \ldots, n \\ \qquad \mathbf{x} \in \mathbf{X}, \quad \mathbf{z} \in \mathbf{Z} \end{cases} \tag{5.5}$$

Problem 5 can be further transformed by introducing penalty functions into:

$$\min_{\mathbf{z},\mathbf{x}} \; \mathbf{f}(\mathbf{x}, \mu_{\mathbf{z}}) + \sum_{i=1}^{n} \Gamma \left(q^R \left(g_i(\mathbf{z}, \mathbf{x})\right)\right) \tag{5.6}$$

with Γ defined, for example, as:

$$\Gamma(x) = \begin{cases} 0 & \text{if} \;\; x \leq 0 \\ kx^2, \; k > 0 & \text{if} \;\; x > 0 \end{cases} \tag{5.7}$$

Considering that $\Gamma \left(q^R = 0\right) = q^R \left(\Gamma \left(g_i(\mathbf{z}, \mathbf{x})\right)\right)$ it is possible to write, finally:

$$\min_{\mathbf{z},\mathbf{x}} \; q^R \left(\mathbf{f}(\mathbf{x}, \mu_{\mathbf{z}}) + \sum_{i=1}^{n} \Gamma \left(g_i(\mathbf{z}, \mathbf{x})\right) \right) \tag{5.8}$$

In the subsequent examples, problem (5.8) is solved using a simple multi-objective genetic algorithm for unconstrained problems. Of course, more sophisticated and efficient algorithms could be used to reduce the computational effort.

The first example considered is taken from [2]. It involves two variables, x and y, and two objectives, f_1 and f_2:

$$\begin{cases} \min & f_1 = x \\ \min & f_2 = \frac{1+y}{x} \\ \text{s. to:} & y + 9x - 6 \geq 0 \\ & -y + 9x - 1 \geq 0 \\ & 0.1 \leq x \leq 1, 0 \leq y \leq 5 \end{cases} \tag{5.9}$$

with both variables uncertain and characterized by a normal distribution with $\sigma = 0.03$.

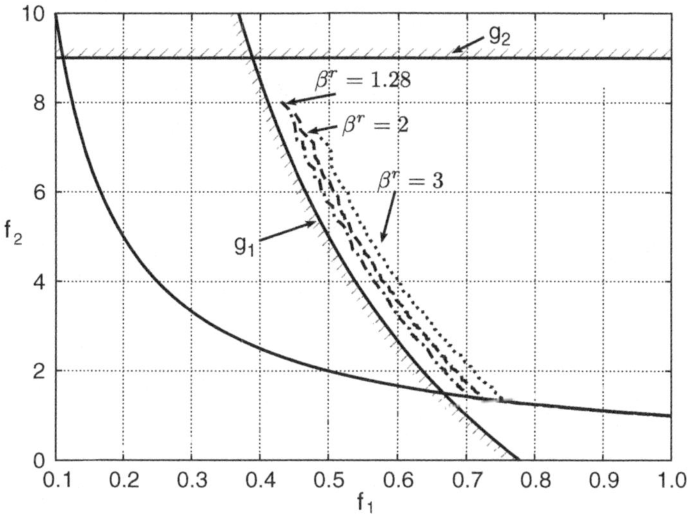

Fig. 5.6 Pareto front for the simple two-objectives reliability problem obtained with three different reliability indexes

The optimization runs have been carried out with three different reliability levels, namely 90 % ($\beta^r = 1.28$), 97.725 % ($\beta^r = 2.00$) and 99.875 % ($\beta^r = 3.00$), where β^r is the reliability index computed according to $R = 50\left(1 + \text{erf}\left(\beta^r / \sqrt{2}\right)\right)$. The results are reported in Fig. 5.6. The multi-objective genetic algorithm crossover and mutation settings are the same that have been used in the previous test case. For each different reliability level, a population of 100 individuals evolved for 400 generations. For $\beta^r = 1.28$ and $\beta^r = 2.00$ the Montecarlo sample size was set to 1,000, while for the $\beta^r = 3.00$ case the sample size was set to 10,000.

The second problem presented is a car side-impact problem related to vehicle crash-worthiness and dealing with vehicle safety rating scores related to human safety issues. The problem is reported in [17] and, in a slightly different form, in [2]. The reader is referred to the references above for more details on the physical nature of the model used and the role and significance of variables and parameters. In extreme synthesis, eleven random variables are involved in the optimization problem, whose characteristics and nature is reported in Table 5.4 for the sake of completeness. The random variables z_{10} and z_{11} are not regarded as design variables, because they are related to aspects that define the type of the problem but which can not be controlled in any way by the designer. The objective function and constraints of the problem are given below:

Table 5.4 Properties of design and random parameters of vehicle side impact model

Random variable	Std dev.	Distr type.	Lower limit	Upper limit
z_1 (B-pillar inner)	0.030	Normal	0.500	1.500
z_2 (B-pillar reinforce)	0.030	Normal	0.500	1.500
z_3 (Floor side inner)	0.030	Normal	0.500	1.500
z_4 (Cross member)	0.030	Normal	0.500	1.500
z_5 (Door beam)	0.030	Normal	0.500	1.500
z_6 (Door belt line)	0.030	Normal	0.500	1.500
z_7 (Roof rail)	0.030	Normal	0.500	1.500
z_8 (Mat. B-pillar inner)	0.006	Normal	0.192	0.345
z_9 (Mat. floor side inner)	0.006	Normal	0.192	0.345
z_{10} (Barrier height)	10.0	Normal	−30.0	30.0
z_{11} (Barrier hitting)	10.0	Normal	−30.0	30.0

$$
\begin{aligned}
f(\mathbf{z}) &= 1.98 + 4.9z_1 + 6.67z_2 + 6.98z_3 + 4.01z_4 + 1.78z_5 + 2.73z_7 \\
g_1(\mathbf{z}) &= 1.16 - 0.3717z_2z_4 - 0.00931z_2z_{10} - 0.484z_3z_9 \\
&\quad + 0.01343z_6z_{10} \le 1 \text{ kN} \\
g_2(\mathbf{z}) &= 0.261 - 0.0159z_1z_2 - 0.188z_1z_8 - 0.019z_2z_7 \\
&\quad + 0.0144z_3z_5 + 0.0008757z_5z_{10} + 0.08045z_6z_9 + 0.00139z_8z_{11} \\
&\quad + 0.00001575z_{10}z_{11} \le 0.32 \text{ m/s} \\
g_3(\mathbf{z}) &= 0.214 + 0.00817z_5 - 0.131z_1z_8 - 0.0704z_1z_9 \\
&\quad + 0.03099z_2z_6 - 0.018z_2z_7 + 0.0208z_3z_8 + 0.121z_3z_9 \\
&\quad - 0.00364z_5z_6 + 0.0007715z_5z_{10} \\
&\quad - 0.0005354z_6z_{10} + 0.00121z_8z_{11} \le 0.32 \text{ m/s} \\
g_4(\mathbf{z}) &= 0.74 - 0.61z_2 - 0.163z_3z_8 \\
&\quad + 0.001232z_3z_{10} - 0.166z_7z_9 + 0.227z_2z_2 \le 0.32 \text{ m/s} \\
g_5(\mathbf{z}) &= 28.98 + 3.818z_3 - 4.2z_1z_2 + 0.0207z_5z_{10} \le 32 \text{ mm} \\
&\quad + 6.63z_6z_9 - 7.7z_7z_8 + 0.32z_9z_{10} \\
g_6(\mathbf{z}) &= 33.86 + 2.95z_3 + 0.1792z_{10} - 5.057z_1z_2 \\
&\quad - 11z_2z_8 - 0.0215z_5z_{10} - 9.98z_7z_8 + 22z_8z_9 \le 32 \text{ mm} \\
g_7(\mathbf{z}) &= 46.36 - 9.9z_2 - 12.9z_1z_8 + 0.1107z_3z_{10} \le 32 \text{ mm} \\
g_8(\mathbf{z}) &= 4.72 - 0.5z_4 - 0.19z_2z_3 - 0.0122z_4z_{10} + 0.009325z_6z_{10} \\
&\quad + 0.000191z_{11}z_{11} \le 4 \text{ kN} \\
g_9(\mathbf{z}) &= 10.58 - 0.674z_1z_2 - 1.95z_2z_8 + 0.02054z_3z_{10} \\
&\quad - 0.0198z_4z_{10} + 0.028z_6z_{10} \le 9.9 \text{ mm/ms} \\
g_{10}(\mathbf{z}) &= 16.45 - 0.489z_3z_7 - 0.843z_5z_6 + 0.0432z_9z_{10} \\
&\quad - 0.0556z_9z_{11} - 0.000786z_{11}z_{11} \le 15.7 \text{ mm/ms}
\end{aligned}
\tag{5.10}
$$

The goal is to minimize the weight of the structure.

Again, the same settings were used for crossover and mutation and, in all cases analyzed, the size of the population and was fixed to 30 individuals that have evolved over 3,000 generations. The Montecarlo sample size was set to 100,000. Figure 5.7

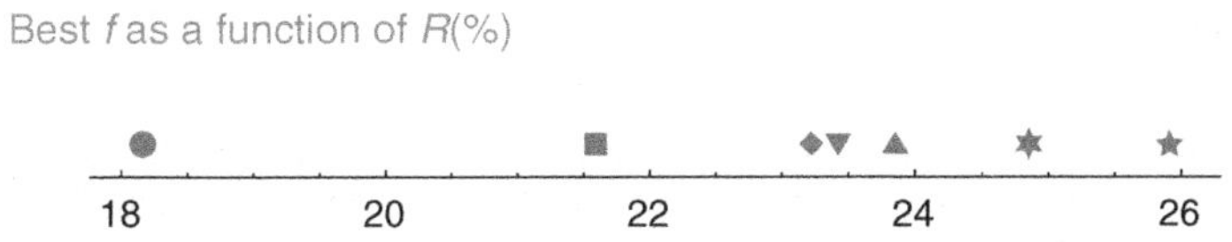

	β^r	R (%)	f		β^r	R (%)	f
●	best outcome		18.15	▲	2.00	77.7250	23.86
■	deterministic		21.60	✱	3.00	99.8650	24.86
◆	1.28	89.9727	23.24	★	4.00	99.9968	25.92
▼	1.50	93.3193	23.43				

Fig. 5.7 Best optimization result as a function of the reliability index for the crash-worthiness vehicle side impact problem

shows, both in tabular and number line format, the values of the objective function for different reliability values (expressed both in percentage and in β^r terms). These values are compared both with the deterministic case, calculated by setting the parameters z_{10} and z_{11} to zero (average value), that with the case here called "best outcome", obtained by considering also the two parameters z_{10} and z_{11} as deterministic variables of the problem.

5.7 Conclusions

An alternative approach to the optimization under uncertainty has been introduced and illustrated with examples related to both robust and reliability based design optimization. Furthermore, a method, based on the well known bootstrap statistical technique, has been introduced to estimate the error introduced by the usage of the ECDF instead of the true CDF.

The algorithm used for optimization is a classical genetic algorithm, but, to further improve the proposed procedure, an optimization algorithm capable of accounting for the errors in the estimation of the CDF has to be conceived. This is a very important topic and it will be subject of next research work. In particular, to reduce the curse of dimensionality, the effect of different sampling methodologies, like stochastic collocation, on the estimation of the ECDF will be considered in future works. Indeed the possibility to use the error on the ECDF estimator to properly refine the probability space using adaptive uncertainty quantification algorithms will be explored.

References

1. Bratley P, Fox BL (1988) Algorithm 659: implementing sobol's quasirandom sequence generator. ACM Trans Math Softw 14(1):88–100. doi:10.1145/42288.214372, http://doi.acm.org/10.1145/42288.214372
2. Deb K, Gupta S, Daum D, Branke J, Mall A, Padmanabhan D (2009) Reliability-based optimization using evolutionary algorithms. Evol Comput IEEE Trans 13(5):1054–1074
3. Diaconis P, Efron B (1983) Computer intensive methods in statistics. Technical Report 83, Division of Biostatistics, Stanford University
4. Efron B (1979) Bootstrap methods: another look at the jackknife. Ann Stat 7(1):1–26. http://projecteuclid.org/euclid.aos/1176344552
5. Filomeno CR, Bouillard P (2011) Multi-objective reliability-based optimization with stochastic metamodels. Evol Comput 19(4):525–560
6. Hughes E (2001) Evolutionary multi-objective ranking with uncertainty and noise. Evolutionary multi-criterion optimization. Springer, Berlin, pp 329–343
7. Poloni C, Padovan L, Parussini L, Pieri S, Pediroda V (2004) Robust design of aircraft components: a multi-objective optimization problem. In: Von Karman Institute for Fluid Dynamics, Lecture Series. Von Karman Institute for Fluid Dynamics, Lecture Series 2004–07
8. Serfling RJ (2008) Approximation theorems of mathematical statistics. Wiley. doi:10.1002/9780470316481, http://dx.doi.org/10.1002/9780470316481.indsub
9. Sobol IM (1994) A primer for the monte Carlo method. CRC Press, Boca Raton
10. Teich J (2001) Pareto-front exploration with uncertain objectives. Evolutionary multi-criterion optimization. Springer, Heidelberg, pp 314–328
11. van der Vaart AW (1998) Asymptotic statistics. Cambridge University Press, Cambridge. http://dx.doi.org/10.1017/CBO9780511802256
12. von Neumann J, Morgenstern O (1953) Theory of games and economic behavior. Princeton University Press, Princeton
13. Vasile M, Minisci E (2011) An evolutionary approach to evidence-based multi-disciplinary robust design optimisation. In: Poloni C, Quagliarella D, Periaux J, Gauger N,Giannakoglou K (eds.) EUROGEN 2011 PROCEEDINGS—Evolutionary and deterministic methods for design, optimization and control with applications to industrial and societal problems, ECCOMAS thematic conference. ECCOMAS, CIRA, Capua, Italy
14. Vicini A, Quagliarella D (1997) Inverse and direct airfoil design using a multiobjective genetic algorithm. AIAA J 35(9):1499–1505
15. Wikipedia: Cumulative distribution function. http://en.wikipedia.org/wiki/Cumulative_distribution_function
16. Woo C, Empirical distribution function (version 4). http://planetmath.org/EmpiricalDistributionFunction.html
17. Youn B, Choi K, Yang RJ, Gu L (2004) Reliability-based design optimization for crashworthiness of vehicle side impact. Struct Multi Optim 26(3–4):272–283. doi:10.1007/s00158-003-0345-0, http://dx.doi.org/10.1007/s00158-003-0345-0

Part II
Theoretical and Numerical Methods and Tools for Optimization: Numerical Methods and Tools

Chapter 6
On the Choice of Surrogates for Multilevel Aircraft Perfomance Models

Manon Bondouy, Sophie Jan, Serge Laporte and Christian Bes

Abstract The objective of this study is to propose a methodology which aims at reducing the memory size of hierarchical multilevel models while satisfying both a given accuracy and a maximum computational time, and keeping the initial multilevel structure. In this paper, we propose to construct a new multilevel model satisfying such requirements, based on a choice among available surrogates associated with each submodel. We show how this metamodel assignments can be formulated in an optimal manner through an integer programming problem. The proposed methodology is illustrated on a drag performance model, with surrogates based on High Dimensional Model Representation (HDMR).

Keywords Hierarchical multilevel models · Surrogates · HDMR

6.1 Description

6.1.1 Introduction

Complex performance models have generally a hierarchical multilevel structure. Hierarchical modeling [1] is used to decompose complex model into multilevel submodels according to their functional attributes, physical structures, or size. Each submodel has multiple entries, which are the outputs of the previous level submodels (the input variables for the top level), and a unique output.

For example, pitching moment or drag coefficients for Airbus models may contain up to thirty submodels spread in about ten levels. Each submodel has at most four input variables.

M. Bondouy (✉) · S. Laporte
Airbus, 316 Route de Bayonne, 31060 Toulouse CEDEX 9, France
e-mail: manon.bondouy@airbus.com

M. Bondouy · S. Jan
IMT, Université de Toulouse, 118 Route de Narbonne, 31062 Toulouse, France

C. Bes
ICA, Université de Toulouse, 118 Route de Narbonne, 31062 Toulouse, France

D. Greiner et al. (eds.), *Advances in Evolutionary and Deterministic Methods for Design, Optimization and Control in Engineering and Sciences*, Computational Methods in Applied Sciences 36, DOI 10.1007/978-3-319-11541-2_6

The strong interest of air transport on environmental impact associated with the growing need to reduce congestion of air traffic and the important economical pressure on airlines [4] induce new challenges on aircraft operations; for example, finding best trajectories allows to reduce the fuel consumption and also decrease the noise pollution close to the airport. This leads to an intensive use of numerical performance models either on ground or in flight. They can then either be part of optimization loops, Monte Carlo sampling or real time computation. Consequently, a crucial need to reduce the **computational time** and **memory size** of the performance models while keeping acceptable **accuracy** appears.

There are different stages for tackling the above problem. These strategies include parallel computing, increasing computer power, eliminating the irrelevant interactions between input variables, reducing the number of input variables, building relevant surrogate models. These strategies are not mutually exclusive and should be used together for a better efficiency.

In this paper, we focus on embedded applications. Therefore, the critical points are the computational time and the structure of the model, strongly constrained by the avionics system capabilities which lead to optimize the memory size. Moreover, most of such functions have to be certified because of their criticity level; this implies a need to preserve the multilevel structure in order to comply with the certification process. Obviously an acceptable level of accuracy has to be fulfilled.

We propose here to construct a global surrogate satisfying requirements of computational time, storage memory and accuracy while keeping the multilevel structure. This paper is organised as follows. In Sect. 6.1.2, we first introduce the mathematical formulation of the submodel multilevel architecture that we have to deal with. Then, from this structure we show in Sect. 6.1.3 how the optimal choice of surrogate model assigments for each submodel can be formulated as an integer programming problem [5]. In Sect. 6.2, we illustrate our method on an industrial case. Finally we conclude and open further perpective.

6.1.2 Notations

$\boldsymbol{X}$ vector of input variables.
d dimension of the input variables vector.
n total number of levels.
F global model.
$\tilde{F}$ global surrogate model.
ε measure of the error, $\tilde{F} - F$
$\|.\|$ norm; the choice of this norm is free.

For each intermediate level i, $1 \leq i \leq n$:

L_i number of submodels; we can notice that $L_n = 1$. By convention $L_0 = d$.
$\boldsymbol{O_i} = (O_{i,1}, \ldots, O_{i,L_i})$ vector representing all outputs at level. By convention $\boldsymbol{O_0} = \boldsymbol{X}$.

For each level i, $1 \leq i \leq n$ and for each submodel j, $1 \leq j \leq L_i$:

$F_{i,j}$ submodel j at level i.
$O_{i,j}$ output of $F_{i,j}$. By convention $O_{0,j} = X_j$ for $1 \leq j \leq d$.
$Q_{i,j}$ number of surrogates of submodel $F_{i,j}$

For each level i, $1 \leq i \leq n$, for each sumodel j, $1 \leq j \leq L_i$ and for each surrogate q, $1 \leq q \leq Q_{i,j}$:

$\tilde{F}_{i,j,q}$ qth surrogate model of the submodel $F_{i,j}$.
$T_{i,j,q}$ local computational time of the surrogate submodel $\tilde{F}_{i,j,q}$.
$S_{i,j,q}$ local memory size of the surrogate submodel $\tilde{F}_{i,j,q}$.
$\varepsilon_{i,j,q}$ measure of the local error, $\tilde{F}_{i,j,q} - F_{i,j}$.
$a_{i,j,q}$ upper-bound of the global error duc thc propagation of the local error between $\tilde{F}_{i,j,q}$ and $F_{i,j}$.
$z_{i,j,q}$ binary decision variable to select or not the surrogate submodel $\tilde{F}_{i,j,q}$.
$\boldsymbol{Z}$ vector composed by the elementary decision variables $z_{i,j,q}$.

Remark L_i's value is generally different at each level i and the number of surrogates $Q_{i,j}$ for each submodel $F_{i,j}$ is not constant, therefore j depends on i and q depends on i and j; nevertheless these dependances are generally obvious and will not be explicitely specified when not necessary.

The data flow of the global model F follows a one way direction from the top level to the bottom level (cf. Fig. 6.1).

The global original model F is composed of submodels $F_{i,j}$ and the global surrogate model $\tilde{F}$ is composed of surrogate submodels $\tilde{F}_{i,j,q}$. We assume that at each $i \in \{1, \ldots, n\}$ and for each $j \in \{1, \ldots, L_i\}$, the submodel $F_{i,j}$ is defined by:

$$\begin{aligned} F_{i,j} : \mathbb{R}^{L_{i-1}} &\longrightarrow \mathbb{R} \\ \boldsymbol{O_{i-1}} &\longmapsto F_{i,j}(\boldsymbol{O_{i-1}}) \end{aligned}$$

and can be approximated by a family of q surrogates $\tilde{F}_{i,j,q}$:

$$\begin{aligned} \tilde{F}_{i,j,q} : \mathbb{R}^{L_{i-1}} &\longrightarrow \mathbb{R} \\ \boldsymbol{O_{i-1}} &\longmapsto \tilde{F}_{i,j,q}(\boldsymbol{O_{i-1}}). \end{aligned}$$

6.1.3 The Optimization Problem

The problem described in the introduction is naturally formulated as a multi-objective optimization problem in which we would like to minimize simultaneously the memory size S, the computational time T and the global error $\|\varepsilon\|$. However, we notice that these three criteria are generally monotonously linked: when the memory size increases, the computational time generally increases as well, while the error

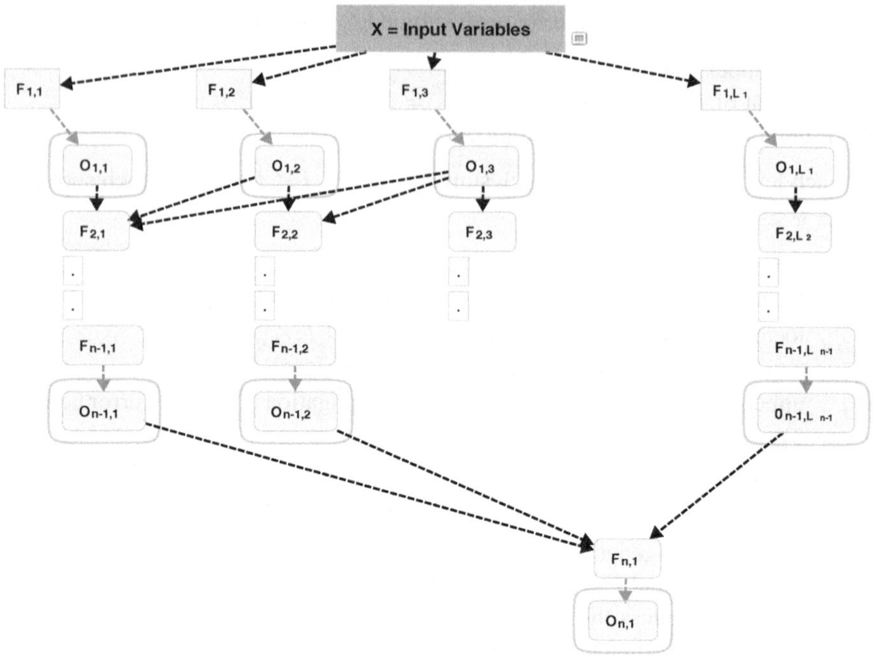

Fig. 6.1 Multilevel architecture description

decreases. The maximum acceptable global error $\overline{\varepsilon}$ is deduced by specialist analysis from the consequences it induces. In the same way, the maximum acceptable global computational time $\overline{T}$ is established in order to be compatible with real time calculation needs during flight. Therefore, these latter quantities are given and can be considered as constraints. The initial multi-criteria problem can then be reformulated as a monocriterium problem subject to constraints, in which the objective is to minimize the memory size.

In order to formalize it, we now introduce a decision variable defined by:

$$\forall i \in \{1, \cdots n\}, \forall j \in \{1, \ldots, L_i\}, \forall q \in \{1, \ldots, Q_{i,j}\}$$
$$z_{i,j,q} = \begin{cases} 1 & \text{if the surrogate model number } q \text{ is selected for submodel } F_{i,j}, \\ 0 & \text{else.} \end{cases}$$

Since there is a unique surrogate selected for each submodel, we have:

$$\forall i \in \{1, \ldots n\}, \forall j \in \{1, \ldots, L_i\}, \sum_{q=1}^{Q_{i,j}} z_{i,j,q} = 1.$$

The global memory size S of the surrogate model $\tilde{F}$ is the sum of the memory sizes of each selected surrogates $\tilde{F}_{i,\,j,q}$:

$$S = \sum_{i=1}^{n}\sum_{j=1}^{L_i}\sum_{q=1}^{Q_{i,\,j}} S_{i,\,j,q} \cdot z_{i,\,j,q}.$$

S is a function of the selected surrogates and is the criterium to minimize.
In the same way, the global computational time T is the sum of the computational times of each selected surrogates:

$$T = \sum_{i=1}^{n}\sum_{j=1}^{L_i}\sum_{q=1}^{Q_{i,\,j}} T_{i,\,j,q} \cdot z_{i,\,j,q},$$

while the global error $\|\varepsilon\|$ is the norm of the output error $\|\varepsilon\| = \|\tilde{F} - F\|$.
Again, T and $\|\varepsilon\|$ are functions of the selected surrogates.

If we note $\boldsymbol{Z}$ the vector composed by the elementary decision variables $z_{i,\,j,q}$, the optimization problem can now be written as follows:

$$\begin{aligned} &\min_{\boldsymbol{Z}} \; S(\boldsymbol{Z}) \\ &\text{subject to} \; \begin{cases} T(\boldsymbol{Z}) \leq \overline{T}, \\ \|\varepsilon(\boldsymbol{Z})\| \leq \overline{\varepsilon}. \end{cases} \end{aligned} \tag{6.1}$$

The exact value of the global error $\|\varepsilon(\boldsymbol{Z})\|$ is not known but we prove that it can be upper bounded if F is differentiable:

$$\|\varepsilon(\boldsymbol{Z})\| \leq \sum_{i=1}^{n}\sum_{j=1}^{L_i}\sum_{q=1}^{Q_{i,\,j}} a_{i,\,j,q} \cdot z_{i,\,j,q}. \tag{6.2}$$

Proof For level one, we have $\forall 1 \leq j_1 \leq L_1$:

$$\begin{aligned} O_{1,\,j_1} &= F_{1,\,j_1}(\boldsymbol{X}), \\ \tilde{O}_{1,\,j_1} &= \sum_{q=1}^{Q_{1,\,j_1}} \tilde{F}_{1,\,j_1,q}(\boldsymbol{X}) \cdot z_{1,\,j_1,q}, \\ \tilde{F}_{1,\,j_1,q}(\boldsymbol{X}) &= F_{1,\,j_1}(\boldsymbol{X}) + \varepsilon_{1,\,j_1,q}(\boldsymbol{X}),\; 1 \leq q \leq Q_{1,\,j_1}. \end{aligned}$$

Since there is only one selection of surrogate model for each submodel, we have:

$$\forall 1 \leq j_1 \leq L_1,\;\; \tilde{O}_{1,\,j_1} = F_{1,\,j_1}(\boldsymbol{X}) + \sum_{q=1}^{Q_{1,\,j_1}} \varepsilon_{1,\,j_1,q}(\boldsymbol{X}) \cdot z_{1,\,j_1,q}.$$

We then obtain:

$$\forall 1 \leq j_1 \leq L_1,\ \tilde{O}_{1,\,j_1} = O_{1,\,j_1} + \sum_{q=1}^{Q_{1,\,j_1}} \varepsilon_{1,\,j_1,\,q}(\boldsymbol{X}) \cdot z_{1,\,j_1,\,q}. \tag{6.3}$$

For level two, we have $\forall 1 \leq j_2 \leq L_2$:

$$\begin{aligned} O_{2,\,j_2} &= F_{2,\,j_2}(\boldsymbol{O_1}), \\ \tilde{O}_{2,\,j_2} &= \sum_{q=1}^{Q_{2,\,j_2}} \tilde{F}_{2,\,j_2,\,q}(\tilde{\boldsymbol{O}}_{\boldsymbol{1}}) \cdot z_{2,\,j_2,\,q}, \\ \tilde{F}_{2,\,j_2,\,q}(\tilde{\boldsymbol{O}}_{\boldsymbol{1}}) &= F_{2,\,j_2}(\tilde{\boldsymbol{O}}_{\boldsymbol{1}}) + \varepsilon_{2,\,j_2,\,q}(\tilde{\boldsymbol{O}}_{\boldsymbol{1}}),\ 1 \leq q \leq Q_{2,\,j_2}. \end{aligned}$$

As for level one, we simplify the equation and have:

$$\forall 1 \leq j_2 \leq L_2,\ \tilde{O}_{2,\,j_2} = F_{2,\,j_2}(\tilde{\boldsymbol{O}}_{\boldsymbol{1}}) + \sum_{q=1}^{Q_{2,\,j_2}} \varepsilon_{2,\,j_2,\,q}(\tilde{\boldsymbol{O}}_{\boldsymbol{1}}) \cdot z_{2,\,j_2,\,q}.$$

By using (6.3), we obtain:

$$\begin{aligned} F_{2,\,j_2}(\tilde{\boldsymbol{O}}_{\boldsymbol{1}}) = F_{2,\,j_2}\Bigg(& O_{1,\,1} + \sum_{q=1}^{Q_{1,\,1}} \varepsilon_{1,\,1,\,q}(\boldsymbol{X}) \cdot z_{1,\,1,\,q}, \dots, \\ & O_{1,\,L_1} + \sum_{q=1}^{Q_{1,\,L_1}} \varepsilon_{1,\,L_1,\,q}(\boldsymbol{X}) \cdot z_{1,\,L_1,\,q}\Bigg). \end{aligned}$$

Since F is differentiable at each point of the line segment $[\boldsymbol{O_1};\, \tilde{\boldsymbol{O}}_{\boldsymbol{1}}]$, then by using the mean value theorem (for several variables) there exists $\zeta_{2,\,j_2} \in [\boldsymbol{O_1};\, \tilde{\boldsymbol{O}}_{\boldsymbol{1}}]$ such that:

$$F_{2,\,j_2}(\tilde{\boldsymbol{O}}_{\boldsymbol{1}}) = F_{2,\,j_2}(\boldsymbol{O_1}) + \sum_{j_1=1}^{L_1} \frac{\partial F_{2,\,j_2}}{\partial O_{j_1}}(\zeta_{2,\,j_2}) \left(\sum_{q=1}^{Q_{1,\,j_1}} \varepsilon_{1,\,j_1,\,q}(\boldsymbol{X}) \cdot z_{1,\,j_1,\,q} \right).$$

Finally we obtain:

$$\begin{aligned} &\forall 1 \leq j_1 \leq L_2, \\ &\tilde{O}_{2,\,j_2} = O_{2,\,j_2} + \sum_{j_1=1}^{L_1} \sum_{q=1}^{Q_{1,\,j_1}} \left(\frac{\partial F_{2,\,j_2}}{\partial O_{j_1}}(\zeta_{2,\,j_2}) \cdot \varepsilon_{1,\,j_1,\,q}(\boldsymbol{X}) \right) \cdot z_{1,\,j_1,\,q} \\ &\qquad + \sum_{q=1}^{Q_{2,\,j_2}} \varepsilon_{2,\,j_2,\,q}(\tilde{\boldsymbol{O}}_{\boldsymbol{1}}) \cdot z_{2,\,j_2,\,q}. \end{aligned} \tag{6.4}$$

For the last output at level n, we obtain:

$$
\begin{aligned}
\tilde{O}_{n,1} = O_{n,1} &+ \sum_{q=1}^{Q_{n,1}} \varepsilon_{n,1,q}(\tilde{\boldsymbol{O}}_{\boldsymbol{n-1}}) \cdot z_{n,1,q} \\
&+ \sum_{j_{n-1}=1}^{L_{n-1}} \sum_{q=1}^{Q(n-1,j_{n-1})} \left(\frac{\partial F_{n,1}}{\partial O_{j_{n-1}}}(\zeta_{n,1}) \cdot \varepsilon_{n-1,j_{n-1},q}(\tilde{\boldsymbol{O}}_{\boldsymbol{n-1}}) \right) \cdot z_{n-1,j_{n-1},q} \\
&+ \cdots \\
&+ \sum_{j_{n-1}=1}^{L_{n-1}} \cdots \sum_{j_3=1}^{L_3} \sum_{q=1}^{Q_{3,j_3}} \left(\frac{\partial F_{n,1}}{\partial O_{j_{n-1}}}(\zeta_{n,1}) \cdots \frac{\partial F_{4,j_4}}{\partial O_{j_3}}(\zeta_{4,j_4}) \cdot \varepsilon_{3,j_3,q}(\tilde{\boldsymbol{O}}_{\boldsymbol{2}}) \right) \cdot z_{3,j_3,q} \\
&+ \sum_{j_{n-1}=1}^{L_{n-1}} \cdots \sum_{j_3=1}^{L_3} \sum_{j_2=1}^{L_2} \sum_{q=1}^{Q_{2,j_2}} \left(\frac{\partial F_{n,1}}{\partial O_{j_{n-1}}}(\zeta_{n,1}) \cdots \right. \\
&\qquad \left. \frac{\partial F_{4,j_4}}{\partial O_{j_3}}(\zeta_{4,j_4}) \cdot \frac{\partial F_{3,j_3}}{\partial O_{j_2}}(\zeta_{3,j_3}) \cdot \varepsilon_{2,j_2,q}(\tilde{\boldsymbol{O}}_{\boldsymbol{1}}) \right) \cdot z_{2,j_2,q} \\
&+ \sum_{j_{n-1}=1}^{L_{n-1}} \cdots \sum_{j_3=1}^{L_3} \sum_{j_2=1}^{L_2} \sum_{j_1=1}^{L_1} \sum_{q=1}^{Q_{1,j_1}} \left(\frac{\partial F_{n,1}}{\partial O_{j_{n-1}}}(\zeta_{n,1}) \cdots \right. \\
&\qquad \left. \frac{\partial F_{4,j_4}}{\partial O_{j_3}}(\zeta_{4,j_4}) \cdot \frac{\partial F_{3,j_3}}{\partial O_{j_2}}(\zeta_{3,j_3}) \cdot \frac{\partial F_{2,j_2}}{\partial O_{j_1}}(\zeta_{2,j_2}) \cdot \varepsilon_{1,j_1,q}(\boldsymbol{X}) \right) \cdot z_{1,j_1,q}
\end{aligned}
\tag{6.5}
$$

By using the triangle inegality on Eq. (6.5) $\left\| \tilde{O}_{n,1} - O_{n,1} \right\|$:

$$
\begin{aligned}
\left\| \tilde{O}_{n,1} - O_{n,1} \right\| \leq & \sum_{q=1}^{Q_{n,1}} a_{n,1,q} \cdot z_{n,1,q} \\
&+ \sum_{j_{n-1}=1}^{L_{n-1}} \sum_{q=1}^{Q_{n-1,j_{n-1}}} a_{n-1,j_{n-1},q} \cdot z_{n-1,j_{n-1},q} + \cdots \\
&+ \sum_{j_3=1}^{L_3} \sum_{q=1}^{Q_{3,j_3}} a_{3,j_3,q} \cdot z_{3,j_3,q} \\
&+ \sum_{j_2=1}^{L_2} \sum_{q=1}^{Q_{2,j_2}} a_{2,j_2,q} \cdot z_{2,j_2,q} \\
&+ \sum_{j_1=1}^{L_1} \sum_{q=1}^{Q_{1,j_1}} a_{1,j_1,q} \cdot z_{1,j_1,q}.
\end{aligned}
$$

with:

$$a_{n,1,q} = \left\|\varepsilon_{n,1,q}\right\|,$$

$$a_{i,j_i,q} = \sum_{j_{n-1}=1}^{L_{n-1}} \cdots \sum_{j_{i+2}=1}^{L_{i+2}} \sum_{j_{i+1}=1}^{L_{i+1}} \left\|\frac{\partial F_{n,1}}{\partial O_{j_{n-1}}}\right\| \cdots \left\|\frac{\partial F_{i+3,j_{i+3}}}{\partial O_{j_{i+2}}}\right\| \left\|\frac{\partial F_{i+2,j_{i+2}}}{\partial O_{j_{i+1}}}\right\| \left\|\frac{\partial F_{i+1,j_{i+1}}}{\partial O_{j_i}}\right\| \left\|\varepsilon_{i,j_i,q}\right\|,$$
$$\forall i \in \{1, \ldots, n-1\}, \forall j_i \in \{1, \ldots, L_i\}, \forall q \in \{1, \ldots, Q_{i,j}\},$$

which is an upper bound for $\|\varepsilon(\mathbf{Z})\|$, see Eq. (6.2).

$\|\varepsilon_{i,j_i,q}\|$ and $\left\|\frac{\partial F_{i,j_i}}{\partial O_{j_{i-1}}}\right\|$ respectively denote the norms of $\varepsilon_{i,j_i,q}(.)$ and $\frac{\partial F_{i,j_i}}{\partial O_{j_{i-1}}}(.)$.

Note that the choice of the norm is free and depends on the criticity of the industrial model we want to reduce (cf Sect. 6.2). □

Finally, the problem (6.1) can be written as the following linear integer programming problem:

$$\min \sum_{i=1}^{n} \sum_{j=1}^{L_i} \sum_{q=1}^{Q_{i,j}} S_{i,j,q} \cdot z_{i,j,q}$$
subject to:
$$\begin{cases} \sum_{i=1}^{n} \sum_{j=1}^{L_i} \sum_{q=1}^{Q_{i,j}} T_{i,j,q} \cdot z_{i,j,q} \leq \overline{T}, & \\ \sum_{i=1}^{n} \sum_{j=1}^{L_i} \sum_{q=1}^{Q_{i,j}} a_{i,j,q} \cdot z_{i,j,q} \leq \overline{\varepsilon}, & \\ z_{i,j,q} \in \{0,1\}, & \forall i \in \{1,\ldots,n\}, \\ & \forall j \in \{1,\ldots,L_i\}, \\ & \forall q \in \{1,\ldots,Q_{i,j}\}, \\ \sum_{q=1}^{Q_{i,j}} z_{i,j,q} = 1, & \forall i \in \{1,\ldots,n\}, \\ & \forall j \in \{1,\ldots,L_i\}. \end{cases} \tag{6.6}$$

6.2 Application on an Industrial Case

The Fuel Consumption Model (FCM) aims at computing a fuel consumption in cruise. It depends on six variables: the load factor (NxAero), the Mach (M), the altitude (Zp), the temperature (ΔISA), the weight (GW) and the center of gravity position (ΔXCG). Its structure complies with the multilevel architecture defined in Sect. 6.1.2 (cf. Fig. 6.1).

This model is derivated from a general performance model. As it is part of an embedded function, some requirements have to be fulfilled:

- the number of input parameters of elementary tabulated submodels must not exceed two,
- the number of values of the tabulated functions is limited due to memory size constraint,
- the computational time is limited due to real time needs.

The figure below shows the multilevel architecture of the FCM. Each of its nine levels is composed of one to five submodels having one, two or three inputs. Nine of them have an analytical expression and therefore do not need to be reduced. They are represented with colored boxes in Fig. 6.2. The eight other ones use interpolation in numerical look-up tables. They are responsible of most of the memory size and computational time. Thus, the proposed optimization process will focus on the reduction of these particular submodels. They are represented with white boxes in Fig. 6.2.

6.2.1 Building High Dimensional Model Representation

The optimization methodology presented in Sect. 6.1.1 will be applied to perform the best choice among a set of surrogates. For the industrial case presented hereafter, each tabulated submodel will be reduced using HDMR decomposition of different orders [2]. Such a decomposition generally achieves a drastic reduction on data to be stored in the numerical tables. More precisely, HDMR methods construct a map of the relationship between the input variables and the output by capitalizing only low order correlations among the input variables having an impact on the output. Therefore HDMR methods provide a means of partitioning the original multivariate input data into low variate data while keeping the dimensionality of the original input space.

Let g be one of the submodels to be reduced and d the dimension of its input variable vector $\boldsymbol{X}$ with $\boldsymbol{X} = (X_1, X_2, \ldots, X_d)$.
The HDMR decomposition gives:

$$g(X) = \tilde{g}_0 + \sum_{i=1}^{d} \tilde{g}_i(X_i) + \sum_{i=1}^{d} \sum_{j=i+1}^{d} \tilde{g}_{i,j}(X_i, X_j) + \ldots + \tilde{g}_{1,2,\ldots,d}(X_1, \ldots, X_d). \quad (6.7)$$

The value $\tilde{g}_0$ is a constant term representing the zeroth-order component function, which is the mean response of the aircraft performance submodel g. The function

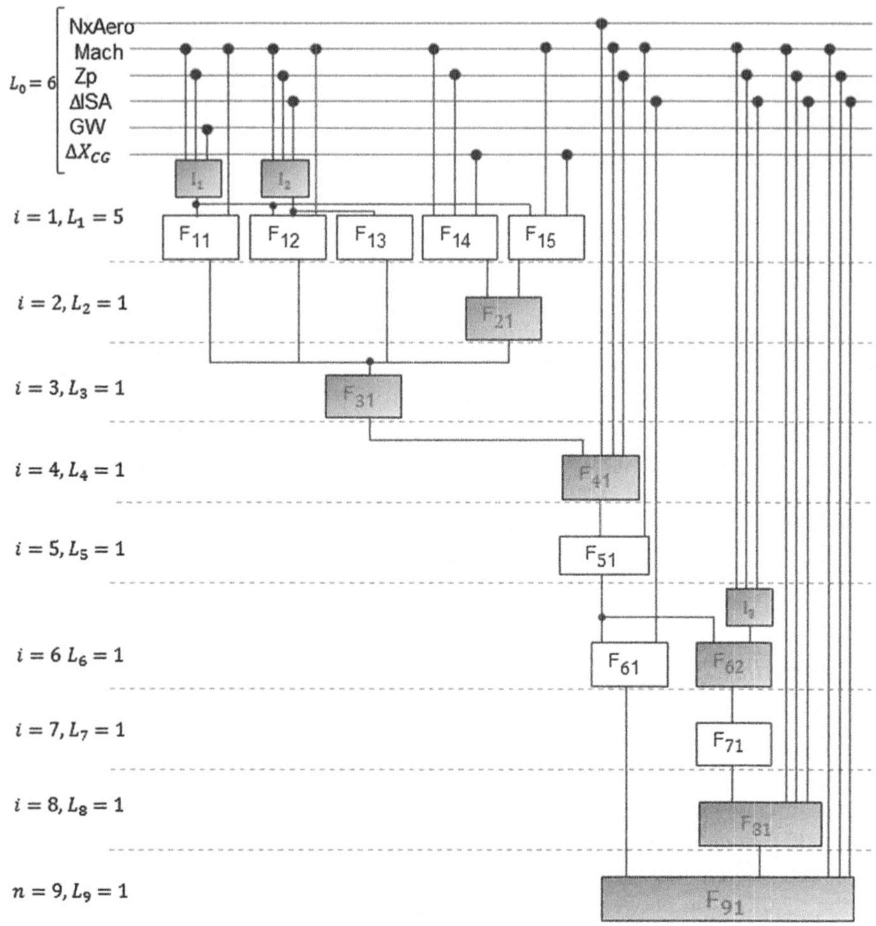

Fig. 6.2 The multilevel architecture of the FCM

$(\tilde{g}_i)_{1 \leq i \leq d}$ is a first-order term expressing the effect of input variable X_i acting alone on g. The function $(\tilde{g}_{i,\,j})_{1 \leq i \leq d, i+1 \leq j \leq d}$ is a second-order term that describes the cooperative effects of the variables X_i and X_j. The higher order terms give the cooperative effects of increasing number of input variables acting together to influence the output g. HDMR is based on the fundamental principle that correlations between input variables affecting the submodels decrease rapidly [3]. Thanks to this method we can build different surrogate models for each submodel by truncating Eq. (6.7) at zeroth-order, first-order or second-order.

An example of such surrogate obtained by truncating HDMR decomposition is given hereafter for submodel F_{12} . This submodel F_{12} represents the Reynolds effect on drag coefficient which is a function of three variables ($d = 3$): Mach, ΔISA and lift coefficient. For this submodel, only first and second-order HDMR are considered. Therefore the associated index q for F_{12} can reach three values:

- $q = 1$: reference submodel (no HDMR built).
- $q = 2$: 2nd-order HDMR. In Eq. (6.7), we keep the constant $\tilde{g}_0$, the first-order terms $\sum_{i=1}^{3} \tilde{g}_i(X_i)$, and the second-order terms $\sum_{i=1}^{3} \sum_{j=i+1}^{3} \tilde{g}_{i,j}(X_i, X_j)$.
- $q = 3$: 1st order HDMR. In Eq. (6.7), we keep the constant $\tilde{g}_0$ and the first order terms $\sum_{i=1}^{3} \tilde{g}_i(X_i)$.

The global characteristics of the eight submodels to be reduced and their candidate surrogates (HDMR of 0th, 1st and 2nd-order) are given in the tables below. Table 6.1 gives the memory size expressed in terms of numbers of tabulated values. Table 6.2 provides the computational time in milliseconds on a personnal computer. Table 6.3 presents the maximum error relatively to the reference model. Finally, Table 6.4 gives the correspondance between the HDMR order and the index q of surrogate.

Remark No value in Tables 6.1, 6.2, 6.3, 6.4 means that the corresponding surrogates are either irrelevant or non-existant.

6.2.2 Surrogate Optimal Choice Results

In this section, the optimization process developped in Sect. 6.1.1 is applied to find out the best choice of surrogates, provided in the previous Sect. 6.2.1. In order to solve this linear integer programming problem, a branch and bound algorithm [5] is used. In this industrial example, the global computational time constraint ($\bar{T}$) has no major effect on the optimal solution, because its value is sufficiently high. Hence, the only active constraint is the maximum global error, which has a major impact on the memory size of the solution; then it is worth to propose a trade-off between these two parameters.

Table 6.1 Size S

Model	Ref	HDMR order		
		2nd	1st	0th
F_{11} 2D	415	–	81	–
F_{12} 3D	504	284	49	–
F_{13} 1D	40	–	–	1
F_{14} 3D	51	83	23	–
F_{15} 3D	474	282	49	–
F_{51} 2D	246	–	61	–
F_{61} 2D	305	–	67	–
F_{71} 1D	12	–	–	1

Table 6.2 Computation time T

Model	Ref	HDMR order		
		2nd	1st	0th
F_{11} 2D	0.4	–	0.3	–
F_{12} 3D	0.8	1.5	0.3	–
F_{12} 3D	0.1	–	–	0
F_{12} 3D	0.8	1.5	0.3	–
F_{12} 3D	0.8	1.5	0.3	–
F_{12} 3D	0.4	–	0.3	–
F_{12} 3D	0.4	–	0.3	–
F_{12} 3D	0.4	–	–	0

Table 6.3 Error $\|\varepsilon\|_\infty$

Model	Ref	HDMR order		
		2nd	1st	0th
F_{11} 2D	0	–	3.88×10^{-4}	–
F_{12} 3D	0	7.83×10^{-7}	2.59×10^{-4}	–
F_{13} 1D	0	–	–	7.05×10^{-5}
F_{14} 3D	0	1.00×10^{-16}	1.60×10^{-1}	–
F_{15} 3D	0	6.70×10^{-5}	4.80×10^{-3}	–
F_{51} 2D	0	–	2.07×10^{-2}	–
F_{61} 2D	0	–	4.72×10^{-2}	–
F_{71} 1D	0	–	–	7.16×10^{-5}

Table 6.4 q values

Model	Ref	HDMR order		
		2nd	1st	0th
F_{11} 2D	1	–	2	–
F_{12} 3D	1	2	3	–
F_{13} 1D	1	–	–	2
F_{14} 3D	1	2	3	–
F_{15} 3D	1	2	3	–
F_{51} 2D	1	–	2	–
F_{61} 2D	1	–	2	–
F_{71} 1D	1	–	–	2

In the first subsection, the impact of the global maximum error constraint value $\bar{\varepsilon}$ on the memory size of the optimal solution is shown. The second subsection describes the choice of surrogates achieved for each optimal solution.

6.2.2.1 Trade-Off Between Error Constraint and Memory Size

The chart below presents the evolution of the optimal memory size in function of the global maximum error constraint value $\bar{\varepsilon}$.

The output of the model is a fuel flow which takes its value between 200 and 9,000 kg/h. Since there is a finite number of candidate surrogates, the set of optimal sizes is discrete. Therefore the evolution of size versus global maximum error constraint is a step curve. Moreover, as the global maximum error constraint tested is also finite (every 100 kg/h), the curve presented in Fig. 6.3 is the conservative one betwen two optimal solutions found. We notice for example that between $100 \leq \bar{\varepsilon} \leq 200$ kg/h the optimal size remains at 1,398, thus between these two values the exact size is visualized. On the same step, between $200 \leq \bar{\varepsilon} \leq 300$ kg/h, the size of the FCM model is at most 1,398 but it could be 1,206 or any other value between 1,206 and 1,398. Because we can not test all the $\bar{\varepsilon}$ value, we made the decision of keeping the conservative size, which is 1,398 in this case.

Obviously, it should be noticed that the memory size is a decreasing function of global maximum error constraint.

6.2.2.2 Evolution of the Optimal Choice of Surrogates

In this subsection, we show the evolution of the optimal HDMR decomposition for each of the eight tabulated functions vs maximum error constraint value of the global corresponding model (see Fig. 6.4).

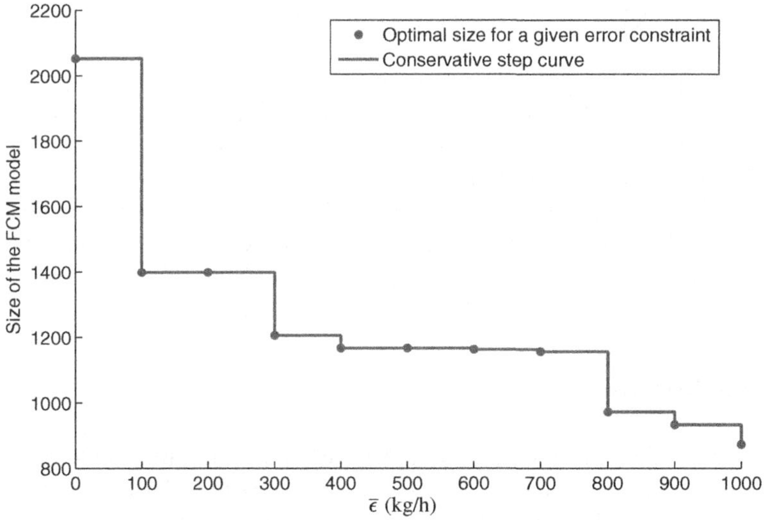

Fig. 6.3 Visualisation of the optimal solutions

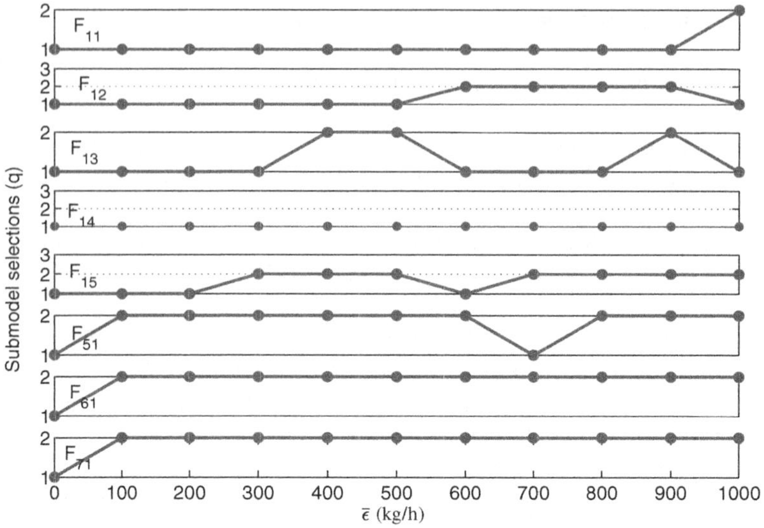

Fig. 6.4 Visualisation of the changes in submodel selections

We can notice in Fig. 6.4 that the submodel selection is not a monotonous function of $\bar{\varepsilon}$. Several comments of these evolutions are given hereafter:

- For $\bar{\varepsilon} = 0$, all the original submodels are selected (no reduction).
- The error due to submodel F_{11} has a strong influence on the output because the reference submodel is kept except for high maximum error constraint.
- The error due to the submodels F_{61} and F_{71} have low influence on the output of the global model FCM: the highest possible reduction is always choosen for both of them.
- F_{12}, F_{13}, F_{15}, F_{51} are the key elements of the reduction. When the global maximum error constraint evolves, their HDMR decomposition order change to obtain the best memory size solution. Roughly speaking, the optimization process found the best balance between these four submodels.
- For the submodel F_{14}, since the 1st order HDMR is not enough accurate and the 2nd order HDMR has a larger size (see Table 6.1), the reference original submodel is always selected.

6.3 Conclusion

For embedded hirarchical multilevel model, we define a new methodology which is able to find the best combination of available surrogate submodels, in order to build a global surrogate satisfying requirements of computational time, storage memory and accuracy while keeping a multilevel structure. We show how this optimal combination method can be formalized as a linear integer programming problem.

The feasability and the efficiency of this method has been demonstrated on a real aircraft performance function, the Fuel Consumption Model. In this application, the set of surrogate submodels have previously been constructed with the HDMR decomposition of different orders.

Our future work will be focused on improving the estimation of the different error propagations in order to reduce the conservatism of the global maximum error. Moreover, to refine the decision about the choices of surrogates, multi-criteria optimization methods will be explored.

References

1. Liu Y, Yin X, Arendt P, Chen W, Huang H (2010) A hierarchical statistical sensitivity analysis method for multilevel systems with shared variables. J Mech Design 132:1–2
2. Rabitz H, Alis O, Shorter J, Shim K (1999) Efficient input-output model representations. Comput Phys Commun 117(1–2):11–20
3. Sobol IM (2003) Theorems and examples on high dimensional model representation. Reliab Eng Syst Saf 79:187–193
4. Talgorn B, Bès C, Laporte S, Segonds S (2009) Optimization of payload multi-step strategy. In: Proceedings of the 9th AIAA aviation technology, integration, and operations (ATIO) conference
5. Wolsey LA (1998) Integer programming, Wiley-Interscience Series In Discrete Mathematics And Optimization. Wiley, New York

Chapter 7
Multi-objective Design Optimization Using High-Order Statistics for CFD Applications

Pietro M. Congedo, Gianluca Geraci, Remi Abgrall and Gianluca Iaccarino

Abstract This work illustrates a practical and efficient method for performing multi-objective optimization using high-order statistics. It is based on a Polynomial Chaos framework, and evolutionary algorithms. In particular, the interest of considering high-order statistics for reducing the number of uncertainties is studied. The feasibility of the proposed method is proved on a Computational Fluid-Dynamics (CFD) real-case application.

Keywords High-order statistics · Dimension reduction · Genetic algorithms · Robust optimization

7.1 Introduction

Optimization and design in the presence of uncertain operating conditions, material properties and manufacturing tolerances poses a tremendous challenge to the scientific computing community. In many industry-relevant situations the performance metrics depend in a complex, non-linear fashion on those factors and the construction of an accurate representation of this relationship is difficult. Probabilistic uncertainty quantification (UQ)approaches represent the inputs as random

P.M. Congedo (✉) · G. Geraci
INRIA Bordeaux Sud-Ouest, 200 Avenue de la Vieille Tour, 33405 Talence, France
e-mail: pietro.congedo@inria.fr

G. Geraci
e-mail: gianluca.geraci@inria.fr

R. Abgrall
Institut für Mathematik, Universität Zürich, Winterthurerstrasse 190 CH-8057, Zürich, Switzerland
e-mail: remi.abgrall@math.uzh.ch

G. Iaccarino
Mechanical Engineering Department, Bld 500, Stanford, CA 94305-3035, USA
e-mail: jops@stanford.edu

D. Greiner et al. (eds.), *Advances in Evolutionary and Deterministic Methods for Design, Optimization and Control in Engineering and Sciences*, Computational Methods in Applied Sciences 36, DOI 10.1007/978-3-319-11541-2_7

variables and seek to construct a statistical characterization of few quantities of interest. Several methodologies are proposed to tackle this problem; polynomial chaos (PC) methods [1] can provide considerable speed-up in computational time when compared to MC. In realistic situations however, the presence of a large number of uncertain inputs leads to an exponential increase of the cost thus making these methodologies unfeasible [2]. This situation becomes even more challenging when robust design optimization is tackled [3]. Robust optimization processes may require a prohibitive computational cost when dealing with a large number of uncertainties and a highly non-linear fitness function. Efforts in the development of numerical method are directed mainly to reduce the number of deterministic evaluations necessary for solving the optimization problem and for the uncertainty quantification (UQ) of the performances of interest. The overall cost is typically the product of the cost of the two approaches because the stochastic analysis and the optimization strategy are completely decoupled. Decoupled approaches are simple but more expensive than necessary.

Several UQ methods have been developed with the objective of reducing the number of solution required to obtain a statistical characterization of the quantity of interest. An alternative solution is based on approaches attempting to identify the relative importance of the input uncertainties on the output. A well known methodology is based on a decomposition of the variance of the quantity of interest in contributions closely connected to each of the input uncertainties (main effects) or combined inputs [4]. Recently, a practical way to decompose high-order statistical moments has been also proposed [5].

In this work, we illustrate an efficient multi-objective optimization method taking into account high-order statistic moments, such as the third and fourth-order statistic moments, i.e. skewness and kurtosis, respectively. These moments can be easily computed by means of a Polynomial Chaos (PC) method. A simplified test-case is presented that displays the importance of considering high-order moments during the optimization. The efficiency of the method in terms of computational cost and fitness function complexity is verified in a realistic CFD-case optimization problem.

7.2 Computing High-Order Statistics by Using Polynomial Chaos Techniques

This section is devoted to show how formulas of variance, skewness and kurtosis can be obtained using a polynomial chaos framework (for more details see Ref. [5]). If a polynomial chaos formulation is used, an approximation $\tilde{f}$ of the function f is provided

$$f(\boldsymbol{\xi}) \approx \tilde{f}(\boldsymbol{\xi}) = \sum_{k=0}^{P} \beta_k \Psi_k(\boldsymbol{\xi}), \tag{7.1}$$

where P is computed according to the order of the polynomial expansion n_0 and the stochastic dimension of the problem d

$$P + 1 = \frac{(n_0 + d)!}{n_0! d!}. \tag{7.2}$$

Each polynomial $\Psi_k(\boldsymbol{\xi})$ of total degree n_o is a multivariate polynomial form which involve tensorization of 1D polynomial form by using a multi-index $\boldsymbol{\alpha}^k \in \mathbb{R}^d$, with $\sum_{i=1}^d \boldsymbol{\alpha}_i^k \leq n_0$:

$$\Psi_k(\boldsymbol{\xi} \cdot \boldsymbol{m}^{\star,k}) = \prod_{\substack{i=1 \\ m_i^{\star,k} \neq 0}}^{d} \psi_{\alpha_i^k}(\xi_i) \tag{7.3}$$

where the multi index $\boldsymbol{m}^{\star,k} = \boldsymbol{m}^{\star,k}(\boldsymbol{\alpha}^k) \in \mathbb{R}^d$ is a function of $\boldsymbol{\alpha}^k$: $\boldsymbol{m}^{\star,k} = (m_1^{\star,k}, \ldots, m_d^{\star,k})$, with $m_i^{\star,k} = \alpha_i^k / ||\alpha_i^k||_{\neq 0}$.

Remark that, for each polynomial basis, $\psi_0(\xi_i) = 1$ and then $\Psi_0(\boldsymbol{\xi}) = 1$. Then, the first coefficient β_0 is equal to the expected value of the function, i.e. $E(f)$. The polynomial basis is chosen accordingly to the Wiener-Askey scheme in order to select orthogonal polynomials with respect to the probability density function $p(\boldsymbol{\xi})$ of the input. Thanks to the orthogonality, the following relation holds

$$\int_{\Xi} \Psi_i(\boldsymbol{\xi})\Psi_k(\boldsymbol{\xi})p(\boldsymbol{\xi})d\xi = \delta_{ij}\langle\Psi_i(\boldsymbol{\xi}), \Psi_i(\boldsymbol{\xi})\rangle \tag{7.4}$$

where $\langle\cdot, \cdot\rangle$ indicates the inner product and δ_{ij} is the Kronecker delta function.

The orthogonality can be advantageously used to compute the coefficients of the expansion in a non-intrusive PC framework

$$\beta_k = \frac{\langle f(\boldsymbol{\xi}), \Psi_k(\boldsymbol{\xi})\rangle}{\langle\Psi_k(\boldsymbol{\xi}), \Psi_k(\boldsymbol{\xi})\rangle}, \quad \forall k. \tag{7.5}$$

Note that the coefficients of the PC expansion are computed by a quadrature employing the points generated by a full tensorization of monodimensional quadrature rules. In particular, employing uniform distribution for the stochastic variables, a Legendre quadrature is chosen following the so-called Wiener-Askey scheme.

Variance (σ^2), skewness (s) and kurtosis (k) can be computed as follows:

$$\sigma^2 = E(f^2) - E(f)^2 = \sum_{k=1}^{P} \beta_k^2 \langle\Psi_k^2(\boldsymbol{\xi})\rangle. \tag{7.6}$$

$$s = \sum_{k=1}^{P} \beta_k^3 \langle\Psi_k^3(\boldsymbol{\xi})\rangle$$

$$
\begin{aligned}
&+3\sum_{i=1}^{P}\beta_i^2\sum_{\substack{j=1\\ j\neq i}}^{P}\beta_j\langle\Psi_i^2(\boldsymbol{\xi}),\Psi_j(\boldsymbol{\xi})\rangle\\
&+6\sum_{i=1}^{P}\sum_{j=i+1}^{P}\sum_{k=j+1}^{P}\beta_i\beta_j\beta_k\langle\Psi_i(\boldsymbol{\xi}),\Psi_j(\boldsymbol{\xi})\Psi_k(\boldsymbol{\xi})\rangle,
\end{aligned}
\tag{7.7}
$$

$$
\begin{aligned}
k &= \int f^4(\boldsymbol{\xi})p(\boldsymbol{\xi})\mathrm{d}\boldsymbol{\xi} - 4E(f)s - 6\sigma^2E(f)^2 - E(f)^4\\
&= \sum_{k=1}^{P}\beta_k^4\langle\Psi_k^4(\boldsymbol{\xi})\rangle + 4\sum_{i=1}^{P}\beta_i^3\sum_{\substack{j=1\\ j\neq i}}^{P}\beta_j\langle\Psi_i^3,\Psi_j\rangle\\
&+6\sum_{i=1}^{P}\beta_i^2\sum_{j=i+1}^{P}\beta_j^2\langle\Psi_i^2,\Psi_j^2\rangle\\
&+12\sum_{i=1}^{P}\beta_i^2\sum_{\substack{j=1\\ j\neq i}}^{P}\beta_j\sum_{\substack{k=j+1\\ k\neq i}}^{P}\beta_k\langle\Psi_i^2,\Psi_j\Psi_k\rangle\\
&+24\sum_{i=1}^{P}\sum_{j=i+1}^{P}\sum_{k=j+1}^{P}\sum_{h=k+1}^{P}\beta_i\beta_j\beta_k\beta_h\langle\Psi_i\Psi_j,\Psi_k\Psi_h\rangle.
\end{aligned}
\tag{7.8}
$$

7.3 Introducing More Sensitivity Indices

As introduced by Sobol [6], sensitivity indexes for variance can be computed for each conditional contribution as follows:

$$
\sigma_{\boldsymbol{m}_i}^{2,\mathrm{SI}} = \frac{\sigma_{\boldsymbol{m}_i}^2}{\sigma^2}. \tag{7.9}
$$

Here, we introduce new sensitivity indexes, basing on the decomposition of skewness and kurtosis, as follows

$$
\begin{aligned}
s_{\boldsymbol{m}_i}^{\mathrm{SI}} &= \frac{s_{\boldsymbol{m}_i}}{s}\\
k_{\boldsymbol{m}_i}^{\mathrm{SI}} &= \frac{k_{\boldsymbol{m}_i}}{k}.
\end{aligned}
\tag{7.10}
$$

If a total sensitivity index is needed, i.e. it is necessary to compute the overall influence of a variable, it can be computed summing up all the contributions in which the variable is present

$$\mathrm{TSI}_j = \sum_{\xi_j \in (\boldsymbol{\xi} \cdot \boldsymbol{m}_i)} \sigma^{2,\mathrm{SI}}_{\boldsymbol{m}_i}$$
$$\mathrm{TSI}^s_j = \sum_{\xi_j \in (\boldsymbol{\xi} \cdot \boldsymbol{m}_i)} s^{\mathrm{SI}}_{\boldsymbol{m}_i} \tag{7.11}$$
$$\mathrm{TSI}^k_j = \sum_{\xi_j \in (\boldsymbol{\xi} \cdot \boldsymbol{m}_i)} k^{\mathrm{SI}}_{\boldsymbol{m}_i}.$$

7.3.1 Dimensional Reduction in the Truncation Sense

The first test case is represented by the following quadratic g-function:

$$f(\boldsymbol{\xi}) = \prod_{i=1}^{4} \frac{|4\xi_i - 2| + a_i}{1 + a_i} \tag{7.12}$$

where $\xi_i \sim \mathscr{U}(0, 1)$. Two possible choices of the coefficients are considered here

- $a_i = (i - 1)/2$ the so called linear g-function f_{glin}
- $a_i = i^2$ the so called quadratic g-function f_{gquad}.

From the analysis, it is possible to note that the third and fourth variables seem to be meaningless for the variance based indexes. Their total sensitivity indexes sum up to 0.05 for the variance, while exceed 0.15 for both skewness and kurtosis. Considering only the sensitivity indexes computed on the variance the decision-maker could be tempted to neglect the variables ξ_3 and ξ_4. In this case the ANOVA expansion will loose all the terms containing ξ_3 and ξ_4

$$f_{G1} = f_0 + f_1(\xi_1) + f_2(\xi_2) + f_{12}(\xi_1, \xi_2)$$
$$f_{G2} = f_0 + f_1(\xi_1) + f_2(\xi_2) + f_{12}(\xi_1, \xi_2) + f_3(\xi_3) + f_{13}(\xi_1, \xi_3) + f_{23}(\xi_2, \xi_3) + f_{123}(\xi_1, \xi_2, \xi_3), \tag{7.13}$$

where in the first case f_{G1} both are neglected; while for f_{G2} only ξ_4 is neglected. In this case the ANOVA terms can be computed analytically and, from them, the statistics too. In the Table 7.1 the percentage errors, for the first four central moments, have been reported with respect the analytical exact solution for both the reduced models f_{G1} and f_{G2}.

Table 7.1 Percentage $\left(\frac{abs(\mu-\mu_{ex})}{\mu_{ex}} \times 100\right)$ errors related to the reduced g-function f_{G1} and f_{G2}

Function	Variance	Skewness	Kurtosis
f_{G1}	4.7997	29.236	15.039
f_{G2}	1.2369	7.7705	4.0632

The Table 7.1 makes evident as an error of only 5 % on the variance can correspond to a much greater error on the higher moments. The effect of a such behavior is clear looking at the Fig. 7.1 where the probability density function has been computed for both f_{G1} and f_{G2} and compared with the complete function (7.12). In this case the model with only the first two variables looses both the tails while a good approximation is achieved in the central part. However this test case clearly shows that considering only the sensitivity indexes based on the variance could be very dangerous if the reduced model need to be obtained for safety purpose. The information related to the less probable realizations of the system are totally lost and in this case the pdf resuls to be analytically bounded between the values of f equal to 0.4 and 1.8. If the third variable is included in the reduced model both variance and skewness are included with an error lower than 5 %, while the error on the skewness remains lower than 8 %. The total sensitivity indexes associated to the fourth variable is lower than 5 % for the three moments. The improvement of the model on the base of the inclusion of the third variable is evident in Fig. 7.1 where the pdf of the reduced model recover both the tails and better approximate the pdf of the complete function.

From a practical point-of-view the dimensional reduction of the model has commonly obtained freezing the neglected parameters. For an analytical function, as here of interest, it is possible to compute the constant values to choose, for both ξ_3 and ξ_4, to obtain a reduced model that preservers both the expectancy and the variance of the original complete model. Of course the two requirements cannot be satisfied contemporary but a set of values of constant satisfying the mean and the variance can be obtained analytically requiring (having a product tensorial function)

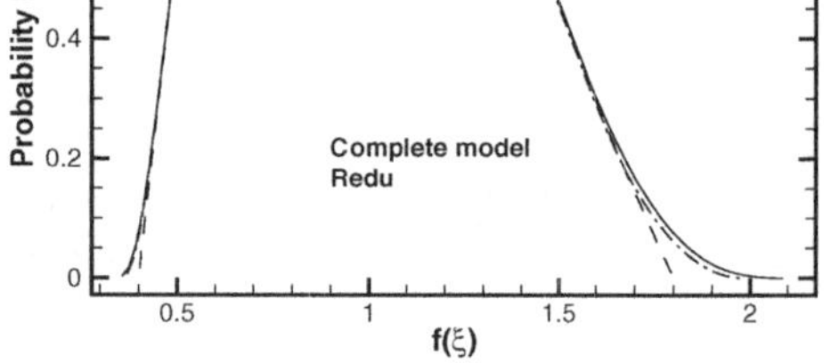

Fig. 7.1 PDFs for the complete g-function and the reduced models (see Eq. 7.13)

$$\frac{|4\bar{\xi}_j - 2| + a_j}{1 + a_j} = 1$$

$$\left(\frac{|4\bar{\xi}_j - 2| + a_j}{1 + a_j}\right)^2 = \int_0^1 \left(\frac{|4\xi_j - 2| + aj}{1 + a_j}\right)^2 d\xi_j. \tag{7.14}$$

The following values can be analytically computed for the two variables: $\xi_3 = \{1/4, 3/4, 91/120, 29/120\}$ and $\xi_4 = \{1/4, 3/4, 77/102, 25/102\}$.

In the Fig. 7.2 the pdf relative to the complete quadratic g-function freezing the parameters ξ_3 and ξ_4 are reported compared to the complete pdf and the totally reduced one.

From Fig. 7.2 it is evident that freezing parameter to assure the correctness of the mean and the variance produces pdf very close to that one obtained neglecting entirely the ANOVA terms. From a practical point-of-view the analysis of the reduced model can be carried out both with the ANOVA reduced model (if it is analytically possible to compute integrals) of by freezing the parameter to neglect satisfying the requirement on the expectancy and variance. In both case the results make in evidence as the analysis on the variance based sensitivity indexes needs to be supplemented by information from sensitivity indexes of higher order to be confident that the reduction of the model can be realized without deteriorate too much the information carried by the reduced model in term of distribution of the realizations, especially in the tails.

In the second part of this section the function f_1 has been analyzed. For the third variable, the level of the TSI 1.55 %, has resulted to be less than the threshold of 2 %, indicated in [7], to detect meaningless parameters. A dimensional reduced model can be obtained freezing the third parameter, or equivalently for what shown in the first part of this section, neglecting all the ANOVA terms in which the variable ξ_3 is present

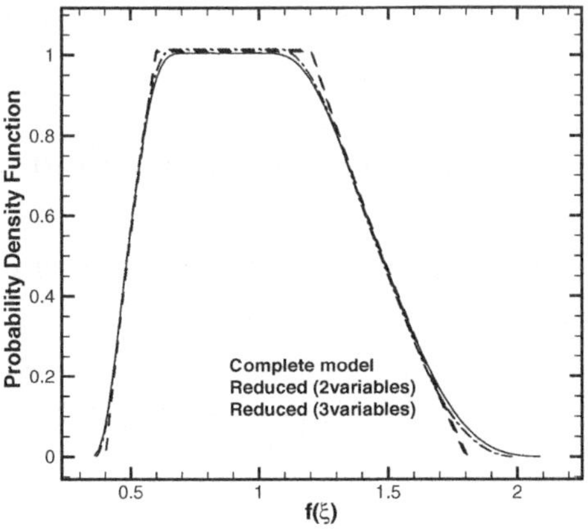

Fig. 7.2 PDFs for the complete g-function and the reduced models freezing the remaining parameters

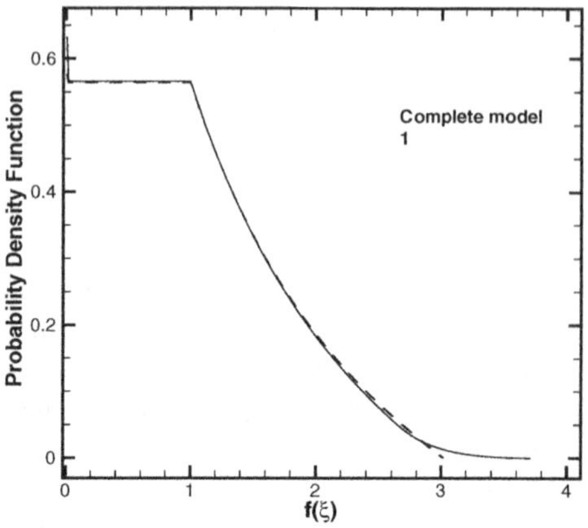

Fig. 7.3 PDFs for the complete f_1 and the reduced models freezing the third parameter

$$\hat{f}_1 = f_0 + f_1(\xi_1) + f_2(\xi_2) + f_{12}(\xi_1, \xi_2). \tag{7.15}$$

If the variable ξ_3 is chosen to recover the exact value of the mean and the variance of the complete model the following values are obtained: $\xi_3 = \{0.4282865945, 0.4166346546\}$. In the Fig. 7.3 the pdf for the complete model and the one freezing the third parameters have been reported. Even in this case, even with a model that it is able to obtain an error on the variance less than the 2 % (that for an engineering application can be very optimistic as goal for a metamodel) the information about the tails of the distributions are, again, totally lost. Of course the intent of this paper and of this section is not to criticize the use of the variance estimator, but to make evident that to obtain a metamodel to employ, for instance, for safety purpose the information relative to high-order sensitivity indexes should be considered. In all the case proposed and in many other, not reported here only for brevity, appear evident that only when even the high-order sensitivity indexes have reached a safety threshold (about 5 %) the model can be really (and more safely) truncated.

7.4 Multi-objective Design Optimization

7.4.1 Importance of Skewness in Decomposition

This paragraph is devoted to show how important is to control the skewness during an optimization process. Let us consider the following polynomial function:

$$f = a\,(xz + xy) + b\left(x^2 + z^2\right) + (cba)\,y^2 \tag{7.16}$$

where x, y and z vary between 0 and 1 with an uniform pdf. Parameters a, b and c are design parameters that vary between -5 and 5. For this function, it is possible

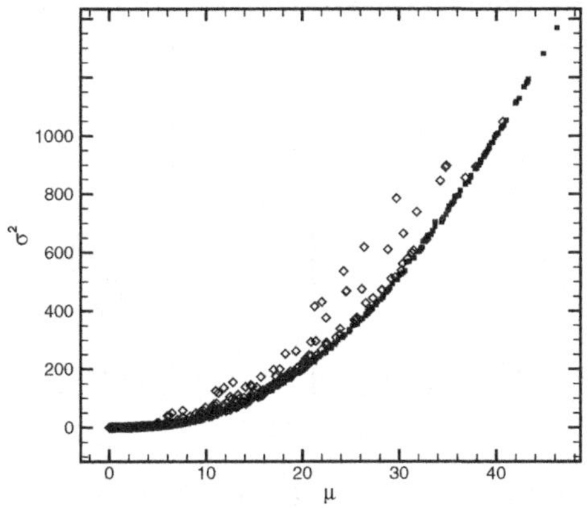

Fig. 7.4 Pareto front in the plan μ–σ^2 for the bi-objectives and three-objectives problem

to compute analytically high-order statistics, as functions of the design parameters. In order to show the importance to take into account also the high-order statistics in the robust optimization, different types of optimization are performed using several objective functions.

First, a classical bi-objective optimization is performed, where the mean of the function is maximized and its variance minimized. The Pareto front is reported in Fig. 7.4. No measures of skewness have been used during the optimization process, then the Pareto front is constituted by various designs displaying a very large variation of skewness.

Now, let us consider a three-objectives optimization, i.e. consisting in the maximization of the mean, the minimization of the variance and the minimization of the absolute value of the conditional skewness s_{xy}. In this case, the Pareto front is no more constituted by a curve, but by a surface in a 3D plan. The Pareto front is represented by means of 2D representation in the Figs. 7.4 and 7.5 with projections on the plans μ–σ^2, μ–s_{xy} and σ^2–s_{xy}, respectively. As shown in Fig. 7.5, designs belonging to the Pareto front display a large variation of the conditional skewness.

Now, let us compare the results obtained with both optimizations. In Fig. 7.4, we show Pareto fronts in the plan μ–σ^2. Designs obtained with the three-objectives optimization are dominated (with respect to only μ and σ^2) by the designs coming from the bi-objectives optimization. This is reasonable seeing that designs from bi-objective optimization are not influenced by the skewness s_{xy} during the optimization.

In Fig. 7.6, curves associated to the three-objectives optimization are obtained by the 3D Pareto front regarding only the designs having a skewness lower than 0.0001. Remark that individuals of this Pareto front take values of μ lower than 3.2 and values of σ^2 lower than 4.4. Moreover, they could be dominated in terms of μ and σ^2 by some individuals of the Pareto front obtained from the bi-objectives optimization.

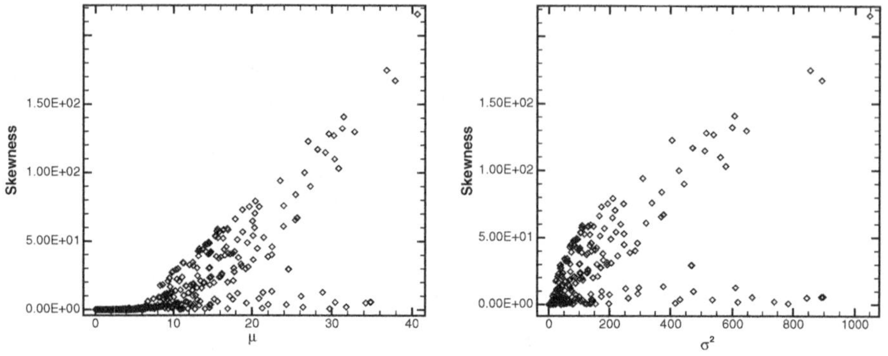

Fig. 7.5 Pareto front in the plan μ–s_{xy} (on the *left*) and σ^2–s_{xy} (on the *right*) for the three-objectives optimization

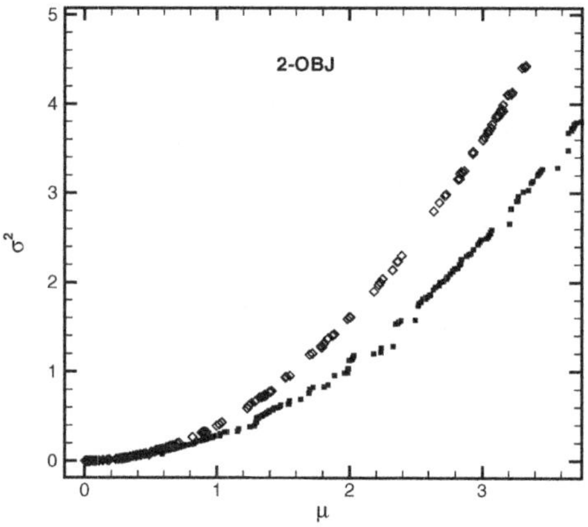

Fig. 7.6 Pareto front in the plan μ–σ^2 for the three-objectives optimization (extracted by the complete one considering only skewness inferior to 0.0001) and the bi-objective optimization

Here, the interest is to get a Pareto front that is not sensitive to large variation in the skewness, since designs obtained from bi-objective optimization could present large skewness values. This displays the great interest to estimate high-order statistics during optimization.

7.4.2 Description of the Algorithm

In this section, the algorithm for multi-objective robust design optimization is described. The strategy is constituted by two steps, that are schematically represented in Figs. 7.7 and 7.8. In the first step, (reported in the Fig. 7.7), a design of experience in the design variables space (called hereafter DOE), i.e. an initial set of design variables y, is generated. For each design variable y of the DOE, a high-order decomposition analysis is performed by computing for each uncertainty j, TSI_j and

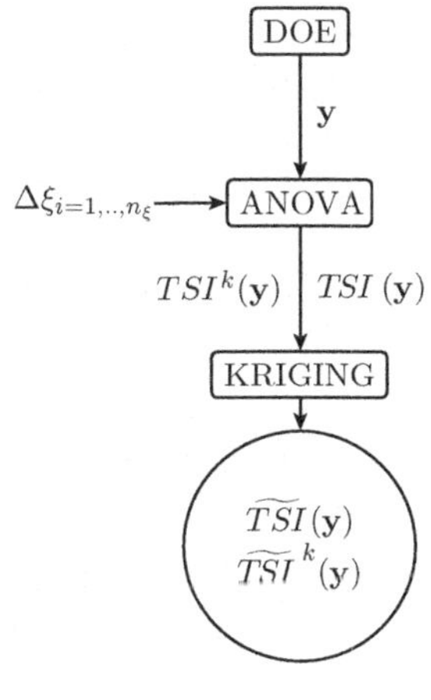

Fig. 7.7 Compact scheme for the kriging procedure

TSI_j^k. Then *TSI* is computed for each y of the DOE and for each uncertainty. A TSI response surface is built for each uncertainty and for each statistical moment as function of design variable space, i.e. $\widetilde{TSI}_j(y)$ and $\widetilde{TSI}_j(y)^k$, by means of a kriging method based on a DACE approach. The advantage of this methodology is the possibility of implement an adaptive response surface in order to minimize the statistical error between the real function and the extrapolated one.

In the second step (represented in Fig. 7.8), basing on the criterion for the *TSI*, the reduced stochastic problem is performed for each point of the DOE, where mean and variance are computed. Genetic algorithms are applied in order to compute new individuals basing on $\mu(y)$ and $\sigma(y)$. The NSGA algorithm [8] is used. The main tuning parameters of the algorithm are the population size, the number of generations, the crossover and mutation probabilities pc, pm and the so-called sharing parameter r used to take into account the relative isolation of an individual along a dominance front. Typical values for pc, pm are, respectively, 0.9 and 0.1; values of r are retained following a formula given in [8] that takes into account the population size and the number of objectives. Then, for the new design, mean and variance are computed for the reduced stochastic problem. Algorithm continues until convergence of genetic

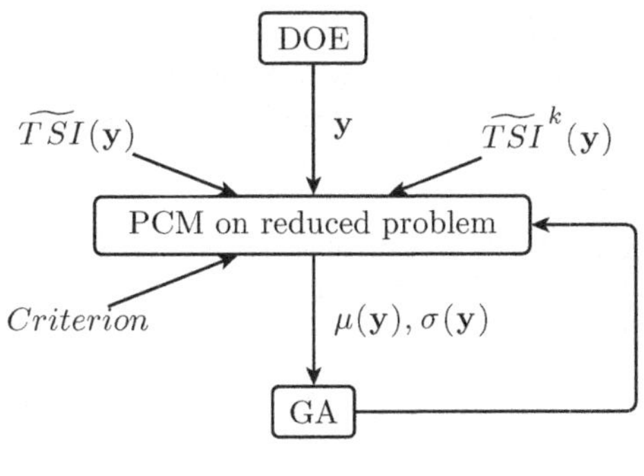

Fig. 7.8 Compact scheme for the overall optimization strategy

algorithms. For further validation of the algorithm, a complete stochastic analysis is performed for each optimal design in order to verify *TSI* for each uncertainty.

The criterion for *TSI* is based on a double verification on TSI_j and TSI^k_j. In particular, when the uncertainty j is considered, the criterion on TSI_j is applied (2 % based on Hestaven criterion) only if the ranking associated to j is the same for TSI_j and TSI^k_j.

7.4.3 Base Configuration and Sources of Uncertainty

In the present work, the turbine blade under consideration is the two dimensional VKI LS-59 cascade, a configuration which has been widely studied [9, 10]. An unstructured CFD dense-gas solver is used to ensure the reliability of the computed results for dense gas flows through a turbine cascade (for more details see Ref. [10]).The two-dimensional flow domain is discretized by a structured C-grid comprised of 192×16 cells. The boundary conditions are imposed as follows: at the inlet and outlet boundaries, non-reflecting boundaries are applied using the method of characteristics; a slip condition is imposed at the wall, which uses multi-dimensional linear extrapolation from interior points to calculate the wall pressure; periodicity conditions are prescribed at the inter-blade passage boundaries.

The siloxane *dodecamethylcyclohexasiloxane* ($C_{12}H_{36}Si_6O_6$), commercially known as D_6, is the fluid considered in this study. The physical properties of D_6 are reported in Table 7.2. The Peng-Robinson (PRSV) equation is used as thermodynamic model for D6. It depends on the following parameters, the fluid acentric factor ω, the isobaric specific heat in the ideal gas state, i.e. $c_{v\infty}$, and a fluid-dependent parameter n (the mean values of these parameters for D_6 are defined in Table 7.3).

Performance of the turbine cascade can be evaluated by using several output criteria. Here, the power output per unit depth (PO) expressed as $\Delta h \cdot \dot{m}/w_{mol}$ [W] is taken into account, where Δh is the enthalpy variation through turbine stage, $\dot{m}$ is the mass flow rate and w_{mol} is the molecular weight.

Three main sources of uncertainties are considered in this study (globally eight uncertainties): (i) the uncertainties on the operating conditions, i.e. inlet total temperature, T_{in}/T_c, inlet total pressure, p_{in}/p_c, angle of incidence β and the stagger angle θ, (ii) the uncertainties on the thermodynamic model, i.e. ω, $c_{v\infty}$ and n, and uncertainties on geometrical parameters, i.e. the blade thickness ϕ. Basing on [11], the 3.0 % of uncertainty for the temperature and pressure levels at the inlet conditions has

Table 7.2 Thermodynamic data for D_6, where M is the percentage molecular weight, and T_b is the boiling temperature at 1 atm

M (g/mole)	T_c (K)	P_c (kPa)	T_b (K)
444.9	645.8	961	518.1

Table 7.3 Thermodynamic constants for D6, PRSV equation of state, mean and min/max values for the uniform probability density function, data taken from [4]

	n	$c_{v\infty}$	ω
Mean	0.5729	105.86	0.7361
Range	0.5385–0.6073	99.50–112.20	0.7214–0.7508

been taken into account. The PRSV thermodynamic model is considered as a good trade-off between the accuracy of thermodynamic properties and the functional complexity since it depends on a limited number of parameters, hence a reduced number of uncertainty sources [11]. The following uncertainties are retained for this model (see the Table 7.3 and Ref. [11]), listed with their associated error bars: the acentric factor ω (2 %), the isobaric specific heat in the ideal gas state and a fluid-dependent parameter n (6 %). For the other parameters, it is assumed an uncertainty of 3 % for the angle of incidence β and the stagger angle θ, and an uncertainty of 2 % for the thickness ϕ.

7.4.4 Problem Definition

The objective is to find the optimal values for T_{in}/T_c, p_{in}/p_c, β and θ (four design variables) in order to maximize the mean of power output, $\mu(PO)$, and to minimize its standard deviation, $\sigma(PO)$ (two objective-optimization problem). Ranges for each design variable are defined in Table 7.4. Remark that the lower limit for the temperature is given by the saturation curve limit (SCL). Seeing that CFD code can compute only 1-phase flows, it has to be verified that the uncertainty region does not cross the maximal saturation curve (that can be computed as the upper limit of the 100 % confidence intervals when uncertainties on thermodynamic model are taken into account).

Finally, the optimization problem consists in finding the optimal values for four design variables where the output to maximize is dependent from eight uncertainties.

7.4.5 Optimization

A design of experiment (DOE) of 50 elements in the four design variable space is generated. Then, for each design, a quasi-Montecarlo plan (based on Sobol sequences) of two hundred individuals in the stochastic plan is generated and TSI is computed for

Table 7.4 Ranges of design variables in the optimization plan

p_{in}/p_c	T_{in}/T_c	β	θ
0.7–0.98	SCL–1.15	25°–35°	29°–39°

each uncertainty. The convergence of TSI indexes for each uncertainty and design is verified by increasing the number of individuals until five hundred. TSI for variance and kurtosis have been computed, displaying very small differences.

In Fig. 7.9, TSI (based on variance) contours are reported for each uncertainty in the plan p-T, where the point in the plan p-T is associated to the couple (p_{in}, T_{in}) of inlet thermodynamic conditions. As shown in Fig. 7.9a, b, TSI associated to the uncertainty on p_{in} vary from 8 to 44 % while vary from 39 to 83 % for uncertainty on T_{in}. For the uncertainties on two geometrical parameters, θ and ϕ (see Fig. 7.9c, d), TSI vary from 7 to 25 % and from 0.7 to 2.9 %, respectively. TSI associated to the uncertainties on thermodynamic model, i.e. ω, $c_{v\infty}$ and n, and on the geometrical parameter ϕ, are less than 0.29 %, then they are negligible with respect to the TSI criterion.

For each design of the DOE that has been previously computed for kriging metamodel, the reduced stochastic problem is performed and the statistics are computed in terms of mean and standard deviation for PO.

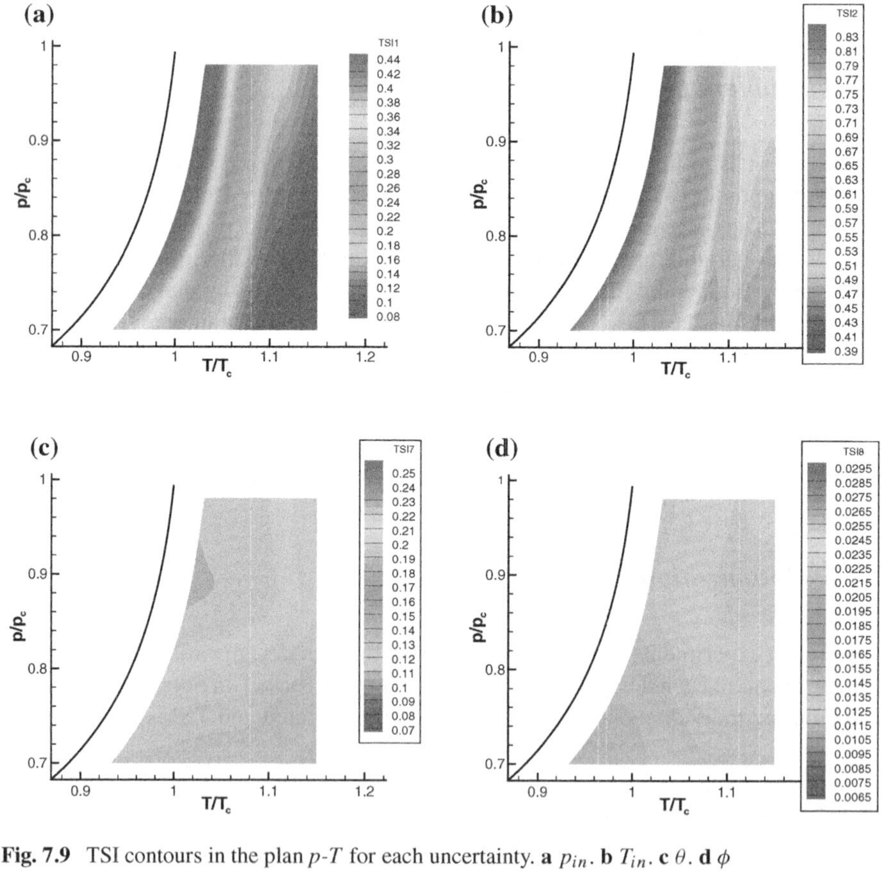

Fig. 7.9 TSI contours in the plan p-T for each uncertainty. **a** p_{in}. **b** T_{in}. **c** θ. **d** ϕ

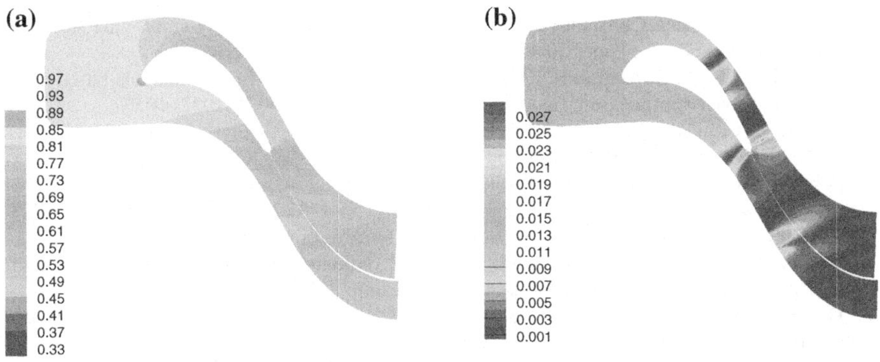

Fig. 7.10 Mean (**a**) and standard deviation (**b**) for pressure for some optimal individuals

Then, twenty individuals evolved during forty generations. The converged Pareto front is represented in Fig. 7.4. Various configurations are obtained with a large variation of the PO, going from 0.91 to 1.46.

The proposed method, in terms of criterion for reduction problem, has been compared with the one proposed in [4]. In particular, the Pareto front is different with the exclusion of several profiles. This is due to a lack of accuracy when only variance is considered in the reduction strategy. With respect to the old version, the introduction of a more strict criterion (based on kurtosis) increases the global computational cost, even if with a greater accuracy.

In Fig. 7.10, the mean pressure is shown in the computational domain. Generally, high inlet turbine pressures are associated to high mean of PO, displaying a strong dependence of turbine performances from thermodynamic inlet conditions. In a similar way, standard deviation of the pressure is reported in Fig. 7.10. Variance is concentrated around the compression shock location near the trailing edge.

7.5 Conclusions

In this work, the interest of using high-order decomposition for reducing the number of uncertainties in a robust optimization problem is assessed. In particular, sensitivity of different problems with respect to the variance or kurtosis decomposition is illustrated. Finally, a well-known optimization algorithm is modified for including this adding information in the reduction loop.

References

1. Xiu D, Karniadakis GE (2002) The Wiener-Askey polynomial Chaos for stochastic differential equations. SIAM J Sci Comput 24(2):619–644. doi:10.1137/S1064827501387826
2. Foo J, Karniadakis GE (2010) Multi-element probabilistic collocation method in high dimensions. J Comput Phys 229:1536–1557. doi:10.1016/j.jcp.2009.10.043, http://linkinghub.elsevier.com/retrieve/pii/S0021999109006044

3. Eldred MS, Webster, Constantine PG (2008) Design under uncertainty employing stochastic expansion methods. In: Paper 20086001. AIAA
4. Congedo P, Geraci G, Abgrall R, Pediroda V, Parussini L (2013) Tsi metamodels-based multi-objective robust optimization. Eng Comput (Swansea, Wales) 30(8):1032–1053
5. Abgrall R, Congedo P, Geraci G, Iaccarino G (2012) Decomposition of high-order statistics. INRIA Research Report (RR-8193), pp 1–40
6. Sobol I (2001) Global sensitivity indices for nonlinear mathematical models and their Monte Carlo estimates. Math Comput Simul 55(1–3):271–280. doi:10.1016/S0378-4754(00)00270-6, http://linkinghub.elsevier.com/retrieve/pii/S0378475400002706
7. Gao Z, Hesthaven JS (2010) Efficient solution of ordinary differential equations with high-dimensional parametrized uncertainty. Commun Comput Phys 10(2):253–278
8. Deb K, Pratap A, Agarwal S, Meyarivan T (2002) A fast and elitist multiobjective genetic algorithm: NSGA-II. IEEE Trans Evol Comput 2(6):619–644
9. Abgrall R, Congedo PM, Santis DD, Razaaly N (2014) A non-linear residual distribution scheme for real-gas computations. Comput Fluids 102:148–169. doi:10.1016/j.compfluid.2014.06.031
10. Congedo P, Corre C, Cinnella P (2011) Numerical investigation of dense-gas effects in turbo-machinery. Comput Fluids 49:290–301
11. Cinnella P, Congedo PM, Pediroda V, Parussini L (2011) Sensitivity analysis of dense gas flow simulations to thermodynamic uncertainties. Phys Fluids 23, 116, 101:1–20

Chapter 8
Extension of the One-Shot Method for Optimal Control with Unsteady PDEs

Stefanie Günther, Nicolas R. Gauger and Qiqi Wang

Abstract The one-shot approach has proven to be very efficient in optimization and control with steady partial differential equations (PDEs) which are solved by fixed-point iterations. The purpose of this paper is to extend the one-shot method to unsteady problems and to make it as efficient as in steady cases. We derive a framework for optimization and control with unsteady PDEs, whose structure is the same as in the steady one-shot method. First results in the direction of one-shot optimization with unsteady Reynolds-averaged Navier-Stokes equations (URANS) are presented. With the Van der Pol oscillator as a generic model problem, we investigate an adaptive time scaling approach, which demonstrates the classical one-shot efficiency on unsteady problems.

Keywords Simultaneous analysis and design · One-shot method · Unsteady optimization · Adaptive timescale

8.1 Introduction

Optimal control of partial differential equations (PDEs) has gained a considerable amount of attention in recent years. Gradient-based methods are typically used for finding optimal design or control parameters [11]. If the number of design parameters is rather large, the adjoint method is preferred since the cost of computing the sensitivities is then independent of the number of design parameters [16].

S. Günther (✉) · N.R. Gauger
Computational Mathematics Group, Department of Mathematics and CCES,
RWTH Aachen University, Schinkelstraße 2, 52062 Aachen, Germany
e-mail: guenther@mathcces.rwth-aachen.de

N.R. Gauger
e-mail: gauger@mathcces.rwth-aachen.de

Q. Wang
Department of Aeronautics and Astronautics, Massachusetts Institute of Technology,
77 Massachusetts Avenue, Cambridge, MA 02139, USA
e-mail: qiqi@mit.edu

D. Greiner et al. (eds.), *Advances in Evolutionary and Deterministic Methods for Design, Optimization and Control in Engineering and Sciences*, Computational Methods in Applied Sciences 36, DOI 10.1007/978-3-319-11541-2_8

For optimization problems where the underlying PDE is steady-state and can be solved iteratively, Ta'asan [19] suggested to solve the PDE simultaneously with the optimization problem. In this so called *one-shot* approach, the necessary optimality conditions, including the state, the adjoint and the design equations, are solved simultaneously. Since then, the one-shot method has been further developed and successfully applied to various optimization tasks especially in the field of aerodynamics [3, 4, 10]. The benefit over classical, so called *sequential* optimization algorithms, where the state and adjoint equations have to be solved accurately in each optimization step, is, that the cost of an one-shot optimization is just a small multiple of the effort of a simulation for the underlying PDE [4]. It has been proven in [9] that the one-shot method converges to a stationary point of the optimization problem, provided that a certain preconditioner for the design updates is used. The one-shot approach for optimization with steady PDEs is recalled shortly in Sect. 8.2.

Along with increasing computer capacities optimal control of time-dependent PDEs has become an active field of research [12, 13, 17, 18, 21]. In computational fluid dynamics, where complex and accurate simulation tools have been developed over years, the computational cost for solving unsteady PDEs is still enormous. Extending the one-shot methodology for unsteady problems such that the state and adjoint equations can be solved approximately during the optimization process, is therefore crucial. Simulation tools for solving unsteady PDEs typically use an implicit time marching scheme and solve the resulting implicit equations by applying an iterative fixed point solver at each time step [2]. In Sect. 8.3 we derive a new framework for solving the time-dependent PDE iteratively, that can be used in a one-shot method. The new method uses the above mentioned fixed point iterations in a black box manner making the unsteady one-shot method applicable for a wide range of existing unsteady PDE solvers.

Section 8.4 presents numerical results for solving the state and the adjoint equation simultaneously within the new framework. The piggy-back iteration is applied to an optimal contol problem solving the unsteady Reynolds-averaged Navier-Stokes equations (URANS). In order to further improve the state iterations and accelerate the convergence behavior, an adaptive time scaling approach is investigated in Sect. 8.5. The Van der Pol oscillator, written as a system of two ODEs, is used as a generic model problem to study the convergence of the unsteady one-shot method including adaptive time scales in Sect. 8.6.

8.2 One-Shot Optimization for Steady PDEs

We consider the general optimization problem

$$\min_{y,u} f(y,u) \quad s.t. \quad c(y,u) = 0 , \tag{8.1}$$

where the constraint represents a steady-state PDE for a state variable $y \in Y$ and a design variable $u \in U$. The objective function $f : Y \times U \to \mathbb{R}$ to be minimized as well as the state contraint $c\colon Y \times U \to Y$ are assumed to be twice continuously differentiable. Y and U are Hilbertspaces with finite dimensions $m = \dim(Y)$ and $n = \dim(U)$ such that elements can be identified with coordinate vectors in $\mathbb{R}^m$ and $\mathbb{R}^n$.

We make the general assumption, that every design uniquely defines a state that satisfies the state constraint. Moreover we assume, that this state can be computed numerically by a contractive fixed-point iteration of the form

$$y_{k+1} = G(y_k, u). \tag{8.2}$$

To ensure convergence of iteration (8.2) we assume, that

$$\left\| \frac{\partial G(y,u)}{\partial y} \right\| \leq \rho < 1 \quad \text{for all } (y,u) \in Y \times U \tag{8.3}$$

for some vector induced matrix norm $\| \cdot \|$. Hence, the fixed-point iterator G is a contractive function and it follows from Banach's fixed point theorem [1] that (8.2) converges to the unique fixed point $y_* = G(y_*, u)$ where $c(y_*, u) = 0$ holds. Replacing the state constraint by the fixed-point equation, the optimization problem reads

$$\min_{y,u} f(y,u) \quad \text{s.t.} \quad y = G(y,u). \tag{8.4}$$

We define the corresponding Lagrangian function L and the shifted Lagrangian N by

$$L(y, \bar{y}, u) := f(y,u) + (G(y,u) - y)^T \bar{y} =: N(y, \bar{y}, u) - y^T \bar{y}\,, \tag{8.5}$$

where the Lagrange multiplier $\bar{y} \in Y$ is called the adjoint variable.

8.2.1 Steady One-Shot Algorithm

The first order necessary optimality conditions for the optimization problem (8.4) are, that the gradients of the Lagrange function L with respect to $\bar{y}$, y and u vanish:

$$\nabla_{\bar{y}} L = 0 \quad \Leftrightarrow \quad y = G(y,u) \tag{8.6}$$
$$\nabla_y L = 0 \quad \Leftrightarrow \quad \bar{y} = N_y(y, \bar{y}, u)^T \tag{8.7}$$
$$\nabla_u L = 0 \quad \Leftrightarrow \quad 0 = N_u(y, \bar{y}, u)^T \tag{8.8}$$

where subscripts denote partial derivatives. To reach such a point, the following one-shot algorithm is suggested [8, 9]

$$\begin{bmatrix} y_{k+1} \\ \bar{y}_{k+1} \\ u_{k+1} \end{bmatrix} = \begin{bmatrix} G(y_k, u_k) \\ N_y(y_k, \bar{y}_k, u_k)^T \\ u_k - B_k^{-1} N_u(y_k, \bar{y}_k, u_k)^T \end{bmatrix}, \tag{8.9}$$

where B_k is a carefully chosen preconditioner that ensures convergence of the coupled iteration (see Sect. 8.2.2). In this approach the state, the adjoint and the design variables are updated simultaneously in one iteration step.

Since G is contractive in its first argument, iterating only in the state and the adjoint variable for a fixed design u converges linearly to the solution of (8.6) and (8.7). Thus, both variables can be updated simultaneously in the so called *piggy-back* iteration without any preconditioning [5]. Although both variables converge at the same convergence rate ρ, the adjoint variable lags behind the convergence of the state variable which was measured in numerical results and analyzed theoretically in [6].

8.2.2 Choosing a Preconditioner B

In order to ensure convergence of the one-shot iteration (8.9) a preconditioner B_k has to be chosen such that contractivity of the coupled iteration is achieved. To this end, Griewank et al. [8, 9] suggest to look for descent on a doubly augmented Lagrangian function

$$L^a(y, \bar{y}, u) := \frac{\alpha}{2} \|G(y, u) - y\|^2 + \frac{\beta}{2} \left\| N_y(y, \bar{y}, u)^T - \bar{y} \right\|^2 + L(y, \bar{y}, u), \tag{8.10}$$

where weighted residuals of the state and the adjoint equations are added to the Lagrangian function with $\alpha, \beta > 0$.

It has been proven in [8] that if the weights α and β are chosen such that

$$\sqrt{\alpha\beta}(1 - \rho) > 1 + \frac{\beta}{2} \|N_{yy}\| \tag{8.11}$$

holds, the augmented Lagrangian L^a is an exact penalty function, i.e. the stationary points of L and L^a coincide. Furthermore, the step increment vector s of the coupled one-shot iteration defined as

$$s(y, \bar{y}, u) := \begin{bmatrix} G(y, u) - y \\ N_y(y, \bar{y}, u)^T - \bar{y} \\ -B^{-1} N_u(y, \bar{y}, u)^T \end{bmatrix} \tag{8.12}$$

is a descent direction for L^a for any preconditioner B satisfying

$$B = B^T \succeq B_0 := \frac{1}{\sigma}(\alpha G_u^T G_u + \beta N_{yu}^T N_{yu}), \tag{8.13}$$

where $\sigma = 1 - \rho - \frac{(1+\frac{\|N_{yy}\|}{2}\beta)^2}{\alpha\beta(1-\rho)}$ (see [8] for a detailed derivation of the above expressions).

A particular preconditioner B can be derived from minimizing a quadratic approximation of L^a with respect to the design changes. It is suggested in [9] to choose

$$B = \frac{1}{\sigma}(\alpha G_u^T G_u + \beta N_{yu}^T N_{yu} + N_{uu}) \tag{8.14}$$

which approximates $\nabla_{uu} L^a$ in a neighborhood of a minimization point. Since (8.13) is satisfied for this particular choice, the increment vector s decreases L^a. Thus applying the one-shot iteration converges to a local minimum of (8.4).

Numerical applications of the one-shot algorithm choosing B as in (8.14) yields a bounded retardation, i.e. the cost for an one-shot optimization is just a small multiple of the cost of a single simulation for the underlying PDE. The factor typically varies between 2 and 8, as various numerical applications have shown [4, 9].

8.3 Extension of the One-Shot Optimization for Unsteady PDEs

For time-dependent PDEs the state variable varies with time and thus is a function $y\colon [0, T] \to Y$. The objective function to be minimized is typically given by some time-averaged quantity. The optimization problem with unsteady PDEs reads

$$\min_{y,u} \frac{1}{T}\int_0^T f(y(t), u)\,\mathrm{d}t \quad \text{s.t.} \quad \begin{cases} \frac{\partial y(t)}{\partial t} + c(y(t), u) = 0 \quad \forall t \in [0, T] \\ y(0) = y_*^0 \end{cases}, \tag{8.15}$$

where f and c are defined as in Sect. 8.2 and $y_*^0 \in Y$ describes some initial state.

Numerical methods for solving unsteady PDEs often discretize the time domain with $t_0 = 0 < t_1 < \cdots < t_N = T$ and approximate the transient term by an implicit scheme. This results in an implicit equation for each time step. For simplicity, we approximate the transient term by the implicit Backward Euler method resulting in the following implicit residuum equations:

$$R(t_{i+1}) := \frac{y^{i+1} - y^i}{t_{i+1} - t_i} + c(y^{i+1}, u) = 0 \quad \forall i = 0, \ldots, N-1, \tag{8.16}$$

where $y^i = y(t_i)$ denotes the state variable evaluated at the discrete time step t_i. Using the first order Backward Euler time marching scheme is not a restriction since modification to any other implicit multistep method is straightforward. The residuum equation at time t_{i+1} then also contains the evaluation of the state variable at other previous time steps $y^{i-1}, y^{i-2}, \ldots$, depending on the order of the transient term approximation.

It is assumed, that the implicit equations can be solved one after another forward in time using an iterative fixed-point solver which converges at each time step to a pseudo-steady-state solution:

$$\text{for } i = 0, \ldots, N-1: \qquad \text{iterate} \quad y_{k+1}^{i+1} = G(y_k^{i+1}, y_*^i, u) \overset{k\to\infty}{\longrightarrow} y_*^{i+1} \tag{8.17}$$

where y_*^{i+1} denote the converged pseudo-steady-states at time steps t_{i+1} which satisfy the residuum equations (8.16) and are fixed points of G, i.e. $y_*^{i+1} = G(y_*^{i+1}, y_*^i, u)$ $\forall i = 0, \ldots, N-1$. In contrast to the steady-state iterations in Sect. 8.2, the unsteady fixed-point iterator now also depends on the state at the previous time step y_*^i.

Convergence of the above fixed-point iterations is ensured, if G is contractive with respect to its first argument:

$$\left\| \frac{\partial G(y^{i+1}, y^i, u)}{\partial y^{i+1}} \right\| \leq \rho < 1 \quad \forall i = 0, \ldots, N-1 \tag{8.18}$$

for all point of interest.

In order to extend from simulation to one-shot optimization, where one incorporates design updates already during the primal flow computation, the time marching scheme (8.17) is modified as

$$\text{iterate } k = 0, 1 \ldots: \qquad y_{k+1}^{i+1} = G(y_k^{i+1}, y_{k+1}^i, u) \quad \text{for} \quad i = 0, \ldots, N-1 \tag{8.19}$$

with $y_k^0 := y_*^0 \; \forall k \in \mathbb{N}$. In contrast to (8.17), where fixed point iterations are performed at each time step to reach the converged states y_*^i one after another, in the one-shot framework (8.19) the complete trajectory of the unsteady solution is updated within one iteration. Interpreting the time-dependent state variable as a discrete vector from the product space $y = (y^1, \ldots, y^N) \in Y^N := Y \times \cdots \times Y$ we can write (8.19) in terms of an update function $y_{k+1} = H(y_k, u)$ where $H: Y^N \times U \to Y^N$ performs the update formulas (8.19) and is defined as

$$H(y, u) := \begin{pmatrix} G(y^1, y_*^0, u) \\ G(y^2, G(y^1, y_*^0), u) \\ \vdots \\ G(y^N, G(y^{N-1}, G(y^{N-2}, \ldots, G(y^1, y_*^0, u), u) \ldots, u), u) \end{pmatrix}. \tag{8.20}$$

Theorem 1 *For a fixed design $u \in U$, the iteration*

$$y_{k+1} = H(y_k, u), \tag{8.21}$$

where $H\colon Y^N \times U \to Y^N$ is defined in (8.20)*, converges locally to the fixed point $y_* = H(y_*, u)$ with linear convergence rate bounded by ρ. Furthermore, H is contractive in a neighborhood of y_* with respect to a suitable norm $\|\cdot\|_H$.*

Proof Let $y_*^{i+1} = G(y_*^{i+1}, y_*^i, u)$ denote the fixed points of G that can be computed with (8.17). Then $y_* = (y_*^1, \dots, y_*^N) \in Y^N$ is by construction a fixed point of H. The partial derivatives of H are given by

$$\frac{\partial H^i(y_*, u)}{\partial y^j} = \begin{cases} 0 & : j > i \\ \partial_1 G_*^i & : j = i \\ \partial_2 G_*^i \cdot \partial_2 G_*^{i-1} \cdots \partial_2 G_*^{j+1} \cdot \partial_1 G_*^j & : j < i \end{cases} \tag{8.22}$$

where $G_*^i := G(y_*^i, y_*^{i-1}, u)$ for $i = 1, \dots, N$ and ∂_k denotes partial derivatives of G with respect to the kth argument with $k \in \{1, 2\}$. Thus, the Jacobian matrix of $H(y_*, u)$ is a tridiagonal block matrix whose diagonal blocks coincide with the derivatives of the fixed point iterator G with respect to its first argument:

$$\frac{\partial H(y_*, u)}{\partial y} = \begin{pmatrix} \partial_1 G(y_*^1, y_*^0, u) & 0 & 0 \\ * & \ddots & 0 \\ * & * & \partial_1 G(y_*^N, y_*^{N-1}, u) \end{pmatrix} \tag{8.23}$$

Using the contractivity of G in (8.18) we can bound the spectral radius of the Jacobian:

$$\operatorname{spr}\left(\frac{\partial H(y_*, u)}{\partial y}\right) = \max_{i \in \{1, \dots, N\}} \operatorname{spr}\left(\partial_1 G(y_*^i, y_*^{i-1}, u)\right) \le \rho < 1 \tag{8.24}$$

It follows from Ostrowski's theorem [15, Proposition 10.1.3 and 10.1.4], that the iteration $y_{k+1} = H(y_k, u)$ converges locally to the fixed point $y_* = H(y_*, u)$ with linear convergence rate bounded by ρ.

Furthermore, since for all $\epsilon > 0$ there exists a norm $\|\cdot\|_\epsilon$ such that $\operatorname{spr}\left(\frac{\partial H(y_*, u)}{\partial y}\right) \le \|\frac{\partial H(y_*, u)}{\partial y}\|_\epsilon \le \operatorname{spr}\left(\frac{\partial H(y_*, u)}{\partial y}\right) + \varepsilon$ [15, Proposition 2.2.8], it follows, that

$$\left\|\frac{\partial H(y_*, u)}{\partial y}\right\|_\epsilon \le \rho + \epsilon < 1 \tag{8.25}$$

for ϵ small enough. Thus, H is contractive in a neighborhood of y_* with respect to a suitable norm which we will refer to as $\|\cdot\|_H$. □

By construction of H and G, the components of the fixed point $y_* = H(y_*, u)$ satisfy the residuum equations in (8.16). We find an approximate solution function $y\colon [0, T] \to Y$ of the unsteady PDE by applying a spline interpolation with the grid $\{t_i : i = 0, \dots, N\}$ for the discrete state vector $y_* \in Y^N$ and initial state y_*^0.

The unsteady PDE-constrained optimization problem can be written in terms of the fixed-point equation as

$$\min_{y,u} J(y,u) \quad \text{s.t.} \quad y = H(y,u) \tag{8.26}$$

where the objective function J approximates the time-averaged quantity that is to be minimized:

$$J(y,u) := \frac{1}{N}\sum_{i=1}^{N} f(y^i,u) \approx \frac{1}{T}\int_0^T f(y(t),u)\,\mathrm{d}t\,. \tag{8.27}$$

We define the Lagrangian function L and the shifted Lagrangian N corresponding to the optimization problem (8.26) by

$$L(y,\bar{y},u) := J(y,u) + (H(y,u)-y)^T\bar{y} =: N(y,\bar{y},u) - y^T\bar{y}\,, \tag{8.28}$$

where the Lagrange multiplier $\bar{y} \in Y^N$ is the adjoint variable.

This formulation has the same structure as the definition of the Lagrangian in (8.5) for optimization with steady PDEs. Thus, the one-shot approach for steady PDEs as in Sect. 8.2 can be applied in the same way by considering the state variable as a vector from the product space Y^N and exchanging the iterator G with H and the objective function f with J. The unsteady one-shot iteration is then given by

$$\begin{bmatrix} y_{k+1} \\ \bar{y}_{k+1} \\ u_{k+1} \end{bmatrix} = \begin{bmatrix} H(y_k,u_k) \\ N_y(y_k,\bar{y}_k,u_k)^T \\ u_k - B_k^{-1}N_u(y_k,\bar{y}_k,u_k)^T \end{bmatrix} \tag{8.29}$$

Since the condition of the one-shot method, i.e. the constraints are solved by a contractive fixed-point iterator, are satisfied for the mapping H, the existing theory for converging the state and adjoint variables in a piggy-back iteration as well as finding a preconditioner B_k that ensures convergence of the one-shot method, also applies for the unsteady optimization problem. The advantages of the steady one-shot method, that have been measured in terms of a bounded retardation factor, are expected for the unsteady one-shot method.

8.4 Piggy-Back Iteration for the Unsteady State and Adjoint Equations

For a fixed design $u \in U$ the piggy-back iteration for the state and adjoint variable is given by

$$\begin{bmatrix} y_{k+1} \\ \bar{y}_{k+1} \end{bmatrix} = \begin{bmatrix} H(y_k, u) \\ N_y(y_k, \bar{y}_k, u)^T \end{bmatrix}. \tag{8.30}$$

Since H is contractive, its adjoint counterpart is also contractive. Thus, both variables converge simultaneously to the solution of the unsteady state and adjoint equations without any preconditioning. In contrast to the steady piggy-back iterations, one update of the state and the adjoint variables now each contains a loop over all time steps. While one evaluation of H performs an update for each state component y^i one after another forwards in time, one evaluation of N_y^T performs an update for each adjoint component $\bar{y}^i$ one after another backwards in time.

The piggy-back approach for solving the state and the adjoint equations simultaneously is applied for two test cases. After setting up the iterator H and the objective function J as explained in Sect. 8.3, Automatic Differentiation in reverse mode [7] is applied to H and J in order to compute the derivative of the shifted Lagrangian with respect to the state variable.

8.4.1 Piggy-Back Iteration Solving the Unsteady RANS Equations

The piggy-back iteration is implemented for an optimal active flow control problem of unsteady flow around a 2D cylinder. Eight actuation slits are installed on the surface of the cylinder where sinusoidal blowing and suction is applied in order to reduce vorticity downstream the cylinder. Amplitude, phase shift and frequency of the actuation are used as design variables. The governing incompressible unsteady Reynolds-averaged Navier-Stokes equations (URANS) are solved by applying the new approach described in Sect. 8.3 to the second order implicit finite volume code ELAN [20]. The original code performs a pressure-velocity correction loop in each time step. In order to apply the piggy-back iteration, these inner correction loops are reduced to perform only one update in each time step and are therefore identified with the fixed point iterator G.

To study the convergence behavior of the piggy-back approach, the L_2-norm of the state and the adjoint residuals $\|y_k - H(y_k, u)\|_2$, $\|\bar{y}_k - N_y(y_k, \bar{y}_k, u)^T\|_2$ are computed. From Fig. 8.1 it can be observed, that both variables converge at the same convergence rate while the adjoint residuum exhibits the typical time lag behind the convergence of the state variable. An initial phase of approximately 150 iterations is needed before the primal variable starts to converge rapidly. It was observed, that this phase enlarges for numbers of time steps, making the convergence behavior dependent on the number of time steps.

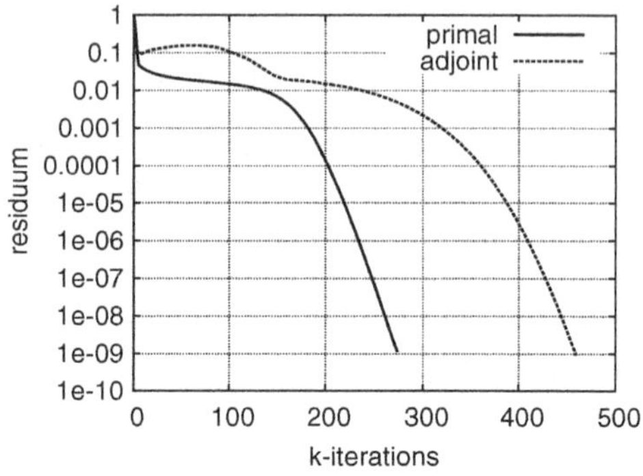

Fig. 8.1 Convergence history of primal and adjoint states for solving incompressible URANS in piggy-back framework ($N = 200$)

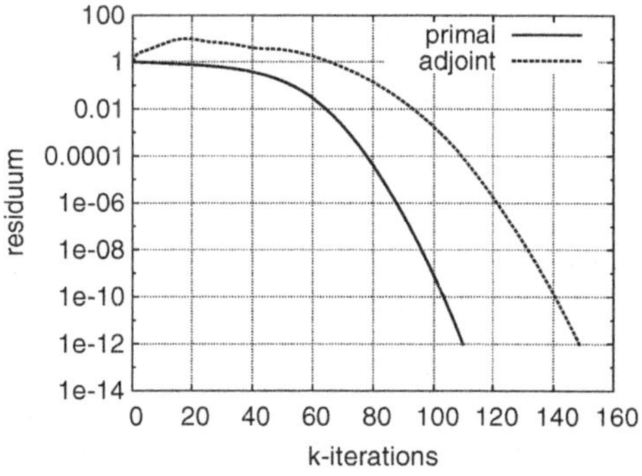

Fig. 8.2 Convergence history of primal and adjoint states solving Van der Pol oscillator in piggy-back framework ($N = 1.000$)

8.4.2 Piggy-Back Iteration Solving the Van der Pol Oscillator

The Van der Pol oscillator is a nonlinear oscillator where a damping factor $\mu \geq 0$ controls the magnitude of the nonlinear term. It can be written as a system of two first order ordinary differential equations (ODEs)

$$\begin{pmatrix} \dot{x} \\ \dot{v} \end{pmatrix} = \begin{pmatrix} v \\ -x + \mu(1 - x^2)v \end{pmatrix} . \tag{8.31}$$

Since any unsteady PDE transforms into a system of ODEs after spatial discretization, the Van der Pol oscillator is used as a simple model problem to study the convergence of the piggy-back approach and test the unsteady one-shot method. The transient term of the ODE is approximated by the implicit Backward Euler method. The resulting implicit equations are solved using a Quasi-Newton method. According to Sect. 8.3, the contractive function H is set up to converge the primal variable at all time steps, while G represents one step of the Quasi-Newton solver.

The L_2-norm of the state and the adjoint residuals during the piggy-back iterations are plotted in Fig. 8.2. About 50 iterations are needed before the state variable starts to converge rapidly.

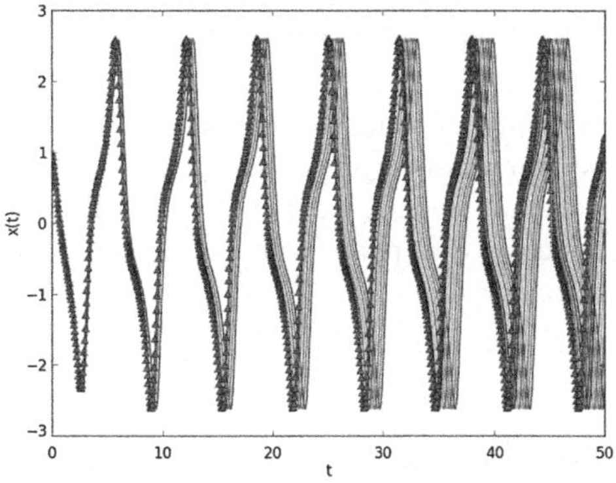

Fig. 8.3 x-component of the Van der Pol oscillator for 16 different k-iterations and final solution marked by *triangles* ($N = 1.000$, $\mu = 2$)

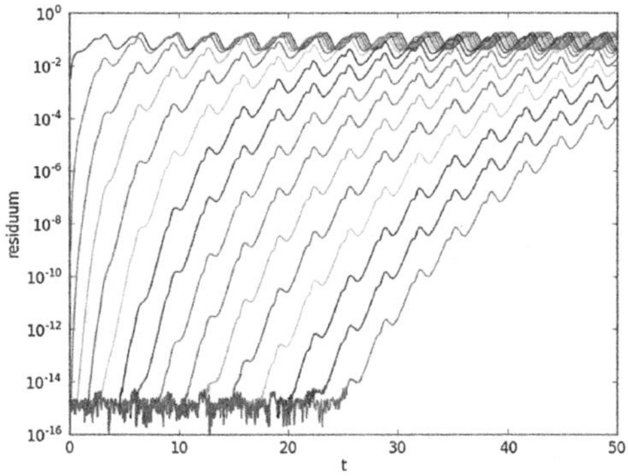

Fig. 8.4 Residuum of the Van der Pol oscillator at each time step for 16 different k-iterations ($N = 1.000$, $\mu = 2$)

To clarify the reason for the slowly converging initial phase, the x-component of the Van der Pol oscillator is plotted in Fig. 8.3 for 16 primal iterations. The magnitudes of each k-iteration follow a similar trajectory, but the first few k-trajectories exhibit an artificial time dilation compared to the final solution, marked by triangles. The time dilation enlarges for increasing number of time steps. The corresponding residuals $\|R_k(t_i)\|_2$ at each time step are plotted in Fig. 8.4. While the residuum decreases rapidly for t close to zero, the magnitude of the residuum at increasing times stays almost unchanged during the first iterations.

Figures 8.3 and 8.4 demonstrate, that the slow convergence comes from this artificial time dilation. Concerning the primal update formulation for a trajectory as in (8.19) one can see, that every update y^i_{k+1} is performed on the basis of an inexact state y^{i-1}_{k+1} at the previous time step. This error propagates to all following time steps. Thus, enlarging the time domain leads to larger accumulated errors at the last time steps leading to larger residuals and increasing number of iterations to reduce the errors. Figure 8.3 shows, that the dominating part of the error occurs in the direction of time while the error in the amplitude of the inexact trajectories are rather marginal.

8.5 Adaptive Time Scaling

To reduce the time dilation observed in the previous section and therefore improve the primal convergence speed, we introduce an adaptive time scaling approach. After each state update, we assign a trajectory y_{k+1} to a scaled time $\tilde{t}$ such that the new trajectory is in phase with the physical solution. More precisely, we define the new trajectory as

$$\tilde{y}_{k+1}^{i} := y_{k+1}(\tilde{t}_i) \quad \forall i = 1, \ldots, N \tag{8.32}$$

where $\tilde{t}_i$ is chosen such that the residuals R defined in (8.16) are minimized:

$$\min_{\tilde{t}_i} \|R(\tilde{t}_i)\|_2 \quad \forall i = 1, \ldots, N \tag{8.33}$$

In this adaptive time scaling approach, we rescale the time parametrization after each state update in such a way, that the new trajectory mimics the solution of the PDE. Since we eliminate the error component that points in the direction of time by minimizing the residuum with respect to time scales, the time dilation effect, that was observed numerically, vanishes. Obviously, this rescaling approach leads to an accelerated convergence speed of the primal iteration since we have

$$\|\tilde{y}_{k+1} - y_*\|_H \leq \|y_{k+1} - y_*\|_H \tag{8.34}$$

$$= \|H(y_k, u) - y_*\|_H \overset{k\to\infty}{\longrightarrow} 0 \tag{8.35}$$

for a design $u \in U$ and the fixed point $y_* = H(y_*, u) \in Y^N$.

8.6 Numerical Results: Unsteady One-Shot Optimization Including Adaptive Time Scales

In this section we present numerical results that demonstrate the efficiency of the one-shot approach including adaptive time scales for optimization with unsteady PDEs.

The unsteady one-shot approach is applied for an inverse design problem subject to the Van der Pol oscillator, while the design parameter μ controls the magnitude of the nonlinear term. With $y = (x, v)^T$, the minimization problem under consideration is defined as

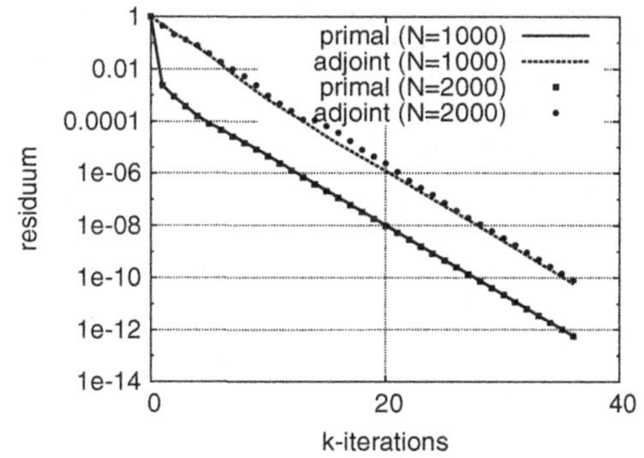

Fig. 8.5 Residuum of state and adjoint variable during the piggy-back iteration for the Van der Pol oscillator including adaptive timescales for $N = 1.000$ and $N = 2.000$ time steps

$$\min_{y,\mu} \frac{1}{T}\int_0^T \|y(t) - y_{ref}(t)\|_2^2\, dt + \gamma\|\mu\|_2^2 \tag{8.36}$$

$$\text{s.t.} \quad \begin{pmatrix} \dot{x}(t) \\ \dot{v}(t) \end{pmatrix} = \begin{pmatrix} v(t) \\ -x(t) + \mu(1 - x(t)^2)v(t) \end{pmatrix} \quad \forall t \in [0, T] \tag{8.37}$$

$$\begin{pmatrix} x(0) \\ v(0) \end{pmatrix} = \begin{pmatrix} x_0 \\ v_0 \end{pmatrix} \tag{8.38}$$

where y_{ref} denotes a target solution for some given reference parameter. A Tykhonov regularization is added in the objective function with $\gamma = 0.0001$.

First, we investigate the accelerated convergence speed of the piggy-back iteration including adaptive time scales. Similar to Sect. 8.4.2, we compute the L_2-norm of the state and the adjoint variable during the piggy-back iteration for $N = 1.000$ and $N = 2.000$ time steps. From Fig. 8.5 we observe, that the initial phase of slow convergence has vanished completely and a fast piggy-back convergence independent of the number of time steps is achieved.

Figure 8.6 shows the residuals $\|R_k(t_i)\|_2$ of the Van der Pol oscillator at each time step for 16 different k-iterations. Comparing Fig. 8.6 with Fig. 8.4 from Sect. 8.4.2 it can be seen, that the adaptive time scaling approach yields a more or less constant accuracy over all computed times leading to a fast convergence of the state variable for all time steps.

Enhancing the fast converging piggy-back iteration with a preconditioned design update, the one-shot algorithm is applied to the optimization problem (8.36)–(8.38). Instead of computing the exact preconditioner B_k as derived in Sect. 8.2 we find an approximate preconditioner by using BFGS-updates [14] on the gradient of the augmented Lagrangian with respect to the design variable. A stopping criterion for the one-shot iteration on the norm of the gradient is implemented as $\|N_u\|_2 < \epsilon$ with $\epsilon = 10^{-9}$.

The convergence behavior of the one-shot algorithm including adaptive time scales can be monitored in Fig. 8.7. The state and the adjoint residuum converge to zero with a linear convergence rate. Moreover the gradient reaches the stopping

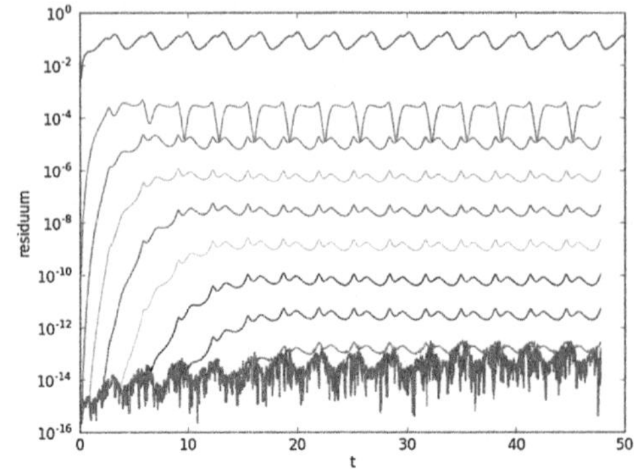

Fig. 8.6 Primal residuals of the Van der Pol oscillator at each time step for 16 k-iterations including adaptive timescales ($N = 1.000$)

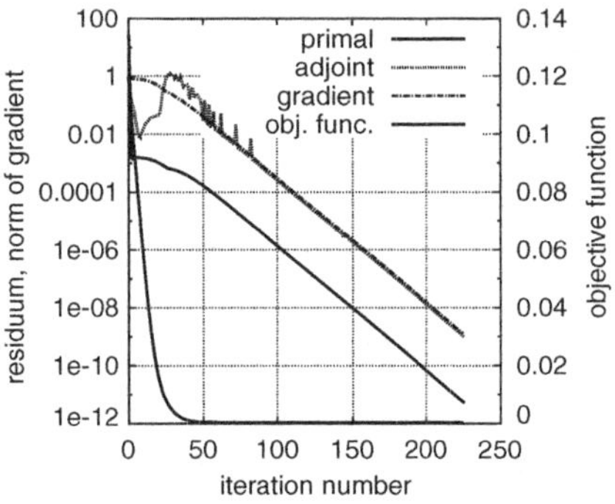

Fig. 8.7 Convergence history of one-shot iterations including adaptive time scales for solving the Van der Pol oscillator ($N = 1.000$)

criterion and thus the optimization stops at a local minimum after approximately 225 iteration steps where the objective function is fully converged and reduced to zero. Comparing Fig. 8.5 with Fig. 8.7 it can be seen that a retardation factor of approximately 7.0 is achieved.

The numerical results demonstrate, that the unsteady one-shot approach enhanced with adaptive time scales yields a similar efficiency as observed in the steady one-shot applications. In the current test case, the cost for an optimization is only a small multiple of the cost of a primal simulation which is in good agreement with numerical results for steady one-shot optimization.

8.7 Conclusion and Outlook

The theory for steady one-shot optimization has been recalled. The one-shot methodology has been extended to unsteady optimization problems by interpreting the time-dependent solution of the PDE as a discrete state vector from the product space Y^N. We introduced an iterative fixed point solver by rearranging the classical implicit time marching scheme. The updating scheme has been further improved by introducing an adaptive time scaling approach. After each iteration step for solving the state equation, the time domain is rescaled such that the new trajectory is in phase with the final solution. The one-shot algorithm including adaptive time scales has been implemented for the Van der Pol oscillator as a first simple model problem. As in the steady case, a bounded retardation was achieved, i.e. the cost of the one-shot optimization is just a small multiple of the cost of a single simulation. The one-shot method will be applied to optimal control of the unsteady RANS equations in future work.

References

1. Barner M, Flohr F (1995) Analysis 2, 3rd edn. Walter de Gruyter, Berlin
2. Ferziger JH, Perić M (2002) Computational methods for fluid dynamics, 3rd edn. Springer, Berlin
3. Gauger NR, Ötzkaya E (2009) Single-step one-shot aerodynamic shape optimization. Int Ser Numer Math 158:191–204
4. Gauger N, Griewank A, Hamdi A, Kratzenstein C, Özkaya E, Slawig T (2012) Automated extension of fixed point PDE solvers for optimal design with bounded retardation. In: Leugering G, Engell S, Griewank A, Hinze M, Rannacher R, Schulz V, Ulbrich M, Ulbrich S (eds) Constrained optimization and optimal control for partial differential equations. Springer, Basel, pp 99–122
5. Griewank A, Faure C (2002) Reduced functions, gradients and Hessians from fixed-point iterations for state equations. Numer Algorithms 30:113–139
6. Griewank A, Kressner D (2004) Time-lag in derivative convergence for fixed point iterations. Proceedings of CARI, vol 4, pp 295–304
7. Griewank A, Walther A (2008) Evaluating derivatives: principles and techniques of algorithmic differentiation. SIAM, Philadelphia
8. Griewank A, Hamdi A (2010) Properties of an augmented Lagrangian for design optimization. Optim Methods Softw 25:645–664
9. Griewank A, Hamdi A (2011) Reduced quasi-Newton method for simultaneous design and optimization. Comput Optim Appl 49:521–548
10. Hazra SB, Schulz V, Brezillon J, Gauger NR (2005) Aerodynamic shape optimization using simultaneous pseudo-timestepping. JCP 204:46–64
11. Lions JL (1971) Optimal control of systems governed by partial differential equations. Springer, New York
12. Nadarajah SK, Jameson A (2007) Optimum Shape design for unsteady flows with time-accurate continuous and discrete adjoint methods. AIAA J 45:1478–1491
13. Nemili A, Özkaya E, Gauger NR, Thiele F, Carnarius A (2013) Optimal control of unsteady flows using a discrete and a continuous adjoint approach. In: Hömberg D, Tröltzsch F (eds): Systems modeling and optimization, pp 318–327. Springer, Heidelberg
14. Nocedal J, Wright SJ (2006) Numerical optimization, 2nd edn. Springer Science+Business Media, New York
15. Ortega JM, Rheinboldt WC (1970) Iterative solution of nonlinear equations in several variables. Academic Press, New York
16. Pironneau O (1974) On optimum design in fluid mechanics. J Fluid Mech 64:97–110
17. Potschka A, Mommer MS, Schlöder JP, Bock HG (2012) Newton-Picard-based preconditioning for linear-quadratic optimization problems with time-periodic parabolic PDE constraints. SIAM J Sci Comput 34:1214–1239
18. Rumpfkeil MP, Zingg DW (2010) The optimal control of unsteady flows with a discrete adjoint method. Optim Eng 11:5–22

19. Ta'asan S (1991) One-shot methods for optimal control of distributed parameter systems I: finite dimensional control. ICASE 91-2 (1991)
20. Xue L (1998) Entwicklung eines effizienten parallelen Lösungsalgorithmus zur dreidimensionalen Simulation komplexer turbulenter Strömungen. Technische Universität Berlin, PhD-Thesis
21. Yamaleev NK, Diskin B, Nielsen EJ (2010) Local-in-time adjoint-based method for design optimization of unsteady flows. J Comput Phys 229:5394–5407

Chapter 9
Adaptive Aerodynamic Design Optimization for Navier-Stokes Using Shape Derivatives with Discontinuous Galerkin Methods

L. Kaland, M. Sonntag and N.R. Gauger

Abstract We state and analyze one-shot optimization methods in a function space setting for optimal control problems, for which the state equation is given in terms of a fixed-point equation. Further, we concentrate on the application of a design optimization problem incorporating the solution of the compressible Navier-Stokes equations using a discontinuous Galerkin method. For the given primal fixed-point solver an appropriate adjoint solver is constructed. For the following design update we compute the shape derivative analytically based on the weak formulation of the governing equations. The primal, adjoint and design updates are performed in a one-shot manner, i.e., the corresponding equations are not fully solved, instead only a few iteration steps are performed. Finally, we add an additional adaptive step. During the optimization routine we refine or coarsen the grid to obtain a better accuracy.

Keywords One-shot method · Function space analysis · Design optimization · Adaptivity · Shape derivative

9.1 Introduction

Many algorithms for solving optimization problems governed by partial differential equations (PDE) aim to solve the necessary first-order optimality system, also called Karush-Kuhn-Tucker (KKT) system. This can be either done consecutively, i.e., after a complete solve of the state equation the corresponding adjoint equation is solved and finally the control or design parameter is updated. On the other hand, this can be

L. Kaland · M. Sonntag (✉) · N.R. Gauger
Computational Mathematics Group, Department of Mathematics and CCES, RWTH Aachen University, Schinkelstr. 2, D-52062 Aachen, Germany
e-mail: sonntag@mathcces.rwth-aachen.de

L. Kaland
e-mail: kaland@mathcces.rwth-aachen.de

N.R. Gauger
e-mail: gauger@mathcces.rwth-aachen.de

D. Greiner et al. (eds.), *Advances in Evolutionary and Deterministic Methods for Design, Optimization and Control in Engineering and Sciences*, Computational Methods in Applied Sciences 36, DOI 10.1007/978-3-319-11541-2_9

done simultaneously or in *one-shot*, where state and adjoint equation are not fully converged. Sequential quadratic programming (SQP) methods are an example for a simultaneous treatment [1, 2], where the new iterate is immediately used in the subsequent computations. The necessary second-order sensitivity informations are often approximated by a BFGS update. Additionally, some variations include the approximation of the linearized state equation components [3]. Instead of incorporating the new iterate immediately into the iteration cycle as for SQP type methods, in [4] state, adjoint and design update depend on the previous iteration. This enables a complete parallel treatment of the KKT system. This so-called *single-step one-shot method* as discussed and analyzed in [4] relies on the fact, that the state equation is solved in terms of a fixed-point solver. A convergence proof is provided for finite dimensions [4].

Different versions of the one-shot method have been tested successfully for many problems, especially in aerodynamics [5, 6]. A further extension of one-shot methods can be achieved by adding adaptivity. In more detail, that means that during the optimization the grid is adapted simultaneously. This procedure yields the problem of handling several dimensions in the optimization algorithm. In other words, before dealing with adaptivity, the optimization method, here the one-shot method, needs to be formulated and analyzed in a function space setting [7, 8].

The paper is structured as follows. We review the main aspects of the one-shot method, its setting in function spaces as well as the principle steps of the convergence proof as given in [8]. We continue by stating the framework with all its components necessary for the design optimization problem in aerodynamics. In particular, we review the compressible Navier-Stokes equations, its discretization by the discontinuous Galerkin method as well as the iterative solution by the backward Euler-Newton method. These are the main steps of the simulation code PADGE (Parallel adaptive discontinuous Galerkin environment), that was developed and provided by the German Aerospace Center (DLR), see [9]. The relevant gradient information for the design step is obtained by the shape gradient approach. Here, we compute the gradient based on a weak formulation of the Navier-Stokes equations as in [10]. Finally, some computational results regarding the optimization of a NACA0012 airfoil are shown.

9.2 General Setting of the One-Shot Method

We consider the optimization problem

$$\min\ J(\mathbf{u}, q) \quad \text{s.t.} \quad e(\mathbf{u}, q) = 0, \tag{9.1}$$

where $e(\mathbf{u}, q) = 0$, $e : U \times Q \to U'$ denotes the governing equations in form of a PDE. Here, $\mathbf{u} \in U$, $q \in Q$ denote the state and control or design variable with appropriate Hilbert spaces U and Q as well as the dual space U' of U. We further assume that the PDE can be reformulated in terms of a fixed-point equation, i.e.,

$$\mathbf{u} = G(\mathbf{u}, q). \tag{9.2}$$

The fixed-point operator G needs to have a contraction factor less than 1. Therefore, it needs to hold

$$\|G_{\mathbf{u}}(\mathbf{u}, q)\| \leq \rho_G < 1, \tag{9.3}$$

where $\|\cdot\|$ denotes the appropriate operator norm. Overall, this leads to the equivalent optimization problem

$$\min\ J(\mathbf{u}, q) \quad \text{s.t.} \quad \mathbf{u} = G(\mathbf{u}, q). \tag{9.4}$$

In order to define the Lagrange function for (9.4) in a Hilbert space setting correctly, we need to consider the transition from the PDE to the fixed-point formulation. This is given according to [8] in terms of a linear, bounded and bijective operator $F(\mathbf{u}) : U \to U'$ such that

$$e(\mathbf{u}, q) = F(\mathbf{u})[\mathbf{u} - G(\mathbf{u}, q)].$$

We note, that for the sake of simplicity we assume $F(\mathbf{u})$ to be independent of q. We can now define the Lagrangian incorporating the fixed-point formulation

$$\begin{aligned} L(\mathbf{u}, q, \lambda) &= J(\mathbf{u}, q) + \langle \lambda, e(\mathbf{u}, q)\rangle_{U,U'} \\ &= J(\mathbf{u}, q) + \langle F(\mathbf{u})^* \lambda, \mathbf{u} - G(\mathbf{u}, q)\rangle_{U',U}. \end{aligned} \tag{9.5}$$

Computing the KKT system based on (9.5) yields a fixed-point formulation of the optimality system according to

$$\mathbf{u} - G(\mathbf{u}, q) = 0 \tag{9.6a}$$

$$\Phi(\mathbf{u}, q, \lambda) - \lambda = 0 \tag{9.6b}$$

$$L_q(\mathbf{u}, q, \lambda) = 0. \tag{9.6c}$$

The operator $\Phi(\mathbf{u}, q, \lambda)$ in (9.6b) is the fixed-point operator of the adjoint equation and defined by (see [8])

$$\begin{aligned} \langle F(\mathbf{u})^* \Phi(\mathbf{u}, q, \lambda), w\rangle_{U',U} :=& \langle J_{\mathbf{u}}(\mathbf{u}, q), w\rangle_{U',U} \\ &- \langle \lambda, F_{\mathbf{u}}(\mathbf{u}) w[\mathbf{u} - G(\mathbf{u}, q)]\rangle_{U,U'} \\ &+ \langle F(\mathbf{u})^* \lambda, G_{\mathbf{u}}(\mathbf{u}, q) w\rangle_{U',U} \end{aligned}$$

for all $w \in U$. Note that it holds

$$\langle L_{\mathbf{u}}(\mathbf{u}, q, \lambda), w\rangle_{U',U} = \langle F(\mathbf{u})^* \Phi(\mathbf{u}, q, \lambda), w\rangle_{U',U} - \langle F(\mathbf{u})^* \lambda, w\rangle_{U',U}. \tag{9.7}$$

A standard optimization method solves (9.6a) to obtain the state $\mathbf{u}$. Afterwards, (9.6b) is solved for the adjoint λ and finally the design is updated according to some optimization routine. Instead, the one-shot method is based on a simultaneous update of state, adjoint and design according to [4]

$$\mathbf{u}^{k+1} = G(\mathbf{u}^k, q^k) \tag{9.8a}$$
$$\lambda^{k+1} = \Phi(\mathbf{u}^k, q^k, \lambda^k) \tag{9.8b}$$
$$q^{k+1} = q^k - (B^k)^{-1} L_q(\mathbf{u}^k, q^k, \lambda^k), \tag{9.8c}$$

with an appropriate preconditioner B^k. In [8] a convergence proof is given for the general case and specified for model problems including the viscous Burgers equations. Additionally, [7] considers the incompressible Navier-Stokes equations in a Hilbert space setting. In the following, we only mention the leading steps of the general convergence proof.

Therefore, consider the augmented Lagrangian defined as

$$L^a(\mathbf{u}, q, \lambda) = L(\mathbf{u}, q, \lambda) + \frac{\alpha}{2}\|G(\mathbf{u}, q) - \mathbf{u}\|_U^2 + \frac{\beta}{2}\|\Phi(\mathbf{u}, q, \lambda) - \lambda\|_U^2, \tag{9.9}$$

with the penalty parameters $\alpha, \beta > 0$. Further, we define the increment vector

$$s = \begin{pmatrix} s_1 \\ s_2 \\ s_3 \end{pmatrix} := \begin{pmatrix} G(\mathbf{u}, q) - \mathbf{u} \\ \Phi(\mathbf{u}, q, \lambda) - \lambda \\ -B^{-1} L_q(\mathbf{u}, q, \lambda) \end{pmatrix}. \tag{9.10}$$

The convergence proof follows the idea to show that the augmented Lagrangian acts as an exact penalty function, i.e., that every local minimum of the original optimization problem (9.1) is also a local minimum of L^a. Further, we show that the one-shot method yields descent on L^a. The next theorem is the main result in this procedure and ensures the equivalence of the stationary points as well as the descent condition.

Theorem 1 *If there exist constants $\alpha > 0$ and $\beta > 0$ such that the following conditions are fulfilled*

$$\alpha(1 - \rho_G) - \frac{\alpha^2}{2\tilde{\gamma}}\|G_q\|^2 > \|F(\mathbf{u})\| + \frac{\beta}{2}\|\Phi_{\mathbf{u}}\|, \tag{9.11}$$
$$\beta(1 - \rho_G) > \|F(\mathbf{u})\| + \frac{\beta}{2}\|\Phi_{\mathbf{u}}\|, \tag{9.12}$$
$$\gamma > \frac{\tilde{\gamma}}{2}, \tag{9.13}$$

for a positive preconditioner B with $(Bh, h)_Q \geq \gamma\|h\|_Q^2$, $\|\Phi_\lambda\| \leq \rho_G < 1$ and a constant $\tilde{\gamma} > 0$, then a point is a stationary point of L^a if and only if it is a solution of *the optimality system* (9.6a), (9.6b) *and* (9.6c). *Additionally s is a descent direction for* L^a.

Proof For the proof we refer to [8]. □

9.3 Design Optimization for Navier-Stokes

In our application the PDE is given by the steady compressible Navier-Stokes equations with the flow field $\mathbf{u}$ around an airfoil q. The vector of conservative variables in two dimensions is given by

$$\mathbf{u} = \begin{bmatrix} \rho \\ \rho v_1 \\ \rho v_2 \\ \rho E \end{bmatrix},$$

with the density ρ, the total energy E and the velocity components v_1 and v_2. We further define the convective fluxes

$$f_1^c(\mathbf{u}) = \begin{bmatrix} \rho v_1 \\ \rho v_1^2 + p \\ \rho v_1 v_2 \\ \rho h v_1 \end{bmatrix} \quad \text{and} \quad f_2^c(\mathbf{u}) = \begin{bmatrix} \rho v_2 \\ \rho v_1 v_2 \\ \rho v_2^2 + p \\ \rho h v_2 \end{bmatrix},$$

as well as the viscous fluxes

$$f_1^v(\mathbf{u}, \nabla\mathbf{u}) = \begin{bmatrix} 0 \\ \tau_{11} \\ \tau_{21} \\ \tau_{11} v_1 + \tau_{12} v_2 + \kappa \frac{\partial T}{\partial x_1} \end{bmatrix} \quad \text{and} \quad f_2^v(\mathbf{u}, \nabla\mathbf{u}) = \begin{bmatrix} 0 \\ \tau_{12} \\ \tau_{22} \\ \tau_{21} v_1 + \tau_{22} v_2 + \kappa \frac{\partial T}{\partial x_2} \end{bmatrix},$$

with the pressure p, the enthalpy h with $h = E + p/\rho$, the thermal conductivity κ, the temperature T and the viscous stress tensor τ. Omitting external volume forces, we finally obtain the compressible Navier-Stokes equations for laminar flow in a two-dimensional domain Ω

$$\sum_{i=1}^{2} \frac{\partial}{\partial x_i} f_i^c(\mathbf{u}) - \frac{\partial}{\partial x_i} f_i^v(\mathbf{u}, \nabla\mathbf{u}) = 0. \tag{9.14}$$

For a Newtonian fluid the stress tensor is expressed by

$$\tau = \mu(\nabla v + (\nabla v)^T) - \frac{2}{3}\mu \; div(v) I,$$

with the dynamic viscosity μ. For an ideal gas it holds $e = c_v T$ for the internal energy e and the temperature is related to the viscosity by

$$\kappa T = \frac{\mu \gamma}{Pr}\left(E - \frac{1}{2}|v|^2\right)$$

where γ is the ratio of specific heat capacity at constant pressure c_p and constant volume c_v, i.e., $\gamma = c_p/c_v$, and $Pr = \mu c_p/\kappa$ is the Prandtl number. The compressible Navier-Stokes equations are completed by appropriate subsonic or supersonic inflow and outflow boundary conditions. Additionally, we have no-slip boundary as well as adiabatic or isothermal boundary conditions on the wall, see [11] for details.

The cost functions under consideration are the lift and drag coefficients given by

$$J(\mathbf{u}, \Omega) = \frac{1}{C_\infty} \int_{\Gamma_W} (pn - \tau n) \cdot \psi \, ds, \tag{9.15}$$

where Γ_W is the wall boundary of the domain Ω, n is the normal vector and ψ is either $\psi_l = (-\sin(\alpha), \cos(\alpha))^\top$ for the lift or $\psi_d = (\cos(\alpha), \sin(\alpha))^\top$ for the drag coefficient at a given angle of attack α. Further, it is $C_\infty = \frac{1}{2}\rho_\infty |v_\infty|^2 L$, where the subscript ∞ denotes the values on the farfield and L is the characteristic length scale.

9.3.1 The Backward Euler-Newton Method

The system of PDEs (9.14) is discretized by the discontinuous Galerkin method. It is based on a weak formulation of the system and on integration by parts. Since the discrete variables are not assumed to be continuous over the element edges in the discontinuous Galerkin method, we obtain a weak formulation inhabiting jumps and numerical flux functions. We refer to [11] for details. We denote the discrete variable by $\mathbf{u}_h$ and obtain the discrete weak formulation

$$\mathcal{N}(\mathbf{u}_h, \mathbf{v}_h) = 0 \tag{9.16}$$

for $\mathbf{u}_h \in V_h^p$ and all $\mathbf{v}_h \in V_h^p$, where V_h^p denotes the finite element space consisting of discontinuous vector-valued polynomial functions of degree p.

The simulation code PADGE offers the iterative solution of (9.16) with the backward Euler-Newton method that will be explained in the following. Incorporating a pseudo-time leads to the following system

$$\left(\frac{d}{dt}\mathbf{u}_h(t), \mathbf{v}_h\right) + \mathcal{N}(\mathbf{u}_h(t), \mathbf{v}_h) = 0 \qquad \text{for all} \qquad \mathbf{v}_h \in V_h^p, t \in (0, T),$$

$$\mathbf{u}_h(0) = \mathbf{u}_h^0.$$

The time is discretized with the backward Euler method. Linearizing the nonlinear term together with $\Delta t^k = t^{k+1} - t^k$, we obtain

$$\left(\frac{\mathbf{u}_h^{k+1} - \mathbf{u}_h^k}{\Delta t^k}, v_h \right) + \mathcal{N}'(\mathbf{u}_h^k, \mathbf{v}_h)(\mathbf{u}_h^{k+1} - \mathbf{u}_h^k) = -\mathcal{N}(\mathbf{u}_h^k, \mathbf{v}_h). \tag{9.17}$$

Here, $\mathcal{N}'(\mathbf{u}, v)w$ denotes the derivative of $\mathcal{N}(\mathbf{u}, v)$ with respect to $\mathbf{u}$ in the direction w. Introducing a finite dimensional basis of V_h^p the fixed-point iteration reads in system notation

$$\mathbf{u}_h^{k+1} = \mathbf{u}_h^k - \left(\frac{M}{\Delta t^k} + N'(\mathbf{u}_h^k) \right)^{-1} N(\mathbf{u}_h^k), \tag{9.18}$$

with the mass matrix M, the Jacobian $N'(\mathbf{u}_h^k)$ and the residual vector $N(\mathbf{u}_h^k)$. We note that by interpreting $M/\Delta t^k + N'(\mathbf{u}_h^k)$ as an augmented Jacobian, the scheme corresponds to the Newton scheme, i.e., $\Delta t^k \to \infty$ leads to a standard Newton update.

9.3.2 Adjoint Navier-Stokes Equations

The discrete adjoint equations to (9.16) are given by

$$\mathcal{N}'(\mathbf{u}_h, \lambda_h)\mathbf{w}_h = J_{\mathbf{u}}(\mathbf{u}_h, \Omega) \tag{9.19}$$

for all $\mathbf{w}_h \in V_h^p$. The control q now corresponds to the domain Ω. In order to satisfy an adjoint consistency condition, the target function J needs to be modified at the boundary. To keep notation shortly, we still denote this function by J and refer to [11] for details. The same arguments as above, i.e., choosing a finite dimensional basis of V_h^p, lead to the linear adjoint system

$$N'(\mathbf{u}_h)^T \lambda_h = J_{\mathbf{u}}(\mathbf{u}_h, \Omega) \tag{9.20}$$

for the discrete adjoint variable λ_h. The discrete linear system (9.20) can be solved directly with a standard iterative scheme. Nevertheless, in the spirit of the one-shot method it is important to update the adjoint variable simultaneously to the state with an appropriate fixed-point iteration. Therefore, we add a pseudo-time and discretize it by backward Euler. The adjoint fixed-point iteration is finally given by

$$\lambda_h^{k+1} = \lambda_h^k - \left(\frac{M}{\Delta t} + N'(\mathbf{u}_h^k) \right)^{-T} \left(N'(\mathbf{u}_h^k)^T \lambda_h^k - J_{\mathbf{u}}(\mathbf{u}_h^k, \Omega) \right). \tag{9.21}$$

The primal-dual update based on the backward Euler-Newton method fits into the general framework of the one-shot method as introduced in Sect. 9.2 in a discrete and simplified manner. This follows directly by defining the preconditioner $R := \left(M/\Delta t + N'(\mathbf{u}_h^k)\right)^{-1}$ and omitting its dependency on the state, so that we obtain the adjoint update with $F(\mathbf{u}_h) := R^{-1}$ and $G(\mathbf{u}_h, \Omega) := \mathbf{u}_h - R\, N(\mathbf{u}_h)$.

9.3.3 Shape Derivative

We present a derivation of the shape gradient, that is based on the variational formulation of the state equations. Using shape and adjoint calculus, we compute the shape gradients of lift and drag coefficient g_d and g_l in the so-called Hadamard form. Compared to an approach using a strong form, where the state equations are fulfilled pointwise, our approach leads to numerically more accurate results (see Fig. 9.1). We summarize the concept of shape derivatives and especially the Hadamard Theorem as stated in [12, 13] for $J(\Omega) := J(\mathbf{u}, \Omega)$. Therefore, let D be an open set in $\mathbb{R}^d$ and the domain Ω be a measurable subset of D. Using the perturbation of identity

$$T_t[V] : D \times [0, \delta) \to \mathbb{R}^d, \quad (x, t) \mapsto x + tV(x),$$

which describes the deformation of the domain Ω by a vector field V, we can define the following Eulerian derivative

$$dJ(\Omega; V) := \lim_{t \searrow 0} \frac{J(\Omega_t) - J(\Omega)}{t}$$

as the shape derivative of the target functional in the direction V. The function J is called shape differentiable at Ω if the Eulerian derivative exists for all directions V and the mapping $G(\Omega) : C_0^k(D; \mathbb{R}^d) \to \mathbb{R}$ defined by $V \mapsto dJ(\Omega; V)$ is linear and continuous. In the following, let $C_0^k(\Omega, \mathbb{R}^d)^*$ denote the dual space of $C_0^k(\Omega, \mathbb{R}^d)$ and $C_0^k(\Gamma)^*$ the dual space space of $C_0^k(\Gamma)$ appropriately.

Theorem 2 (Hadamard Theorem) *For every domain $\Omega \subset D$, let $J(\Omega)$ be shape differentiable of class C^k. Furthermore, let the boundary Γ be of class C^{k-1}. Then, there exists the following scalar distribution $g(\Gamma) \in C_0^k(\Gamma)^*$, such that the shape gradient $G(\Omega) \in C_0^k(\Omega, \mathbb{R}^d)^*$ of $J(\Omega)$ is given by*

$$G(\Omega) = \gamma_\Gamma^*(g \cdot n),$$

where $\gamma_\Gamma \in L\left(C_0^k(\Omega, \mathbb{R}^d), C_0^k(\Gamma, \mathbb{R}^d)\right)$ is the trace operator and γ_Γ^* its adjoint. Hence, it follows

$$dJ(\Omega; V) = dJ(\Gamma; V \cdot n) = \langle g, V \cdot n \rangle_{(C_0^k(\Gamma,\mathbb{R}^d))^* \times C_0^k(\Gamma,\mathbb{R}^d)}.$$

If $g(\Gamma)$ is integrable over Γ, than the Hadamard Formula

$$dJ(\Omega; V) = \int_{\Gamma} g\,(V \cdot n)\,ds$$

is fulfilled. In the following, we denote terms which are written in this form to be in Hadamard form.

Proof A proof can for example be found in [12] or in [13]. □

The Hadamard Formula implies that the shape derivative only depends on the normal component of the vector field V at the boundary of the domain. For the optimization the scalar distribution $g(\Gamma)$, which corresponds to the shape gradient, has to be found for the drag and lift coefficient.

Theorem 3 (Preliminary shape derivative of the cost functional) *If the vector field of the perturbation of identity fulfills $V = 0$ in the neighborhood of the farfield boundary Γ_∞, then the shape derivative of the lift and drag coefficient according to* (9.15) *is given by*

$$dJ(\Omega; V) = \frac{1}{C_\infty} \int_{\Gamma_W} (p'n - \tau'n) \cdot \psi + (V \cdot n)\ div(p\psi - \tau\psi)ds, \tag{9.22}$$

where f' denotes the local shape derivative of a function $f(t, x)$, defined as the partial derivative $f' := \frac{\partial}{\partial t} f(t, x)$.

Proof A proof can be found in [14]. □

Since this preliminary shape derivative is not yet in Hadamard form, one has to use adjoint calculus to eliminate the local shape derivatives p' and τ'. This can be done on the one hand for the Navier-Stokes equations in pointwise form as stated in equation (9.14), see [14, 15]. On the other hand the shape gradient can be computed for the Navier-Stokes equations in variational form

$$\left(\sum_{i=1}^{2} \frac{\partial}{\partial x_i} f_i^c(\mathbf{u}) - \frac{\partial}{\partial x_i} f_i^v(\mathbf{u}, \nabla \mathbf{u}), \mathbf{v} \right)_{\Omega} = 0 \quad \forall \mathbf{v} \in H, \tag{9.23}$$

where H represents an appropriate broken Sobolev space necessary for the weak discontinuous Galerkin formulation, see [10, 16] for details. Computing the shape derivative based on the latter approach leads to additional terms compared to the shape derivative based on the strong form. This fact is caused by the reversed treatment of shape differentiation and integration. Using technical reformulations, particularly exploiting the no-slip condition and its local shape derivative at the wall boundary, leads to the Hadamard form of the drag and lift coefficient.

Theorem 4 *The shape derivative of the drag and lift coefficient* (9.15) *in Hadamard form is given by*

$$
\begin{aligned}
dJ(\Omega;V) = &\int_{\Gamma_W} \frac{\partial v}{\partial n} \cdot (n \cdot \Sigma)(V \cdot n) ds \\
&- \int_{\Gamma_{iso}} \frac{\partial T_W - T}{\partial n} \kappa n \cdot \nabla \lambda_4 (V \cdot n) ds \\
&+ \int_{\Gamma_W} \frac{\partial v}{\partial n} \cdot (\rho n \lambda_1 + (\rho h n - \tau n) \lambda_4)(V \cdot n) ds \\
&- \int_{\Gamma_{adia}} \left(\frac{\partial^2 T}{\partial n^2} \kappa \lambda_4 + \mathrm{div}_\Gamma (\nabla T \kappa \lambda_4) \right)(V \cdot n) ds \\
&- \boxed{\int_{\Gamma_W} \rho (\nabla \cdot v) \lambda_1 (V \cdot n) ds} \\
&- \boxed{\int_{\Gamma_{adia}} \left(\rho h (\nabla \cdot v) - \sum_{i,j} \tau_{ij} \frac{\partial v_j}{\partial x_i} - \kappa \Delta T \right) \lambda_4 (V \cdot n) ds},
\end{aligned}
$$

where Σ *is the adjoint stress tensor and the last two integrals, inside the boxes, only occur for the variational approach of the Navier-Stokes equations.*

Proof A detailed proof is given in [10] or [17]. □

We use all discretization points of the shape as parameters for the one-shot optimization. This is often called free-node parameterization.

Figure 9.1 shows the shape gradient of the drag and lift coefficient based on the variational as well as the pointwise formulation, computed with a flow solution of

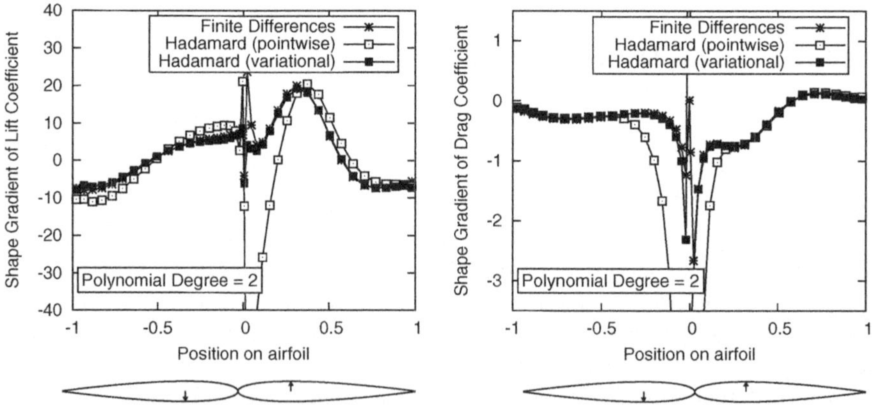

Fig. 9.1 Shape gradient for lift and drag in Hadamard form for variational/strong approach with $p = 2$

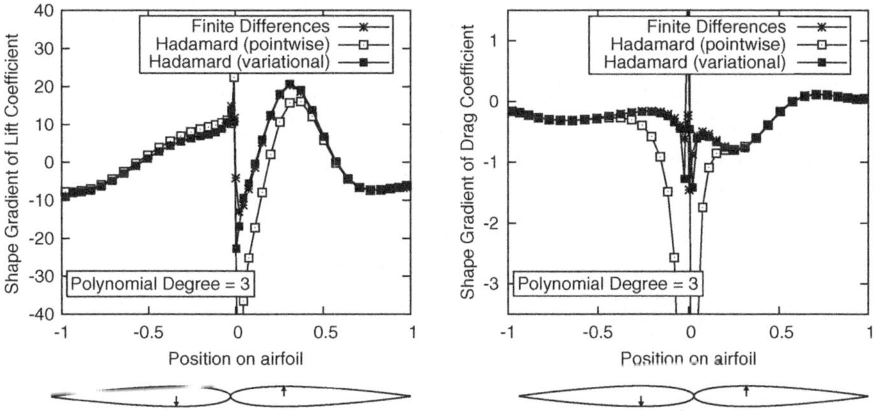

Fig. 9.2 Shape gradient for lift and drag in Hadamard form for variational/strong approach with $p = 3$

polynomial degree 2. For a verification of our implementation we also computed the shape gradients with finite differences. While the Hadamard form of the variational approach shows a good agreement with the computation by finite differences for both cases, the shape derivative of the pointwise approach deviates from the finite difference results significantly. Solving the flow solution more accurately we are getting closer to a pointwise solution of the Navier-Stokes equations and can therefore expect the Hadamard form of the pointwise approach to match the finite differences. Hence, an increase of the polynomial degree of the flow and adjoint solution, see Fig. 9.2, let the Hadamard form based on the pointwise approach match the finite differences better in total. Nevertheless, the difference around the leading edge of the airfoil, where the magnitude of the gradients of the flow solution is big, is still noticeable. However, the agreement of the shape derivative based on the variational approach compared to finite differences is nearly perfect.

Naturally, the gradient $g(\Gamma)$ belongs to $C_0^k(\Gamma)^*$ and lacks the smoothness necessary for the optimization. Therefore, as in [14] we additionally solve the following PDE for $\tilde{g}$ to smooth the gradient

$$(I - \varepsilon \Delta_\Gamma)\tilde{g} = g. \tag{9.24}$$

Here, ε is a smoothing parameter and Δ_Γ the so-called Laplace-Beltrami operator. Numerical experiments show good results for $\varepsilon = 5$. The PDE (9.24) is solved with homogeneous Dirichlet boundary conditions, which corresponds to fixing the leading and trailing edge.

9.3.4 Adaptive One-Shot Method

Putting all the steps together, we obtain the one-shot method. Motivated by the formulation of the algorithm in function spaces in Sect. 9.2, we extent the method by an adaptive step. The complete adaptive one-shot algorithm then reads:

---

repeat

1. **if** (res < tol_1): refine/coarsen the mesh
2. perform n coupled steps in primal and adjoint solver:
   $\mathbf{u}_h^k \to \mathbf{u}_h^{k+1}$
   $\lambda_h^k \to \lambda_h^{k+1}$
5. compute shape gradient g^{k+1}
6. compute the smoothed gradient $\tilde{g}^{k+1}$
7. update the geometry Γ_W due to
   $\Gamma_W^{k+1} = \Gamma_W^k - \tau^k\, \tilde{g}^{k+1}$
9. deform the mesh

until res < tol_2

Therefore, whenever the residual reaches a given tolerance tol_1, we refine or coarsen the mesh. After every adaptive step or design update, the solution vectors need to be interpolated to the new mesh. An interpolation error is introduced and hence we perform n simultaneous primal-dual updates before we continue. We therefore call this method n-step one-shot method. The design corresponds to the boundary wall of the airfoil Γ_W. After computing and smoothing the shape gradient, we can finally update the design via

$$\Gamma_W^{k+1} = \Gamma_W^k - \tau^k\, \tilde{g}^{k+1} \tag{9.25}$$

with the step size parameter $\tau^k \in (0, 1]$, which might improve the convergence behavior. We proceed until we reach a prescribed tolerance tol_2.

9.4 Computational Results

We consider the drag minimization of a NACA0012 airfoil at $M = 0.5$, $Re = 5{,}000$ and an angle of attack of 2. The leading as well as the trailing edge of the airfoil are kept fixed. As design parameters all surface grid points of the shape are chosen. The initial mesh has only 1,600 cells and is given in Fig. 9.3.

This is sufficient, because we adapt the mesh during the one-shot optimization. Furthermore, the contraction rate on coarse meshes is usually better than on fine meshes, which yields further speed-up at the beginning of the optimization. For the initial state we choose the given freestream solution with subsonic inflow and outflow as well as adiabatic wall boundary conditions. The initial adjoint is set to zero. For

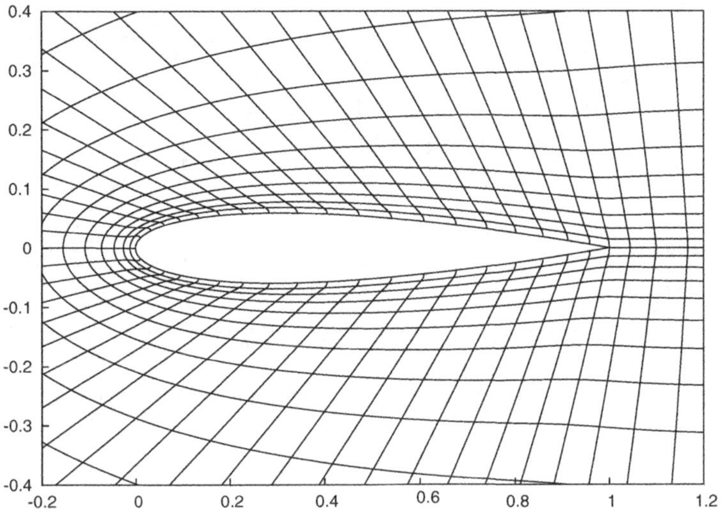

Fig. 9.3 Initial mesh for NACA0012 airfoil

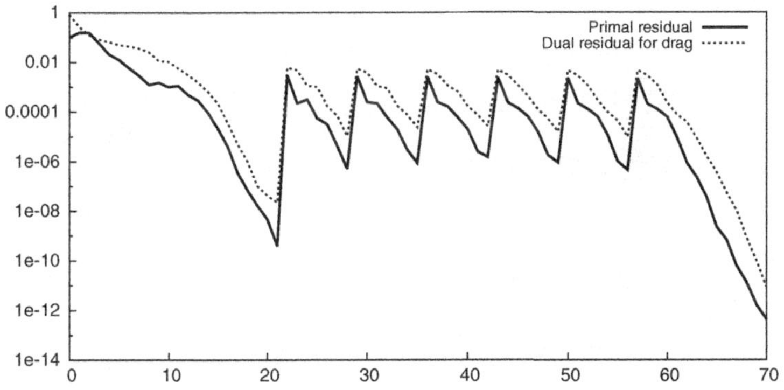

Fig. 9.4 Primal and dual residual for the one-shot optimization

the discretization of the primal the polynomial degree 2 is used. The adjoint variable is discretized with degree 3. We need the higher polynomial degree for the adjoint in order to compute the dual weighted residual (DWR) error estimator with PADGE, see [18] for details.

Figure 9.4 shows the primal and dual residual of the one-shot method without adaptive steps. After 20 iterations in the beginning we obtain a feasible starting iterate for the optimization. The choices $n = 7$ and $\tau^k = 0.5$ yield good results. The similar behavior of primal and dual iteration can be seen clearly. After 6 deformation steps the drag was reduced significantly. As expected, the shape of the airfoil appears to become thinner. Whereas the initial drag is $J_d^0 = 5.586 \times 10^{-2}$, after the optimization

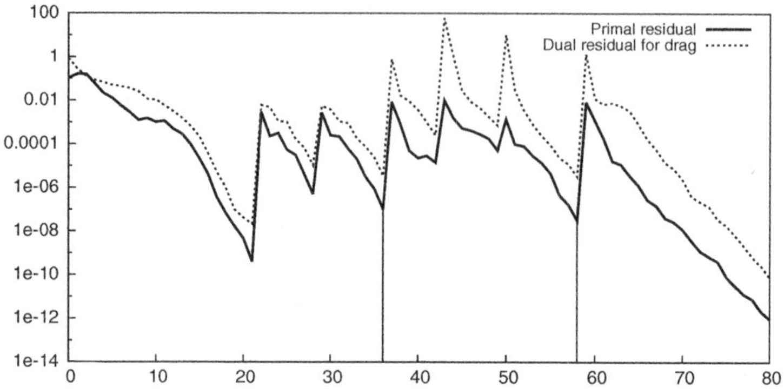

Fig. 9.5 Primal and dual residual for the adaptive one-shot optimization with refinement at step 36 and 58

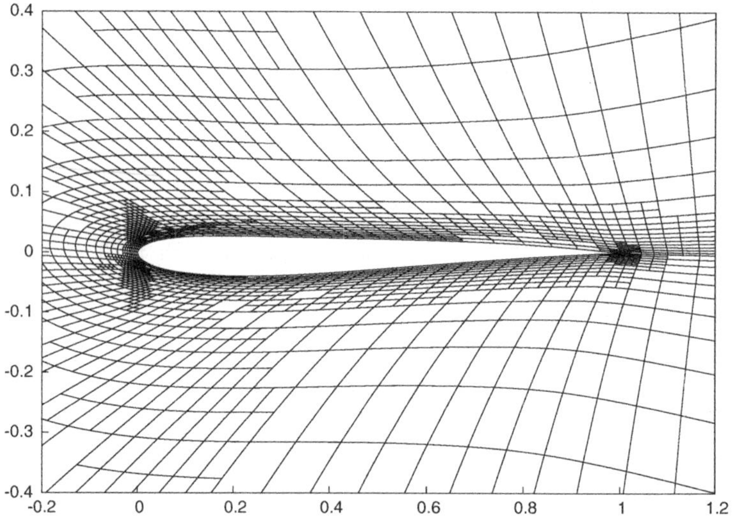

Fig. 9.6 Deformed and refined mesh after 4 deformation and 2 refinement steps

it is $J_d^* = 4.939 \times 10^{-2}$. Due to fixing the leading and trailing edge, the airfoil develops a curved behavior and the lift increases.

We finally add two adaptive steps. After 2 deformation steps 20 % of the cells with the largest DWR error indicator are refined. After the first refinement the mesh has 2,656 cells, after 2 refinement steps it has 4,333 cells. The residuals are plotted in Fig. 9.5. Since the accuracy is improved by the adaptivity and the shape gradients can therefore be evaluated more accurately, the drag reaches $J_d^* = 4.728 \times 10^{-2}$ already after 4 deformation steps. The airfoil becomes much thinner, the deformed and refined mesh is plotted in Fig. 9.6.

9.5 Conclusions

In this paper we studied the extension of the one-shot method by adaptivity. Since the method needs to be capable of handling several dimensions during the iteration, we presented the main steps of the convergence proof in a function space setting. In this framework, we considered the shape optimization of a NACA0012 airfoil in more detail. We introduced the compressible Navier-Stokes equations, which were discretized by the discontinuous Galerkin method, as well as their solution by the backward Euler-Newton method. An appropriate iterative solver for the adjoint equations was set up. The relevant gradient information for the optimization was computed based on a weak formulation of the Navier-Stokes equations. The computational results underlined the positive effect of adaptivity during the optimization with respect to accuracy and efficiency.

Acknowledgments The authors gratefully acknowledge the support of BMBF Grant 03MS632D/ "DGHPOPT".

References

1. Biros G, Ghattas O (2005) Parallel Lagrange-Newton-Krylov-Schur methods for PDE-constrained optimization. Part I: the Krylov-Schur solver. SIAM J Sci Comput 27:687–713
2. Ziems JC, Ulbrich S (2011) Adaptive multilevel inexact SQP methods for PDE-constrained optimization. SIAM J Optim 21:1–40
3. Ito K, Kunisch K, Schulz V, Gherman I (2010) Approximate nullspace iterations for KKT systems. SIAM J Matrix Anal A 31:1835–1847
4. Hamdi A, Griewank A (2010) Properties of an augmented Lagrangian for design optimization. Optim Methods Softw 25:645–664
5. Gauger NR, Griewank A, Hamdi A, Kratzenstein C, Özkaya E, Slawig T (2012) Automated extension of fixed point PDE solvers for optimal design with bounded retardation. In: Leugering G et al (eds) Constrained optimization and optimal control for partial differential equations. Springer, Basel, pp 99–122
6. Hazra SB, Schulz V, Brezillon J, Gauger NR (2005) Aerodynamic shape optimization using simultaneous pseudo-timestepping. J Comput Phys 204:46–64
7. Kaland L (2013) The one-shot method: function space analysis and algorithmic extension by adaptivity. PhD thesis, RWTH Aachen University, Aachen, Germany, in preparation
8. Kaland L, De Los Reyes JC, Gauger NR (2013) One shot methods in function space for PDE-constrained optimal control problems. Optim Methods Softw doi:10.1080/10556788.2013.774397
9. Hartmann R, Held J, Leicht T, Prill F (2010) Discontinuous Galerkin methods for computational aerodynamics 3D adaptive flow simulation with the DLR PADGE code. Aerosp Sci Technol 14:512–519
10. Kuhn M (2011) Herleitung von Formableitungen aerodynamischer Zielgrößen für die laminaren kompressiblen Navier-Stokes Gleichungen und deren Implementierung in diskontinuierlichen Galerkin-Verfahren. Masters thesis, Humboldt-Universität zu Berlin, Berlin, Germany
11. Hartmann R (2008) Numerical analysis of higher order discontinuous Galerkin finite element methods. In: Deconinck H (ed) VKI LS 2008-08: CFD— ADIGMA course on very high order

discretization methods, 13–17 October 2008, Von Karman Institute for Fluid Dynamics, Rhode Saint Genèse, Belgium
12. Sokolowski J, Zolésio J-P (1992) Introduction to shape optimization. Springer, Berlin
13. Delfour MC, Zolésio J-P (2001) Shapes and geometries. Society for Industrial and Applied Mathematics, Philadelphia
14. Schmidt S (2010) Efficient large scale aerodynamic design based on shape calculus. PhD thesis, University of Trier, Trier, Germany
15. Castro C, Lozano C, Palacios F, Zuazua E (2007) Systematic continuous adjoint approach to viscous aerodynamic design on unstructured grids. AIAA 45:2125–2139
16. Ito K, Kunisch K, Gunther G, Peichl H (2008) Variational approach to shape derivatives. ESAIM Control Optim Calc Var 14:517–539
17. Sonntag M, Gauger NR, Schmidt S (2013) Shape derivatives for the compressible Navier-Stokes equations in variational form. J Comput Appl Math (submitted)
18. Hartmann R (2005) Discontinuous Galerkin methods for compressible flows: higherorder accuracy, error estimation and adaptivity. In: Deconinck H, Ricchiuto M (eds.), VKI LS 2006-01:CFD-Higher Order Discretization Methods, Nov. 14-18, 2005, Von KarmanInstitute for Fluid Dynamics, Rhode Saint Genèse, Belgium

Chapter 10
Optimal Flow Control and Topology Optimization Using the Continuous Adjoint Method in Unsteady Flows

Ioannis S. Kavvadias, George K. Karpouzas, Evangelos M. Papoutsis-Kiachagias, Dimitris I. Papadimitriou and Kyriakos C. Giannakoglou

Abstract This paper presents the development and application of the unsteady continuous adjoint method to the incompressible Navier-Stokes equations and its use in two different optimization problems. The first is the computation of the optimal setting of a flow control system, based on pulsating jets located along the surface of a square cylinder, in order to minimize the time-averaged drag. The second is dealing with unsteady topology optimization of a duct system with four fixed inlets and a single outlet, with periodic in time inlet velocity profiles, where the target is to minimize the time-averaged viscous losses. The presentation of the adjoint formulation is kept as general as possible and can thus be used to other optimization problems governed by the unsteady Navier-Stokes equations. Though in the examined problems the flow is laminar, the extension to turbulent flows is doable.

Keywords Unsteady continuous adjoint · Flow control optimization · Topology optimization · Level-set

---

I.S. Kavvadias · G.K. Karpouzas · E.M. Papoutsis-Kiachagias · D.I. Papadimitriou
K.C. Giannakoglou (✉)
Laboratory of Thermal Turbomachines, Parallel CFD and Optimization Unit,
School of Mechanical Engineering, National Technical University of Athens (NTUA),
Iroon Polytechniou 9, 15780 Athens, Greece
e-mail: kgianna@central.ntua.gr

I.S. Kavvadias
e-mail: kavvadiasj@hotmail.com

G.K. Karpouzas
e-mail: grgkarpou@yahoo.gr

E.M. Papoutsis-Kiachagias
e-mail: vaggelisp@gmail.com

D.I. Papadimitriou
e-mail: dpapadim@mail.ntua.gr

D. Greiner et al. (eds.), *Advances in Evolutionary and Deterministic Methods for Design, Optimization and Control in Engineering and Sciences*, Computational Methods in Applied Sciences 36, DOI 10.1007/978-3-319-11541-2_10

10.1 Introduction

Adjoint methods [1–4] are successfully used to compute the gradient of the objective function with respect to the design variables and support gradient-based optimization methods. This paper is dealing with the unsteady continuous adjoint [5–8] methods, where the adjoint PDEs are firstly derived and, then, discretized. The primal problem is governed by the unsteady flow equations and time-averaged performance metrics are used as objective functions. Two optimization problems are solved: an active flow control and a topology optimization problem.

Active flow control [9, 10], based on suction of blowing jets, steady or unsteady, may control the boundary layer of the flow by preventing or delaying separation or controlling other flow phenomena, such as the Karman vortices generated behind a cylinder. The case examined here is the unsteady flow developed around a square cylinder, which is controlled by a set of pulsating jets at fixed locations around the cylinder, with their amplitudes as the design variables.

On the other hand, topology optimization is a shape parameterization-free design method, which is used to identify which parts of an extended domain should be solidified, so as to minimize the objective function which quantifies the quality of the fluid flow in the remaining, non-solidified, part of the domain. Two major variances of topology optimization exists, the porosity [11–13] and the level-set [14, 15] methods.

The porosity-based class of topology optimization algorithms compute a real-valued porosity field, a, over an extended domain, which minimizes the objective function. Domain areas corresponding to the fluid flow are identified as those with nodal values $a = 0$ or, practically, $a \leq \varepsilon$ where ε is an infinitesimally small positive number. Remaining areas where $a \neq 0$ or, practically, $a > \varepsilon$ define the part of the domain to be solidified. In contrast, the level-set method is based on the signed distance φ from the sought solid-fluid interface. If locally $\varphi < 0$, this cell is a fluid cell, whereas cells with $\varphi > 0$ must be solidified. The isolines $\varphi = 0$ define the solid walls.

In this paper, the topology optimization is based on the level-set method and aims at designing a duct system with four fixed inlets and a single fixed outlet, for minimum time-averaged viscous losses. Unsteadiness is caused by the time-varying inlet velocity profiles.

In the unsteady adjoint method, a major issue is the storage of the primal solution fields, at different time steps. When solving the unsteady adjoint equations, information travels backwards in time. Also, to solve the adjoint equations at a given time instant, the primal fields for the same instant must be available. The full-storage of the primal field evolution in time is memory-wise too expensive and alternatives must be used. The check-pointing technique [16, 17], which is used in this paper, stores the primal solution at a number of instants and recomputes the solution at all other time instants, starting from the closest check-point.

10.2 Flow Model and Objective Functions

The flow is modeled by the Navier-Stokes equations for the unsteady laminar flow of an incompressible fluid. The last term in Eq. 10.1, which is activated only in topology optimization problems, is added. The primal equations are

$$\begin{aligned} R_i^v &= \frac{\partial v_i}{\partial t} + v_j \frac{\partial v_i}{\partial x_j} - \frac{\partial}{\partial x_j}\left[\nu\left(\frac{\partial v_i}{\partial x_j} + \frac{\partial v_j}{\partial x_i}\right)\right] \\ &+ \frac{\partial p}{\partial x_i} + \alpha H(\varphi)\, v_i = 0, \quad i = 1, 2(3) \end{aligned} \tag{10.1}$$

$$R^p = -\frac{\partial v_j}{\partial x_j} = 0 \tag{10.2}$$

where v_i and p stand for the velocity components and the static pressure divided by the density, respectively. φ is the signed distance (d or $-d$) from the solid walls used in the level set method,

$$\varphi(x) = \begin{cases} d, & \text{if x} \in \text{solid region}(\Omega_s), \\ -d, & \text{if x} \in \text{fluidic region}(\Omega_f), \\ 0, & \text{if x is on the interface.} \end{cases} \tag{10.3}$$

H(φ) is the Heaviside function and α is a penalty multiplier.

In view of the derivation of the primal equations, needed for the adjoint equations, the non-differentiable Heaviside function is replaced with the sigmoid function

$$\widehat{H}(\varphi, h) = \begin{cases} 1 & , \quad \text{if } \varphi \geq h \\ \frac{1}{2} + \frac{15\varphi}{16h} - \frac{5\varphi^3}{8h^3} + \frac{3\varphi^5}{16h^5} & , \quad \text{if } |\varphi| < h \\ 0 & , \quad \text{if } \varphi \leq -h \end{cases}$$

and its derivative is

$$\tau(\varphi, h) = \frac{\partial \widehat{H}(\varphi, h)}{\partial \varphi} = \begin{cases} +\frac{15}{16h}\left(1 - \frac{\varphi^2}{h^2}\right) & , \quad \text{if } |\varphi| < h \\ 0 & , \quad \text{if } |\varphi| \geq h \end{cases}$$

where h takes on a very small positive value determining the shape of the sigmoid function.

For the numerical solution of the primal equations the SIMPLE algorithm [18] was used, with a cell-centered, finite-volume discretization scheme.

The first problem examined is the design of an optimal flow control system. It is about the optimal configuration of a set of pulsating jets activated at fixed locations along the perimeter of a square cylinder, to minimize the time-averaged drag. The velocity components of each jet are given by

$$v_\lambda^m = \left(A^m \sin\left(2\pi f^m \left(t - f_0^m\right)\right) - A^m\right) n_\lambda, \quad \lambda = 1, 2(3) \tag{10.4}$$

where m is the jet counter. A^m is the amplitude, f^m the frequency and f_0^m the phase of each jet. Jets are aligned with the outwards, normal to the wall, unit vector. Positive A^m corresponds to blowing and negative A^m to suction. The frequencies f^m and phases f_0^m of all jets are the same and fixed to $f^m = \frac{v_\infty}{d}$ [8] and $f_0^m = 0$, where v_∞ is the infinite flow velocity and d is the side length of the square cylinder. The only design variables are the amplitudes A^m. In this case, the period of pulsating jets is about 6 times shorter than that of the Karman vortices.

The time-averaged (squared) drag force is expressed as

$$J_1 = \frac{1}{2T} \int_T D^2(t)dt \tag{10.5}$$

where T is the flow period. In the uncontrolled case, the flow period is the Karman vortices' period whereas in the optimally controlled case T stands for the jets' period. D is the time-dependent drag force

$$D(t) = \int_{S_w} \left[pn_i - \nu \left(\frac{\partial v_i}{\partial x_j} + \frac{\partial v_j}{\partial x_i} \right) n_j - \left| v_j n_j \right| v_i \right] r_i dS \tag{10.6}$$

where r_i are the components of the unit vector aligned with the farfield velocity and S_w stands for the solid wall boundary. The last term in Eq. 10.6 stands for the contribution of jets on the forces acting upon the body, at the jets locations. A similar study, for a circular cylinder, can be found in [8].

The second problem is concerned with the design of an optimal duct system connection fixed inlets with a single outlet. The flow is unsteady since time-varying inflow conditions are imposed. The velocity at each inlet is still given by Eq. 10.4, where $A^m(>0)$, f^m and f_0^m have fixed values. The problem is handled as a topology optimization problem on an extended domain where the level set values φ at each cell center are the design variables.

The objective function to be minimized is the time- and mass-averaged total pressure losses between the inlets S_I and the outlet S_O. This is mathematically expressed as

$$J_{p_t} = -\frac{1}{T} \int_T \int_{S_{I,O}} \left(p + \frac{1}{2} v_j v_j \right) v_i n_i dS dt \tag{10.7}$$

with a term expressing a volume constraint being added to it. The constraint function is

$$c = \left[\frac{\int_\Omega H(\varphi) d\Omega}{V_{all}} - V_{tar} \right]^2 = \left(\frac{V_{solid}}{V_{all}} - V_{tar} \right)^2 \tag{10.8}$$

where V_{all} is the volume of Ω, V is the volume occupied by fluid, V_{solid} that of the solidified part of the domain and V_{tar} gives the desired percentage of Ω to be solidified. After including the equality constraint of Eq. 10.8, the objective function becomes

$$J_2 = J_{p_t} - \lambda c + wc^2 \tag{10.9}$$

where λ is a Lagrangian multiplier and w a weight associated with the constraint. During the optimization loop, both are updated according to the Augmented Lagrange Multiplier (ALM) algorithm [19]. λ is initialized with a zero value and w with a small positive value. At the end of each optimization cycle, w is multiplied by a user-defined positive factor $\gamma > 1$ (unless it exceeds w_{max}) and λ is updated as $\lambda^{new} = \lambda^{old} - 2w^{old}c$.

10.3 The Continuous Unsteady Adjoint Method

10.3.1 Field Adjoint Equations

The augmented objective function L_k is defined as the sum of J_k and the time-space $(T - \Omega)$ integrals of the products of the state equations and the corresponding adjoint fields. So,

$$L_k = J_k + \int_T \int_\Omega u_i R_i^v \, d\Omega dt + \int_T \int_\Omega q R^p \, d\Omega dt, \; k = 1, 2 \tag{10.10}$$

where u_i and q are the adjoint velocities and pressure, respectively.

The derivatives of L_k w. r. t. the design variables b_m, after applying the Leibniz theorem, become

$$\frac{\delta L_k}{\delta b_m} = \frac{\delta J_k}{\delta b_m} + \int_T \int_\Omega u_i \frac{\partial R_i^v}{\partial b_m} d\Omega dt + \int_T \int_\Omega q \frac{\partial R^p}{\partial b_m} d\Omega dt \tag{10.11}$$

By applying the Green-Gauss theorem and eliminating the integrals which depend on variations in the flow variables w. r. t. b_m, the field adjoint equations are derived. These are

$$R^q = \frac{\partial u_i}{\partial x_i} = 0 \tag{10.12}$$

$$R_i^u = -\frac{\partial u_i}{\partial t} - v_j \frac{\partial u_i}{\partial x_j} + u_j \frac{\partial v_j}{\partial x_i} + \frac{\partial q}{\partial x_i} - \frac{\partial}{\partial x_j}\left[\nu \left(\frac{\partial u_i}{\partial x_j} + \frac{\partial u_j}{\partial x_i}\right)\right] + \alpha H(\varphi)\, u_i = 0 \tag{10.13}$$

Since both objective functions, J_1 and J_{pt}, comprise only boundary integrals, their derivatives $\frac{\delta J_1}{\delta b_m}$ and $\frac{\delta J_{pt}}{\delta b_m}$ are defined only at these boundaries and do not contribute to the adjoint field equations. The c constrain term, added to J_{p_t} to form J_2, contains a volume integral which contributes only to the sensitivity derivatives. Since none of the objective functions contributes to the field adjoint equations, the same field adjoint equations are valid for both. Of course, in the flow control optimization problem, the last term in Eq. 10.13 vanishes.

After eliminating the integrals depending on the variations in flow quantities, the sensitivity derivatives are expressed, in the most general form, as

$$\frac{\delta L_k}{\delta b_m} = \frac{\delta J_k}{\delta b_m} + \int_{\Omega} \left[v_i \frac{\partial v_i}{\partial b_m} \right]_0^T d\Omega + \int_T \int_S D_i^u \frac{\partial v_i}{\partial b_m} dSdt + \int_T \int_S D^q \frac{\partial p}{\partial b_m} dSdt$$
$$+ \int_T \int_S E_i^u \left[\frac{\partial}{\partial x_j} \left(\frac{\partial v_i}{\partial b_m} \right) + \frac{\partial}{\partial x_i} \left(\frac{\partial v_j}{\partial b_m} \right) \right] n_j dSdt \tag{10.14}$$

where $S = S_I \cup S_O \cup S_w$ or $S = S_\infty \cup S_w$ is the boundary and $D_i^u = u_i v_j n_j + \nu \left(\frac{\partial u_i}{\partial x_j} + \frac{\partial u_j}{\partial x_i} \right) n_j - q n_i$, $E_i^u = -\nu u_i$ and $D^q = u_j n_j$.

By substituting the derivative of each objective function into Eq. 10.14, the elimination of the boundary integrals which depend on the variation in the flow variables gives the adjoint boundary conditions. The remaining terms give the expression of the sensitivity derivatives to be used in the descent algorithm.

10.3.2 Boundary Conditions and Sensitivity Derivatives for J_1

The derivative of the 'mean drag' objective function w. r. t. b_m is

$$\frac{\delta J_1}{\delta b_m} = \frac{1}{T} \int_T \int_{S_w} D \left(-\nu \left[\frac{\partial}{\partial x_j} \left(\frac{\partial v_i}{\partial b_m} \right) + \frac{\partial}{\partial x_i} \left(\frac{\partial v_j}{\partial b_m} \right) \right] n_j \right.$$
$$\left. + \frac{\partial v_i}{\partial b_m} |v_j n_j| + \frac{v_j n_j}{|v_j n_j|} \frac{\partial v_j}{\partial b_m} n_j v_i + \frac{\partial p}{\partial b_m} n_i \right) r_i dSdt \tag{10.15}$$

After substituting Eq. 10.15 into Eq. 10.14, the elimination of the boundary integrals depending on the variation of the flow variables w. r. t. b_m gives the adjoint boundary conditions at every time step.

The adjoint boundary conditions along S_w, S_I and S_O, at every time-step, are S_w: $u_i = -\frac{D(t)}{T} r_i$ and S_∞: $u_i = 0$; for the whole domain Ω, the initial condition at $t = T$ is $u_i|_{t=T} = 0$.

The incoming or outgoing adjoint velocity is proportional to the instantaneous value of drag $D(t)$; this is the origin of the unsteady adjoint flow.

Finally, the sensitivities of J_1 w. r. t. the control variables $b_m = A^m$ are given by

$$\frac{\delta J_1}{\delta b_m} = \int\limits_T \int\limits_{S_w} \left[u_i v_j n_j - u_i \left| v_j n_j \right| + \nu \left(\frac{\partial u_i}{\partial x_j} + \frac{\partial u_j}{\partial x_i} \right) n_j - q n_i - \frac{v_j n_j}{\left| v_j n_j \right|} u_j v_j n_i \right]$$
$$(sin(2\pi f^m (t - f_0^m)) - 1) n_i dS dt \quad (10.16)$$

10.3.3 Boundary Conditions and Sensitivity Derivatives for J_2

For the time-averaged total pressure losses, used in topology optimization, the adjoint boundary conditions are derived by substituting the derivative of J_2 w. r. t. b_m in Eq. 10.14 and eliminating terms depending on the derivatives of the flow fields. The derivative of J_2 w. r. t. b_m is

$$\frac{\delta J_2}{\delta b_m} = -\frac{2}{T} \int\limits_T \int\limits_{S_{I,O}} \left[v_i n_i \frac{\partial p}{\partial b_m} + \left(v_i v_j n_j + \left(p + \frac{1}{2} v_j^2 \right) \right) \frac{\partial v_i}{\partial b_m} \right] dS dt$$
$$+ (4wc - \lambda) \left(\frac{V_{solid}}{V_{all}} - V_{tar} \right) \frac{\int\limits_\Omega \tau(\varphi) d\Omega}{V_{all}} \quad (10.17)$$

The adjoint boundary conditions are S_w: $u_i = 0$, S_I: $u_{(n)} = \frac{v_{(n)}}{T}$, $u_{(t)} = 0$, S_O: $q = u_{(n)} v_{(n)} + \nu \left(\frac{\partial u_{(n)}}{\partial n} + \frac{\partial u_{(n)}}{\partial n} \right) + \left(p + \frac{1}{2} v^2 \right) + v_{(n)} v_{(n)}$ and $u_{(t)} v_{(n)} + \nu \left(\frac{\partial u_{(t)}}{\partial n} + \frac{\partial u_{(n)}}{\partial t} \right) + v_{(t)} v_{(n)} = 0$ and the initial condition for the adjoint field, at $t = T$, is $u_i|_{t=T} = 0$. Indices (n) and (t) stand for the normal and tangent components to the boundary.

The sensitivity derivatives of J_2 w. r. t. b_m, where b_m are the φ values at the cell-center, are

$$\frac{\delta J_2}{\delta b_m} = \frac{\delta J_2}{\delta \varphi} = \int\limits_\Omega [\alpha (v_i u_i) \tau(\varphi)] d\Omega + 2(-\lambda + 2wc) \left(\frac{V_{solid}}{V_{all}} - V_{tar} \right)$$
$$\times \frac{\int_\Omega \tau(\varphi) d\Omega}{V_{all}} \quad (10.18)$$

10.3.4 Check-Pointing

For the solution of the unsteady adjoint equations, the primal fields at all time steps must be available. Since the adjoint information travels backwards in time, in order to use the primal fields at each time step, these should have been stored during the solution of the primal equations. Due to memory limitations, this is replaced by the binomial check-pointing technique.

The check-pointing technique is a compromise between memory consumption and CPU cost. Instead of storing the primal solutions at every time step, which is very memory consuming, only those at a predefined number of time-instances, called check-points, are stored; from them, the primal solution in every other time-instant is re-computed.

The binomial check-pointing method uses a binomial distribution of check-points in time, for which it can be proved [17] that the number of flowfield recomputations is minimal for given numbers of check-points and time-steps. The distribution of check-points is dynamically updated as time progresses, so that, at any given time-step, each check-point is always in the optimal position in time, as dictated by the binomial distribution.

10.4 Results

In both optimization problems, the steepest descent method [19]

$$b_m^{new} = b_m^{old} - \eta \frac{\delta J_k}{\delta b_m} \tag{10.19}$$

is used to update the design variables values, after solving the adjoint equations.

10.4.1 Flow Control Optimization

The Reynolds number of the flow around the cylinder is $Re = 100$. Five jets were equi distributed along each side of the square cylinder, the placement of which can be seen in Fig. 10.1. All 20 jets share the same frequency $f^m = v_\infty/d = 10\,\text{Hz}$ and phase $f_0^m = 0$. Recall that the optimization variables are the amplitudes A^m of the jets and the minimization of J_1 is targeted.

The time step for the simulation is $\Delta t = 4 \times 10^{-4}$ sec. Two variants were tried. In the first variant, the so-called "full-in-time" approach, the flow computation was performed for 11 periods of time in each optimization circle. It was decided to discard the solution during the first 5 periods, so as to get rid of the transient phase of the primal problem and do the same for the last 5 to also avoid the transient phase of the adjoint problem. Only the intermediate period, which is considered representative of the periodic primal and adjoint phenomena, was used to calculate sensitivity derivatives and the value of the objective function. In the second variant, to be referred to as the "fast-in-time" approach, only one period of time is simulated in each optimization cycle. After the numerical solution of the primal equations for a single period of time, the adjoint equations were solved for this period. Then, the

Fig. 10.1 Flow control optimization—mean drag minimization of a square cylinder at $Re = 100$. The optimal solution is shown. Snapshot of the vorticity field at a random time instant. Primal (*top*) and adjoint (*bottom*) vorticity fields

sensitivity derivatives were computed using the solution to the primal and adjoint equations, for this single period, and the design variables were updated.

Because the transient effects were not discarded, the computed sensitivities were not exact but the CPU cost per optimization cycle was lower. In order to reduce transient effects, the results at the last time-step of the previous period/cycle were used as initialization for the primal flow in the next optimization cycle. For the adjoint equations, the first time-instant was used instead, since time goes backwards. At the end of the optimization process, both the primal and the adjoint equations were converged to a periodic solution.

In this case, 400 check-points were used and enough optimization cycles were performed for both variants to converge. The convergence of both approaches is presented in Fig. 10.2. The "fast-in-time" simulation appears to be twice as fast as the "full-in-time" one. As such, the "fast-in-time" approach was exclusively used in the second problem.

As both variants converged to a similar mean drag value and, except convergence, only the outcome of the "fast-in-time" approach is shown.

The time variation in the drag and lift coefficients for the uncontrolled and controlled cases are shown in Fig. 10.3 and 10.4. The resulted reduction in the amplitude of the oscillating lift force is nothing more than a by-product of the optimization process and is attributed to the controlled flow field symmetry.

A snapshot of the optimized flow, at an arbitrary time instant, is presented in Fig. 10.1. The Karman vortices of the uncontrolled flow were suppressed and both the primal and adjoint flows are symmetric in space. Also, the wake in the adjoint flow, developed in the upwind direction, is visible.

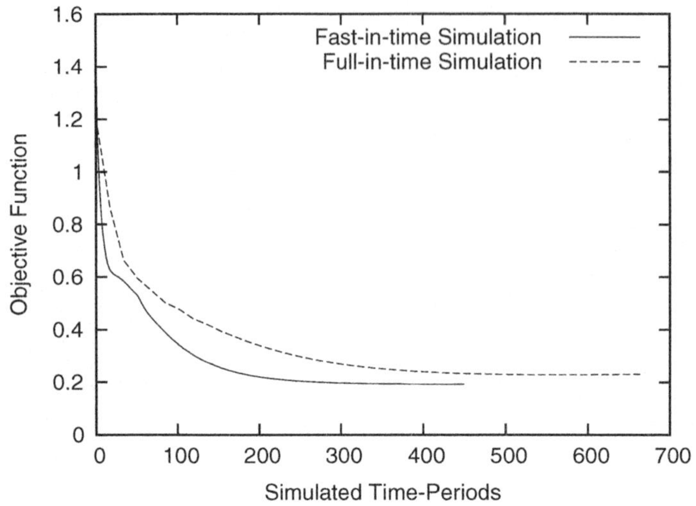

Fig. 10.2 Flow control optimization—mean drag minimization of a square cylinder, at $Re = 100$. Convergence of both "-in-time" approaches. The x-axis corresponds to simulated periods of time. For the "full-in-time" approach, each optimization cycle solves 11 primal and 6 adjoint periods, while the "fast-in-time" approach solves for only 1 primal and 1 adjoint period per cycle. From this case, a speed-up of about ×2 was achieved by using the "fast-in-time" approach

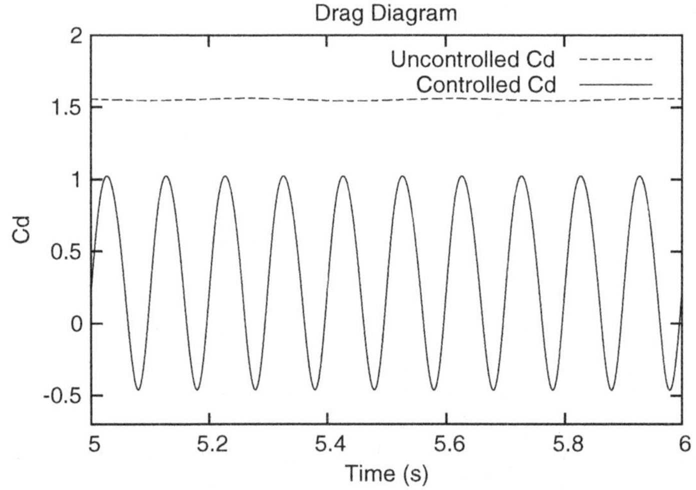

Fig. 10.3 Flow control optimization—mean drag minimization of a square cylinder, at $Re = 100$. Drag coefficient C_d ($= \frac{D(t)}{0.5d^2v_\infty^2}$) for the uncontrolled case and the optimally controlled configuration. The mean drag coefficient was reduced from ∼1.6 to ∼0.3

The computed optimal jet amplitudes are shown in Fig. 10.5. Slots 4–10 create symmetric vortices above and below the cylinder and slots 1–3 push them away. These vortices do not allow the Karman street to be developed and produce a symmetric flow field around the horizontal axis.

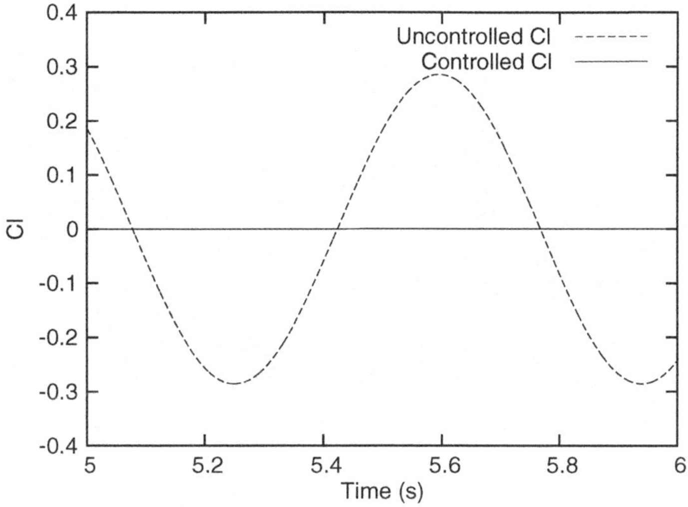

Fig. 10.4 Flow control optimization—mean drag minimization of a square cylinder, at $Re = 100$. Lift coefficient C_l for the uncontrolled and the optimally controlled configuration. The lift was almost stabilized to zero, though this was not included in the objective function

Slot	*Amplitude*
1	0.0437
2	0.0792
3	0.0836
4	0.0488
5	- 0.0416
6	- 0.0374
7	- 0.0163
8	- 0.0127
9	0.0756
10	0.0577
11	0.0034

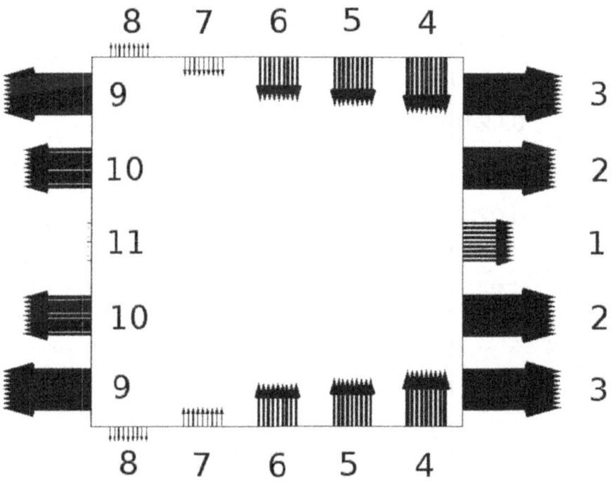

Fig. 10.5 Flow control optimization—mean drag minimization of a square cylinder, at $Re = 100$. Jet locations and slot widths are shown. The computed optimal amplitudes of the pulsating jets are listed and sketched

10.4.2 Unsteady Topology Optimization

The topology optimization was carried out on an empty square box, $2 \times 2\,\mathrm{m}^2$, with four inlets of 5 cm each and a single 15 cm outlet. The set-up of the inlets and outlets is shown in Fig. 10.6. The velocity at each inlet is expressed by Eq. 10.4. All inlet

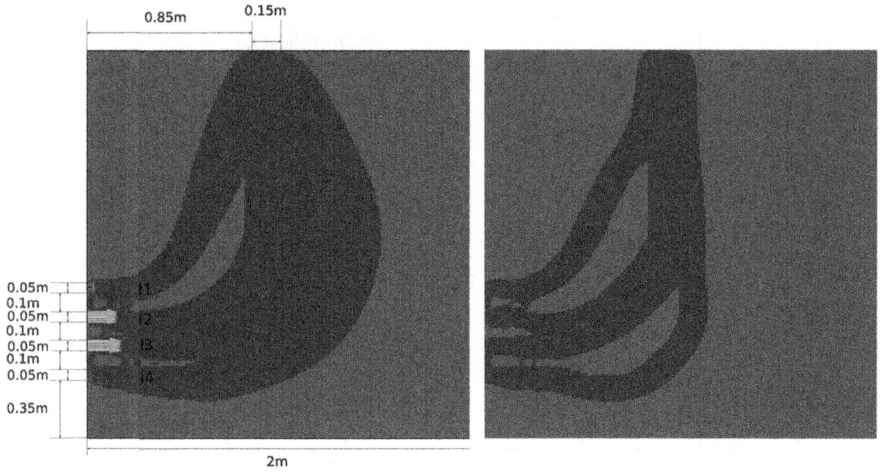

Fig. 10.6 Unsteady topology optimization. The $\widehat{H}(\varphi)$ field showing the optimal duct computed by the optimization loop. *Left* $V_{tar} = 0.6$, *Right* $V_{tar} = 0.8$, *Red areas* ($\widehat{H}(\varphi) \approx 1$) indicate the solidified part of the domain whereas the *blue* one ($\widehat{H}(\varphi) \approx 0$) is the fluid. *Left* the inlet velocity vectors at this instant are also shown, so as to make clear that the four incoming mass flow rates are not in phase

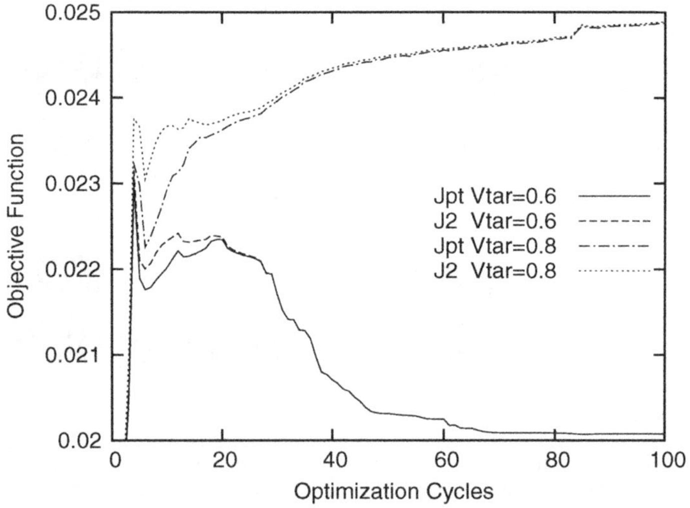

Fig. 10.7 Unsteady topology optimization. The mean total pressure losses as the optimization algorithm. Evolves the presented results correspond to two runs with different constraints, $V_{tar} = 0.6$ and $V_{tar} = 0.8$. The number of optimization cycles might appears to be high but the CPU cost per cycle is quite low since both the primal and adjoint equations were solved for a single period of time only. The constrained and unconstrained value of the objective function are plotted. Once the solidified part of the domain reaches the desired percentage of the overall volume, the terms added to J_{p_t} become zero and the two curves coincide. As expected, increasing the solidified part of the domain led to a narrow fluid passage and increased pressure losses

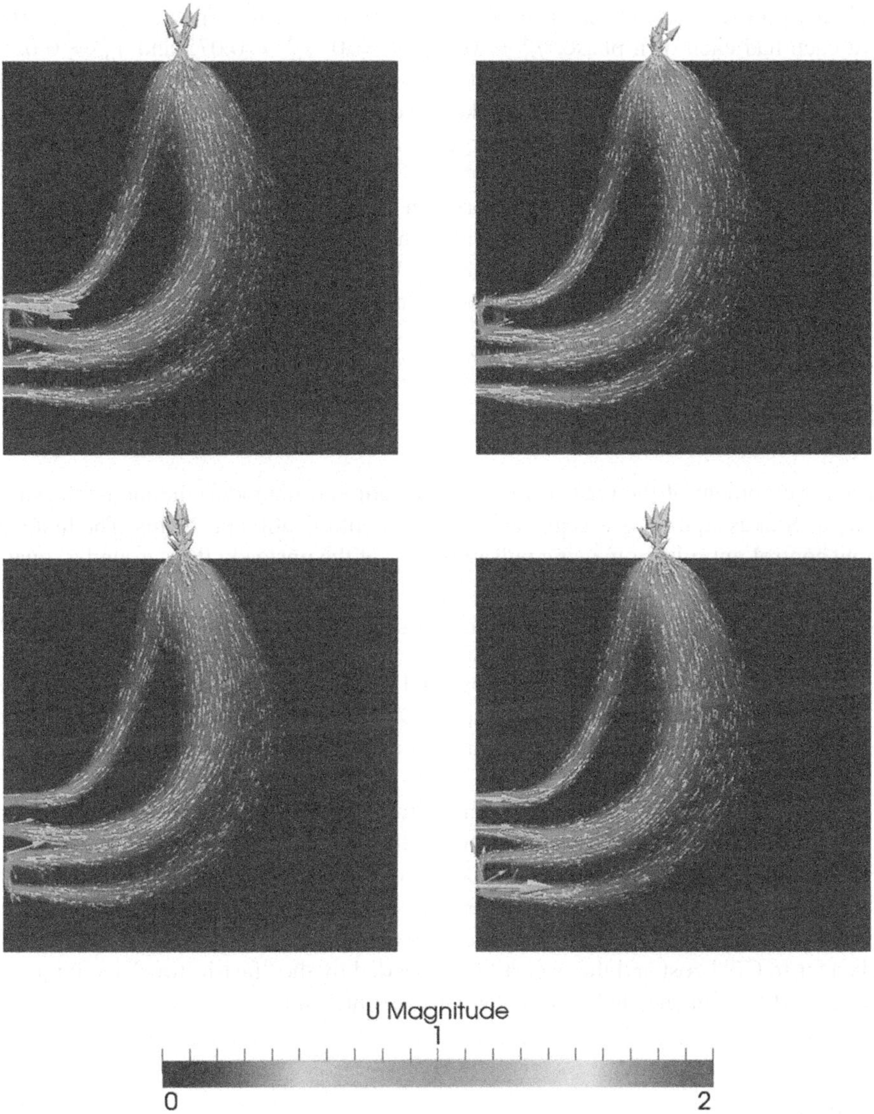

Fig. 10.8 Unsteady topology optimization. The velocity flow field at 4 different time instants, corresponding to maximum velocities at inlet 1 (*top-left*), inlet 2 (*top-right*), inlet 3 (*bottom-left*) and inlet 4 (*bottom-right*). Strong swirl effects are present close to the flow outlet. Should these be undesirable, they could be controlled or even eliminated [13] by adding a second constraint to the objective function

velocity profiles share the same amplitude $A^m = 1\,\text{m/s}$ and frequency $f^m = 10\,\text{Hz}$, but each had each own phase, $f_0^1 = 0.05$, $f_0^2 = 0$, $f_0^3 = 0.075$ and $f_0^4 = 0.025$. The percentage of the square box volume to be solidified was initially set at 60 % or, in Eq. 10.8, $V_{tar} = 0.6$. For the sake of comparison, a second optimization was carried out using $V_{tar} = 0.8$. The target is to minimize J_2. In this case, only the "fast-in-time" approach was used.

The optimal shape of the duct, for each constraint, is shown in Fig. 10.6 and the progress of the optimization algorithm in Fig. 10.7. Also, four snapshots of the velocity field are presented in Fig. 10.8, for $V_{tar} = 0.6$, each corresponding to the time instant at which the velocity of each inlet jet is at its maximum value.

10.5 Conclusions

The development of the unsteady continuous adjoint method to the incompressible Navier-Stokes equations was presented for two optimization problems. The first is a flow control optimization, using pulsating jets, of the unsteady flow around a square cylinder and the second is a level-set optimization problem to design an optimal duct system in a box with four inlets, a single outlet and unsteady inlet boundary conditions, under a volume constraint.

On the flow control problem the optimal amplitude for each jet, as well as its type (blowing or suction) were identified. On the topology optimization problem, the optimal duct systems were identified for different volume constraints, based on the level-set topology method.

In all cases, the binomial check-pointing method was used to overcome the memory requirements of the unsteady adjoint method.

The "fast-in-time" technique was formulated, where the optimization is based on an approximation of the sensitivity derivatives, due to transient effects. Using this technique, though more optimization cycles are needed, each one of them is much cheaper in CPU cost and the over-all time needed by the "fast-in-time" technique is about half the time needed by its "standard" counterpart.

References

1. Pironneau O (1974) On optimum design in fluid mechanics. J Fluid Mech 64:97–110
2. Jameson A (1988) Aerodynamic design via control theory. J Sci Comput 3:233–260
3. Shubin G, Frank PA (1991) Comparison of the implicit gradient approach and the variational approach to aerodynamic design optimization. Boeing computer services report AMS-TR-163
4. Burgreen G, Baysal O (1996) Three-dimensional aerodynamic shape optimization using discrete sensitivity analysis. AIAA J 34(9):1761–1770
5. Anderson W, Venkatakrishnan V (1997) Aerodynamic design optimization on unstructured grids with a continuous adjoint formulation. In: 35th aerospace sciences meeting and exhibit, Reno, NV, AIAA-1997-0643

6. Papadimitriou D, Giannakoglou K (2008) Aerodynamic shape optimization using first and second order adjoint and direct approaches. Arch Comput Methods Eng (State of the Art Reviews) 15(4):447–488
7. Wang Q (2008) Uncertainty quantification for unsteady fluid flow using adjoint-based approaches. PhD thesis, Standford
8. Carnarius A, Thiele F, Ozkaya E, Nemili A, Gauger N (2011) Optimal control of unsteady flows using a discrete and a continuous adjoint approach. Syst Model Optim IFIP Adv Inf Commun Technol 391:318–327
9. Bewley T (2001) Flow control: new challenges for a new renaissance. Prog Aerosp Sci 37:21–58
10. Zymaris A, Papadimitriou D, Papoutsis-Kiachagias EM, Giannakoglou K (2013) The continuous adjoint method as a guide for the design of flow control systems based on jets. Eng Comput 30(4):494–520
11. Borvall T, Peterson J (2003) Topology optimization if fluids in stokes flow. Int J Numer Meth Fluids 41:77–107
12. Gersborg-Hansen A, Sigmund O, Haber R (2005) Topology optimization of channel flow problems. Struct Multi Optim 30:181–192
13. Kontoleontos EA, Papoutsis-Kiachagias EM, Zymaris AS, Papadimitriou DI, Giannakoglou KC (2013) Adjoint-based constrained topology optimization for viscous flows, including heat transfer. Eng Optim 45(8):941–961
14. Sethian J (2001) Evolution, implementation, and application of level set and fast marching methods for advancing fronts. J Comput Phys 169:503–555
15. Chang Y, Hou T, Merriman B, Osher S (1996) A level set formulation of eulerian interface capturing methods for incompressible fluid flows. J Comput Phys 124:449–494
16. Griewank A, Walther A (2000) Algorithm 799: revolve: an implementation of checkpointing for the reverse or adjoint mode of computational differentiation. ACM Trans Math Softw (TOMS) 26(1):19–45
17. Wang Q, Moin P, Iaccarino G (2008) Minimal repetition dynamic checkpointing algorithm for unsteady adjoint calculation. SIAM J Sci Comput 31(4):2549–2567
18. Caretto L, Gosman A, Patankar S, Spalding D (1972) Two calculation procedures for steady three-dimensional flows with recirculation. In: Proceedings of the third international conference on numerical methods in fluid mechanics, Paris
19. Nocedal J, Wright S (1999) Numerical optimization. Springer, Heidelberg

Part III
Engineering Design and Societal Applications: Turbomachinery

Chapter 11
Design Optimization of the Primary Pump of a Nuclear Reactor

T. Verstraete and L. Mueller

Abstract Engineers are often challenged by designing new equipment without any prior knowledge or guidance from an existing similar product. The large degree of freedom that this generates can become a bottleneck as it could lead to a loss of global oversight and may even lead to wrong, uninformed choices. It is essential to have a large exploration of the design space to allow for innovative solutions, on the other hand it is important to introduce a high level of detail as early as possible in the design process to increase the reliability of the model predictions, which drive the decision process. This leads to a well-known conflict where more knowledge is needed upfront in the design process in the early stages of the design, and a larger degree of freedom is needed near the end of the design process where typically more knowledge is available. In this work it is demonstrated how modern design optimization tools can be effectively used to integrate the preliminary with the detailed design process. The key to achieve a good balance between design exploration and detailed design is obtained by reducing the parameters that are fixed during the preliminary design to an absolute minimum, such that the detailed design phase has still a large degree of freedom. The parameters that are fixed in the preliminary design phase are moreover those parameters that have a pronounced influence on the design performance and can be reliably predicted by a lower detail analysis code. Both preliminary and detailed design processes rely heavily on optimization techniques. Due to the larger computational cost in the detailed design phase, a surrogate model based optimization is used opposed to an evolutionary algorithm in the preliminary design phase. The application within this paper is the design of a liquid-metal pump for the primary cooling system of the advanced nuclear reactor MYRRHA conceived by the Belgian research center (SCK·CEN). This single stage axial-flow pump has unique design requirements not met by any previously designed pump, and hence demands for a novel approach.

T. Verstraete (✉) · L. Mueller
Turbomachinery and Propulsion Department, von Karman Institute for Fluid Dynamics, Waterloosesteenweg 72, B-1640 Sint-Genesius-Rode, Belgium
e-mail: tom.verstraete@vki.ac.be

L. Mueller
e-mail: lasse.mueller@vki.ac.be

D. Greiner et al. (eds.), *Advances in Evolutionary and Deterministic Methods for Design, Optimization and Control in Engineering and Sciences*, Computational Methods in Applied Sciences 36, DOI 10.1007/978-3-319-11541-2_11

Keywords Turbomachinery · Design optimization · Computational fluid dynamics

11.1 Introduction

The MYRRHA project, initiated by the Belgian nuclear research center (SCK·CEN), aimes at demonstrating a new generation of nuclear reactors. Indeed, fourth generation fast reactors show considerable improvements in fuel efficiency, safety and nuclear waste generation compared to the current installed reactors. The primary coolant is either sodium or lead/lead-bismuth eutectic (LBE). The use of sodium has more safety hazards as it reacts explosively with water and ignites in air, and has a lower boiling point. In contrast, the main problem with LBE is related to erosion.

The MYRRHA reactor is conceived as an accelerator driven system (ADS), able to operate in sub-critical and critical modes [7] and will allow the demonstration and performance assessment of the transmutation concept and associated technologies starting in 2023.

The primary system of the MYRRHA research reactor is a pool-type design, as illustrated in Fig. 11.1. All components of the primary loop, i.e. the pumps, heat exchangers, fuel handling tools, experimental rigs, etc., are inserted from the top and immersed in the reactor vessel, which is filled with lead-bismuth eutectic (LBE) as primary coolant. The relatively high boiling temperature of LBE of 1,670 °C leads

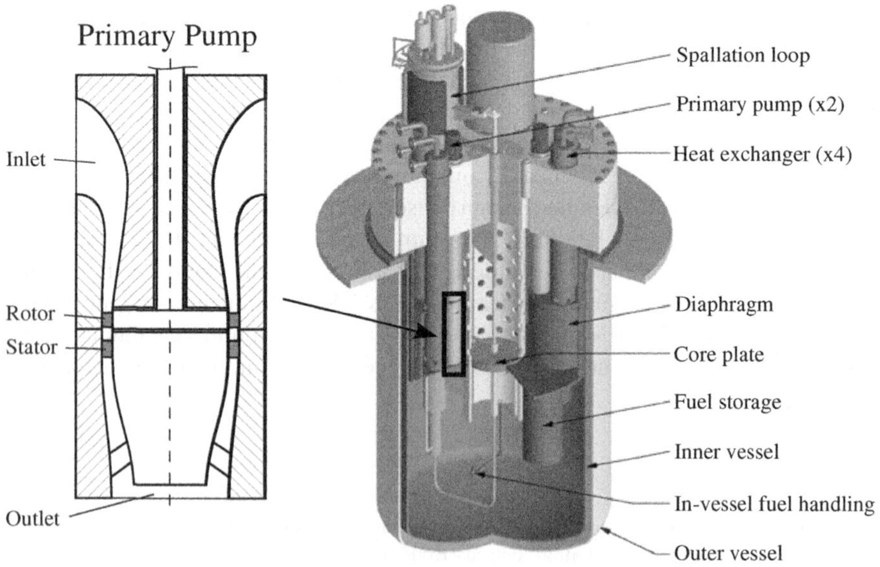

Fig. 11.1 Schematic assembly of the MYRRHA reactor [7] with a close up of the primary pump (dimensions not to scale)

to a passively safe design regarding a loss of coolant accident (LOCA), as it allows the operation without pressurizing the reactor even at high temperatures. However, the high density of 10,000 kg/m^3 and the corrosive properties of LBE add additional challenges to the design, which regard mainly a restriction in flow velocity. Lower flow velocities will for instance reduce the convective heat transfer, hence resulting in large heat exchangers. But more significantly, low flow velocities have a major impact on the pump design reducing dramatically the realizable head. The maximum relative velocity in the pump needs indeed to be limited to rather low values to avoid exessive blade erosion, e.g. it is reported that pumps developed by OKBM [6] for nuclear submarines operate at maximum velocities of $W_{max} = 25–30$ m/s.

11.2 Design Procedure

Reducing the maximum velocity to limit erosion is an essential and unique objective to be addressed in the design of the primary pump. Up to this date, such pumps have only been designed for Soviet Alfa class nuclear submarines during the 1970s and no detailed reports on the design of such pumps exists. Hence one cannot rely on any previous experience for the design, necessitating a thorough study of how the particular requirements can be met.

As a design process evolves over time, more and more design parameters are fixed limiting the remaining degrees of freedom. However, knowledge on the design is gathered over the course of the design process and builds up mainly after the preliminary design stage, as shown graphically in Fig. 11.2. This means that important decisions have to be taken at early stages in the design process where only limited information is available, which may lead to poorly informed decisions. It is therefore advisable to limit as much as possible the decisions made early in the design process while gathering as much knowledge as possible.

In this work it was therefore intended to limit the degree of freedom as much as possible in the early design phase. To this end, the design process was split in two sequential phases, each relying on modern optimization techniques. The first stage consisted of a preliminary design phase where the type of pump and global dimensions were fixed. The following phase consisted in a detailed design optimization of the pump shape, keeping only few parameters fixed as defined from the preliminary design phase. Such approach allows to prevent premature conclusions to be drawn in the early design phase where only limited models are used, but on the other hand still allows for a reasonably manageable optimization problem in the detailed design phase where the main interactions between the different design parameters is limited. Indeed, it is well known that the complexity of optimization problems increases the higher the degree of interaction is between the different design variables. Within this work, the most interfering design parameters have been fixed during the preliminary analysis, which is possible since they can be well represented by lower accuracy analysis tools.

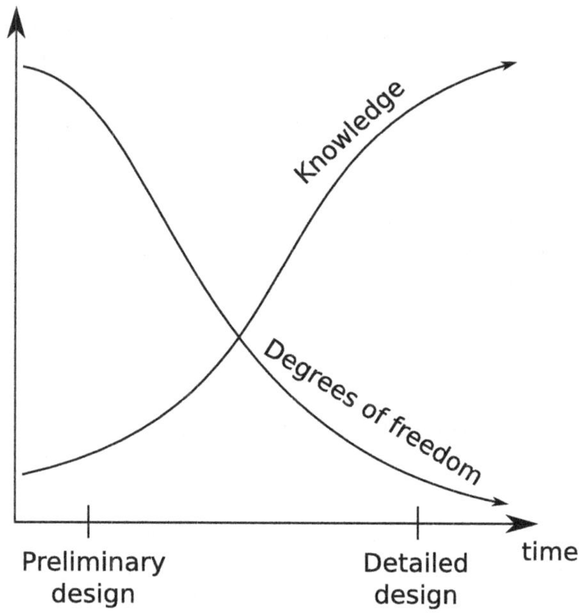

Fig. 11.2 Degrees of freedom and acquisition of knowledge during design

A fully detailed design optimization from the start, without any preliminary design phase, is practically impossible. The large degree of freedom, combined with high order interactions between several parameters, lead to a very complicated design space that requires a high level of evaluations to accurately cover the myriad of design options.

11.3 Preliminary Design Optimization

11.3.1 Machine Type

The preliminary design phase starts by the selection of the machine type. For pumps a wide choice of different machine types exists. In the present work volumetric (or positive displacement) pumps are ruled out for maintenance and other reasons. The most attractive pump type is a rotodynamic pump which still has a large choice of machine type (axial flow, mixed flow, or radial flow). This choice is mainly related to the design requirements of the pump, i.e. to the flow rate Q and the total head ΔH. A classical parameter in the selection of hydraulic machinery is the specific speed

$$N_s = \frac{RPM\sqrt{Q}}{\Delta H^{3/4}} \,. \tag{11.1}$$

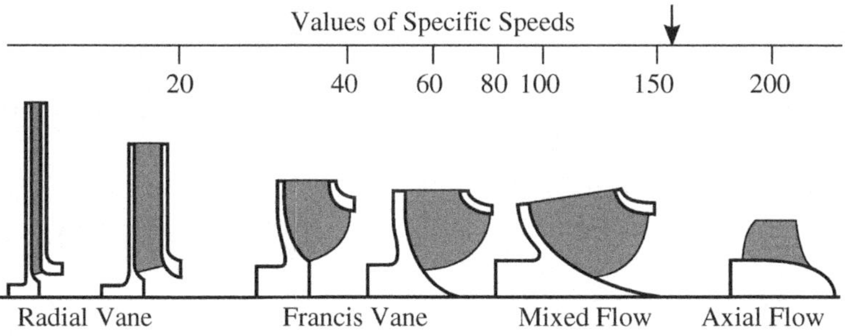

Fig. 11.3 Specific speed chart

This specific speed is used as a guideline for the selection of the machine type based on a large database of existing designs. It suggests a type of machine that should deliver the highest efficiency for the required specific speed. This should be viewed as a strong suggestion, however one may deviate from the suggested type with the risk of having a lower efficiency.

Based on the design requirements of the MYRRHA pump provided by SCK·CEN the specific speed is in the range of mixed to axial-flow pumps $N_s \approx 156$ (see Fig. 11.3). Although towards lower RPM a mixed-flow configuration might provide higher efficiency, an axial-flow pump is regarded as the best trade-off solution due its lower mechanical complexity and simplified manufacturing.

The use of charts as Fig. 11.3 is very helpful in the preliminary design phase, as it gives a clear indication for the choice of machinery in terms of global performance parameters. Such charts do however not guarantee the identification of a global optimum and provide no detailed shape. Moreover, no chart exists for the erosion induced requirement of low relative velocity with respect to the walls.

With respect to this requirement, it was decided to analyze various design options in this preliminary design phase. For that reason a simple one-dimensional model has been applied in the hub, mid, and tip section of the rotor to assess the meridional flow path and its rotational speed (Fig. 11.4, left). This rather simple model based on velocity triangles up- and downstream of the rotor blade allows a rapid screening of several design parameters, such as blade angles, rotational speed (RPM), and hub and tip radii, R_{hub} and R_{tip}, respectively. Although it does not include complex flow features, e.g. flow deviation or even flow separation, it provides a good estimation of the initial rotor design as an input for the following detailed 3D-optimization.

Input parameters to the 3D-optimization are the hub and tip radii (R_{hub} and R_{tip}) and the RPM of the rotor as illustrated in Fig. 11.4 (right). These parameters were unchanged in the subsequent detailed high fidelity 3D-optimization, in which the three dimensional design of the pump comprising the inlet section, the rotor and stator rows, and the diffuser has been performed. It is important to note that these three design variables have a large effect on all other design parameters for the detailed design process. Flow angles will change for instance with different rotational speeds, as well as with higher or lower hub and tip radii. To account for such changes, a much

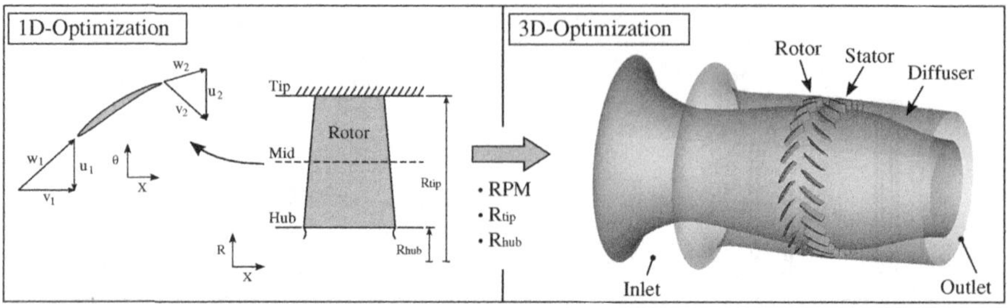

Fig. 11.4 Design procedure

wider range for the blade metal angles will be needed, which is increasing the design space. Additionally, the optimal value for blade angle will depend largely on these three parameters, and hence will require a much larger number of design evaluations with the used methodology to find the optimal value. This can be easily understood as follows: Suppose a sampling technique is used to find the optimum, i.e. the design space is probed in discrete points. When two design variables have no interaction, it is sufficient to put experiments on the diagonal of the design space spanning these two design variables to find the optimum. However, if a strong interaction exists, only sampling the diagonal is not sufficient as it would not allow to see the effect of changing one design variable while keeping the other constant (and necessary if the optimum is in an opposite corner of the diagonal). As such, the entire 2D space spanned by both design variables will need to be sampled. It can be reasonably assumed that most parameters in a 3D design of the axial pump, as will be shown later, have only a low level of interaction with other parameters, and hence result in a relatively easy optimization process. This situation would dramatically change if hub and tip radii and RPM would be added as design variables.

11.3.2 Computational Model

The 1D computational model is illustrated in Fig. 11.5. Input parameters are the hub and tip radii, R_{hub} and R_{tip}, and the RPM, next to the pump requirements (ΔH and Q). The axial velocity V_x is derived from the hub and tip radius for the specified flow rate Q. It is assumed at this stage that the axial velocity upstream of the rotor is constant from hub to tip although the pump will have a radial inlet bend (see Fig. 11.1).

The peripheral speed U is computed at three sections (hub, mid and tip) from the rotational speed and the respective radii. The mid section is positioned in the middle of the hub and tip radius: $R_{mid} = 0.5 \cdot (R_{hub} + R_{tip})$.

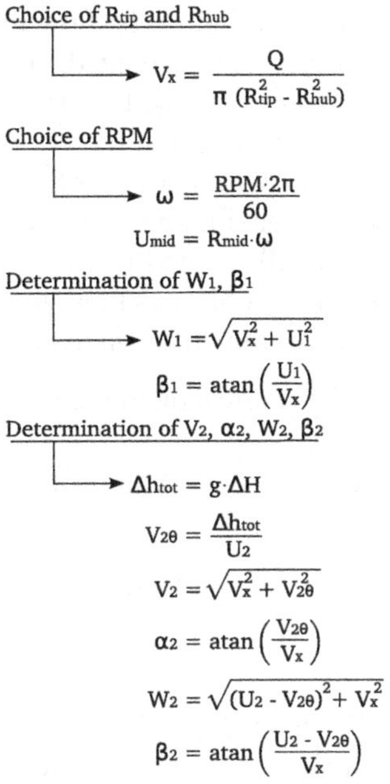

Fig. 11.5 1D computational model

The inlet absolute velocity V_1 is assumed to be axial (no pre-rotation), which allows to compute the relative velocity W_1 and the relative flow angle β_1.

The quantities downstream of the rotor (index 2) are computed based on the required total head ΔH and assuming no losses ($\eta_{hyd} = 1.0$) using the Euler-equation for pure axial inlet flow ($V_{1\theta} = 0$)

$$\frac{g \cdot \Delta H}{\eta_{hyd}} = U_2 V_{2\theta} - U_1 V_{1\theta} = U_2 V_{2\theta} \,. \tag{11.2}$$

and assuming the axial velocity V_x to maintain constant through the rotor (i.e. free vortex design). The absolute velocity downstream of the rotor V_2, the absolute flow angle α_2, the relative velocity W_2, and the relative flow angle β_2 are computed as illustrated in Fig. 11.5.

11.3.3 Objectives and Constraints

Three design parameters (hub radius R_{hub}, tip radius R_{tip}, and RPM) can be chosen to obtain the required total head ΔH and flow rate Q. However, additional requirements need to be imposed:

- The turning of the flow in the rotor ($\beta_2 - \beta_1$) needs to be limited to reduce the losses, not accounted for in this preliminary design phase. According to [9] the turning is limited to 30–25°.
- The diffusion in the rotor (W_2/W_1) needs to remain feasible. A limit of 0.72 is often used, known as the de Haller number [8].
- The maximum relative velocity in the rotor W_{max} needs to remain at low values (10–20 m/s) to limit erosion.
- The absolute exit flow angle (α_2) needs to be small (from axial direction) to perform a feasible diffusion in the subsequent stator [9].

The corresponding optimization problem is formulated as:

$$\text{Minimize:} \quad Obj_1 = W_1^{tip} \tag{11.3}$$

$$Obj_2 = -W_2^{hub}/W_1^{hub} \tag{11.4}$$

$$\text{Subject to:} \quad Constr_1 = abs(\beta_2^{hub} - \beta_1^{hub}) \leq \Delta\beta_{max} \tag{11.5}$$

$$Constr_2 = V_x \leq V_{x,max} \tag{11.6}$$

The first objective Obj_1 (Eq. 11.3) reduces the maximum relative velocity in the rotor, which is in the tip section at the inlet. A value in the range of $W_1^{tip} = 10$–20 m/s is considered as feasible, although lower values are preferred, as they would increase the lifetime with respect to erosion. The second objective Obj_2 (Eq. 11.4) maximizes the diffusion ratio near the hub to prevent an excessive diffusion. For a free vortex design the highest diffusion occurs in the hub section. The first constraint (Eq. 11.5) restricts the turning at the hub, where the largest relative flow turning will take place, while the second constraint (Eq. 11.6) puts an additional limitation on the axial velocity to limit the erosion risk in the meridional passage.

11.3.4 Methodology

Three different design variables are allowed to be changed during the preliminary design phase, i.e. the hub and tip radii (R_{hub} and R_{tip}) and the rotational speed RPM. Each design variable is allowed to change within a specified range. For each choice of the hub radius R_{hub}, tip radius R_{tip}, and RPM, the velocity triangles can be computed by the 1D model and the requirements can be evaluated. In order to find the three design parameters that give a suitable compromise for the diffusion W_2/W_1, turning $\Delta\beta$, and maximum velocity W_{max} requirements, a Differential Evolutionary (DE) algorithm [10] has been used. Figure 11.6 illustrates schematically the optimization flowchart.

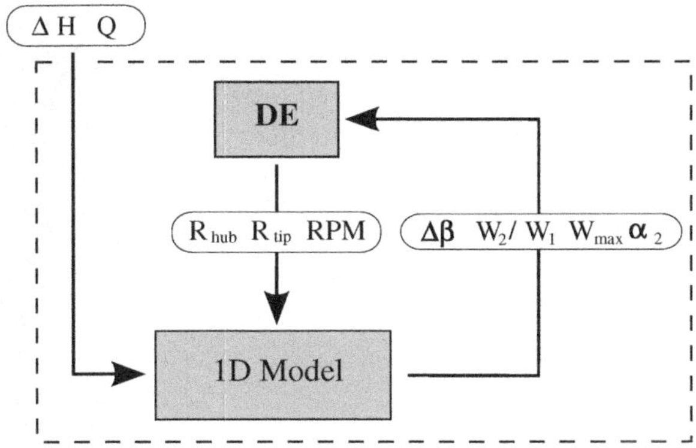

Fig. 11.6 1D optimization flowchart

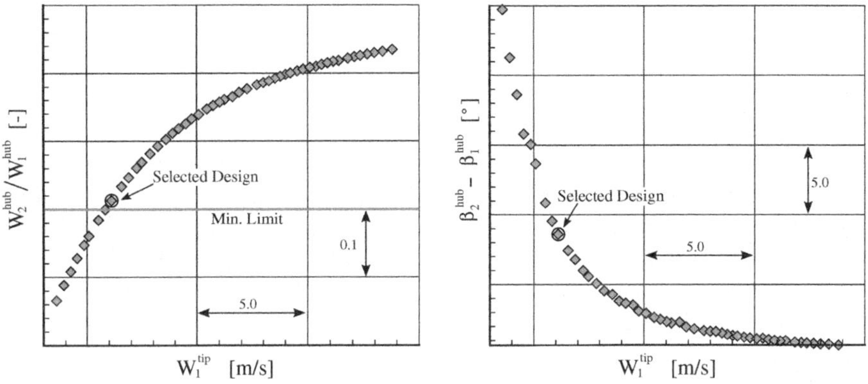

Fig. 11.7 Diffusion in the hub versus relative velocity in the tip section (*left*), turning in the hub section versus relative velocity in the tip section (*right*)

11.3.5 Results

The results of the 1D optimization are illustrated in Fig. 11.7, showing both the diffusion in the hub section W_2^{hub}/W_1^{hub} (left) and the flow turning $\beta_2^{hub} - \beta_2^{hub}$ (right) with respect to the relative velocity in the rotor tip section W_1^{tip}. A Pareto front is found with non-dominated designs, in which one objective cannot be improved without worsening the other.

In Fig. 11.7 (left), a clear Pareto front (towards the upper left hand corner) is visible indicating that a lower maximum velocity W_1^{tip} comes at the expense of a larger diffusion near the hub (lower W_2^{hub}/W_1^{hub}). This evidently results also in a higher flow turning near the hub, as can be seen in Fig. 11.7 (right). The resulting Pareto front allows the designer to select designs depending on the weight given to each objective. In the present case, it was decided to select the design with a hub diffusion ratio (W_2^{hub}/W_1^{hub}) close to the minimum limit as indicated in Fig. 11.7 (left).

Fig. 11.8 NASA axial pump rotor for liquid rocket application (hub-to-tip ratio $\nu = 0.9$) [11]

This design has a specific speed of $N_s \approx 70$ with a high hub-to-tip ratio of $\nu = R_{hub}/R_{tip} = 0.88$, which is larger than commonly used for axial (propeller-like) pumps in the range of $\nu = 0.3$–0.7 [9]. However, this is due to the very low velocity requirement (i.e. low V_x and W_1^{tip}) and high diffusion (i.e. low W_2^{hub}/W_1^{hub}) for a single stage pump. Similar conclusions were drawn within a design study of axial pump rotors for liquid rocket application [1, 3, 11]. Several rotors (see Fig. 11.8) were designed and tested with hub-to-tip ratios between $\nu = 0.4$–0.9 for a specified diffusion, which confirms the results obtained from this preliminary design optimization.

11.4 Detailed Design Optimization

11.4.1 Methodology

The 3D design of the primary pump is performed with the optimization algorithm developed at the von Karman Institute (VKI) with special focus on turbomachinery applications. The system (Fig. 11.9) makes use of a Differential Evolution algorithm (DE), a metamodel based on an Artificial Neural Network (ANN), a database, and high fidelity simulation tools for the flow analysis (CFD).

The basic approach of this method is that the Artificial Neural Network substitutes the computational expensive tools for the CFD in the *Generation Loop* (see Fig. 11.9) and provides less accurate but very fast performance predictions to evaluate the large number of geometries necessary by the DE during its search for the optimum. However, the metamodel requires a validation which is then performed in the *Iteration Loop* according to Fig. 11.9. After a specified number of generations, the optimum

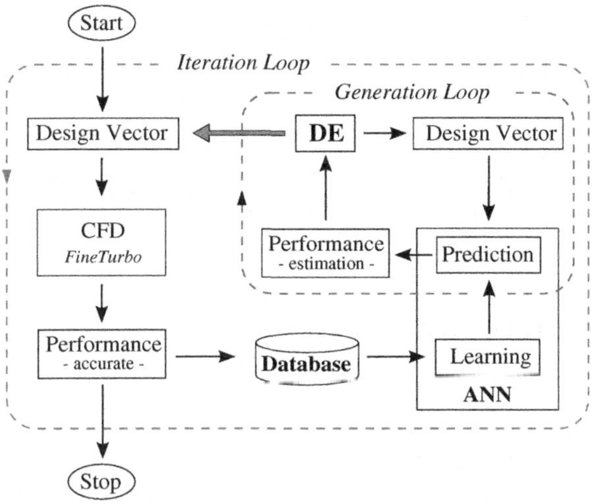

Fig. 11.9 VKI optimization algorithm

geometries according to the ANN predictions are analyzed by the more accurate but much more computationally expensive CFD calculations to verify the accuracy of the metamodel. The results of the accurate performance analysis are added to the database and a new *Generation Loop* is started after a new training of the metamodel on the enlarged database. In this way the whole system is self-learning, resulting in a more accurate ANN.

The Differential Evolution algorithm used was developed by Storn and Price [10]. In present optimization, 1,000 generations are created with a constant population size of 40 individuals. To validate the ANN predictions in each *Iteration Loop* eight individuals were selected and reassessed by the high fidelity tools.

The initial sampling of the database was performed by means of a Design of Experiments (DOE) [4]. The DOE is based on statistical methods and considers, that k design variables can take two values fixed at a specified position in the design space (here: 20 and 80 % plus one central case = 50 %). Further, to reduce the number of required evaluations, the fractional factorial design approach [4] is used: $2^{(k-p)}$ with $(k - p) = 6$, thus resulting in 64 experiments plus the central case, which are the initial sampling of the database. This clearly illustrates that this method is only working properly if a low level of interactions between the different parameters is present. This justifies the choice of keeping the R_{hub}, R_{tip} and RPM parameters constant after the preliminary analysis.

11.4.2 Parametrization

The 3D model of the rotor of the pump (cf. Fig. 11.1) is based on Bézier and B-spline curves and surfaces and is defined by:

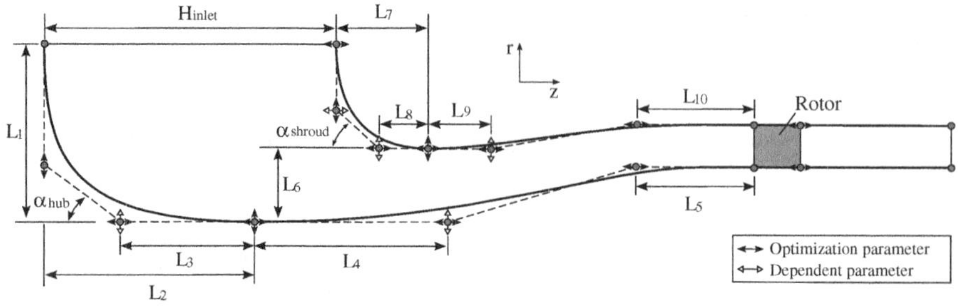

Fig. 11.10 Meridional parametrization of the inlet section and the rotor

1. the meridional contour,
2. the blade camber line at hub and tip,
3. the thickness distribution, which is added normal to the camber line at hub and tip, and
4. the number of blades.

The definition of the meridional contour is shown in Fig. 11.10. The meridional flow path is subdivided into different patches: an inlet patch (90°-bend and swan neck), a blade patch, and an outlet patch. The blade patch corresponds to where the rotor blade is located in the meridional plane. The coordinates of the control points are the geometrical parameters which can be modified by the optimization program and the possible variation in axial and radial direction is indicated by arrows. In order to have a good control about the geometry and to avoid undesirable shapes, lengths and angles are used (e.g. L_1 to L_{10}, H_{inlet}, α_{hub} and α_{shroud}). In total 15 parameters define the meridional flow path.

The rotor blade is defined by the blade camber line of the hub and tip section, each separately defined by the blade angle β_m-distribution with respect to the meridional plane. The β_m-distribution is parametrized by a Bézier curve with three control points, one at the leading and trailing edge and one intermediate control point. This results in a total of 6 degrees of freedom.

The final blade with the pressure and suction side surfaces is created by adding a thickness distribution normal to the camber line at hub and tip. In the present optimization study, a NACA 65 thickness distribution at hub and tip has been chosen, which shape is kept constant during the optimization process.

The number of blades, which defines the full rotor, is introduced as an optimization parameter within this design study. This parameter has a major effect on solidity, i.e. on the blade loading and on the blockage of the flow in case of accidents with no pump rotation, in which cooling needs to be guaranteed by natural convection in the reactor.

In total 22 parameters define the geometry of the pump and will be subject to optimization.

11.4.3 Flow Analysis

Every geometry in the *Iteration Loop* according to Fig. 11.9 has been analyzed at the design operating point using the commercial flow solver fineTurbo™. The incompressible Reynolds-Averaged Navier-Stokes (RANS) equations are solved using a Runge-Kutta scheme in conjunction with accelerating techniques such as variable-coefficient implicit residual smoothing and a multi-grid scheme. Discretization is based on finite volumes with a cell-centered scheme stabilized by artificial dissipation. For the turbulence closure the one equation model of Spalart and Allmaras [12] is used with the assumption of fully turbulent flow with an inlet Reynolds number of

$$Re_{inlet} = \frac{V_{inlet} R_{inlet} \rho}{\mu} = 6.5 \cdot 10^5 \tag{11.7}$$

where V_{inlet} and R_{inlet} are the absolute inlet velocity and the inlet radius, respectively. The flow properties of lead-bismuth eutectic (LBE) according to [5] have been used for the CFD simulations.

11.4.4 Objectives and Constraints

The optimization of the inlet section and the rotor has two objectives:

1. Maximizing the hydraulic efficiency

$$Obj_1 = \eta_{hyd} = \frac{\Delta P_{tot}}{\rho \cdot \Delta(V_\theta U)} \tag{11.8}$$

where ρ is the density of LBE, ΔP_{tot} the mass-flow averaged absolute total pressure rise, and $\Delta(V_\theta U)$ the difference of the mass flow averaged angular momentum between inlet and outlet.

2. Reducing the maximum isentropic velocity $(W/W_0)_{max,SS}$ on the blade suction side at 90 % span. The isentropic velocity objective aims at reducing the maximum relative velocity in the tip section (90 %-span) as shown in Fig. 11.11 and is computed as follows:

$$Obj_2 = (W/W_0)_{max,SS} = \sqrt{1 - C_p|_{min,SS}} \tag{11.9}$$

with the pressure coefficient C_p:

$$C_P = \frac{P - P_{ref}}{0.5 \cdot \rho \cdot W_{ref}^2} \tag{11.10}$$

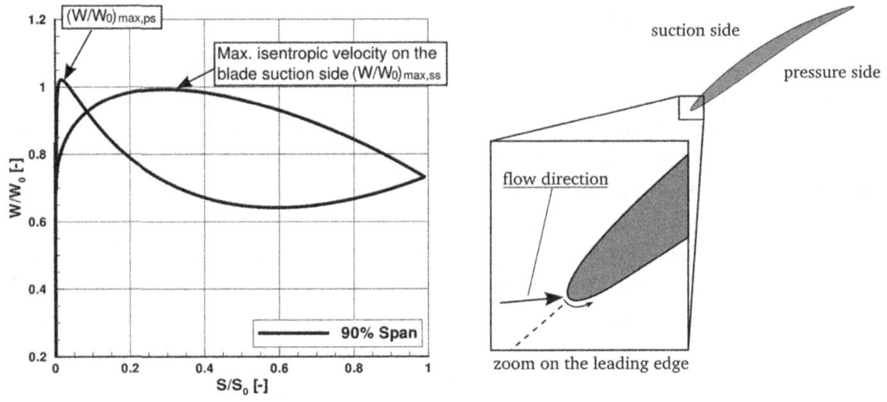

Fig. 11.11 Illustration of negative incidence resulting in a velocity peak on the pressure side

The two objectives are subject to two constraints:

$$Constr_1 = \Delta H \geq \Delta H_{min} \tag{11.11}$$

and

$$Constr_2 = \left[(W/W_0)_{max,ps} - (W/W_0)_{max,ss}\right] \leq 0\,. \tag{11.12}$$

The first constraint (Eq. 11.11) ensures that the designs generated by the optimization program supply the required total head for the design flow rate. The second constraint (Eq. 11.12) prevents that the isentropic velocity objective Obj_2 is not improved at the expense of a higher velocity on the pressure side at the leading edge due to negative incidence, i.e. if the flow impinges on the suction side, which results in a velocity peak on the pressure side (Fig. 11.11).

11.4.5 Results

The results of the optimization are presented in Fig. 11.12, showing both the total head ΔH (Fig. 11.12, left) and the maximum isentropic velocity on the suction side $(W/W_0)_{max,SS}$ (Fig. 11.12, right) with respect to the hydraulic efficiency η_{hyd}. Each symbol in Fig. 11.12 represents one design that has been analyzed by CFD. The square symbols represent the geometries analyzed for the initial database (DOE) prior to the optimization to train the metamodel. The designs generated during the optimization process are the diamond shape symbols. From the initial scattered distribution of the DOE (Fig. 11.12, left), the optimizer generated a large number of designs with high hydraulic efficiency η_{hyd} and minimum required total head (indicated with the horizontal line in Fig. 11.12, left). The selected design of this optimization

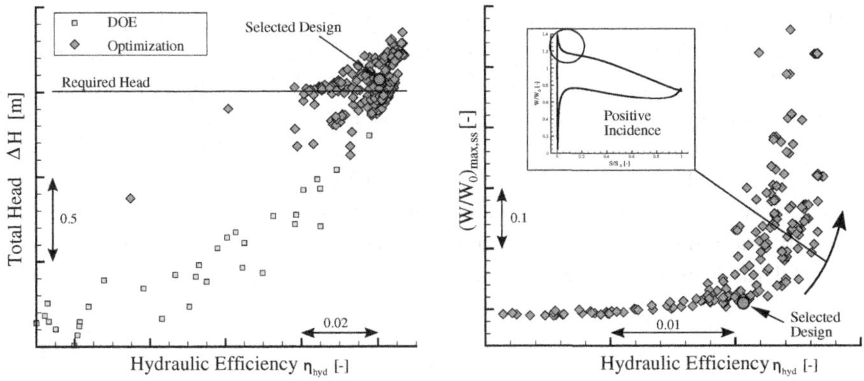

Fig. 11.12 Total Head versus hydraulic efficiency (*left*) 2D objective space: $(W/W_0)_{max,SS}$ versus hydraulic efficiency (*right*)

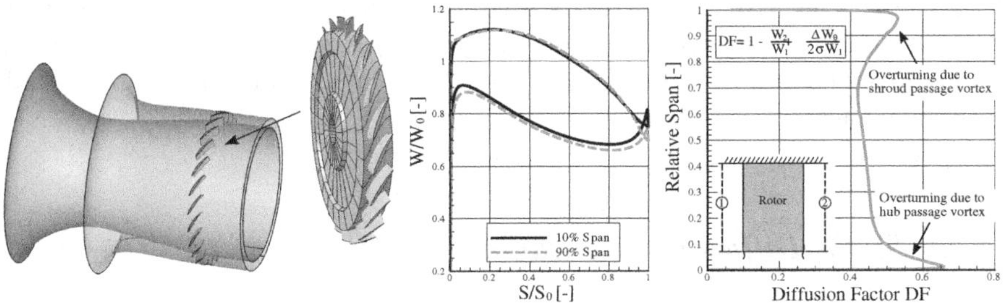

Fig. 11.13 Selected design indicated in Fig. 11.12 with a close up on the rotor (*left*), isentropic velocity distribution at 10 and 90 % span (*middle*), span-wise distribution of rotor diffusion factor (*right*)

supplies a total Head, which is above the minimum required value, with high hydraulic efficiency. The reason why this particular design has been chosen is more obvious from Fig. 11.12 right, showing the two-dimensional objective space, i.e. the maximum isentropic velocity on the suction side $(W/W_0)_{max,SS}$ with respect to the hydraulic efficiency η_{hyd}.

In this plot only designs which satisfy the constraints according to Eqs. (11.11) and (11.12) are presented such that the apparent front of designs towards the lower right hand corner is the Pareto front comprising non dominated designs, i.e. no other designs outperform the Pareto optimal designs with respect to both higher efficiency and lower relative velocity. Higher hydraulic efficiency comes at the expense of higher relative velocity due to positive incidence resulting in a velocity peak on the suction side leading edge. In this optimization reducing the isentropic velocity on the suction side, i.e. the relative velocity to limit erosion had a higher priority than improving the efficiency, resulting in the selected design, which is considered as the best trade-off solution of both objectives.

The selected design, which is presented in Fig. 11.13 left, is a high staggered, low aspect ratio rotor ($h/c_{tip} \approx 0.36$) with a solidity in the tip section of $(c/p)_{tip} \approx 1.26$. It is a mid-loaded blade with an equal loading in the hub and tip section and with a smooth diffusion on the suction side as illustrated in Fig. 11.13 middle, showing the isentropic velocity distribution at 10 and 90 % span. Additionally, as illustrated already in the 2-dimensional objective space in Fig. 11.12 right, this design has a rather low velocity on the suction side which limits the erosion risk. Figure 11.13 (right) illustrates the mass-flow averaged span-wise distribution of the rotor Diffusion Factor (DF) at design operating conditions. Although the Diffusion Factor introduced by [2]

$$DF = 1 - \frac{W_2}{W_1} + \frac{\Delta W_\theta}{2\sigma W_1} \tag{11.13}$$

using the relative velocities up- and downstream of the rotor and the solidity σ is strictly valid for two-dimensional flows, it is a suitable parameter to estimate off-design tendencies. Except close to the side walls where the flow is overturned due to secondary flows (i.e. due to the hub and shroud passage vortices), the computed rotor Diffusion Factor is below DF ≤ 0.5, which indicates a sufficient margin towards lower flow rates.

11.4.6 Stator-Diffuser Optimization

Subsequently to the inlet-rotor design optimization, the stator-diffuser has been designed and optimized. A similar parametrization as for the inlet-rotor optimization is used with 18 degrees of freedom, comprising the meridional flow path and blade shape. The design of the rotor is kept constant during this process, but the inlet and rotor are modeled in the CFD during this design effort. The objectives of the optimization are to diffuse as much as possible inside the diffuser with limited losses, while reducing as much as possible the velocity peak on the suction side of the stator vane to limit erosion. The outcome of this optimization is illustrated in Fig. 11.14, showing the meridional velocity with a span-wise distribution upstream of the rotor and the isentropic velocity distribution at 10 and 90 % span of the stator vane.

11.4.7 Performance Map

The selected design (Fig. 11.14) has been analyzed further regarding its off-design performance. Although the pump was designed primarily for the design point, it has good off-design characteristics, which is illustrated in Fig. 11.15, showing the Head coefficient

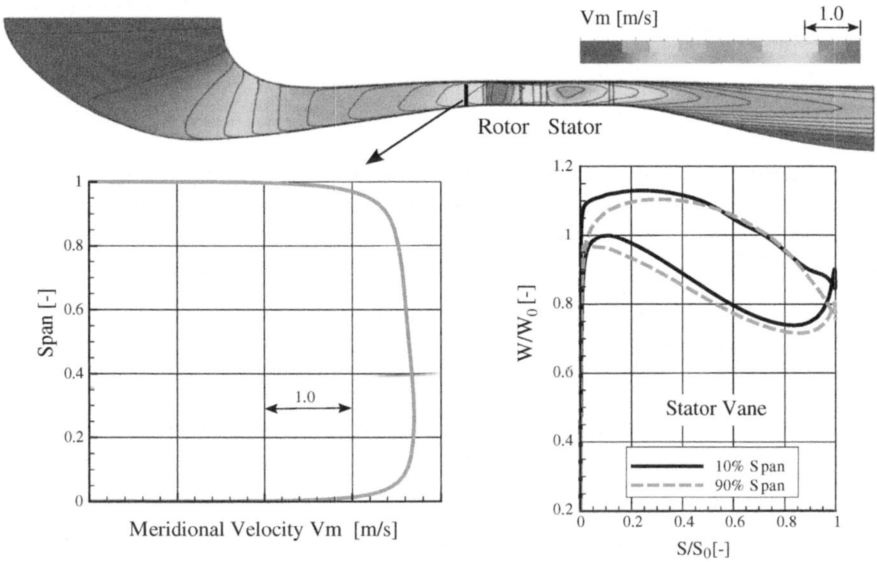

Fig. 11.14 Meridional velocity in the final pump design and isentropic velocity distribution at 10 and 90 % span of the stator vane

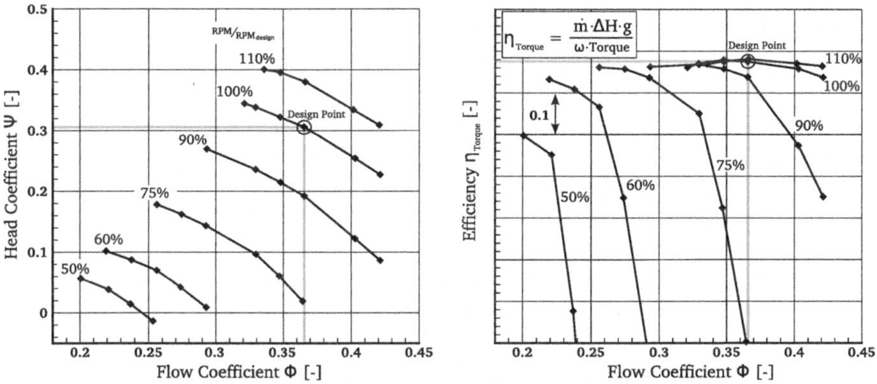

Fig. 11.15 Performance map of the pump

$$\psi = \frac{g \,.\, \Delta H}{U_{tip}^2} \tag{11.14}$$

and the efficiency based on the torque

$$\eta_{Torque} = \frac{\dot{m} \Delta H g}{\omega \cdot Torque} \tag{11.15}$$

with respect to the flow coefficient

$$\phi = \frac{V_x}{U_{tip}}. \tag{11.16}$$

The reason for the good off-design behavior is related to the chosen design strategy. The optimization was aiming at limiting the diffusion in the rotor and stator and reducing the incidence of both blade rows resulting in a robust design with an entirely sufficient negative slope ($\equiv \Delta\psi/\Delta\phi$) of the characteristics curve.

11.5 Conclusions

A hydrodynamic optimization based on evolutionary methods is used to design the primary pump of the MYRRHA nuclear reactor. The design approach is aimed at

- maximizing the hydraulic efficiency of the rotor,
- minimizing the pressure losses in the stator and diffuser, and
- reducing the relative velocity in the rotor tip section.

Based on the preliminary 1D optimization, the rotor and stator have been designed in two successive steps keeping the number of design parameters and objectives to feasible values in each optimization run. The outcome is a pump, which supplies the required Head and is respecting the design target of low relative velocity to limit erosion.

The pump has been realized as a moderate specific speed, high hub-to-tip ratio axial pump ($N_s = 70$, $R_{hub}/R_{tip} = 0.88$), although a mixed-flow pump would provide higher hydraulic efficiency according to classical specific speed charts. However, due to the additional complexity of a mixed flow configuration in the manufacturing process, an axial configuration is more attractive.

Although the pump was designed primarily for the design operating point, it has a good off-design behavior, which is the result of the chosen design strategy. The optimization was aiming at limiting the diffusion in the rotor and stator and reducing the incidence of both blade rows resulting in a robust design.

Finally, the use of a two phase approach consisting of a preliminary design phase and a detailed 3D design optimization allows for a strong reduction of the degree of interactions of the design variables, which significantly reduces the design optimization problem at hand. The use of a preliminary search with optimization techniques on a reduced model allows a rapid selection of optimal design variables that are well predicted by the reduced model. The authors advise that a well-thought use of physical models in the early stages of the design process is a prerequisite for a successful detailed design optimization.

References

1. Crouse JE, Sondercock DM (1964) Blade-element performance of 0.7 hub-tip radius ratio axial-flow pump rotor with tip diffusion factor of 0.43. In: NASA TN D-2481, Lewis Resarch Center, Cleveland Ohio, 1954
2. Lieblein S, Schwenk FC, Broderick RL (1953) Diffusion factor for estimating losses and limiting blade loadings in axial-flow-compressor blade elements. In: NACA RM E53D01, Lewis Flight Propulsion Laboratory, Cleveland Ohio, 1953
3. Miller MJ, Crouse JE (1965) Design and overall performance of an axial-flow pump rotor with a blade-tip diffusion of 0.66. In: NASA TN D-3024, Lewis Resarch Center, Cleveland Ohio, 1965
4. Montgomery D (2006) Design and analysis of experiments. Wiley, New York
5. NEA: Handbook on lead-bismuth eutectic alloy and lead properties, materials compatibility, thermal-hydraulics and technologies. Nuclear energy agency. http://www.oecd-nea.org/science/reports/2007/nea6195-handbook.html. 2008
6. OKBM: Experimental designing bureau of machine building. http://www.okbm.nnov.ru/ 5 Dec 2013
7. SCKCEN: MYRRHA: Multi-purpose hybrid research reactor for high-tech applications. http://myrrha.sckcen.be/ 5 Dec 2013
8. Saravanamuttoo H, Rogers GFC, Cohen H, Straznicky PV (2001) Gas turbine theory. Pearson Prentice Hall, Harlow
9. Stepanoff AJ (1957) Centrifugal and axial flow pumps. Wiley, New York
10. Storn R, Price K (1997) Differential evolution - a simple and efficient heuristic for global optimization over continuous spaces. J Global Optim 11:341–359
11. Urasek DC (1971) Design and performance of a 0.9 hub-tip-ratio axial-flow pump rotor with a blade-tip diffusion factor of 0.63. In: NASA TM X-2235, Lewis Resarch Center, Cleveland Ohio 1971
12. Wilcox DC (1993) Turbulence modelling for CFD. DCW Industries Inc, California

Chapter 12
Direct 3D Aerodynamic Optimization of Turbine Blades with GPU-Accelerated CFD

Philipp Amtsfeld, Dieter Bestle and Marcus Meyer

Abstract Secondary flow features of turbine blade flows are only assessable by 3D computational fluid dynamics (CFD) which is a time-consuming task. In this paper a fast automatic optimization process for the aerodynamic improvement of three-dimensional turbine blades is described and applied to a two-stage turbine rig. Basically, standard tools are used where the 3D CFD analysis, however, is significantly accelerated by a novel CFD solver running on graphics processing units (GPU) and the entire blade is parameterized in 3D. This approach shows that three-dimensional optimization of turbine blades is feasible within days of runtime and finds an improved blade design.

Keywords Turbine blade design · GPU flow solver · Shape optimization · Aerodynamic optimization · Parameter reduction

12.1 Introduction

Although turbomachinery aerodynamics has been already extensively improved over recent years, further optimization is still desirable to reduce emissions and fuel consumption. Typically, the aerodynamic design process of turbine blades is split into two phases: optimal design of 2D blade sections and then stacking them optimally along a three-dimensional stacking line. This separation of section design and stacking

P. Amtsfeld (✉) · D. Bestle
Engineering Mechanics and Vehicle Dynamics, Brandenburg University of Technology, Siemens-Halske-Ring 14, 03046 Cottbus, Germany
e-mail: philipp.amtsfeld@tu-cottbus.de

D. Bestle
e-mail: bestle@tu-cottbus.de

M. Meyer
Rolls-Royce Deutschland Ltd & Co KG, Eschenweg 11, 15827 Blankenfelde-Mahlow, Germany
e-mail: marcus.meyer@rolls-royce.com

D. Greiner et al. (eds.), *Advances in Evolutionary and Deterministic Methods for Design, Optimization and Control in Engineering and Sciences*, Computational Methods in Applied Sciences 36, DOI 10.1007/978-3-319-11541-2_12

is due to high computational costs of 3D flow simulation. Splitting the two phases reduces the number of design parameters and eases the design tasks. However, some loss mechanisms like secondary flow can only be assessed in 3D. Therefore, turbine blade design will be tackled as a real 3D problem in this paper.

Section design by automatic optimization processes is well-established since more than one decade [3, 4, 6, 13]. Various section parameterization approaches have been presented which are mainly based on physical design parameters and/or free form curves [5]. Flow simulation on stream surfaces is usually performed by Euler or Navier-Stokes solvers, where the former can be enhanced with a boundary layer model. Many kinds of optimization algorithms such as Evolutionary Algorithms have been used. Often the optimizer is coupled with a surrogate model to reduce the number of function evaluations.

There are also many reports about stacking optimization [1, 14]. Usually the sections are arranged in radial direction along the stacking line which is parameterized as a polynomial or free form curve. Keskin et al. [10] report on stacking optimization in an industrial environment with GPU-accelerated CFD.

Of course, also fully three-dimensional aerodynamic optimization has already been performed [9, 11, 12]. Blade parametrization or modification can be based on stacked sections, free form surfaces or free form deformation (FFD). These approaches are mainly limited by large number of design parameters and huge 3D evaluation time. If indicated, runtime of 3D blade optimization so far is in the order of many weeks.

In this work we present a fully 3D optimization of a turbine blade as part of a two-stage turbine research rig, where runtime is just a few days although hundreds of 3D design evaluations are performed. Aerodynamic design evaluation is significantly accelerated by a rapid flow solver running on GPUs. All computations run on a workstation equipped with four high-performance NVIDIA Tesla C2050 GPUs. Each of these GPUs offers 448 stream processors with up to 515 GFLOPs double precision peak computational power and 3 GB on-board memory. Apart from that the design process is built up from standard components. Design parameters are chosen according to a sensitivity analysis. The following sections describe the optimization problem, CFD analysis, parameterization, parameter reduction and optimization results.

12.2 Problem Formulation

The main goal in aerodynamic turbine blade design is to maximize the component efficiency η:

$$\max_{\mathbf{p}_l \leq \mathbf{p} \leq \mathbf{p}_u} \eta. \tag{12.1}$$

However, this is constrained by keeping given operation conditions and engine properties like inlet capacity $\dot{m}_{\text{cor}}^{\text{ref}}$ and degrees of reaction ρ_1^{ref} and ρ_2^{ref} of both stages of the considered two-stage turbine which are set in preliminary design phases:

$$\begin{aligned} \dot{m}_{\text{cor}} &= \dot{m}_{\text{cor}}^{\text{ref}} \\ \rho_1 &= \rho_1^{\text{ref}} \\ \rho_2 &= \rho_2^{\text{ref}}. \end{aligned} \tag{12.2}$$

Inlet capacity is associated with the mass flow at turbine inlet. Usually the inlet capacity determines the mass flow through the whole turbomachine and is kept fixed in this design phase. The degree of reaction of a turbomachinery stage is defined as fraction of fluid temperature drop across the rotor in relation to the temperature drop of the corresponding stage. This is a measure for the fluid work on the rotor and related to axial bearing loads. Thus, the reactions might affect other engine components which is why they may not be changed.

More constraints, e.g. on geometry or aerodynamics would be applicable, too. However, this is omitted here to keep problem formulation and CFD post-processing as simple as possible. For simplification of numerical optimization, constraints (12.2) are only enforced to be fulfilled within small tolerances ε_i, i.e.,

$$\begin{aligned} \left|\dot{m}_{\text{cor}} - \dot{m}_{\text{cor}}^{\text{ref}}\right| &\leq \varepsilon_0, \\ \left|\rho_1 - \rho_1^{\text{ref}}\right| &\leq \varepsilon_1, \\ \left|\rho_2 - \rho_2^{\text{ref}}\right| &\leq \varepsilon_2. \end{aligned} \tag{12.3}$$

The constrained scalar optimization problem (12.1), (12.3) may then be solved with a penalty strategy. The blade parameterization and choice of design parameters is explained in more detail in Sect. 12.4.

12.3 CFD Analysis with a GPU-Accelerated Flow Solver

In this work, CFD analyses are performed with the fast flow solver Turbostream [2] which is a massively parallel re-implementation of an existing CFD solver for flows in turbo-machinery to optimally run on modern multi-core hardware like high-performance GPUs. Turbostream solves the Reynolds-averaged Navier-Stokes (RANS) equations with the finite volume method in cylindrical polar coordinates. Turbulence is modeled by a mixing length model and wall shear stresses are computed with a wall function. Unsteady multi-stage effects are modeled with a mixing plane approach.

Boundary conditions are specified by radial distributions of total pressure $P_I(r)$, total temperature $T_I(r)$, yaw angle $\alpha_t(r)$ and pitch angle $\alpha_r(r)$ at the inlet and by

Fig. 12.1 Computational domain of the investigated turbine with boundary conditions at inlet and exit

static pressure distribution $p_E(r)$ at the exit, see Fig. 12.1. Extra inlet patches are used for film cooling and cooling flows at the hub cavities and trailing edge slots. The initial flow field is generated from rough aerodynamic data based on preliminary design.

The structured mesh has a multi-block topology with arbitrary patches. In general, a HOH mesh topology is used for each blade. On the blade surface, y+ values are in the order of 20. Furthermore, geometric features like fillets, stub cavities, trailing edge slots and tip gaps are represented by the mesh. The resulting mesh generated by the rapid Rolls-Royce in-house meshing system PADRAM [15] has 2.6 million cells for four blade rows. This mesh size fits into the memory of one GPU card used for this project. Thus four CFD analyses may be performed in parallel on the machine described in Sect. 12.1.

12.4 Parametric Blade Model

Parameterization is a key driver for successful optimization. The concept of the three-dimensional blade description chosen here is based on sections which are defined on stream surfaces at several radial heights from hub to casing, see Fig. 12.2a. Each section is defined by circular arc segments for the leading and trailing edge, respectively, and B-splines for suction and pressure side, respectively. Typical section parameters are the chord length c, stagger angle ξ, metal angles β_j, wedge angles μ_j, circle radii r_j, tangential spline control points t_j^k and free spline control points $(s_i^k;\ n_i^k)$, Fig. 12.2b, where $j \in \{I, E\}$ denotes inlet or exit, $k \in \{S, P\}$ marks suction or pressure side and i is a numbering of spline control points. This concept is implemented as an industrial parametric blade design tool [7]. In this case, the suction and pressure side are defined with four and two free control points, respectively.

The radial height of every section is determined by the stream surface it is defined on. To account for optimal stacking, a section can be moved on its stream surface

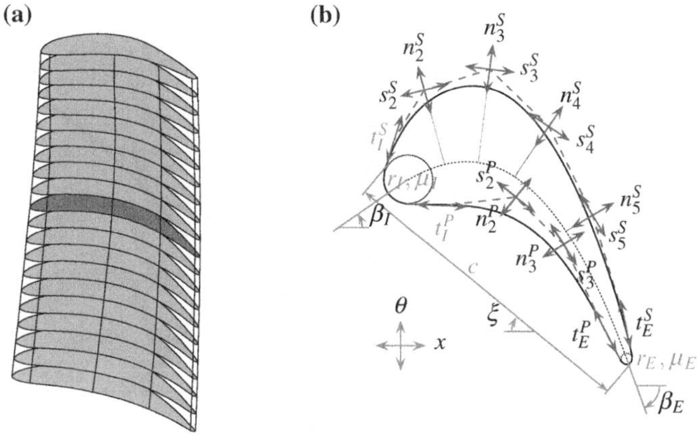

Fig. 12.2 Turbine blade consisting of stacked sections (**a**) and a parameterized section (**b**). Labels of section parameters which are not modified in this investigation are colored in *gray*

in axial and circumferential directions by axial shift x and circumferential shift θ, respectively. The three-dimensional blade is then obtained by arranging the sections above each other and interpolating the section shapes.

So far, section design and stacking would be described separately. For a three-dimensional parametric blade model the section parameters need to be coupled in a proper way to fulfill two goals:

- the output of the parametric model shall be compatible with an existing blade and section storage system
- the resulting blade shape shall be smooth.

In order to achieve this, the change of every section parameter $\Delta \bullet (\tilde{r})$ along the normalized radial coordinate $\tilde{r}$ relative to a reference design is parameterized with a second-order B-spline with an arbitrary number of control points. The first and last control point can move only along hub ($\tilde{r} = 0$) and casing ($\tilde{r} = 1$), respectively, whereas the others are free to move in both directions, Fig. 12.3. Usually only three control points are used for each section parameter described above. However, circumferential shift is parameterized with five control points as it is considered to be important and to allow for more design freedom.

These moves of the control points are used as design variables summarized in the design vector $\mathbf{p}$ of problem formulation (12.1) and limited by lower and upper bounds $\mathbf{p}_l$ and $\mathbf{p}_u$, respectively. The parametric model requires that the reference design is already smooth and roughly adapted to flow conditions. Changed design parameters are translated to modified sections which are saved, converted to a three-dimensional blade and passed to the meshing system and CFD analysis (Sect. 12.3).

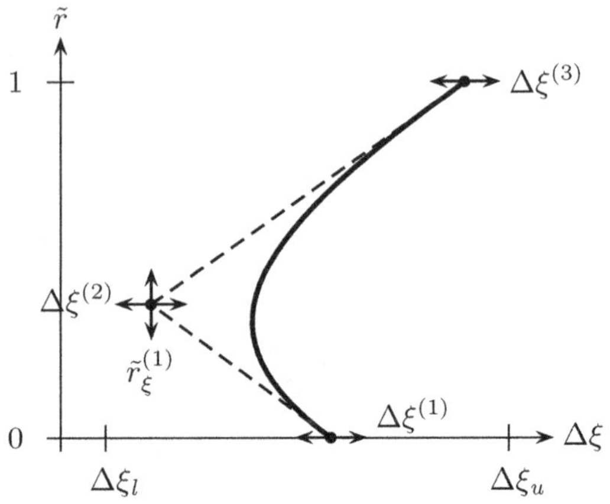

Fig. 12.3 Exemplary radial parameterization of the change of stagger angle $\Delta\xi$ relative to the reference design by a B-spline with three control points

12.5 Parameter Reduction

The used 3D parametric model has much more degrees of freedom than could reasonably be handled in a direct optimization. In order to decide on the most important parameters, a design of experiments (DoE) with Optimal Latin Hypercube sampling is performed. The varied parameters are the control points of the radial B-splines determining changes of axial and circumferential shifts, stagger angle, metal angles, free control points of suction and pressure side and tangential control points at the trailing edge. In total this yields 76 design parameters. 500 design evaluations are performed within 2.4 days, where four evaluations are computed in parallel in approximately 40 min and 82 % of the designs are analysed successfully.

Based on these DoE results a sensitivity analysis is performed to assess the influence of every design parameter on the objective and the constraints. Due to the large number of parameters and relatively small number of samples, the Spearman correlation coefficient is used. It provides a measure for the monotone dependence between two sampled variables.

According to Fig. 12.4a, turbine efficiency shows the strongest correlations with changes to circumferential and axial shifts and free control points on suction side. Free control points on pressure side and exit metal angle have smaller influence. Inlet metal angle and tangential control points only have insignificant correlations. Inlet capacity is mostly influenced by variations of stagger angle, free control points on suction side near the throat and exit metal angle, Fig. 12.4b. Changes of tangential shift and others have small influence on inlet capacity.

A similar interpretation can be done for the degrees of reaction of both stages, Fig. 12.5. The reaction of stage 1 also has the strongest correlation with changes to stagger angle, free control points on suction side near the throat and exit metal angle. Reaction of stage 2 is mostly influenced by changes of tangential shift, free control

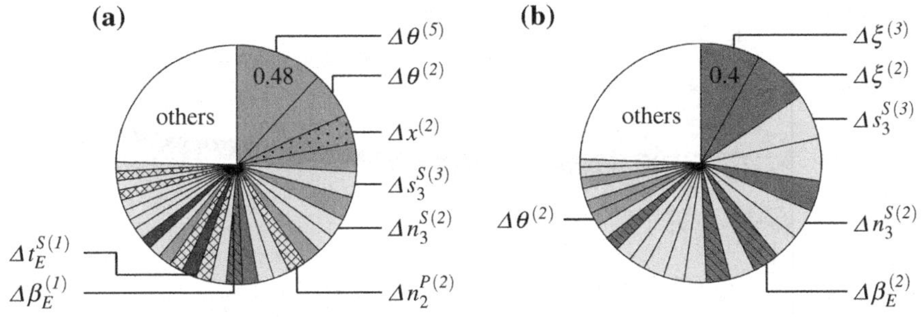

Fig. 12.4 Normalized absolute values of Spearman correlation coefficients as measure of correlation between exemplary design parameters and (**a**) turbine efficiency and (**b**) inlet capacity

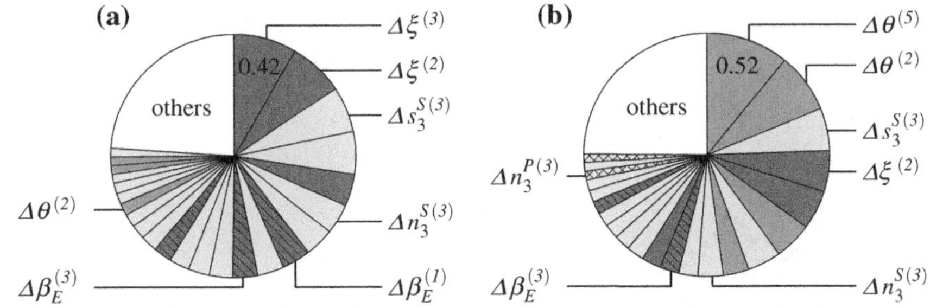

Fig. 12.5 Normalized absolute values of Spearman correlation coefficients as measure of correlation between exemplary design parameters and reaction of (**a**) stage 1 and (**b**) stage 2

points on suction side and stagger angle. A small correlation can be noticed for exit metal angle.

Consequently, parameters associated to changes of inlet metal angle, tangential control points and control points on the pressure side seem to have only small influence on the objective and constraints. Therefore, the 28 parameters associated with these quantities will not be used as design parameters in the following optimization runs.

12.6 Optimization

The automated aerodynamic design evaluation can be employed in an optimization process. For optimization the algorithm CMA-ES [8] is used which is a stable, derivative-free, black box evolution strategy for unconstrained single-objective problems. Its most important property is the de-randomized change of strategy parameters which means that they are adapted deterministically. The step-size σ is adapted based on the evolution path of previous generations. Themulti-variate

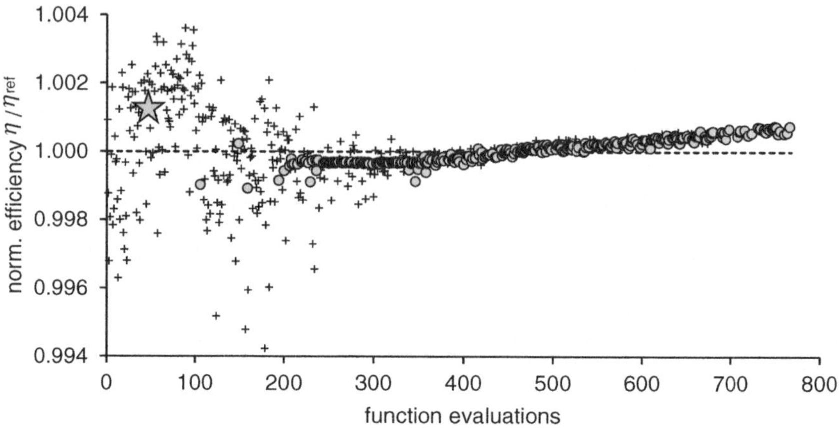

Fig. 12.6 Efficiency of optimal (⋆), feasible (○) and infeasible (+) designs relative to baseline

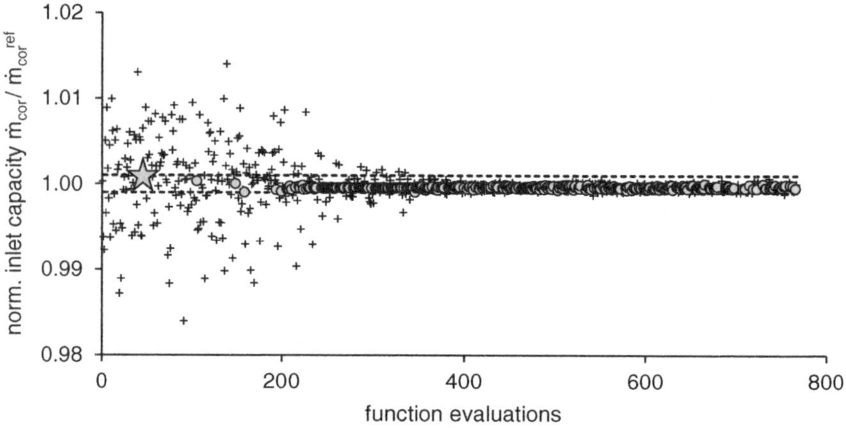

Fig. 12.7 Inlet capacity of optimal (⋆), feasible (○) and infeasible (+) designs relative to baseline

normal search distribution is controlled by an adapted covariance matrix of successful previous designs. Furthermore, CMA-ES has some invariance properties which make it very robust.

Based on the sensitivity analysis in Sect. 12.5, the selected design parameters for optimization are associated with axial and circumferential shifts, stagger angle, exit metal angle and free control points of suction side. This yields 48 design parameters of the 3D blade model. The population size of CMA-ES is reduced from the default value of 13 to 12 to optimally exploit parallel evaluation capability of the used GPU workstation. Several optimization runs were performed where Figs. 12.6 and 12.7 represent a typical result.

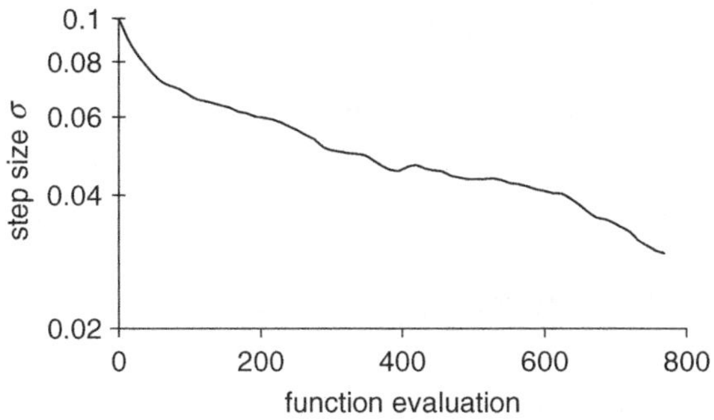

Fig. 12.8 Step size σ of CMA-ES during optimization

In total 768 design evaluation respectively 64 generations are performed in four and a half days. The optimization is terminated prematurely when the maximum number of generations is reached. All design evaluations converge where 40.7 % are feasible w.r.t. the optimization constraints (12.3). In the first 400 iterations large variations in objective and constraints can be observed. After that the optimization algorithm has reduced its step-size to a more successful order (Fig. 12.8) and begins to converge which results in a further decreasing step size, increased number of feasible designs and a continuous improvement of efficiency. The finally best feasible design ($\star$) is found in the early phase already. However, if the optimization runs longer, the optimization will converge and reach such an efficiency level again.

The optimal design is shown in Fig. 12.9. It fulfills the constraints (12.3) and has a slightly improved efficiency. Major design changes are that the tip section is moved

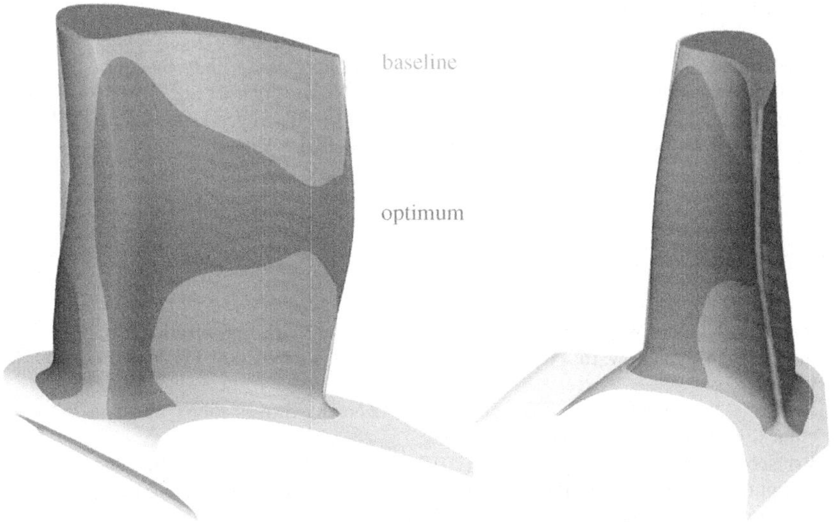

Fig. 12.9 Two views of a comparison between the blade shapes of baseline (*light grey transparent*) and optimal (*dark gray*) design

in circumferential direction of the suction side whereas the midsection is moved towards the opposite direction. In addition to the modifications of the stacking there are also significant modifications of the section parameters like a varying increase of the exit metal angle along the blade height or an decrease of the stagger angle.

12.7 Conclusions

A fast optimization process for the aerodynamic improvement of three-dimensional turbine blades is achieved by assessing the aerodynamic quality of a multi-stage turbine by a novel CFD solver running on GPUs. Wall-clock time of one 3D CFD is reduced to approximately 40 min, however, further speed-up is achieved by parallel design evaluation using multiple GPUs. Only standard tools are used for optimization, mesh generation and blade modification, where the three-dimensional blade parameterization is based on a well-known section parameterization. This allows to implement the proposed design approach in an industrial design environment.

A consequence of three-dimensional blade parameterization is the large number of design parameters. Therefore, a DoE is performed on all appropriate blade parameters and the design vector for optimization can be reduced to 48 design variables showing most significant influence on objective and constraints assessed by Spearman correlation coefficients. The optimization is characterized by a large variation of constraint values at the beginning and a steady improvement of efficiency after that. All optimization runs are terminated prior to convergence when the maximum number of generations is reached. Nevertheless, this paper shows that a fully three-dimensional optimization of turbine blades is feasible within days of runtime and finds improved blade designs.

Acknowledgments This work has been carried out in collaboration with Rolls-Royce Deutschland as part of the research project VIT 3 (Virtual Turbomachinery, contract no. 80142272) funded by the State of Brandenburg, the European Community and Rolls-Royce Deutschland. Rolls-Royce Deutschland's permission to publish this work is greatly acknowledged.

References

1. Arabnia M, Ghaly W (2009) A strategy for multi-point shape optimization of turbine stage in three-dimensional flow. In: Proceedings of ASME turbo expo 2009, GT2009-59708, Orlando, Florida
2. Brandvik T, Pullan G (2011) An accelerated 3D Navier-Stokes solver for flows in turbomachines. J Turbomach 133(2):021025
3. Briasco G, Cravero C, Macelloni P (2010) Turbine blade profile optimization using soft-computing techniques. In: Proceedings of 2nd international conference on engineering optimization, Lisbon, Portugal
4. Dennis B, Egorov I, Han ZX, Dulikravich G, Poloni, C (2000) Multi-objective optimization of turbomachinery cascades for minimum loss, maximum loading, and maximum gap-to-chord

ratio. In: Proceedings of 8th AIAA/NASA/USAF/ISSMO symposium on multidisciplinary analysis and optimization, AIAA 2000-4876, Long Beach, California

5. Ghaly W (2010) Shape representation. In: North Atlantic treaty organization, research & technology organization (ed) strategies for optimization and automated design of gas turbine engines, RTO-EN-AVT-167
6. Giannakoglou KC (1999) Designing turbomachinery blades using evolutionary methods. In: Proceedings of 44th ASME gas turbine & aeroengine congress, 9-GT-181, Indianapolis
7. Gräsel J, Keskin A, Swoboda M, Przewozny H, Saxer A (2004) A full parametric model for turbomachinery blade design and optimisation. In: Proceedings of ASME DETC 2004, DETC2004-57467, Salt Lake City, Utah
8. Hansen N, Ostermeier A (2001) Completely derandomized self-adaptation in evolution strategies. Evol Comput 9(2):159–195
9. Hasenjäger M, Sendhoff B, Sonoda T, Arima T (2005) Three dimensional aerodynamic optimization for an ultra-low aspect ratio transonic turbine stator blade. In: Proceedings of ASME turbo expo 2005, GT2005-68680, Reno-Tahoe, Nevada
10. Keskin A, Haselbach F, Meyer M, Janke E, Brandvik T, Pullan G (2011) Aerodynamic design and optimzation of turbomachinery component using a GPU flow solver. In: Proceedings of 9th European conference on turbomachinery fluid dynamics and thermodynamics, Istanbul, Turkey
11. Menzel S, Olhofer M, Sendhoff B (2006) Evolution design optimisation of a turbine stator blade using free form deformation techniques. In: Proceedings of conference on modelling fluid flow (CMFF'06), Budapest, Hungary
12. Nagel MG (2004) Numerische Optimierung dreidimensional parametrisierter Turbinenschaufeln mit umfangsunsymmetrischen Plattformen—Entwicklung, Anwendung und Validierung. PhD thesis, Universität der Bundeswehr München
13. Pierret S, Van den Braembussche RA (1999) Turbomachinery blade design using a navier-stokes solver and artificial neural network. J Turbomach 121(2):326–332
14. Pierret S, Demeulenaere A, Gouverneur B, Van den Braembussche RA (2000) Designing turbomachinery blades with the function approximation concept and the navier-stokes equations. In: Proceedings of 8th AIAA/USAF/NASA/ISSMO symposium on multidisciplinary analysis and optimization, AIAA 2000-4879, Long Beach, California
15. Shahpar S, Lapworth L (2003) Padram parametric design and rapid meshing system for turbomachinery optimisation. In: Proceedings of ASME turbo expo 2003, GT2003-38698, Atlanta, Georgia, pp 579–590

Chapter 13
Evaluation of Surrogate Modelling Methods for Turbo-Machinery Component Design Optimization

Gianluca Badjan, Carlo Poloni, Andrew Pike and Nadir Ince

Abstract Surrogate models are used to approximate complex problems in order to reduce the final cost of the design process. This study has evaluated the potential for employing surrogate modelling methods in turbo-machinery component design optimization. Specifically four types of surrogate models are assessed and compared, namely: neural networks, Radial Basis Function (RBF) Networks, polynomial models and Kriging models. Guidelines and automated setting procedures are proposed to set the surrogate models, which are applied to two turbo-machinery application case studies.

Keywords Surrogate models · Neural networks · Turbo-machinery

13.1 Introduction

This paper is based on a MSc thesis in Mechanical Engineering (University of Trieste) [1]. The main author was a research student at the University and is now employed as an Aerodynamics Methods Engineer at ALSTOM Power UK, facilitating tailored applied research collaboration between the University and the Company. The aim of the research was to evaluate the possibility of employing surrogate modelling methods for turbo-machinery component design optimization. The main idea behind these methods is to replace expensive to compute physical models with surrogate models, in order to speed up the entire design optimization process. These surrogate models

G. Badjan (✉) · A. Pike · N. Ince
ALSTOM Power Ltd, Newbold Road, Rugby CV21 2NH, UK
e-mail: gianluca.badjan@power.alstom.com

A. Pike
e-mail: andrew.pike@power.alstom.com

N. Ince
e-mail: nadir.ince@power.alstom.com

C. Poloni
University of Trieste, Via Valerio 8, 34127 Trieste, Italy
e-mail: poloni@units.it

D. Greiner et al. (eds.), *Advances in Evolutionary and Deterministic Methods for Design, Optimization and Control in Engineering and Sciences*, Computational Methods in Applied Sciences 36, DOI 10.1007/978-3-319-11541-2_13

are required to be cheap to compute and easy to use, whilst providing an adequate representation of the real problem. Four different surrogate models were employed in this study: Feed-Forward Backpropagation Neural Networks (FFBP NN), Radial Basis Function (RBF) Networks, Kriging models and polynomial models.

In the first part of this paper the surrogate modelling methods will be summarized, providing guidelines and automated procedures for their setting. Surrogate models will subsequently applied to two turbo-machinery case studies.

13.2 Neural Networks

The FFBP NNs employed in this study are multilayer networks with a single hidden layer. FFBP NNs are characterized by a very complex setting process, due to the high number of parameters to be set and their multi-modal performance function [2]. A critical choice for the neural network is the number of hidden neurons, since it determines the "flexibility" of the model. This parameter is usually chosen directly by the user. Unfortunately, the complexity of the modelled process is usually unknown, and FFBP NNs are tested for different architectures in order to find the best fitting for a particular dataset. This "Trial and Error" procedure is very time consuming, making it desirable to automate the setting of neural networks.

There are two main approaches to design a FFBP NN in an automatic fashion [1]:

- Constructive Methods
- Pruning Methods

The pruning methods appear to be the most convincing, since they allow a more tailored neural networks setting than the constructive methods. Two pruning techniques are evaluated in this paper: the Optimal Brain Surgeon (OBS) and the MATLAB *trainbr* algorithm.

13.2.1 Optimal Brain Surgeon

The OBS algorithm is a pruning technique developed by Hassibi [3] and implemented by Noorgard [4]. Each pruning session returns a certain number of pruned (partially connected) networks, one for each OBS algorithm iteration. Consequently, a neural network must be chosen according to some criteria, which are provided by Noorgard [4].

In addition, a new criteria was introduced by Badjan [1]:

$$\text{Balanced Valid. Error} = \text{Valid. Error} + \|\text{Valid. Error} - \text{Train. Error}\|_2 \qquad (13.1)$$

called *balanced validation error*, which takes into account both the error on the estimation subset and on the validation subset (note that Valid. and Train. error are ≥ 0 by definition).

According to this criteria, FFBP NNs with a low validation error that show similar performances on both the estimation and the validation subsets are preferred to the other networks.

13.2.2 MATLAB trainbr

In MATLAB *trainbr* [13], the following cost function is implemented:

$$MSE_{reg} = \alpha MSW + \beta MSE \tag{13.2}$$

where MSE is the mean square error, MSW is the sum of the squares of the network weight and biases, α and β are regularization parameters.

Minimizing Eq. (13.2) leads to lower values of the network weights and biases, making the network response smoother and less prone to overfit. In fact, assigning low values to the free parameters may be viewed as equivalent to pruning the neural network.

13.2.3 Dynamic Threshold Neural Networks

The Dynamic Threshold Neural Network (DTNN) was originally proposed by Chiang and Fu [5] for pattern recognition purposes, but it was also successfully applied to function approximation problems by Pediroda [6] and Poloni et al. [7]. The DTNN was designed to employ *Static Threshold Quadratic Sigmoidal Neurons* in the hidden layer and *Dynamic Threshold Quadratic Sigmoidal Neurons* in the output layer.

This network configuration produces outputs in the range [0, 1], which is adequate for pattern recognition purposes, but it could represent a limitation for function approximation purposes. In the view of the Authors, having an output range limited between two fixed values implies that the training-set contains both the minima and the maxima of the objective function. If the training-set targets are normalized between [0, 1], then other new input configurations will always produce target values included between [0, 1]. An example may illustrate the concept: considering a training-set with the maximum objective function value 12 and the minimum objective function value -5, after the data normalization 12 will correspond to 1 and -5 to 0; if there is a maxima (or minima) somewhere in the input domain with a value 15 (-7), then the corresponding output will be again 1 (0). It can be noticed that even if the objective value is saturated, the input configuration might represent the true maxima (minima). In any case, no robust analysis could be performed using the surrogate model on that point, since it would not be possible to approximate in a proper way the shape of the objective function in the saturated zone.

For these reasons, in this study it was decided to rearrange the architecture of the DTNN, employing the Dynamic Threshold Quadratic Sigmoidal neurons directly in the hidden layer and the standard linear transformation in the output layer, thereby removing the output limits.

The setting process of the DTNN presents some differences with respect to classic FFBP NNs. In fact, fewer neurons are generally required to fit a dataset, since they have a higher approximation capability than the neurons of classic FFBP NNs [5]. It was therefore decided to use a constructive methodology to train this type of network.

The proposed setting process for DTNNs consists in:

1. Set the DTNN with n hidden neurons.
2. Train the network m times from different initial configurations.
3. Check the performance on the validation subset.
4. Set a new DTNN with $n + 1$ hidden neurons.
5. Train the new network a couple of times and check the performance on the validation subset.
6. If the validation error increases stop the procedure, otherwise go to point 4.

13.3 RBF Networks

In this paper, Gaussian, multiquadrics and inverse-multiquadrics functions were chosen to build RBF Networks, since they have a shape parameter σ used to control the domain of influence of the radial basis function. There are various strategies in the literature for selecting an appropriate value for the shape parameter σ. The *leave-one-out* (LOO) error is a well known criteria for setting RBF Networks. However, the computational cost can be very high, of order $O(N^4)$, which becomes prohibitively expensive even for problems of modest size. Fortunately, Rippa [8] proposed a technique to reduce the computational cost of the LOO metric to $O(N^3)$, which was here implemented.

Based on the LOO error, an iterative procedure to select the optimal shape parameter for interpolating RBF Network is proposed in this paper:

1. Initialize σ to 1 and evaluate the LOO error.
2. Set σ_{new} to 0.5 and evaluate the corresponding LOO error.
3. If LOO $\sigma_{new} <$ LOO σ then $\sigma = \sigma_{new}$ and $\sigma_{new} = \sigma/a$, otherwise $\sigma = \sigma_{new}$ and $\sigma_{new} = \sigma * b$.
4. Evaluate the LOO error for the RBF Network set with σ_{new}.
5. Repeat the procedure from point 3 until the maximum number of iterations is reached.
6. Return the RBF Network which scored the minimum LOO error.

The parameters a, b can be set by the user, determining how much σ is increased or decreased at each iteration. In addition, there is also the possibility to vary these parameters during the iterations, to gradually reduce or increase the step size of the shape parameter. The Authors suggest to set $a = 1.5$ and $b = 1.8$. The maximum number of iterations should take into account the time required to solve a single LOO measure. However, 20 iterations should be an appropriate number for the majority of the problems.

RBF Networks can also perform a regression of the data, introducing the regularization parameter λ in a similar way as for the Kriging model [9]. Keane and Nair [10] suggest to set λ to the variance of the noise in the response data, but since this information in usually unknown. The remaining option is to add it to the list of parameters to be estimated. In this study, both the shape parameter σ and the regularization parameter λ were searched throughout their domain using a Genetic Algorithm (GA). Suitable upper and lower bounds for the search of λ are 10^{-6} and 1 respectively [10].

13.4 Polynomial Models

Polynomial models can be applied to multi-dimensional problems taking into account interaction terms [9]. In this paper, optimal values for global and interaction orders are found by applying cross-validation.

13.5 Kriging Models

Kriging models are powerful methods based on Gaussian processes. They can perform either interpolation or regression of data. In this paper, Kriging models are set via maximizing the marginal likelihood function [9].

13.6 Assessment Criteria for Surrogate Models

If the observational data are abundant, a randomly selected subset (Hastie et al. [11] recommend around the 25 % of the total $\boldsymbol{x} \rightarrow y$ pairs) should be set aside for model testing purposes. These observations must not be touched during the previous stages, as their sole purpose is to allow us to evaluate the testing error (based on the difference between the true and approximated function values at the test sites) once the model has been built. Standard assessment criteria for surrogate models are Normalized Root Mean Square Error (NRMSE) and Coefficient of Determination (r^2). According to [9], good surrogate models should have NRMSE $<10\,\%$ and $r^2 > 0.8$.

Furthermore, a new criteria is introduced in this paper, called RANKING [1], the aim of which is to assess the capability of surrogate models to replicate the trend expressed by the underlying function.

The RANKING is evaluated using the following procedure:

1. Sort the true solutions of a particular dataset in ascending order.
2. Check if the corresponding approximated solutions increase their values monotonically.
3. A score of 1 is given to the solutions that increase step by step, referring to the previous highest value (absolute RANKING).
4. Finally the score is divided by the number of points in the dataset and multiplied by 100.

A numerical example may illustrate the steps:

- Assuming the following true solutions y: $[12, 43, 2, 33, 30, 31]$ and the surrogate model approximations $\widehat{y}$: $[10, 45, 3, 32, 35, 31]$.
- Now sorting the true solutions y in an ascending order: $[2, 12, 30, 31, 33, 43]$ with the corresponding original index: $[3, 1, 5, 6, 4, 2]$.
- Then the corresponding approximation $\widehat{y}$ will be: $[3, 10, 35, 31, 32, 45]$.
- The scores for each point are $[1, 1, 1, 0, 0, 1]$ and their sum is 4, it should be noticed that this metric is done on the absolute ascending order.

$$\text{RANKING} = \frac{4}{6}100 = 67\,\% \tag{13.3}$$

The higher the value of the RANKING, the better the surrogate model can follow the underlying response trend. However, this criteria in isolation is insufficient to determine the overall accuracy of the model.

13.7 Optimization Case Studies

Two optimization case studies were chosen to evaluate the application of surrogate models in turbo-machinery component design optimization:

- Mono-objective optimization of the operating conditions of a turbine cascade.
- Mono-objective optimization of a turbine labyrinth seal.

The optimization procedure consisted in two consecutive steps:

1. *Global search* of the optima, over all the design space.
2. *Local search* of the optima, refining the result obtained from the global search.

This optimization strategy combines both robustness and accuracy.

In particular, a *Genetic Algorithm* was chosen as the global optimizer, since it is a robust and reliable algorithm widely used in optimization [7, 12]. The subsequent local optimization was done using the *Sequential Quadratic Programming* (SQP) method [13].

13.7.1 Turbine Cascade Case

The performance of a steam turbine cascade [14] was analyzed for different operating conditions employing an ALSTOM in-house CFD code.

The design space was defined by three input variables:

- Incidence Angle
- Inlet Total Pressure (for adjusting Mach Number)
- Fluid Viscosity (for adjusting Reynolds Number)

The objective of the optimization was to maximize the efficiency of the turbine profile. A dataset of 150 points was obtained running an Optimized Latin Hypercube DOE. Each simulation took about three minutes on a PC (Quad Core CPU running @ 2.66 Ghz, 3.25 GB RAM). Afterwards, the dataset was normalized in the range $[-1, 1]$ and randomly split into a training-set of 120 points and a test-set of 30 points.

13.7.1.1 Performance of Surrogate Models

Different setting approaches were adopted for each type of surrogate model. Five FFBP NNs were created using the *trainbr* algorithm and the OBS technique. It is worth reminding that FFBP NNs have a multi-modal performance function, therefore finding the best network configuration for a particular dataset usually requires to train the network from different initial weights/biases configurations. The same concept applies to the OBS technique, since the setting of the first oversized neural network influences the results of the subsequent pruning process. The validation subset was the 20 % of the training-set. The DTNN was built finding the optimal number of hidden neurons via "trial and error" procedure. Eventually, five DTNNs were created with the optimal architecture. As described for neural networks, five polynomial models were built using cross-validation with 10 subsets, in order to investigate how the random splitting affects the setting process. The same global order and interaction order were obtained for all five models, confirming that cross-validation is a robust procedure to set polynomial models. The interpolating RBF Networks were built only once, using the iterative procedure previously described in Sect. 13.3, with 20 steps. However, the setting procedure for regressive RBF Networks was different. In fact, these models were tuned using the GA, which was set with a population of 30 individuals and 30 generations. In this case, five regressive RBF Networks were built for each basis function, resulting in broadly similar performance. Finally, five Kriging models were also built. The likelihood function employed to build the Kriging models was optimized setting the GA with a population of 50 individuals and 100 generations. The adopted setting configuration produced almost

Table 13.1 Surrogate models performance on the blade test-set

	Test RMSE (%)	Test r^2	Test RANKING (%)
FFBP NN OBS	0.505	0.9996	90.00
FFBP NN *trainbr*	3.069	0.9848	56.67
DTNN	0.712	0.9993	93.33
RBF G	1.908	0.9943	73.34
RBF IM	1.668	0.9965	73.34
RBF M	1.803	0.9965	76.67
Reg. RBF G	1.740	0.9954	73.34
Reg. RBF IM	1.620	0.9962	73.34
Reg. RBF M	1.501	0.9967	73.34
Kriging	1.888	0.9960	66.67
Reg. Kriging	1.034	0.9983	76.67
Polynomial	1.389	0.9967	83.34

G Gaussian, *IM* inverse-multiquadrics and *M* multiquadrics

Table 13.2 Optimized and validated results for the blade study case

	Incidence angle (deg)	Tot. inlet pressure (*bar*)	Fluid viscosity (Ns/m^2) (10^{-6})	Optimized solution (−)	Validated solution (−)
FFBP NN OBS	−27.4056	190.10	1.8	0.93319	0.93378
FFBP NN *trainbr*	−8.4097	188.06	1.8	0.93115	0.93360
DTNN	−24.2300	189.31	1.8	0.93330	0.93383
RBF G	3.3989	191.63	1.8	0.93213	0.93304
RBF IM	−32.2721	189.11	1.8	0.93247	0.93376
RBF M	−32.0508	189.01	1.8	0.93250	0.93377
Reg. RBF G	−23.5060	190.59	1.8	0.93240	0.93375
Reg. RBF IM	−22.8310	190.87	1.8	0.93219	0.93372
Reg. RBF M	−32.5122	188.99	1.8	0.93267	0.93376
Polynomial	−14.2838	190.56	1.8	0.93389	0.93364
Reg. Kriging	−28.9457	189.05	1.8	0.93409	0.93382

The maximum efficiency in the DOE dataset is 0.93099, *G* Gaussian, *IM* inverse-multiquadrics and *M* multiquadrics

identical models, the small differences were related to the GA obtaining only the neighbourhood of the maximum as opposed to maximum of the likelihood function.

For each type of surrogate model, only the best performing model was chosen for the comparison summarized below.

It can be noticed from Table 13.1 that all the models performed very well, with low values for the NRMSE and high values for r^2 and RANKING.

Once the surrogate models were built, then it was possible to use them to evaluate all the other input configurations required by the optimizer algorithm. The constraints

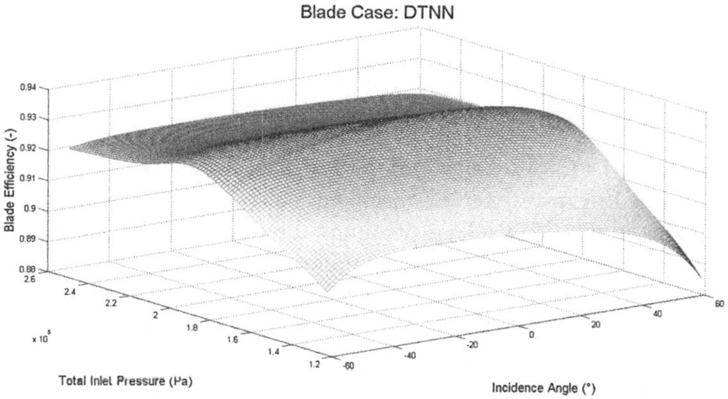

Fig. 13.1 Blade case—DTNN

of the optimization were represented by the design space boundaries, which were defined by the highest and lowest values of each input variable in the DOE.

As can be seen in Table 13.2, all the surrogate models gave very similar validated solutions. In particular, the DTNN produced the best results, which was chosen to plot a graphic representation of the problem, fixing the viscosity to its lowest value in the dataset (see Fig. 13.1). It is also interesting to note that the DTNN had the best RANKING score on the test-set. However, almost the same validated solution was obtained with the Regressive Kriging, which was also definitely far quicker to set than DTNNs and FFBP NNs. Polynomial model and RBF Networks were also quicker and easier to set than neural networks. In the opinion of the authors, the simplicity of the setting process should be always considered in the assessment of surrogate modelling methods. In practical applications, a quick-to-set surrogate model should be preferred to other models with time consuming and non-robust setting processes, especially when the results are almost the same, as in this case.

Finally, it should be considered that the efficiency improvements obtained from the initial DOE were small from the numerical point of view, but very important in engineering design.

13.7.2 Turbine Seal Case

The leakage of a labyrinth seal of the high-pressure stage of a steam turbine was evaluated via CFD simulations, which were performed with the commercial code ANSYS FLUENT [15]. Seven geometric parameters were originally chosen in order to determine the key variables for prediction of leakage, such as fin height, thickness, angle, etc. These input variables were screened using full-factorial DOEs and Pareto Charts (based on polynomial regression). Finally four top parameters were selected to be included in the surrogate modelling and subsequent model based optimization (minimization) of seal leakage:

- a_1, Angle parameter
- a_2, Angle parameter
- L_1, Length parameter
- L_2, Length parameter

A full-factorial DOE of 5 levels per variable (resulting in $5^4 = 625$ points) was originally planned to investigate the problem, but some simulations failed due to technical issues in the CFD solver, obtaining a reduced dataset composed of 517 points. Each simulation took about eight minutes on a PC (12 Core CPU running @ 2.92 Ghz, 24 GB RAM). The dataset was randomly split into a training-set of 414 points and test-set of 103 points. The surrogate models were built adopting the same methodology employed for the turbine cascade case.

The subsequent optimizations were run setting the GA with 100 individuals and 30 generations, and allowing a maximum of 30 iterations for the local optimizer. As for the turbine cascade case, the design space boundaries were defined by the highest and lowest values of each variable in the DOE.

All the values shown in the tables and pictures regarding the seal case were normalized in the range $[-1, 1]$, for the purpose of protecting commercially sensitive information.

13.7.2.1 Performance of Surrogate Models

Table 13.3 shows that the surrogate models did not perform very well in this case, scoring high NRMSE values and low r^2 values. Also the scores for the RANKING were very low. In the opinion of the authors, the RANKING criteria should be applied to the cases where surrogate models perform very similarly, as for the blade case.

Table 13.3 Surrogate models performance on the seal test-set

	Test NRMSE (%)	Test r^2	Test RANKING (%)
FFBP NN OBS pruned	10.490	0.7554	5.83
FFBP NN *trainbr*	10.181	0.7704	9.71
DTNN	11.903	0.6871	5.83
RBF G	13.339	0.6072	8.74
RBF IM	11.880	0.6924	8.74
RBF M	11.530	0.7103	7.77
Reg. RBF G	11.246	0.7267	8.74
Reg. RBF IM	11.420	0.7164	7.77
Reg. RBF M	11.544	0.7093	7.77
Polynomial	11.573	0.7108	7.77
Kriging	14.199	0.5573	9.71
Reg. Kriging	10.749	0.7504	7.77

G Gaussian, *IM* inverse-multiquadrics and *M* multiquadrics

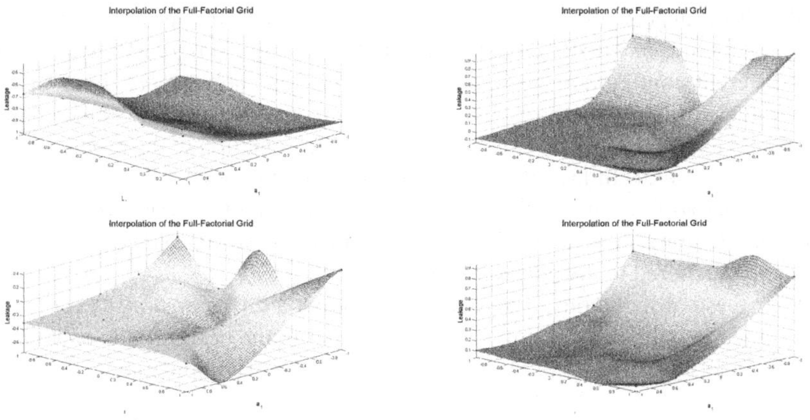

Fig. 13.2 Full factorial cubic spline interpolations: a_2 and L_2 are fixed to different values

The validated solutions were generally worse than the "best solution" contained in the DOE (which was equal to -1 following normalization). However, very small reductions of the leakage were obtained with RBF Networks (surrogate model based optimal solution -1.001, validated CFD code solution -1.0067), but not enough to consider the optimization a success.

It was also found that most of the optimized solutions were found for the lowest value of a_2 and L_2. Recalling that a generic full-factorial DOE consists in a multi-dimensional grid, Fig. 13.2 was generated fixing some variables and using MATLAB cubic spline interpolation [13].

As can be seen from Fig. 13.2, the underlying function shows very different scenarios varying the values of the same fixed variables, making it difficult to be modelled even with a full-factorial DOE of 517 points. This behavior is probably due to the fact that the input variables are highly correlated. However, it should be noted that the cubic spline interpolation does not correspond to the true function, which is obviously unknown, and the underlying function might be even more complex.

In addition, Figs. 13.3 and 13.4 show a comparison between some surrogate models and the corresponding full factorial cubic spline. It is clear from Fig. 13.3 that

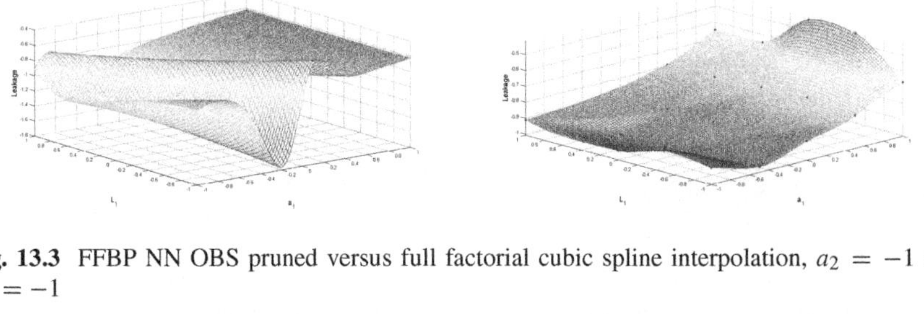

Fig. 13.3 FFBP NN OBS pruned versus full factorial cubic spline interpolation, $a_2 = -1$ and $L_2 = -1$

FFBP NNs pruned with OBS overfitted the data. In fact, it appeared that the high flexible structure of FFBP NNs can fit the data in a large variety of ways (i.e. with very different configurations for weights/biases), generating approximations with good values for NRMSE and r^2 but also with very strange shapes. On the contrary, the FFBP NN trained with *trainbr* gave a good representation of the problem, as can be seen in Fig. 13.4. In fact, the *trainbr* algorithm increases the level of regression of neural networks, making them smooth and less prone to overfit [16]. RBF Networks and Kriging showed less flexibility than neural networks, since their structure is directly anchored to the points in the dataset.

After these observations, it was decided to adopt a different strategy for the seal case, aimed to obtain an optimized solution similar to the best solution contained in the full-factorial DOE of 517 points, but using less CFD computations.

13.7.2.2 Further Investigation with Alternative Datasets

An Optimized Latin Hypercube DOE of 100 points was run with the objective to gather information over all the design space using less points. Again, some points failed to produce a result, obtaining a reduced dataset composed of 94 points. Surrogate models were built with the new dataset, employing all the points as a training-set.

The new validated solutions were not better than the solutions obtained in the previous optimization. However, it can be noticed that the majority of the regressive models gave again the optimized solutions for the lowest value of a_2 and L_2.

It was therefore decided to run a further Optimized Latin Hypercube DOE of 50 points with a_2 and L_2 fixed to their lowest value, in order to reduce the dimensionality of the problem. Finally a dataset composed of 47 points was obtained. The surrogate models were built employing all the 47 points as the training-set.

As can be seen in Table 13.4, the Kriging model and the Regressive RBF Gaussian Network improved the best solution contained in the first dataset of 517 points. Thus, the computational budget was reduced from 600 points to 150 points. Unfortunately the FLUENT solver failed to converge at the optimized solution of the Regressive Kriging.

In addition, a further full-factorial DOE of 400 points (20 levels per variable) was run fixing a_2 and L_2 to their lowest value, in order to investigate in detail the

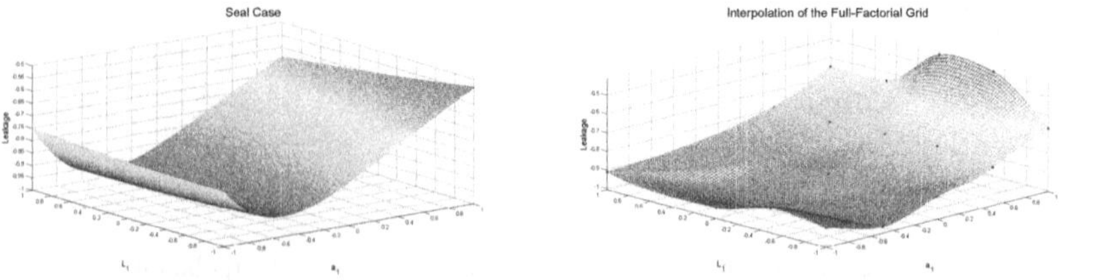

Fig. 13.4 FFBP NN *trainbr* versus full factorial cubic spline interpolation, $a_2 = -1$ and $L_2 = -1$

Table 13.4 Optimized and validated results for the seal study case using the dataset composed of 47 points

	a_1 (−)	L_1 (−)	Optimized solution	Validated solution
OBS pruned	−0.9350	−0.4219	−1.2282	−0.8986
trainbr	−0.5230	0.0042	−1.0246	−0.9818
DTNN	−0.7167	0.0274	−1.0073	−0.9966
RBF G	−0.7374	0.6752	−1.0303	−1.0013
RBF IM	−0.5981	−0.0644	−1.0210	−0.9943
RBF M	−0.6270	−0.0109	−1.0195	−1.0188
Reg. RBF G	−0.7383	0.6797	−1.0292	−1.0196
Reg. RBF IM	−0.5976	−0.0651	−1.0210	−0.9982
Reg. RBF M	−0.3396	0.1970	−1.1035	−0.9693
Kriging	−0.6989	0.5896	−1.1035	−1.0286
Reg Kriging	−0.5494	1	−1.0054	N/A
Polynomial	−0.7461	0.7949	−1.0454	−0.9824

Leakage normalized w.r.t. the first dataset, *G* Gaussian, *IM* inverse-multiquadrics and *M* multiquadrics

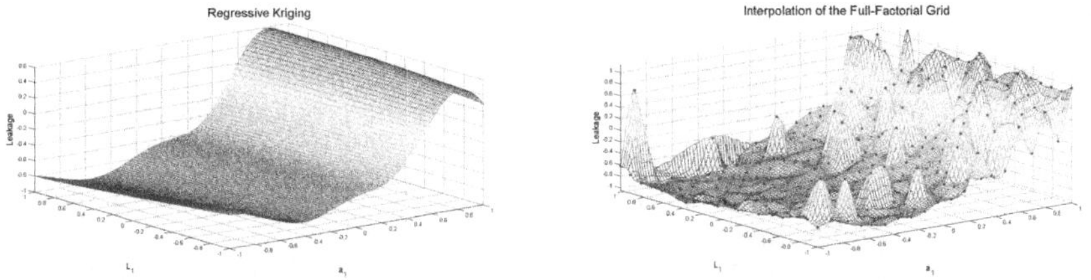

Fig. 13.5 Regressive Kriging versus FF interpolation, DOE 50 points

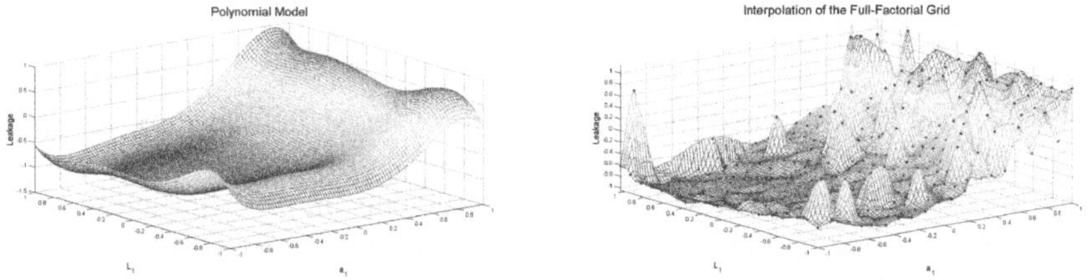

Fig. 13.6 Polynomial model versus FF interpolation, DOE 50 points

morphology of the underlying function. The final dataset was composed of 384 points (with 16 points failing to converge).

As can be seen in Figs. 13.5 and 13.6, regressive models developed with 47 points were capable to detect the general trends of the real problem, which was highly irregular with many peaks (see the full factorial cubic spline interpolations). In particular, the Polynomial model was able to define the "borders" of the underlying function very well. On the contrary, FFBP NN pruned with OBS and DTNN overfitted the data.

In engineering design, the visualization of a problem is extremely useful, since it can provide an indication of promising regions that may yield a robust optimal design. Flat zones with stable performance will be preferred to peaky zones, where small variations of the input variables lead to high variation of the output. For example, the blue valley in Fig. 13.6 represents a stable zone, where small variations of a_1 and L_1 do not particularly affect the leakage. In fact, the geometric parameters defining a labyrinth seal are subject to manufacturing tolerances. Thus, it is clear that an important aspect in industrial design is managing the uncertainties, to find solutions which are insensitive to the stochastic fluctuations of the parameters (*Robust Design*) [10, 17].

Summarizing, the seal case was significantly more challenging than the blade case. FFBP NNs pruned with OBS and DTNNs performed poorly, overfitting the surface. Instead, the FFBP NNs trained with *trainbr* were able to detect the main trends of the underlying function. However, other surrogate models such as RBF Networks, Kriging and Polynomial, gave better results with less training. In particular, the Kriging model produced the best numerical result and the Polynomial model gave the best representation.

In addition, it appeared that a good strategy in optimization assisted by surrogate models may consist in:

1. Run a small global DOE, according to the available computational budget.
2. Build regressive surrogate models and visualize the problem where possible.
3. Validate the optimized solutions.
4. Evaluate the possibility of reducing the dimensionality of the problem, or at least to define a small promising zone in the domain.
5. Run a reduced/local DOE, according to the available computational budget.
6. Validate the new optimized solutions.

13.8 Conclusion

This paper has demonstrated the utility of Surrogate Models in turbo-machinery design optimization. In the first instance, different surrogate modelling methods should be used when dealing with unknown problems, in order to find the model that best fits a particular dataset. In addition, surrogate models should be assessed on the basis of their ease of configuration. From this point of view, FFBP NNs and DTNNs present too many drawbacks to be considered a valid methodology in turbo-machinery component design optimization. They did not show any clear advantage compared to other methodologies in terms of accuracy, but their setting process presented many issues. However, neural networks are widely applied in control engineering and signal processing, where their flexibility represents a benefit in modelling of dynamic systems.

Finally for the considered case studies, Kriging models were assessed as being the most promising surrogate model among those evaluated in this paper, combining high performance with a relatively easy setting process.

References

1. Badjan G (2013) Evaluation of surrogate modelling methods for turbo-machinery component design optimization. MSc Thesis in Mechanical Engineering, University of Trieste, Italy
2. Haykin S (2005) Neural networks: a comprehensive foundation. Pearson Education, Hamilton
3. Hassibi B, Stork DG (1993) Optimal brain surgeon and general network pruning. IEEE Int Conf Neural Netw 169:293–299
4. Noorgard M (2000) Neural network based system identification toolbox. Technical Report 00-E-891, Technical University of Denmark
5. Chiang CC, Fu HC (1994) The classification capability of a dynamic threshold neural network. Pattern Recogn Lett 15:409–418
6. Pediroda V (2001) Utilizzo e sviluppo di tecniche "Soft-Computing" per lo studio e la progettazione di macchine a fluido. Ph.D. thesis in Chemical and Energy Technologies, University of Udine, Italy
7. Poloni C, Giurgevich A, Onesti L, Pediroda V (2000) Hybridization of a multi-objective genetic algorithm, a neural network and a classical optimizer for a complex design problem in fluid dynamics. Comput Methods Appl Mech Eng 186:403–420
8. Rippa S (1999) An algorithm for selecting a good value for the parameter c in radial basis function interpolation. Adv Comput Math 11:193–210
9. Forrester A, Sóbester A, Keane A (2008) Engineering design via surrogate modelling. Wiley, University of Southampton
10. Keane A, Nair P (2005) Computational approaches for aerospace design: the pursuit of excellence. Wiley, New York
11. Hastie T, Tibshirani R, Friedman J (2001) The elements of statistical learning. Springer, New York
12. Goldberg DE (1989) Genetic algorithm in search, optimization and machine learning. Addisons-Wesley
13. MATLAB (2008) User guide. The MathWorks Inc.
14. Perdichizzi A, Dossena V (1993) Incidence angle and pitch-chord effects on secondary flows downstream of a turbine cascade. J Turbomach 115:383–391
15. ANSYS FLUENT (2008) User guide. ANSYS, Inc.
16. Demuth H, Beale M (2000) Neural network toolbox: for use with MATLAB. User's guide, Version 4, The MathWorks Inc.
17. Pediroda V, Poloni C (2006) Approximation methods and self organizing map techniques for MDO problems. Department of Mechanical Engineering, University of Trieste, Italy

Chapter 14
Robust Aerodynamic Design Optimization of Horizontal Axis Wind Turbine Rotors

Marco Caboni, Edmondo Minisci and Michele Sergio Campobasso

Abstract The work reported in this paper deals with the development of a design system for the robust aerodynamic design optimization of horizontal axis wind turbine rotors. The system developed is here used to design a 126-m diameter, three-bladed rotor, featuring minimal sensitivity to uncertainty associated with blade manufacturing tolerances. In particular, the uncertainty affecting the rotor geometry is associated with the radial distributions of blade chord and twist, and the airfoil thickness. In this study, both geometric and operative design variables are treated as part of the optimization. Airfoil aerodynamics and rotor aeroelasticity are predicted by means of XFOIL and FAST codes, respectively, and a novel deterministic method, the Univariate Reduced Quadrature, is used for uncertainty propagation. The optimization is performed by means of a two-stage multi-objective evolution-based algorithm, aiming to maximize the rotor expected annual energy production and minimize its standard deviation. The design optimization is subjected to a single structural constrain associated with the maximum out-of-plane blade tip deflection. The results of this research highlight that a lower sensitivity to uncertainty tied to manufacturing tolerances can be achieved by lowering the angular speed of the rotor.

Keywords Horizontal axis wind turbine design · Multidisciplinary design optimization · Robust design optimization · Uncertainty propagation

M. Caboni (✉)
School of Engineering, University of Glasgow, James Watt Building South,
University Avenue, Glasgow G12 8QQ, UK
e-mail: m.caboni.1@research.gla.ac.uk

E. Minisci
Department of Mechanical and Aerospace Engineering, University of Strathclyde,
75 Montrose Street, Glasgow G1 1XJ, UK
e-mail: edmondo.minisci@strath.ac.uk

M.S. Campobasso
Department of Engineering, Lancaster University, Engineering Building,
Lancaster LA1 4YR, UK
e-mail: m.s.campobasso@lancaster.ac.uk

D. Greiner et al. (eds.), *Advances in Evolutionary and Deterministic Methods for Design, Optimization and Control in Engineering and Sciences*, Computational Methods in Applied Sciences 36, DOI 10.1007/978-3-319-11541-2_14

14.1 Introduction

By the late 20th century, wind power has become one of the most promising new energy sources worldwide, achieving a rapid global growth in installed capacity. Considerable efforts have been put into wind turbine design in order to improve performance and reduce costs, making wind power a competitive energy source. In recent years, several studies have focused on the aerodynamic and structural design optimization of horizontal axis wind turbines (HAWTs) [1–6], encompassing different approaches to increasing the annual energy production (AEP) and reducing the cost of energy. However, an optimized design can become inefficient in the presence of environmental, operation, manufacturing or assembly uncertainties. Therefore, one of the ways to further improve the design of modern wind turbines is to consider the effect of the aforementioned sources of uncertainty throughout the optimization process, leading to the design of more effective devices with minimal sensitivity to uncertainties. The design optimization under uncertainty, aiming to maximize the expected value of one of more objective functions (e.g., AEP), while minimizing the effect of uncertainties, is often denoted by the attribute "robust". Incorporating sources of uncertainty into a robust optimization process implies the use of a suitable technique for uncertainty propagation, which should keep computational costs affordable while maintaining an acceptable accuracy.

Petrone et al. [7] developed a comprehensive multi-physics computational model to study the impact of wind condition variability, manufacturing tolerances and roughness induced by insect contamination on HAWT aerodynamic performance and noise. In this framework, the Latin Hypercube Sampling (LHS) and the Stochastic Simplex Collocation (SSC) methods were successfully used to propagate uncertainties throughout the computational model. In a more recent work, Petrone et al. [8] developed a system for the robust optimization of HAWT rotors under uncertainty represented by insect contamination. Uncertainty was propagated by means of the SSC method. The proposed design strategy was coupled with a multi-objective genetic algorithm.

Minisci et al. [9] demonstrated a methodology for the aerodynamic optimization of HAWT rotors under geometric uncertainty of the blade geometry caused by manufacturing and assembly tolerances. Chord and twist distributions, and the angular speed of the rotor were included in the optimization process. Uncertainty propagation was conveniently performed by means of the Univariate Reduced Quadrature (URQ) approach [10]. The adopted optimization method was based on a two-stages multi-objective optimization strategy.

The work described in this paper aims to improve the current state-of-art in robust aerodynamic design optimization of HAWT rotors by including the effect of a comprehensive range of geometric uncertainties associated with blade manufacturing tolerances in the design process. In this context, blade chord and twist distributions as well as airfoil thickness are considered affected by uncertainty. Design variables include blade chord and twist distributions, airfoil shape, and the angular speed of the rotor. The URQ method for uncertainty propagation, and the two-stages multi-objective optimization strategy developed by Minisci et al. [9] are used to carry out the robust design process.

The first part of the paper provides a general description of the optimization system, including: (a) the blade geometric module, (b) the aeroelastic module, (c) the URQ uncertainty propagation method, and (d) the procedure and algorithms used to carry out the optimizations. In the second part, the optimization process set-up is presented in detail, and the obtained results are described and discussed in the third part. A conclusion section summarizes the key findings of this work, and suggests some directions for future developments.

14.2 Optimization System

The integrated design system developed within the framework of this work consists of four main components, including: a blade geometric module defining airfoil shapes and blade chord and twist distributions, an aeroelastic module including airfoil aerodynamics and wind turbine rotor aeroelasticity, an algorithm for uncertainty propagation, and a two-stage multi-objective evolution-based optimizer.

14.2.1 Blade Geometric Module

Blade geometric module defines airfoil shapes and blade chord and twist distributions. The airfoil shape parametrization is achieved through a composite third order Bezier curve. More specifically, airfoil suction and pressure sides are described by four third order Bezier curves joined with C_0, C_1 and C_2 continuity. This solution provides a flexible airfoil parametrization within the context of the optimization process by actively using a total number of eleven design variables, corresponding to the degrees of freedom of the Bezier curve control points. Figure 14.1 shows the control points of the composite third order Bezier curves used to parametrize the airfoil shape. Along the suction side, points from $p1$ to $p4$ define the first third order Bezier curve, while points from $p4$ to $p7$ define the second one. In the pressure side, points from $p7$ to $p10$ define the third Bezier curve, and points from $p5$ to $p7$ define the fourth one. Thus, a total of 13 control points are used.

The leading and the trailing edges (points $p1$ and $p7$, respectively) are fixed. The degrees of freedom of the Bezier curve control points are the y-coordinate of point $p6$, and the x- and y-coordinates of points $p2$, $p3$, $p5$, $p11$, and $p12$. x and y-coordinates of all the remaining control points are determined by the algorithmic to keep the tangent and the curvature continuity between consecutive Bezier curves.

Blade chord and twist distributions are each defined by four design variables, representing chord and twist at four fixed radial sections. Blade chord and twist distributions are reconstructed by using the MATLAB shape-preserving piecewise cubic (pchip) interpolation function over the four radial stations.

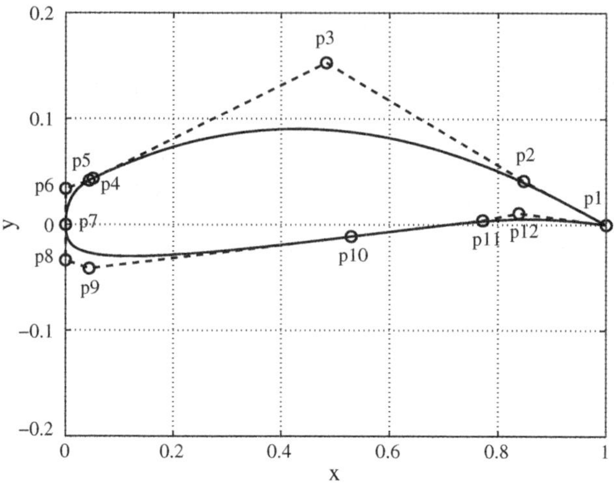

Fig. 14.1 Airfoil shape parametrization through a composite third order Bezier curve

14.2.2 Aeroelastic Module

The aeroelastic module consists of two main components, which are used to compute airfoil aerodynamic loads, as well as rotor power extraction and structural deformations. Lift and drag coefficients of the airfoil, as a function of the Reynolds number and the angle of attack, are calculated using the viscous-inviscid airfoil analysis code XFOIL [11]. XFOIL is a rapid and efficient way of calculating airfoil performance, however it may overestimate lift coefficient, and it does not provide reliable prediction beyond stall. Moreover, two-dimensional (2D) aerodynamic data calculated by means of XFOIL need to be corrected to account for the complex three-dimensional (3D) physics occurring over rotating blades, especially in the stall regime. Based on empirically derived equations, AERODAS [12] provides a method for calculating stall and post-stall lift and drag characteristics of rotating airfoils, using as input a limited amount of pre-stall 2D aerodynamic data of the airfoils used by the turbine under investigation.

The calculation of rotor power extraction and structural deformations is performed by means of the NREL aeroelastic design code FAST [13]. FAST solves the rotor aerodynamics through the AeroDyn code, employing the blade-element momentum (BEM) [14] theory and several corrections including those to account for tip and hub losses, axial induction factors exceeding the maximum theoretical limit of 0.5, and dynamic stall. FAST model uses a linear modal representation to model flexible blades [13]. Blade modes depend on the blade span-variant structural properties, which are tied to the blade external shape and the internal layup of composite laminates.

14.2.3 Method for Uncertainty Propagation

Uncertainty propagation is performed by means of the non intrusive URQ deterministic sampling technique, requiring $2n_u + 1$ evaluations, where n_u is the number of uncertain variables. URQ has a computational cost comparable to that of the linearization method, but allows a higher accuracy. More details about this method can be found in [10], and its validation against Monte Carlo method for wind turbine design can be found in [9].

14.2.4 Optimization Algorithms

Evolutionary Algorithms (EAs) solve optimization problems by making a generation of individuals evolve subject to selection and search operators. In this study, an individual denotes a HAWT rotor configuration, defined by the geometry of the bladed rotor and its rotational speed. This iterative process eventually leads to a population containing the fittest possible individuals (best rotor configuration designs), or individuals who are significantly fitter than those of the starting population. The role of the selection operators is to identify the fittest or most promising individuals of the current population, whereas search operators such as crossover and mutation attempt to generate better offspring starting from suitably selected individuals of the current generation. Each individual is defined by genes, which correspond to design variables in design optimization. The solution of the optimization problems reported in this study is based on a two-stage approach using the Multi-Objective Parzen-based Estimation of Distribution (MOPED) [15] and the Inflationary Differential Evolution Algorithm (IDEA) [16].

MOPED belongs to a subset of EAs and was developed to circumvent certain algorithmic problems of conventional EAs, which can be ineffective when the problem at hand features a high level of interaction among the design variables. This is mainly due to the fact that the recombination operators are likely to disrupt promising sub-structures that may lead to optimal solutions. Additionally, the use of the crossover and mutation operators may result in slow convergence to the solution of the optimization; that is, it may require a large number of generations to obtain very fit individuals. MOPED was developed to circumvent shortfalls of this kind. Its use of statistical tools enables it to preserve promising sub-structures associated with variable interaction from one generation to another (automatic linkage learning). Such statistical tools also replace the crossover and mutation operators of conventional EAs, and they allow a faster convergence of MOPED with respect to the latter class of optimizers. Starting from the individuals of the current population, MOPED builds an approximate probabilistic model of the search space. The role of the crossover and mutation operators is replaced by sampling of this probabilistic model. There exist similar other evolutionary methods that use the aforementioned strategy, and they are called Estimation of Distribution Algorithms (EDAs) [17].

MOPED is a multi-objective optimization EDA for continuous problems that uses the Parzen method [18] to build a probabilistic representation of Pareto solutions, and can handle multivariate dependencies of the variables [15, 19]. MOPED implements the general layout and the selection techniques of the Non-dominated Sorting Genetic Algorithm II (NSGA II) [20], but traditional crossover and mutation search approaches of NSGA-II are replaced by sampling of the Parzen model. NSGA-II was chosen as the base for MOPED mainly due to its simplicity, and also for the excellent results obtained for many diverse optimization problems [21, 22].

The Parzen method utilizes a non-parametric approach to kernel density estimation, and results in an estimator that converges asymptotically to the true Probability Density Function (PDF) over the whole design space. Additionally, when the true PDF is uniformly continuous, the Parzen estimator can also be made uniformly consistent. The Parzen method allocates N_{ind} identical kernels (where N_{ind} is the number of individuals of the current population), each centered on a different element of the sample. A probabilistic model of the promising search space portion is built on the basis of the statistical data provided by the N_{ind} individuals through their kernels, and $\tau_E\ N_{ind}$ new individuals ($\tau_E \leq 1$) are sampled. The variance of each kernel depends on (i) the location of the individuals in the search space and (ii) the fitness value of these individuals, and its construction leads to values that favor sampling in the neighborhood of the most promising solutions.

The features of MOPED often prevent the true Pareto front from being achieved, particularly when the front is broad and the individuals of the population are spread over different areas, which are far apart from each other in the feasible space. This circumstance has prompted coupling MOPED with another EA, which has better convergence properties. To this aim, the Inflationary Differential Evolution Algorithm (IDEA) [16] has been selected. IDEA was first developed for the design optimization of interplanetary trajectories, and it is an improved variant of the differential evolution (DE) algorithms [16]. The IDEA algorithm is based on a synergistic hybridization of a standard DE algorithm and the strategy behind the monotonic basin hopping (MBH) [23]. The resulting algorithm was shown to outperform both standard DE optimizers and the MBH algorithm in the solution of challenging space trajectory design problems, featuring a multiple funnel-like structure. In this paper, a modified version of IDEA has been used to move the individuals of the approximate Pareto front obtained with MOPED closer to the true front.

The main features of the original IDEA algorithm are reported in [16]. The IDEA algorithm works as follows: a DE process is performed several times and each process is stopped when the population contracts below a predefined threshold. At the end of each DE step, a local search is performed in order to get closer to the local optimum. In the case of non-trivial functions, where there is a high likelihood of converging to local optima, the combined DE/local search is usually iterated several times, performing either a local or a global restart on the basis of a predefined scheduling.

The design optimization presented in this study is constrained. Therefore, the DE step has been modified so that the fitness assessment of the individuals during the DE process also takes into account the constraints. The constraint handling technique used herein is one of the approaches that can be adopted in evolutionary computing, and is similar to the approach used by MOPED. In the unconstrained DE algorithm [24], and also in the unconstrained IDEA algorithm [16], each parent solution is compared with its offspring, and the solution with a better value of the objective function is passed to the next generation. In the constrained case, on the other hand, when parents and offspring are compared, the solutions are first evaluated in terms of constraint compatibility *cp*. Its definition is:

$$cp(\mathbf{x}) = \sum_{j=1}^{m} s_j(\mathbf{x}) \tag{14.1}$$

where $\mathbf{x}$ is the array of design variables, m is the number of constraints, and the constraint factor s_j is:

$$s_j(\mathbf{x}) = \max g_j(\mathbf{x}), 0 \tag{14.2}$$

The constraint factor equals 0 when the constraint ($g_j(\mathbf{x}) \leq 0$) is satisfied and is strictly positive when the constraint is violated. The solution with the better values of *cp* is then passed to the next generation. When the *cp* values of parent and offspring are the same, the selection is performed on the basis of the objective function. In the current implementation, MOPED and IDEA are used sequentially. When MOPED has reached a given number of generations, its final population represents a first and good approximation to the sought Pareto front. Then, clustered sub-populations of such a population are used as initial solutions of the single-objective constraint IDEA optimizer. this algorithm "pushes" the individuals of a sub-population of the MOPED front towards a better local approximation of the sought Pareto front. The resulting two-stage optimizer blends the exploratory capabilities of MOPED (global exploration) and the favorable convergence characteristics of IDEA (exploitation of local information).

14.3 Optimization Set-Up

In this study, two optimizations have been performed: a robust optimization, and a deterministic one (i.e., without considering uncertainties throughout the design process) for comparison purposes.

14.3.1 General Settings

All optimizations performed in this research aimed at maximizing the AEP of a three-bladed HAWT based on the NREL 5-MW reference turbine [25]. The yearly wind distribution was represented by a Weibull distribution with scale parameter of 7 m/s and shape parameter of 2 (Fig. 14.2). The wind turbine was regulated through

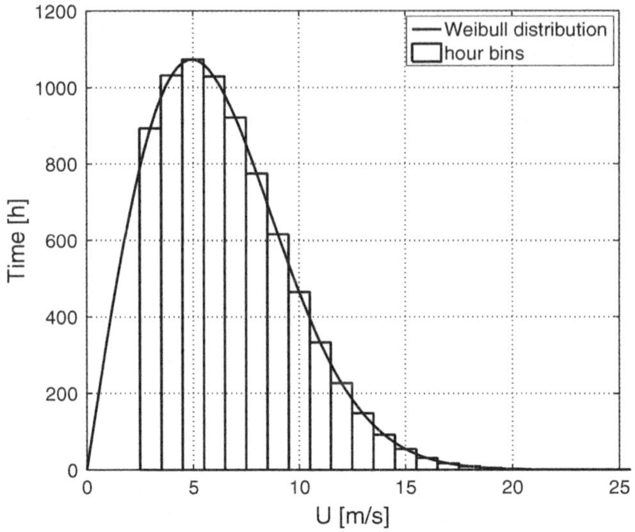

Fig. 14.2 Weibull distribution with scale parameter of 7 m/s and shape parameter of 2, and hour bins for the wind speeds considered between cut-in and cut-out

variable rotational speed before rated wind speed, and variable blade pitch thereafter. Cut-in, rated and cut-out wind speeds were fixed to 3, 12 and 25 m/s, respectively. Blade root radius was set equal to 1.5 m, while blade tip radius was set equal to 63 m. From root to tip, each blade was modeled through seventeen radial sections, including two cylindrical sections near the root, one section transitioning from the last cylindrical section to the first airfoil section, and fourteen airfoil sections over the remainder.

Excluding the cylindrical and transitioning sections near the root, the airfoil distribution along the blade span was defined exclusively by a single aerodynamic shape. Airfoil lift and drag coefficients were computed through XFOIL for a single Reynolds number of $1.2 \cdot 10^7$ over an angle of attack range spanning form -5 to $25°$. XFOIL polars were then extended for angles of attack ranging from -180 to $180°$, and corrected to account for 3D aerodynamic effects by means of the AERODAS model. In XFOIL, transition from laminar to turbulent flow along the airfoil is simulated by the e^N method, through the parameter NCRIT. For all optimizations reported in this paper, NCRIT was fixed to 9. The airfoil shape and the twist and chord distributions at the airfoil sections were treated as part of the optimization. Lift and drag coefficients of the cylindrical sections near the hub were assumed to be equal to 0 and 0.5, respectively. Lift and drag coefficients for the transitioning section were obtained by interpolating between the neighboring sections. Blade chord and twist distributions were defined at four radial sections r_1, r_2, r_3 and r_4 fixed to 11.75, 28.15, 48.65 and 61.6333 m, respectively. The blade shape was parametrized by means of eleven design parameters defining the airfoil shape (x_1 to x_{11}), four design parameters defining the chord distribution (x_{12} to x_{15}), and four design parameters defining the twist distributions (x_{16} to x_{19}). Wind turbine operating conditions were completely determined by one design variable defining the rotational speed associated with the rated wind speed x_{20}. Indeed, the optimal rotational speed at the given rated wind speed defined the constant optimal tip speed ratio at which the turbine operated between cut-in and rated wind speeds. Thus, the total number of design parameters was equal to 20. Design variables and their ranges of variability are shown in Table 14.1. All optimizations were subject to a structural constraint tied to the maximum out-of-plane blade tip deflection (BTD), which was assumed to be equal to 2/3 of the total clearance in unloaded conditions [3]. Tower diameter was assumed to be constant and equal to 6 m. Rotor overhang, rotor shaft tilt angle, and blade precone angle were fixed to 5.0191 m, 5°, and 2.5°, respectively. Maximum BTD allowed was therefore equal to 6.8 m.

FAST is able to account for flexible bodies, including tower, blades and drive shaft. However, since this work focuses only on the design of bladed rotors, tower and drive shaft deflections were neglected. Moreover, during the optimization process, the adopted hypothesis was to change the internal layup of each turbine in such a

Table 14.1 Range of design variables

x_1	$\in [0.001, 0.2]$	$y(p6)$
x_2	$\in [0.001, 0.2]$	$x(p5)$
x_3	$\in [0.001, 0.2]$	$y(p5))$
x_4	$\in [0.3, 0.6]$	$x(p3)$
x_5	$\in [0.1, 0.5]$	$y(p3)$
x_6	$\in [0.7, 0.9]$	$x(p2)$
x_7	$\in [0.001, 0.1]$	$y(p2)$
x_8	$\in [0.6, 0.8]$	$x(p11)$
x_9	$\in [-0.1, 0.01]$	$y(p11)$
x_{10}	$\in [0.8, 0.99]$	$x(p12)$
x_{11}	$\in [-0.05, 0.1]$	$y(p12)$
x_{12}	$\in [4, 5.5]$ m	$c(r_1)$
x_{13}	$\in [3.5, 5]$ m	$c(r_2)$
x_{14}	$\in [2, 3.5]$ m	$c(r_3)$
x_{15}	$\in [0.5, 2]$ m	$c(r_4)$
x_{16}	$\in [11, 16]°$	$\theta_T(r_1)$
x_{17}	$\in [5, 10]°$	$\theta_T(r_2)$
x_{18}	$\in [0, 5]°$	$\theta_T(r_3)$
x_{19}	$\in [0, 3]°$	$\theta_T(r_4)$
x_{20}	$\in [5, 15]$ RPM	$\Omega(U_{\text{rated}})$

$x(p5)$, $x(p3)$, $x(p2)$, $x(p11)$, and $x(p12)$ represent the x-coordinates of control points $p5$, $p3$, $p2$, $p11$ and $p12$, respectively. $y(p6)$, $y(p5)$, $y(p3)$, $y(p2)$, $y(p11)$ and $y(p12)$ represent the y-coordinates of control points $p6$, $p5$, $p3$, $p2$, $p11$ and $p12$, respectively. $c(r_1)$, $c(r_2)$, $c(r_3)$ and $c(r_4)$ represent the chord length at r_1, r_2, r_3 and r_4, respectively. $\theta_T(r_1)$, $\theta_T(r_2)$, $\theta_T(r_3)$, and $\theta_T(r_4)$ are the twist angles at r_1, r_2, r_3 and r_4, respectively. $\Omega(U_{\text{rated}})$ represents the rotational speed at the rated wind speed way that the span-variant structural properties remained constant. Therefore, blade modes were considered constant throughout the optimization process.

To avoid intersection between airfoil suction and pressure sides, and more than one change in their curvatures, two constraints were enforced. For practical purposes tied to blade manufacturing, a monotonicity constraint was enforced on both chord and twist distributions.

For both MOPED optimization process, the size of the population have set to 100 and the fitness parameter α_f and the sampling proportion τ_E have been set to 0.5 and 1, respectively. The maximum number of generation was set 100 for the deterministic process and 300 for the robust one. In both IDEA-based optimizations, the weighting factor F and the crossover probability CR have been set to 0.6 and 0.9, respectively. The IDEA search has used a random population of 40 individuals, and has stopped when the population has contracted to 25 % of the maximum expansion during the evolution.

14.3.1.1 Robust Design Optimization

The 8 design parameters defining chord and twist distributions as well as the airfoil thickness were assumed to be affected by normally distributed uncertainty. The Gaussian distribution of these parameters was centered at their nominal values. Standard deviations were set to 3° for twist, and 30 cm for chord. At a given radial section, standard deviation for thickness was considered equal to 1 % of the chord. The robust optimization described in this paper aimed to maximize the mean value of the AEP and minimize its standard deviation, by varying the twenty aforementioned design variables and propagating the uncertainties affecting the nine variables described above. Robust optimization was achieved by minimizing the following objective functions:

$$F_1 = -\mu_{AEP} \tag{14.3}$$

$$F_2 = \sigma_{AEP}^2 \tag{14.4}$$

where μ_{AEP} is the mean value of the AEP in kWh, and σ_{AEP}^2 is its variance in kWh^2. The robust optimization was subject to the following constraints:

$$C_1 : \mu_{AEP} \geq 5e6 \tag{14.5}$$

$$C_2 : \sigma_{AEP}^2 \leq 1e11 \tag{14.6}$$

$$C_3 : (\mu_{BTD} + 3 \cdot \sigma_{BTD}) \leq 6.8 \tag{14.7}$$

where μ_{BTD} and σ_{BTD} represent maximum BTD main value and standard deviation in m, respectively. AEP was computed for each turbine by integrating its power curve against the given Weibull distribution. In order to save computational sources, power curve of each wind turbine was determined through three FAST runs, by interpolating between the power values computed at cut-in wind speed, rated wind speed, and at one intermediate wind speed of 8 m/s. This interpolation was performed by a cubic spine. For each turbine, one additional FAST run was required to determine the maximum BTD. According to Ghedin [3], maximum blade deflections occur when the wind turbine works at rated conditions under a severe gust. In these conditions, the gust occurs so suddenly that the blades cannot be pitched. The gust intensity a wind turbine can withstand depends on its class. The NREL 5-MW reference turbine belongs to the 1A IEC wind class [3], which means it is able to withstand gusts up to 21 % above its rated wind speed. Therefore, maximum BTD was computed at a wind speed equal to 14.52 m/s.

The robust optimization was performed considering identical geometric errors affecting all blades. Each robust analysis performed by means of the URQ technique required 19 computations of AEP, namely 76 FAST runs.

14.3.1.2 Deterministic Design Optimization

Along with the robust optimization, a deterministic optimization was performed. In the deterministic optimization uncertainty sources where not included, and therefore the following objective function was minimized:

$$F_1 = -AEP \tag{14.8}$$

The enforced constraints were:

$$C_1 : AEP \geq 5e6 \tag{14.9}$$

$$C_2 : BTD \leq 6.8 \tag{14.10}$$

14.4 Results and Discussions

14.4.1 Robust Optimization

The robust design optimization problem led to a Pareto front arising from the trade-off between the mean and the standard deviation of AEP. Figure 14.3 shows the URQ Pareto front obtained by means of MOPED. In Fig. 14.3, the performance of a nominal rotor, labeled "URQ ref.", obtained by using the IDEA local refinement is also represented. Given the final population of the MOPED optimization, the IDEA refinement was performed selecting a sub-population containing a solution with maximum μ_{AEP}, and using it as starting point of the IDEA optimization. This optimization aimed at maximizing μ_{AEP} subject to the constraints (14.5)–(14.7).

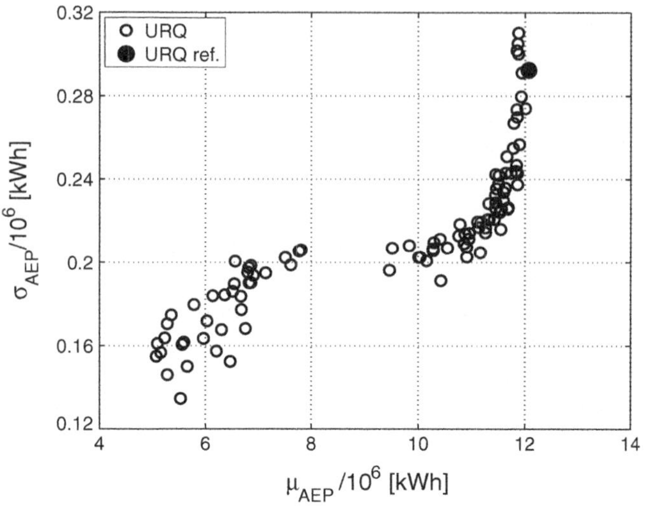

Fig. 14.3 Airfoil shapes of deterministic and robust designs

14.4.2 Robust and Deterministic Optimal Rotors

The deterministic optimization problem led to an optimal rotor which has a nominal AEP of 13.3352 GWh and $\mu_{AEP} = 13.1474$ GWh. This rotor configuration, denoted by "det. opt.", is compared to the "URQ ref." rotor, which has a nominal AEP of 12.1435 GWh, and $\mu_{AEP} = 12.0732$ GWh. The AEP standard deviation of the "det. opt." rotor is $\sigma_{AEP} = 0.2305$ GWh, and is lower than that of the "URQ ref." rotor, which is equal to $\sigma_{AEP} = 0.2924$ GWh. These results do not meet the usual expectations as the AEP standard deviation of the rotor designed taking into account stochastic geometry errors due to manufacturing tolerances is higher than that of the rotor designed neglecting such errors. The structural constrain on the maximum BTD is the explanation for these results. The robust rotor has a nominal maximum BTD equal to 5.6550 m, and maximum BTD mean and standard deviation equal to 5.6533 and 0.3797 m, respectively. The nominal maximum BTD associated with the deterministic design is equal to 6.788 m, while its mean value and standard deviation are 6.7435 and 0.4915 m, respectively. As can be seen, the constraint (14.7) is verified only for the robust optimization. To demonstrate that the obtained results are strongly influenced by the constraint, a new uncertainty based optimization was performed replacing the reliability constraint (14.7) with the following:

$$C_3 : \mu_{BTD} \leq 6.8 \tag{14.11}$$

The rotor obtained through this new optimization, denoted by "URQ ref. 2", has a nominal AEP of 13.2751 GWh, $\mu_{AEP} = 13.1263$ GWh and $\sigma_{AEP} = 0.1816$ GWh. The "URQ ref. 2." rotor has an AEP standard deviation which is lower than that of the "det. opt." rotor by about 20 %.

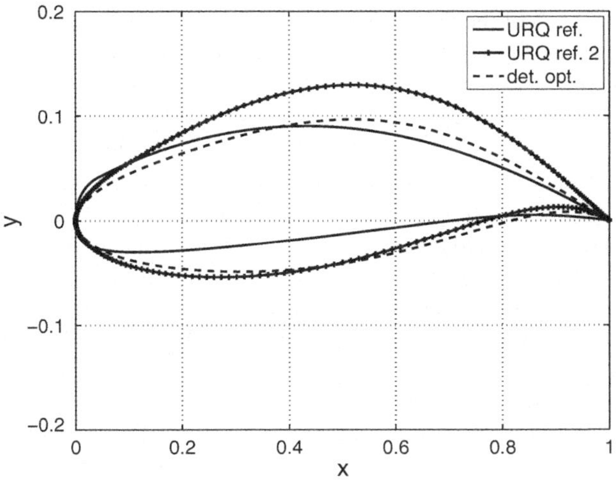

Fig. 14.4 Airfoil shapes of deterministic and robust designs

The airfoil shapes obtained through (i.e., "URQ ref." and "URQ ref. 2") and deterministic (i.e. "det. opt.") optimizations are shown in Fig. 14.4. The radial profiles of the chord c and the pitch angle θ_p of the three rotors are reported in the top left and bottom left subplots, respectively, of Fig. 14.5. Chord distributions of the deterministic and robust designs have the same shape near the root because the cylindrical sections, not treated as part of the optimization, were considered constant.

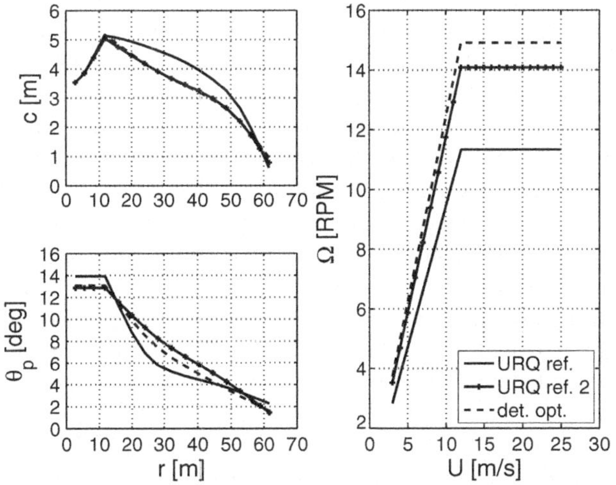

Fig. 14.5 Rotor geometry and control of deterministic and robust designs. *Top left subplot* blade chord c against radius r. *Bottom left subplot* section pitch angle θ_p against radius. *Right subplot* rotational speed Ω against wind speed U

The rotational speed of the three rotors for all considered wind speeds is reported in the right subplot of Fig. 14.5, which highlights that the probabilistically optimized rotors have lower Ω values than the deterministically optimized one.

In light of the above analysis, as it was also reported by Minisci et al. [9], a lower sensitivity of AEP to rotor geometry errors can be achieved by lowering rotational speeds. Further analyses are needed, and have been carrying out by the authors to demonstrate that these results meet the findings obtained in [9]. In this paper it was explained and demonstrated that the reduction of power due to lower circumferential velocities is compensate by an increase of blade aerodynamic loading (i.e., radial lift coefficient distribution), achieved by increasing the angle of attack. This increase is due to the reduced circumferential speed itself, which results in higher values of the relative wind angle. In such conditions, the overall level of angle of attack is in a region where the slope of the angle of attack/lift coefficient curve starts to decrease with respect to the linear part corresponding to lower angles of attack. Therefore, the variation of the lift coefficient caused by a given variation of the angle of attack is smaller for robust rotors.

14.5 Conclusions

A cascade of evolutionary algorithms has been applied to the robust aerodynamic design of a wind turbine rotor to maximize the annual energy production and, at the same time, minimize its variations due to blade manufacturing tolerances. The deterministic URQ sampling approach has been adopted for uncertainty propagation instead of the much more expensive Monte Carlo sampling. The performed robust optimizations and the comparison with the reference deterministic optimization stress the influence of the structural constraint on the achievable results. When the considered constraint limits the magnitude of the mean value of the maximum out-of-plane blade tip deflection, the robust optimization procedure can obtain a rotor producing the same annual energy of the deterministic one, but with a standard deviation which is 20 % lower. On the other hand, if a reliability constraint on the maximum out-of-plane blade tip deflection is considered, the average performance of obtained rotors decrease significantly.

Both robust optimization processes performed in this paper confirm that the search for the lower sensitivity to geometry errors is pursued by adopting lower rotational speeds, and further investigations are needed to demonstrate that robustness is actually obtained by moving to a range of higher values of the angle of attack where the slope of the angle of attack/lift coefficient curve is lower than for lower values of the angle of attack.

For this work, low fidelity models such as XFOIL and FAST have been adopted. If the use of low fidelity models allows preliminary design procedures requiring a huge number of model evaluations, on the other hand, the search space should be heavily bounded, to avoid regions of the design space where the (low fidelity) models can not provide correct results. The next step of this work will regard the integration of the current optimization approach with a multi-fidelity method, which will allow one to achieve a true global design optimization.

References

1. Benini E, Toffolo A (2002) Optimal design of horizontal-axis wind turbines using blade element theory and evolutionary computation. J Sol Energy Eng 124(4):357–363
2. Fuglsang P, Madsen HA (1999) Optimization method for wind turbine rotors. J Wind Eng Ind Aerodyn 80(1–2):191–206
3. Ghedin F (2010) Structural Design of a 5 MW wind turbine blade equipped with boundary layer suction technology. Master's thesis, Eindhoven University of Technology
4. Vesel RW Jr (2012) Aero-structural optimization of a 5 MW wind turbine rotor. Master's thesis, The Ohio State University
5. Jureczko M, Pawlak M, Mężyk A (2005) Optimisation of wind turbine blades. J Mater Process Technol 167(2–3):463–471
6. Xudong W, Shen WZ, Zhu WJ, Sørensen JN, Jin C (2009) Shape optimization of wind turbine blades. Wind Energy 12(8):781–803
7. Petrone G, de Nicola C, Quagliarella D, Witteveen J, Iaccarino G (2011) Wind turbine performance analysis under uncertainty. In: Proceeding of the 49th AIAA aerospace sciences meeting including the new horizons forum and aerospace exposition, January 2011
8. Petrone G, de Nicola C, Quagliarella D, Witteveen J, Axerio-Cilies J, Iaccarino G (2011) Wind turbine optimization under uncertainty with high performance computing. In: Proceeding of the 29th AIAA applied aerodynamics conference, June 2011
9. Minisci E, Campobasso MS, Vasile M (2012) Robust aerodynamic design of variable-speed wind turbine rotors. In: Proceeding of the ASME Turbo Expo 2012 technical conference, June 2012
10. Padulo M, Campobasso MS, Guenov MD (2011) A novel uncertainty propagation method for robust aerodynamic design. AIAA J 49(3):530–543
11. Drela M (1989) XFOIL: an analysis and design system for low Reynolds number airfoils. In: Low Reynolds number aerodynamics, Lecture Notes in Engineering, vol 54. Springer, Heidelberg
12. Spera D (2008) Models of lift and drag coefficients of stalled and unstalled airfoils in wind. Turbines and Wind Tunnels
13. Buhl ML Jr, Jonkman JM (2005) FAST user's guide. Technical report NREL/EL-500-38230, National Renewable Energy Laboratory
14. Manwell J, McGowan J, Rogers A (2002) Wind energy explained. Design and application theory. Wiley, New York
15. Costa M, Minisci E (2003) MOPED: a multi-objective parzen-based estimation of distribution algorithm. In: EMO 2003, Faro, Portugal, Springer, Heidelberg, pp 282–294
16. Vasile M, Minisci E, Locatelli M (2011) An inflationary differential evolution algorithm for space trajectory optimization. IEEE Trans Evol Comput 15(2):267–281
17. Lozano JA, Larranaga P, Inza I, Bengoetxea E (2006) Towards a new evolutionary computation: advances on estimation of distribution algorithms (Studies in fuzziness and soft computing). Springer, Heidelberg
18. Fukunaga K (1972) Introduction to statistical pattern recognition. Academic Press, New York
19. Avanzini G, Biamonti D, Minisci EA (2003) Minimum-fuel/minimum-time maneuvers of formation flying satellites. Adv Astronaut Sci 116:2403–2422
20. Deb K, Pratap A, Agarwal S, Meyarivan T (2002) A fast and elitist multiobjective genetic algorithm: NSGA-II. IEEE Trans Evol Comput 6(2):182–197
21. Datta D, Deb K, Fonseca CM, Lobo FG, Condado PA, Seixas J (2007) Multi-objective evolutionary algorithm for land-use management problem. Int J Comput Intell Res 3(4):371–384

22. Deb K (2008) Scope of stationary multi-objective evolutionary optimization: a case study on a hydro-thermal power dispatch problem. J Glob Optim 41(4):479–515
23. Leary RH (2000) Global optimization on funneling landscapes. J Glob Optim 18(4):367–383
24. Price K, Storn RM, Lampinen JA (2005) Differential evolution: a practical approach to global optimization. Springer, Heidelberg
25. Jonkman J, Butterfield S, Musial W, Scott G (2009) Definition of a 5-MW reference wind turbine for offshore system development. Technical report NREL/TP-500-38060, National Renewable Energy Laboratory

Chapter 15
Horizontal Axis Hydroturbine Shroud Airfoil Optimization

Elia Daniele, Elios Ferrauto and Domenico P. Coiro

Abstract The present work concerns the optimization of the shroud of an horizontal axis hydro turbine (HAHT). The main aim is to improve the hydro-turbine efficiency by designing a new shroud airfoil through an optimization process that maximize, as objective function, the power coefficient. The optimization process is carried out by MATLAB© on the supercomputing infrastructure SCoPE of the University of Naples, "Federico II". Results are obtained with CFD calculations, namely by STARCCM+ for an axisymmetric model, taking advantage of the symmetry of the problem, to minimize the computational time; in addition the HAHT is simulated with an actuator disk that gave reliable results in good agreement with previous works, developed with different software, and with experimental results. The original airfoil was designed for high-lift regimes, so it already gave excellent performance in these kind of applications. For that reason, is not expected a very high increase of the power coefficient. Nevertheless the optimization process results into a power coefficient increase of 4.5 %, with respect to the original airfoil.

Keywords Horizontal axis hydro turbine · Shroud · Optimization · Genetic algorithm · CFD

15.1 Introduction

One of the most promising fields in energy production from renewable sources is related to the intensive exploitation of marine and river currents [5]. The marine current resource has a major advantage over other renewable energy resources in

E. Daniele (✉)
Ammerländer Heerstraße 136, 26129 Oldenburg, Germany
e-mail: elia.daniele@iwes.fraunhofer.de

E. Ferrauto · D.P. Coiro
Department of Industrial Engineering (DII), University of Naples, "Federico II",
Via Claudio 21, 80125 Naples, Italy
e-mail: e.ferrauto@studenti.unina.it

D.P. Coiro
e-mail: domenico.coiro@unina.it

D. Greiner et al. (eds.), *Advances in Evolutionary and Deterministic Methods for Design, Optimization and Control in Engineering and Sciences*, Computational Methods in Applied Sciences 36, DOI 10.1007/978-3-319-11541-2_15

that it is essentially non-intermittent and predictable over long time periods. Having a density of approximately 830 times that of air, water allows for greater kinetic energy extraction over that of wind at similar flow speeds and rotor size. Given a suitable site of concentrated flow and high speed velocities, ocean currents can offer up to four times the energy density of a good wind site, and 30 times the energy density of a solar plant in the Sahara Desert [17]. Ocean currents are consistent, and their strength and directional frequency can be predicted. Flow speed velocities in ocean currents vary little from average flow velocities, and the source availability differentiates it from other renewable technologies. Many configurations have been proposed so far to extract amounts of energy from water currents.

This paper is related to the optimization of an innovative system configuration for clean energy applications whose concept has been already tested on a preliminary prototype model at the University of Naples. The research activities, upon which the present work is based, directly follows the preliminary testing phase oriented to explore the feasibility of plants based on such concept [6]. The original system configuration consists of a submerged floating body linked to the seabed by means of a tether. Electrical generators and auxiliary systems are housed within the hulls of the floating bodies. Two turbines are installed on a support structure exposed to the flow. The system under development is suitable for applications in sea and rivers. According to what has been observed at this stage of development, it possesses also the desirable characteristics of constructive simplicity and ease of operation. This is the basic concept of what we call here "GEM system" (Generatore Elettrico Marino, Marine Electrical Generator) or "Ocean's kite" configuration, and starting from such patented concept a scale model has been realized and tested (see Figs. 15.1 and 15.2).

The proposed configuration, due to a relatively safe and easy self-orienting behavior, is a good candidate to solve some problems involved with oscillating and reversing streams. An additional major advantage of this configuration is related to the possibility of avoiding the use of expensive submarine foundations on the seabed. The use of diffuser augmented turbines is intended to reduce the dimensions of actuator disks for a given rated power and to increase the rotational speed, respect of a rotor with higher diameter, reducing the torque. To extract energy from the wind, a thrust force

Fig. 15.1 A CAD image of GEM hydro-turbine

Fig. 15.2 A picture of GEM hydro-turbine full-scale prototype installed in Venice's lagoon. The prototype has been sponsored by a consortium of Venetian companies and by the Veneto regional authority (*source* http://www.adag.unina.it)

T directed downstream must be generated. On a horizontal axis turbine the thrust is obtained by rotating blades, which create a pressure drop across the rotor disk. The extracted power is the product of the airflow through the disk (m^3/s) and the pressure drop across the disk. Increasing the thrust increases the pressure drop but decreases the airflow through the disk. An optimum exists for the power coefficient C_p, which is the ratio between the extracted power and the available power. This optimum for a bare turbine (without a diffuser) is known as the Betz limit [24],

$$C_{P_{max}} = \frac{16}{27}. \tag{15.1}$$

However, if a mechanism is used to increase the airflow through the rotor disk, the Betz limit can be exceeded. According to [13], a ring vortex of the appropriate sign around the rotor plane would induce (by the Biot-Savart law) a velocity vector to increase the mass flow through the disk. This effect can be obtained by placing a diffuser around the rotor. If the cross-section of the diffuser is shaped as an aerofoil, the generated lift will give the circulation of a ring vortex. The more lift that can be achieved, the more the air will be sucked through the disk. The amount of lift is limited by separation of the boundary layer on the diffuser wall, so the geometry of the diffuser plays an important role in the overall performance of the shrouded wind turbine (see Fig. 15.3 on the following page). The radial distribution of the axial flow is also important in order to ensure a uniform speed-up factor.

Conversely, Van Bussel [25] asserts that, with a simple momentum theory, developed along the lines of momentum theory for bare turbines, power augmentation is proportional to the mass flow increase generated at the nozzle of the diffuser augmented turbine. Such mass flow augmentation can be achieved through two basic principles: increase in the diffuser exit/inlet ratio and/or by decreasing the negative back pressure at the exit. The power increase could results in a misleading overcoming of the Betz limit. As shown by Van Bussel this is due to an incorrect choice of the reference area. In this case the suitable reference area should be the shroud exit area. From this momentum theory, it can be seen that the achievable power

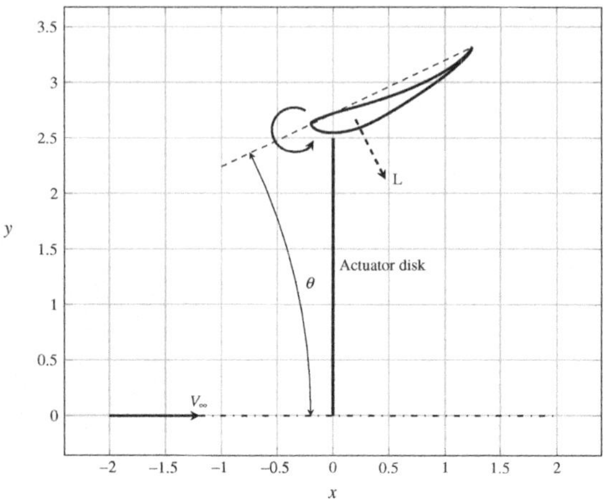

Fig. 15.3 Lift force on the shroud airfoil and pitch angle θ

is comparable with the power of a normal Horizontal Axis Wind Turbine (HAWT) having a diameter equal to the exit diameter of the diffuser. But from this momentum model it can also be seen that larger performances are possible when a substantial low *back pressure level* can be achieved at the diffuser exit. As shown by Tognaccini in [24] the power augmentation is proportional to the thrust exerted by the flow on the diffuser. The shroud design criteria were based on the maximization of this thrust. An high lift airfoil has been opportunely chosen, and the working angle of attack has been numerically estimated by Scherillo et al. [18]. The experimental results achieved in that work, both in the wind tunnel and in the towing tank, show a remarkable increase of the thrust for the shrouded configurations. Besides, according to the theory, the power increases remarkably with the diffuser. The power coefficient of the shrouded configuration, computed referring to the turbine area is almost 0.8, and shows an increase of about twice compared to the bare turbine, that has a $C_P = 0.4$. Referred the power coefficient of the shrouded configuration to the diffuser exit area, the C_P increase is about 7 %.

The main target of this work is to exploit this limit with the use of an optimization process for the shape of the airfoil on which is based the annular diffuser. In Sect. 15.2 the numerical methodology is described, in Sect. 15.3 the results are summarized, then in Sect. 15.4 the conclusions and future works are highlighted.

15.2 Method

Both in aircraft design and in turbine design, the choice of airfoils is critical because it affects overall project performance. Often, an *ad hoc* designed airfoil is used. The aim of this section is to apply numerical optimization concepts to the airfoil design problem. One of the most important ingredients in numerical optimization is the choice of design variables and the parametrization of the system by using these variables. In general, an airfoil is given by its coordinates, typically a set of 150–200 points for panel codes; evidently, it is not suitable to use directly the airfoil's coordinates as design variables, because, even if it is the easier method to implement, there is a large number of design variables to represent 2D or 3D geometries, and displacement of a single mesh point can lead to unsmooth shapes and cause the flow solver to become ill-conditioned, as observed in [4, 27].

In order to reduce the number of parameters to take into account necessary to describe the airfoil's shape, but without geometrical information loss, several mathematical formulations have been proposed in literature as Bezier [12], PARSEC [21, 22] and Legendre polynomials as suggested by Hicks and Van der Plaats in [14, 15]. The latter one has been chosen among the others because of its capability of describing only limited change in the airfoil shape by summing polynomials on the original airfoil coordinates. A Legendre polynomial is a function that satisfies the Legendre's differential equation whose expression is shown in Eq. 15.2:

$$\frac{\mathrm{d}}{\mathrm{d}x}\left[(1-x^2)\frac{\mathrm{d}}{\mathrm{d}x}P(x)\right]+n(n+1)P(x) \tag{15.2}$$

The ordinary differential equation is quite frequent in mathematics and physics since it allows to solve Laplace's equation in spherical coordinates and several partial derivative differential equation. Legendre's differential equation can be solved trough standard methods applying power series so that converging solutions are obtained if $|x| < 1$. Converging solutions are obtained also if $x = \pm 1$ and n is a natural integer (i.e. $n = 0, 1, 2, \ldots$). In such cases the solutions according to n form a polynomial succession called Legendre's polynomials succession. The generic Legendre's polynomial $P_n(x)$ of n degree can be expressed trough the following Eq. 15.3:

$$P_n(x) = \left(2^n n!\right)^{-1}\frac{\mathrm{d}^n}{\mathrm{d}x^n}\left[\left(x^2-1\right)^n\right] \tag{15.3}$$

Several artifices have been adopted in order to use Legendre's polynomials for the optimization process according to [14]. Airfoil thickness distributions are given by summing a perturbation on the original geometry as shown in Eq. 15.4

$$y(x)_{new} = y(x)_{old} + \Delta y(x)_{(up/low)} \tag{15.4}$$

where $y(x)_{old}$ represents the original airfoil y coordinate and Δy the perturbation term for the upper and lower surfaces ordinates. This term is evaluated according to Eqs. 15.5 and 15.6

$$\Delta y(x)_{up} = (1-x)^3\Big[\sqrt{a_1 x} + a_2\,(P_2+1) + a_3\,(P_3-1) + a_4\,(P_4+1) + a_5\,(P_5-1) + a_6\,(P_6+1)\Big] \tag{15.5}$$

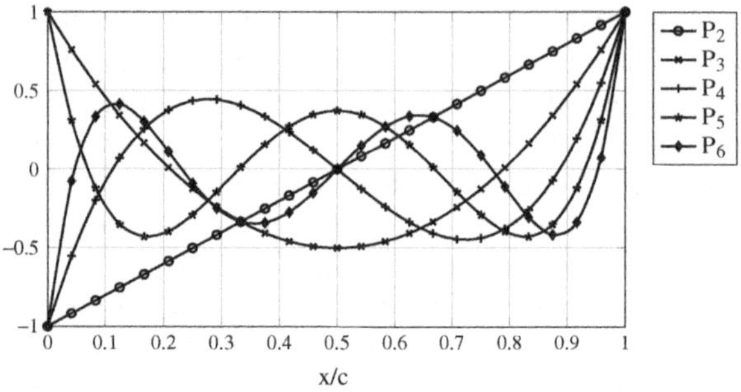

Fig. 15.4 Legendre polynomials

$$\Delta y(x)_{low} = (1-x)^3\Big[\sqrt{b_1 x} + b_2\,(P_2+1) + b_3\,(P_3-1) + b_4\,(P_4+1) + b_5\,(P_5-1) + b_6\,(P_6+1)\Big] \quad (15.6)$$

where and $P_2, \ldots, P_6$, are Legendre polynomials given by Eq. 15.7 and shown in Fig. 15.4. The coefficients $a_1, \ldots, a_6$ and $b_1, \ldots, b_6$ are the design variables perturbed by the optimization program to achieve optimum design. The square root term in Eqs. 15.5 and 15.6 allows a blunt leading edge and assures matching of upper-surface and lower-surface derivatives of all orders at the leading edge. The term $(1-x)^3$ allows to optimize the airfoil in a chosen chord range. This term could be removed by specify that whole the airfoil wants be optimized.

$$\begin{aligned}
P_2 &= 2\,(x/c) - 1\\
P_3 &= 6\,(x/c)^2 - 6\,(x/c) + 1\\
P_4 &= 20\,(x/c)^3 - 30\,(x/c)^2 + 12\,(x/c) - 1\\
P_5 &= 70\,(x/c)^4 - 140\,(x/c)^3 + 90\,(x/c)^2 - 20\,(x/c) + 1\\
P_6 &= 252\,(x/c)^5 - 630\,(x/c)^4 + 560\,(x/c)^3 - 210\,(x/c)^2 + 30\,(x/c) - 1
\end{aligned} \quad (15.7)$$

During years, a lot of optimization methods have been proposed and developed, often starting from theoretical concepts and logics very far each from each other. In general it is very difficult to state which method is the best because each one has several advantages and, at same time, disadvantages; just referring to a particular application, or problem, it is possible to operate this choice. Genetic algorithms (GA) is a heuristic search method derived from natural selection and evolution. At the start of a GA optimization, a set of decision variable solutions are encoded as members of a population. There are multiple ways to encode elements of solutions including binary, value, and tree encodings. Crossover and mutation, operators based on reproduction, are used to create the next generation of the population. One of advantages of GA is that multiple areas of the search space are explored to find a

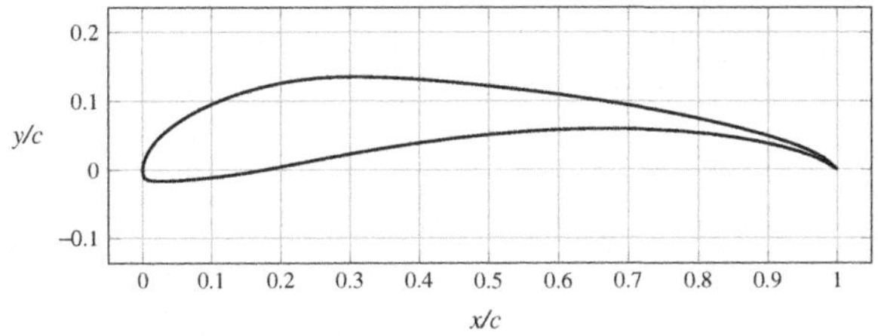

Fig. 15.5 Base airfoil used for the shroud

global minimum. Through the use of the crossover operator, GA are particularly strong at combining the best features from different solutions to find one global solution. Through observation of these crossover combinations, the user gains insight about how parts of the simulation interact. The non-dominated Sorting Genetic Algorithm [1] is a Multiple Objective Optimization (MOO) algorithm and is an instance of an Evolutionary Algorithm from the field of Evolutionary Computation [8]. NSGA is an extension of the Genetic Algorithm for multiple objective function optimization. There are two versions of the algorithm, the classical NSGA and the updated and currently canonical form NSGA-II [9]. The objective of the NSGA algorithm is to improve the adaptive fit of a population of candidate solutions to a Pareto front constrained by a set of objective functions. The algorithm uses an evolutionary process with surrogates for evolutionary operators including selection, genetic crossover, and genetic mutation. The population is sorted into a hierarchy of sub-populations based on the ordering of Pareto dominance. Similarity between members of each sub-group is evaluated on the Pareto front, and the resulting groups and similarity measures are used to promote a diverse front of non-dominated solutions. NSGA-II [8, 9] is an improved version of NSGA which alleviates main criticisms of the NSGA approach. NSGA-II is here used in its single objective functionality. Parametrization and optimization setup for this numerical optimization will be now presented in detail.

The base airfoil selected for the optimization process is a high-lift airfoil for low Reynolds number (see Fig. 15.5 [19]). It is a quite suitable airfoil for a shroud that maximize the hydroturbine power coefficient [11, 16]. In this type of problem, a suitable method of reconstruction is that based on Legendre polynomials, because, as explained before, it performs geometric reconstruction by applying small perturbation on the original geometrical coordinates of the airfoil. Three cases have been analysed, with different abscissa variation ranges in percent of the chord; for each of them variation on 90 % of chord length are imposed: $BC = \pm 0.01$, ± 0.05 and ± 0.10. No geometric and aerodynamic constraints are setted in these analysis. Concerning numerical optimization, the chosen method is NSGA-II [20], executed by use of MATLAB©. A single-objective optimization was performed, the aim of which is maximize the hydro-turbine power coefficient.

The population size of the algorithm was setted to 40, while the generation limit is 20. Three analysis are carried out, by changing the airfoil pitch angle, that is defined

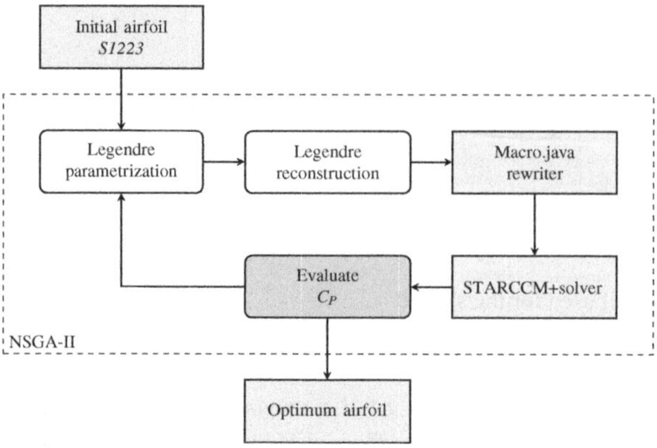

Fig. 15.6 Numerical optimization routine

as the angle between the turbine axis and the chord of the airfoil, as explained in Fig. 15.3. Starting from a pitch angle of 26° (that is the best angle in this analysis conditions for the base airfoil, as proved by [26]) a variation of $\pm 3°$ was provided for the optimization process.

Finally three pitch angles are implemented: $\theta = 23,\ 26,\ 29°$; considering the fact that the airfoil resulting from the optimization process could work better with different pitch angle. Moreover, an airfoil with good performance in a range of ± 3 deg can have good performance also in off-design conditions.

The analysis solver is STAR-CCM+, and, to carry out the analysis, the axisymmetric model was chosen. As a matter of fact, this model is the less computationally expensive one, but results are consistent with previous thesis works [26] and exhibit the same ratio between shrouded and non-shrouded configuration (slightly more than 2) as measured experimentally in [18]. Hence it is the most suitable model to carry out a very expensive calculation as an optimization analysis. The logic diagram of the optimization algorithm used is shown in Fig. 15.6.

The analysis was performed with the parameters illustrated in Table 15.1. The Turbulent Viscosity Ratio (TVR) has the default value present in STAR-CCM+ [3] simulation set-up, and it represents the sole parameter that characterizes the turbulence using the Spalart-Allmaras model [23].

In Table 15.2 mesh parameters are shown. Base size is setted to have the same chord order of magnitude. Prism layer thickness is calculated by *XFOIL* software [10], and the other prism layer parameters are such as to have a y^+ value in the first laminar sublayer less than one. Figure 15.7 shows the mesh around the airfoil with the prism layer mesh. Physical boundary conditions are setted as shown in Table 15.3.

The pressure difference in the fan interface is setted as a constant value to simulate the nominal condition of the hydroturbine, as done in previous simulations based on different CFD solver [26].

Table 15.1 Axisymmetric model: physics data of the simulation

Parameter	Value or type
Fluid	Water
Density	997.561 (kg/m^3)
Dynamic viscosity	8.8871×10^{-4} (Pa · s)
Turbulence model	Spalart–Allmaras
TVR	10.0
T_u	0.52 (%)
C_T	0.89
ΔP	1137.15 (Pa)
Shroud airfoil	*Selig 1223*
Chord	1.6 (m)
Gap	0.05 (m)
Pitch angle	26°
Reynolds number	2.87×10^6
V_∞	1.6 (m/s)

Table 15.2 Axisymmetric model: mesh characteristics, general parameters

Parameter	Value or type
Model	Polyhedral
Base size	1.0 (m)
Number of prism layers	40
Prism layer stretching	1.2
Prism layer thickness	0.0060 (m)
Mesh cells	50449

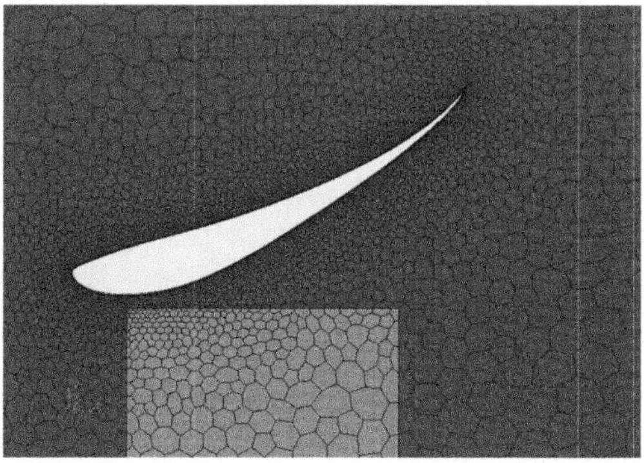

Fig. 15.7 Axisymmetric model: mesh around the airfoil

Table 15.3 Axisymmetric model: physical description of the boundaries

Boundary	Type	Physical quantity	Value
Inlet	Velocity inlet	Velocity	1.6 (m/s)
Outlet	Pressure outlet	Pressure	0.0 (Pa)
Wall	Velocity inlet	Velocity	1.6 (m/s)
Fan	Fan interface	Pressure difference	−1137.15 (Pa)

Table 15.4 Axisymmetric model: physical quantities for the original diffuser airfoil *Selig 1223*

Velocity at disk (m/s)	$C_{P_{ref}}$	θ (deg)	C_P (Δ_{ref})	$C_{P_{exit}}$ (Δ_{ref})
2.1494	0.615	23	1.197 (+95 %)	0.707 (+15 %)
2.2028	0.615	26	1.226 (+99 %)	0.696 (+13 %)
1.8911	0.615	29	1.054 (+71 %)	0.575 (−6.5 %)

15.3 Results

Most of the analysis has been executed on the supercomputing infrastructure SCoPE, in the University of Naples Federico II. This infrastructure consists of a grid of computer, where is possible to use hundreds of parallel processors. Further details about SCoPE can be found in [7], while concerning the programming language, it can be consulted in [2].

The objective function of the analysis is the turbine power coefficient, defined, in the hypothesis of actuator disk, by the Eq. 15.8, and evaluated with respect to the diffuser exit area, defined in the Eq. 15.9 [26], where the exit area is a function of the airfoil pitch angle. This latter value is the most important to evaluate the real advantage of the shrouded configuration with respect to a not-shrouded HAHT with an area equal to the exit area of the shroud [25].

$$C_P = C_T \left(\frac{V_d}{V_\infty} \right), \tag{15.8}$$

$$C_{P_{exit}} = C_P \left(\frac{A_d}{A_{exit}} \right), \tag{15.9}$$

where the subscript d stands for disk, C_T is the disk thrust coefficient, V_∞ is the free-stream or inlet velocity, A_{exit} is the shroud exit area. In Table 15.4 the pressure and velocity values, evaluated as surface average on the actuator disk interface, are indicated. The power coefficients are also specified, respectively, the reference $C_{P_{ref}}$ of the not-shrouded case, and the C_P for shrouded configuration evaluated using both the actuator disk and shroud exit area (see Eqs. 15.8 and 15.9) for the original diffuser airfoil *Selig 1223* at three different pitch angles. The pitch angle of the prototype installed in Venice lagoon is 26° because of the greater extracted power for a given

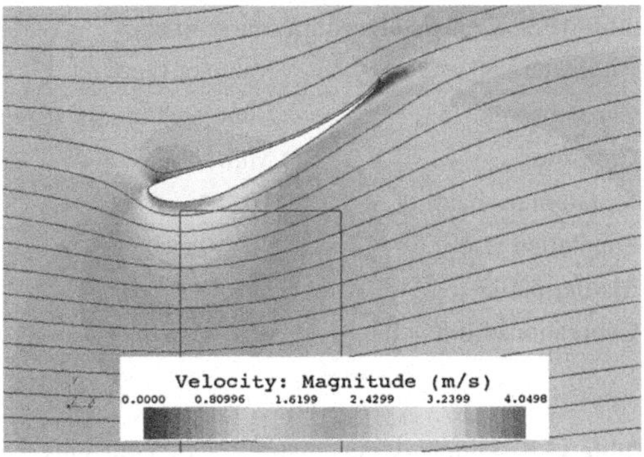

Fig. 15.8 Axisymmetric model: velocity field around the original airfoil for a pitch angle $\theta = 26°$. The streamlines are indicated by the continuos *solid lines*. Actuator disk is located at the throat

pressure drop across the actuator disk. The power coefficient values suggest that using a shroud implies an increasing in power generated of 99 %, if evaluated with respect to the throat area, but of only 13 %, with respect to the outlet area. By using the shroud there is a little zone of separation flow, and not a complete stall of the airfoil, despite the high pitch angle. This is due to two factors:

- the presence of the actuator disk (or of the turbine, in the real case), deviates upwards the streamlines, reducing the effective angle of attack of the airfoil (see Fig. 15.8 in which the velocity field around the original airfoil for a pitch angle θ equal to 26° is shown). Without actuator disk, there would be a complete separation of the flow behind the disk along the diffuser surface.
- The gap between shroud and actuator disk generates a channel that energizes the flow and avoids separation. By reducing the gap, the actuator disk interacts with the shroud boundary layer, the velocity profile is altered, and the separation of the flow occurs, as proven in [26].

In this section the optimization results are shown, for each of the three boundary conditions analysed: $BC = \pm 0.02$, ± 0.05 and ± 0.10. The values of the power coefficient with respect to the generations show a tendency of the C_P to compact the region with elevate values, by increasing the generations, because of the automatized procedure that avoid replication of bad performing airfoil or bad reconstructed ones (the CAD modeler within STARCCM+ is not always capable of correctly reconstruct the parametrized shape produced by Legendre polynomials). The genetic algorithm used for the optimization process, namely the NSGA-II, has been employed with the characteristics summarized in table Table 15.5.

This trend is evident in all of the cases, thanks to the generations limit imposed in the optimization process. The most interesting case is the one with boundary condition amounting to 2 % of the airfoil chord for which after 10 generations the algorithm stabilizes the population's objective function range (see Fig. 15.9).

In Table 15.6 are summarized the results of the optimization process. The choice of the *Selig 1223* as starting airfoil derives from its own already excellent performance. Thus, the optimization process, conducted for different diffuser pitch angle

Table 15.5 Genetic algorithm characteristics

Parameter	Value or type
Chromosome	Simulated binary string
Crossover	Multi-cut
Mutation probability	10(%)
Population size	40
Mating-pool	50(%)
Generations limit	20

Table 15.6 Summary table with power coefficients

θ (deg)				
Cp_{exit} (ΔCp_{exit})		23	26	29
BC	0.02	0.712 (+0.7 %)	0.727 (+4.2 %)	0.695 (+17.3 %)
	0.05	0.713 (+0.8 %)	0.725 (+4.0 %)	0.699 (+17.8 %)
	0.10	0.711 (+0.5 %)	0.723 (+3.8 %)	0.699 (+17.7 %)
Cp_{exit}	*Selig 1223*[a]	0.707	0.696	0.575

[a] See Table 15.4

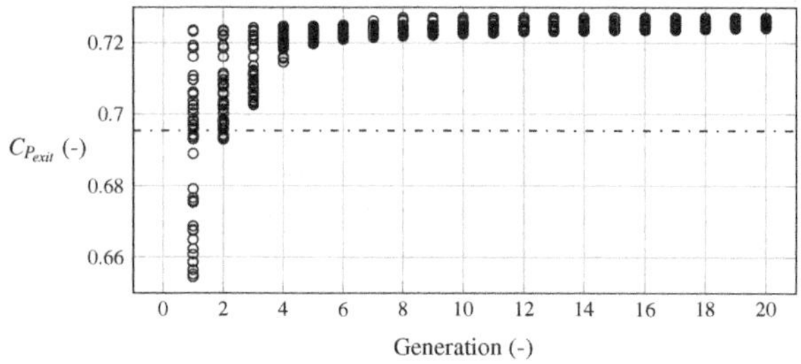

Fig. 15.9 $BC = \pm 0.02$: variation of the power coefficient, $\theta = 26°$

and boundary coefficient for Legendre polynomials to be added on the original shape (see Figs. 15.10, 15.11 and 15.12), returns as best solution the one characterized by a pitch angle equal to 26° and the minimum modification of the airfoil shape, i.e. limiting it to the ±0.02 % of the airfoil chord. The maximum improvement with respect to the initial condition is obtained for a pitch angle of 29° and modification of the airfoil shape fixed to the ±0.05 % of the airfoil chord (Fig. 15.12).

It is important to remarks that all the results refer to an axisymmetric model with an actuator disk modelling the turbine. This means that all the power coefficient should be scaled for taking in consideration the difference between actuator disk and real turbine. From previous experimental tests [18] and forthcoming numerical simulations of the authors, is has been observed that this scaling factor is about 2/3, having finally that from a bare turbine maximum power coefficient of about 0.41,

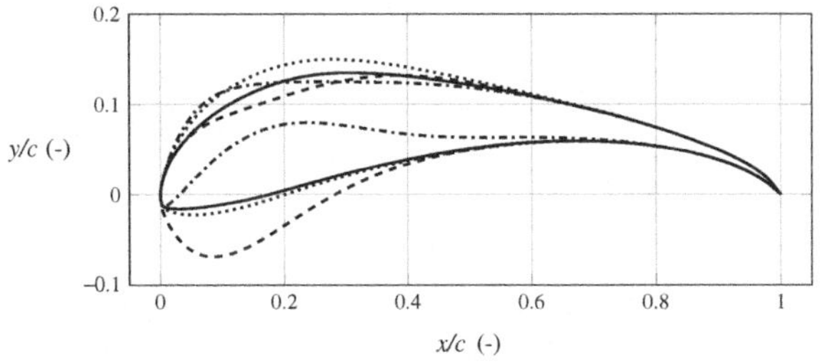

Fig. 15.10 Optimized airfoils, $\theta = 23°$: — Original airfoil, $\cdots$ ±0.02, – – ±0.05, – · – ±0.10

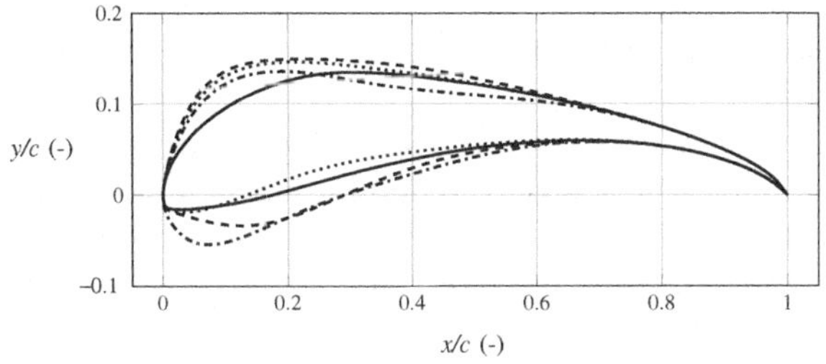

Fig. 15.11 Optimized airfoils, $\theta = 26°$: — Original airfoil, $\cdots$ ±0.02, – – ±0.05, – · – ±0.10

with the old diffuser shape based on *Selig 1223* airfoil grows up to about 0.46, and it reaches with the optimized diffuser shape the value of 0.48.

Concerning the computational time the following considerations could be done. It should be divided into two phases: the meshing and simulation. The meshing phase is performed in serial mode, having the STARCCM+ 7.06 version here used a not very reliable management of parallel meshing tools, expected to be improved in a future version. It takes about 7.5 min, while the simulation phase for a parallel run on 16 cpus lasts about 3.5 min. Since the number of evaluations for the objective functions is 800 with a serial use of the NSGA-II code the total optimization time is about 6 days for a single angle of pitch and a single boundary condition relative to Legendre polynomials. This computational time is abruptly reduced by virtue of the *JobCollection* utility available on the SCoPE distributed grid computing resource: it is an object with the main purpose of allowing the execution of collective operations on sets of independent jobs, being just a logical container, in which both not yet submitted and already submitted jobs can be inserted in. A job collection is somehow orthogonal with respect to a job cluster being a set of dependent jobs (e.g. all jobs spawned by the same father process). The parallel use of the NSGA-II code leads to a total optimization time of only 3.7 h. Further reduction in the computational time could be obtained by means of a different meshing technique as suggested in Sect. 15.4.

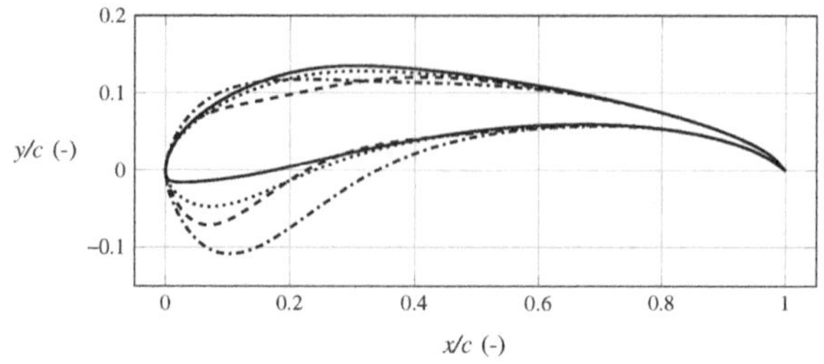

Fig. 15.12 Optimized airfoils, $\theta = 29°$: —Original airfoil, ··· ±0.02, – – ±0.05, – · – ±0.10

15.4 Conclusions and Future Works

The main objective of this work has been the optimization process of an airfoil shape for the shroud of a horizontal axis hydro turbine that that would increase the power extracted from marine currents. To achieve the target, an optimization routine has been developed: starting from the input base airfoil, a geometry parametrization method is chosen, in this case Lagrange polynomial, and a numerical optimization method as NSGA evaluates the objective function, namely the C_P, with a CFD solver giving in output, through an automated process, a new airfoil that improves the objective function. The base airfoil is designed for high-lift regimes, so it gives excellent performance in these kind of applications. For that reason, was not expected a very high increase of the power coefficient. The airfoil obtained as a result of the optimization process shows a higher power coefficient of 4.5 %. This result is quite in agreement with the expectations, so it is a satisfying result. The CFD analysis, in axisymmetric flow field and actuator disk hypothesis, gave reliable results, in agreement with previous works, developed with different software, and with experimental results. Future works will aim to reduce computational time in optimization process, and to enhance the general robustness of the developed numerical code in order to increase its versatility and use in more complex contexts. In particular, by virtue of the use of the overlapping grid tool, recently incorporated in the STARCCM+ distribution, it is possible to design a single grid for more than one diffuser position and or orientation, so that a multi-objective optimization (considering the power coefficient at several pitch angle as objective functions) could be implemented saving the time consuming fraction related to mesh generation.

In addiction, further analysis on curved plate would be performed, following the preliminary promising results of Reinecke [17] for an even more efficient diffuser shape, not more constrained to a classical airfoil shape, exploiting the expansion ratio and simplifying the manufacturing process. At the end of this process two candidates would be ready for testing and then prototyping a new diffuser solution for GEM.

References

1. Brownlee J (2011) Clever algorithms: nature-inspired programming recipes. 1st edn. Lulu enterprises. ISBN 978-1-4467-8506-5
2. Burke S et al (2011) gLite-3-UserGuide 1.4.1. In: Experiment integration and distributed analysis, Andrea Sciabà. https://www.isragrid.org.il/documents/gLite-3-UserGuide.pdf
3. CD-Adapco (2012) Star-CCM+ Version 7.02.011 User Guide
4. Castonguay P, Nadarajah S (2007) Effect of shape parameterization on aerodynamic shape optimization. In: 45th AIAA aerospace science meeting and exhibit, Reno, Nevada, January 8–11
5. Coiro DP, Melone S, Montella F (2004–2005) Energia pulita dalle correnti marine: aspetti tecnici ed economici. In: Department of aerospace engineering, University of Naples Federico II, internal report, in Italian
6. Coiro DP, De Marco A, Scherillo F, Maisto U, Familio R, Troise G (2009) Harnessing marine current energy with tethered submerged systems: experimental tests and numerical model analysis of an innovative concept. In: IECCP, international conference on clean energy production, June, Capri, Italy
7. Cucco D (2011) Analisi numeriche del flusso intorno a microgeneratori eolici ad asse verticale nell'infrastruttura di supercomputing S.Co.P.E. Master thesis in Aerospace Engineering, Department of Aerospace Engineering, University of Naples Federico II, in Italian
8. Deb K (2001) Multi-objective optimization using evolutionary algorithms. Wiley. ISBN 978-0471873396
9. Deb K, Pratap A, Agarwal S, Meyarivan T (2002) A fast and elitist multiobjective genetic algorithm: NSGA-II. IEEE Trans Evol Comput 6(2), http://www.iitk.ac.in/kangal/Deb_NSGA-II.pdf
10. Drela M (1989) XFOIL: an analyse and design system for low reynolds number airfoils. In: Low reynolds number aerodynamics, Springer Verlag Lecture Notes in Engineering, vol. 54
11. Ferrauto E (2013) Horizontal axis hydroturbine Shroud airfoil analysis and optimization. Master thesis in aerospace engineering, department of industrial engineering, University of Naples Federico II
12. Grasso F (2008) Multi-objective numerical optimization applied to aircraft design. PhD thesis in Aerospace Engineering, Department of Aerospace Engineering, University of Naples Federico II http://www.fedoa.unina.it/3188/1/thesis_grasso_francesco.pdf
13. Hansen MOL, Sørensen NN, Flay RGJ (2000) Effect of placing a diffuser around a wind turbine. In: Wind energy, vol. 3
14. Hicks RM, Vanderplaats GN (1975) Application of numerical optimization to the design of low-speed airfoils. In: NASA TM X-3213, Ames Research Center, Moffet Field California
15. Hicks RM, Henne PA (1978) Wing design by numerical optimization. J Aircr 15(7):407–412
16. Popov A (2005) Genetic algorithms for optimization. Programs for MATLAB, http://p0p0v.com/science/downloads/Popov05a.pdf Version 1.0
17. Reinecke J, von Backström TW, Venter G, (2010) Effect of a diffuser on the performance of an ocean current turbine. First annual CRSES student conference, 11–12 November 2010 Sustainability Institute. Stellenbosch University, South Africa
18. Scherillo F, Maisto U, Troise G, Coiro DP, Miranda S (2011) Numerical and experimental analysis of a shrouded hydroturbine. International conference on clean electrical power (ICCEP), 14–16 June, Ischia, Italy, IEEE Catalog Number: CFPI036B-USB, ISBN: 978-1-4244-8928-2
19. Selig MS, Guglielmo JJ (1997) High-lift low reynolds number airfoil design. J Aircr, 34(1), http://www.ae.illinois.edu/m-selig/pubs/GuglielmoSelig-1997-JofAC-S1223.pdf
20. Seshadri A (2009) NSGA-II: a multi-objective optimization algorithm. In: http://www.mathworks.com/matlabcentral/fileexchange/10429-nsga-ii-a-multi-objective-optimization-algorithm
21. Sobieczky H (1997) Geometry generator for CFD and applied aerodynamics. In: Courses and Lecture International

22. Sobieczky H (1998) Parametric airfoils and wings. In: Fujii K, Dulikravich GS (eds) Notes on numerical fluid mechanics, vol. 68. Vieweg Verlag, pp 7–88
23. Spalart PR, Allmaras SR (1994) A one-equation turbulence model for aerodynamic flows. In: Recherche aerospatiale, No. 1, pp 5–21
24. Tognaccini R (2004–2005) Lezioni di aerodinamica dell'ala rotante. Department of Aerospace Engineering, University of Naples Federico II http://wpage.unina.it/rtogna/Aerodinamica_del_rotore.pdf, in Italian
25. Van Bussel JW (2007) The science of making more torque from wind: diffuser experiments and theory revisited. J Phys: Conf Ser 75
26. Voria P (2008) Analisi numerica di una turbina marina dotata di diffusore e slat. Bachelor thesis in Aerospace Engineering, Department of Aerospace Engineering, University of Naples Federico II, in Italian
27. Yang S, Wu H, Liu F (2003) Comparison of three geometric representations of airfoils for aerodynamic optimization. 16th AIAA computational fluid dynamics conference 23–26 June Orlando, FL

Chapter 16
Parametric Blending and FE-Optimisation of a Compressor Blisk Test Case

Kai Karger and Dieter Bestle

Abstract Due to raising demands from aviation industry concerning weight reduction and increased efficiency, compressor front stages of jet engines are designed as blade integrated disks (blisks). However, a major drawback of blisks is that small cracks from foreign object impacts occurring in service may propagate into the whole disk causing burst at worst case which is unacceptable. As a damaged blade of a blisk cannot easily be replaced, there is a need for repair. For example, borescope blisk blending may be applied on-wing to ensure safe on-going operation. To determine best solutions for the blending shape, process integration and optimisation tools are used which modify a parametric model and examine its impact on fatigue criteria by FEM.

Keywords Blisk · Blending · FEM · Process integration · Optimisation

16.1 Introduction

Actually, modern civil jet engines are developed towards higher efficiency and lower weight. Especially at the front stages of compressors, blade integrated disks (blisks) may contribute to both demands since they have less leakage flow and lower weight than bladed disks. However, such blisks are characterised by low vibration damping and they need a higher foreign object damage resistance to ensure rotor integrity. Therefore, stricter design criteria for vibration resistance and static stresses are required. Further, single blades cannot be replaced easily in case of damage demanding repair strategies such as blending which is investigated in this paper within an industrial test case. A fictional elliptical dent at the leading edge of an

K. Karger · D. Bestle (✉)
Chair of Engineering Mechanics and Vehicle Dynamics, Brandenburg University of Technology, P.O.BOX 101344, 03013 Cottbus, Germany
e-mail: bestle@tu-cottbus.de

K. Karger
e-mail: kai.karger@tu-cottbus.de

D. Greiner et al. (eds.), *Advances in Evolutionary and Deterministic Methods for Design, Optimization and Control in Engineering and Sciences*, Computational Methods in Applied Sciences 36, DOI 10.1007/978-3-319-11541-2_16

aerofoil is used to represent the damage causing a strong decay of the fatigue criteria below demanded lower bounds. By smoothly cutting out the damaged area, the fatigue criteria shall be recovered to original conditions.

Existing guidelines for borescope blending repair use circular scallops to be applied perpendicularly to the blades if a damage occurs at the mid height area [6]. Rules for geometric limits have been determined within aerodynamic investigations [5, 12] where the strongest limitation is that the fillet region must not be blended. In order to provide fast results for maintenance, typically the creation of a database for standard blending procedures is suggested [3].

The scope of this work is to find optimal blending shapes for a given damage by an automated optimisation process where a parametric blending shape is modified, relevant endurance measures for a blisk are evaluated by FEM, and optimisation objectives and constraint criteria are computed. To speed up optimisation, Kriging-based surrogate models are used which allow to use genetic algorithms.

16.2 Parametric Model and Evaluation Process

Borescope blending is an in-situ-abrade procedure to repair aerofoils suffering from small foreign object damages. This can be done during regular visual engine inspections using borescopes without time-consuming disassembly of the engine and by supervising the repair process with cameras. Due to space limitations simple cutout geometries are realised on-wing. A flexible motor driven shaft is inserted through borescope ports. At the end of the shaft diverse tool kits can be mounted for nick, dent or crack removal. They are then positioned for blend repair according to engine manufacturer guidelines. In this paper, the resulting contour for damage removal is chosen as a D-shape (Fig. 16.1), where a tool with radius R starts at an adjustment height A with an attack angle α, permeates the aerofoil with depth D and ends at height H. These blending shape parameters may be summarised in the design vector

$$\mathbf{p} = [D\ A\ H\ R\ \alpha]^T \tag{16.1}$$

to be determined by an optimisation process.

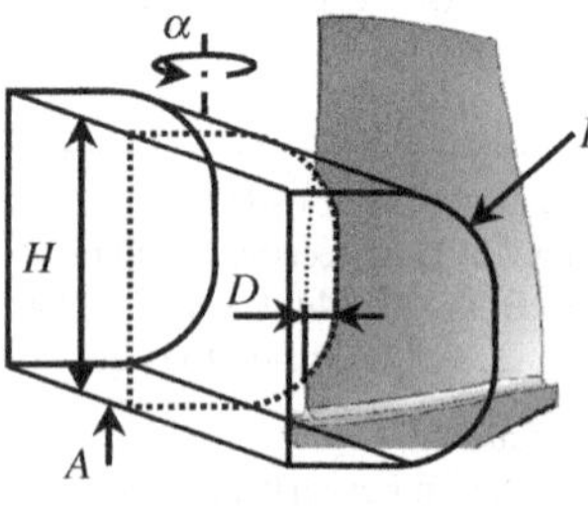

Fig. 16.1 Parametric blending shape model

A major goal of the repair process is to ensure structural feasibility. Here, fatigue criteria [10, 11] for regular conditions are used as constraints where the maximum stresses $\sigma_{\max}$ of the blade must not exceed the yield stress R_m, i.e.,

$$\sigma_{\max} \leq R_m, \tag{16.2}$$

to prevent it from plastic deformation. Another endurance measure is flutter stability. Flow excited vibrations of the blades can be avoided most likely if the ratio λ of natural blade frequency $2\pi f$ and flow excitation frequency $v_{rel,75\,\%}/c_{75\,\%}$is above a mode dependent lower bound λ^c found from experimental tests [11]:

$$\lambda_j := \frac{2\pi f_j \cdot c_{75\,\%}}{v_{rel,75\,\%}} \geq \lambda_j^c, \quad j \in \{1\text{F}, 1\text{T}\}. \tag{16.3}$$

Especially critical are the first flap (1F) and the first torsion (1T) mode resulting in constraints $\lambda_{1\text{F}} \geq \lambda_{1\text{F}}^c, \lambda_{1\text{T}} \geq \lambda_{1\text{T}}^c$. Equation (16.3) considers the relative flow velocity $v_{rel,75\,\%}$ and the chord length $c_{75\,\%}$ at 75 % radial blade height.

Besides static stresses, also dynamic stresses due to vibration must be taken into account. The *af*-strength is a combination of static and dynamic stresses according to the Goodman-diagram regarding fatigue stress R_f, dynamic stresses σ_{dyn}, static stresses σ_{stat}, and yield stress R_m, which has to be kept above an experience based level af^c:

$$af_j := \frac{R_f}{\sigma_{j,\text{dyn}}} \left(1 - \frac{\sigma_{j,\text{stat}}}{R_m}\right) \geq af_j^c, \quad j \in \{1\text{F}, 1\text{T}, 1 \ldots 4\}. \tag{16.4}$$

Again, first flap and first torsion modes are identified as critical using specific lower bounds $af_{1\text{F}}^c, af_{1\text{T}}^c$, whereas four other modes use a common lower bound $af_1^c = \cdots = af_4^c := af^c$.

In addition to the regular load case, also gas loads of a numerical surge event representing worst case running conditions are taken into account. Here, maximum stresses during surge should not exceed yield stress R_m:

$$\sigma_{\max}^{\text{surge}} \leq R_m. \tag{16.5}$$

Flutter stability and *af*-strength from Eq. (16.3) respectively Eq. (16.4) are not used as constraints for numerical surge conditions, since surge events happen for a short period of time only. Both values define endurance levels and it is not intended to run an engine permanently at surge.

In order to get the necessary information on constraint functions (16.2)–(16.5) for specific design values (16.1), the blended rotor blisk is evaluated by the commercial FE-software ANSYS 13.0 [1]. Typically, aerofoil shapes are defined at running conditions, and a hot-to-cold transformation is required to determine the unloaded aerofoil shape before blending can be performed and various loads can be applied. This cold manufacturing geometry needs to be computed only once and was delivered by an

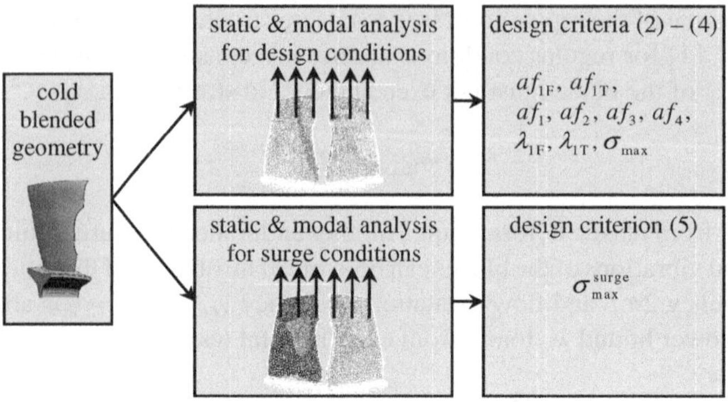

Fig. 16.2 Process flow for FE blisk evaluation

industrial partner. The cold blended blisk model is then used as input for non-linear static analyses and subsequent pre-stressed modal analyses to deliver relevant fatigue criteria [8]. The second step is an FE-analysis for surge conditions resulting in a different stress distribution, and hence changed fatigue criteria. A rough scheme of this FE-evaluation process is shown in Fig. 16.2.

In order to get some reference results, the FE-evaluation is firstly carried out for the undamaged and the damaged rotor blisk, respectively. The damaged blade is simulated by an elliptical damage of depth 2 mm and height 1 mm at 25 % radial blade height. For the damaged blade many of the endurance measures decay below the undamaged values, see Fig. 16.3 which shows normalised endurance measures $\overline{af}_j := af_j/af_j^c$, $\bar{\lambda}_{1F} := \lambda_{1F}/\lambda_{1F}^c$, $\bar{\lambda}_{1T} := \lambda_{1T}/\lambda_{1T}^c$, $\bar{\sigma}_{max} := R_m/\sigma_{max}$, and $\bar{\sigma}_{max}^{surge} := R_m/\sigma_{max}^{surge}$. Especially the maximum stress for surge loads is highly increased which would lead to serious blade damage. Obviously also the required values of *af*-strength for lower modes $af_{1F}, af_{1T}, af_1, af_2$ cannot be fulfilled by the damaged blade anymore, whereas the values of flutter stability $\lambda_{1F}, \lambda_{1T}$ and *af*-strengths af_3, af_4 change only slightly and still meet the required limits. The question arises if proper blending can bring up the violated criterion values above the necessary level.

16.3 Optimisation Problem and Automated Design Approach

In order to find a proper repair geometry, different blending shapes have to be created and evaluated within an optimisation process where each iteration step involves several subtasks to be coupled. Such a process integration may be done e.g. within the software package Isight 4.0 [2] which eases automation of computational tasks. Each design evaluation starts with a modification of the blending shape parameters (16.1) in an expression file which acts as input to the CAD programme Unigraphics NX 6.0. The CAD programme is used to adapt the model geometry and to determine the blade mass. The blended geometry is then updated in a template of the FE-model and analysed with ANSYS 13.0. Based on the results, the objective and constraint values are written to files which are then parsed to the optimisation algorithm which suggests a new design vector for the next loop. In the following, an overview on the design objectives and the implemented optimisation strategy is given.

To ensure minimal impact on aerodynamic performance and minimal rotor imbalance, the blending volume should be as small as possible. However, structural feasibility has to be met as well. Regarding fatigue criteria, a bigger cutout leads to lower blade mass and greater notch radii which may result in lower stresses and better fatigue criteria. In order to meet both design goals, two conflicting targets are defined. For keeping the cutout size low, the ratio between undamaged blade mass m^{ref} and blade mass after blending m^{blend} is minimised:

$$f_1 = \frac{m^{\text{ref}}}{m^{\text{blend}}}. \tag{16.6}$$

The second objective aims to improve the surge vibration resistance for constant surge conditions. Of course, in reality surge conditions are not steady-state, since the reversed flow is a shockwave which interrupts the regular flow and stops right after reaching pressure balance between the combustion chamber and the inlet of the engine. However, also a comparison of undamaged and blended objective values for steady conditions can show the right trend of improvement or degradation. Here, the surge *af*-strengths for the six modes in Fig. 16.3 are normalised with respect to the lower bounds used in (16.4), respectively. For minimisation these ratios are inverted and averaged resulting in the objective

$$f_2 = \frac{1}{6}\sum_{i=1}^{6} \frac{af_i^c}{af_{i,\text{surge}}}. \tag{16.7}$$

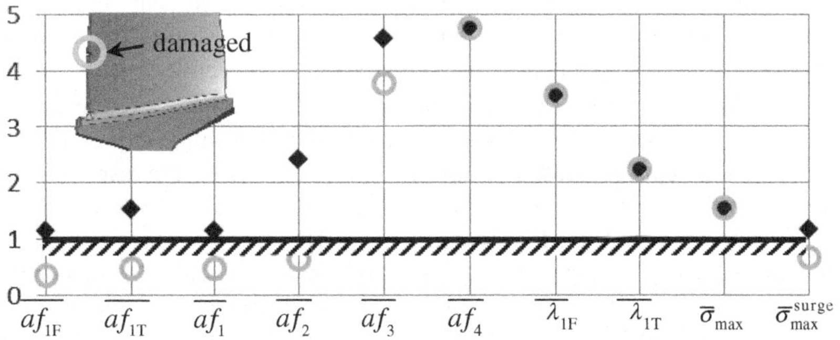

Fig. 16.3 Degradation of normalised endurance measures from undamaged (♦) to damaged (○) blisk aerofoil

Simultaneously, the endurance measures (16.2)–(16.5) have to be fulfilled, which may be summarised in a vector of implicit constraints:

$$\mathbf{h} := \begin{bmatrix} af_j^c/af_j - 1 \\ \sigma_{\max}/R_m - 1 \\ \\ \sigma_{\max}^{\text{surge}}/R_m - 1 \\ \\ \lambda_{1\text{F}}^c/\lambda_{1\text{F}} - 1 \\ \lambda_{1\text{T}}^c/\lambda_{1\text{T}} - 1 \end{bmatrix} \leq \mathbf{0}. \quad (16.8)$$

Finally, a multi-criterion optimisation problem based on (16.1) and (16.6)–(16.8) may be formulated as

$$\min_{\mathbf{p} \in P} \begin{bmatrix} f_1 \\ f_2 \end{bmatrix} \quad \text{where } P = \left\{ \mathbf{p} \in \mathbb{R}^5 | \mathbf{h}(\mathbf{p}) \leq \mathbf{0}, \mathbf{p}^l \leq \mathbf{p} \leq \mathbf{p}^u \right\}. \quad (16.9)$$

Such type of problems may be solved by multi-objective genetic algorithms. However, these algorithms usually require a huge number of design evaluations and a single FE-evaluation already takes about 10–90 min depending on the mesh size at the cutout and hence on the cutout size. Response surface methods help to resolve this problem. They are based on only few evaluated supporting points and are computationally much cheaper than direct FE-analyses. Therefore, the optimisation problem (16.9) is not solved directly, but adaptive, Kriging-based response surfaces are applied which are implemented in the DACE toolbox [9] of MATLAB. Such strategies based on surrogate models are always recommendable in case of computationally expensive problems [4].

A Kriging response surface models a deterministic response function value $f(\mathbf{p})$ as normally distributed random number characterised by mean value $\mu_f(\mathbf{p})$ and standard deviation $s_f(\mathbf{p})$. Then, the most probable realisation of the unknown true function value $f(\mathbf{p})$ is $\mu_f(\mathbf{p})$; however, if minimising $\mu_f(\mathbf{p})$ only, the algorithm may get stuck in a local minimiser. Therefore, the strategy "Minimising a Statistical Lower Bound" [7] is implemented where the artificial objective $\tilde{f}(\mathbf{p}) = \mu_f(\mathbf{p}) - \kappa s_f(\mathbf{p})$, $\kappa \in \mathbb{R}$, is minimised instead. Design points minimising $\tilde{f}(\mathbf{p})$ are worth to be evaluated since they either are points with low function values expressed by low μ_f or high uncertainty expressed by large s_f values. Therefore, the response surface is refined iteratively at these points, where in the following $\kappa = 3$ will be used.

To illustrate this strategy, Fig. 16.4a shows a 1D-example where the unknown original function $f(p)$ is approximated by estimated mean values $\mu_f(p)$ found from a set of supporting points (○). In combination with the predicted standard deviation $s_{f(p)}$, the implemented strategy delivers the goal function $\tilde{f}(p)$ in Fig. 16.4b with minimiser $p*$ which serves as additional supporting point in the next iteration step. This strategy is applied to both objectives f_i resulting in statistical estimates $\mu_{f,i}$, $s_{f,i}$ and constraints h_j resulting in $\mu_{h,j}$, $s_{h,j}$.

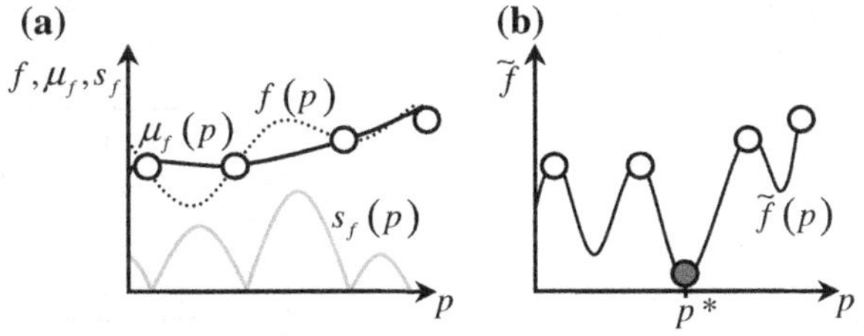

Fig. 16.4 1D-example of Kriging models (**a**) and "Minimising a Statistical Lower Bound" (**b**)

Before solving the optimisation problem (16.9) by the mentioned Kriging strategy, another modification needs to be done to obtain an unconstrained optimisation problem

$$\min_{\mathbf{p} \in P} \begin{bmatrix} \hat{f}_1(\mathbf{p}) \\ \hat{f}_2(\mathbf{p}) \end{bmatrix}. \tag{16.10}$$

Within a penalty strategy, both objectives $\tilde{f}_i(\mathbf{p})$ described above are artificially degraded with the same penalties $w_h(\mathbf{p})$ and $w_{dist}(\mathbf{p})$ as

$$\hat{f}_i(\mathbf{p}) := \mu_{f,i}(\mathbf{p}) - 3s_{f,i}(\mathbf{p}) + (w_h(\mathbf{p}) + w_{dist}(\mathbf{p}))^2 \tag{16.11}$$

where the first penalty term

$$w_h(\mathbf{p}) = \sum_j w_{h,j}(\mathbf{p}),$$

$$w_{h,j}(\mathbf{p}) = \begin{cases} \mu_{h,j}(\mathbf{p}) - 3s_{h,j}(\mathbf{p}) + 1{,}000 & \text{if} \quad \mu_{h,j}(\mathbf{p}) - 3s_{h,j}(\mathbf{p}) > 0 \\ 0 & \text{else} \end{cases} \tag{16.12}$$

deals with constraints (16.8). If the statistical lower bound $\mu_{h,j}(\mathbf{p}) - 3s_{h,j}(\mathbf{p})$ violates the constraint, the real value $h_j(\mathbf{p})$ is unlikely to fulfil it and a high penalty value of 1,000 is added. This value is chosen to guarantee that even feasible designs with poor objective values are rated better than infeasible designs with good objective values.

The second penalty term is intended to avoid designs within the FE-evaluation process where the associated blade cannot be created or analysed. That is why the term $w_{dist}(\mathbf{p})$ in Eq. (16.11) deteriorates the objectives if the distance of a new design suggestion $\mathbf{p}$ is too close to any already known non-converged design $\mathbf{p}_k$ which is stored in an archive:

$$w_{dist}(\mathbf{p}) = \begin{cases} 1{,}000 & \text{if} \quad \min_k |\mathbf{p} - \mathbf{p}_k| \le \varepsilon \\ 0 & \text{else} \end{cases}. \tag{16.13}$$

The unconstrained surrogate optimisation problem (16.10) is solved within the Isight process shown in Fig. 16.5. Firstly, a design of experiments (DoE) defines an initial set of designs which are evaluated by direct analysis. The obtained data

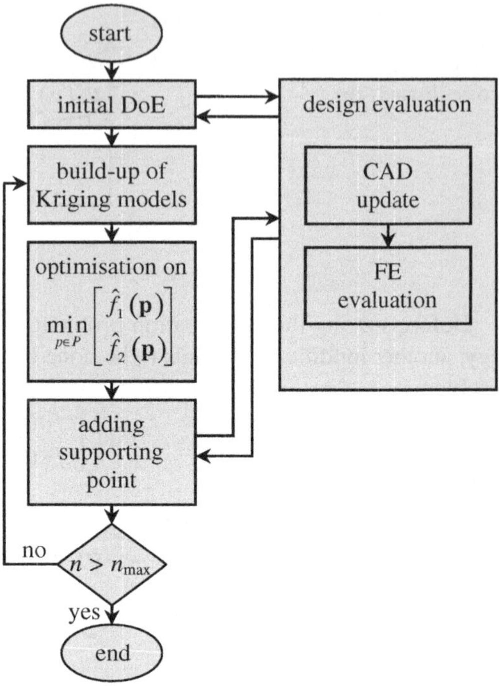

Fig. 16.5 Scheme of optimisation process flow

are used for building up first Kriging models for both criteria (16.6), (16.7) and for the constraints (16.8). Based on these surrogate models a genetic algorithm determines Pareto-optimal solutions in MATLAB for the Kriging-based problem (16.10)–(16.13). The model is then refined iteratively by evaluating and adding the Pareto-optimal solutions with the biggest distance to already converged designs in a user-defined number of loops.

16.4 Optimisation Results

In the following, the settings and the obtained results of the optimisation strategy are described. The initial DoE for creating the first set of Kriging-based surrogate models of objective and constraint functions consists of a Latin hypercube sample of 150 random points. After analysing the function values for these designs, the MATLAB DACE toolbox builds up Kriging models with zero-order polynomial regression models and cubic spline correlation models [9]. The converged points are used as input data whereas non-converged designs cannot contribute since they provide no useful information on objective and constraint functions. The latter points are stored in an archive of non-converged designs to be avoided. The Kriging-based optimisation problem (16.10)–(16.13) is solved by the optimisation algorithm NSGA-II

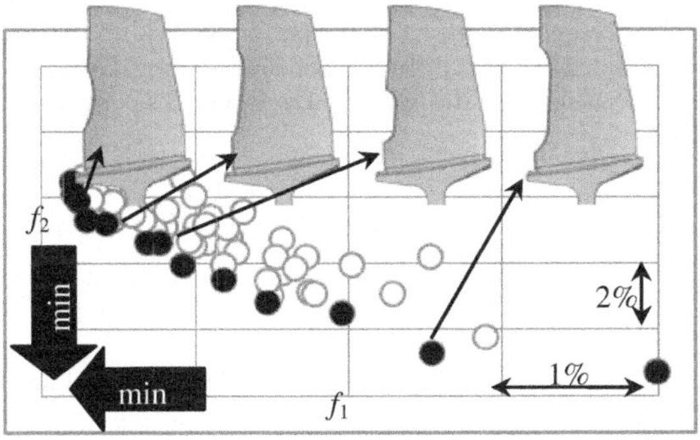

Fig. 16.6 Admissible designs in criterion space (○), Pareto-optimal solutions (●), and examples for optimally blended aerofoil shapes

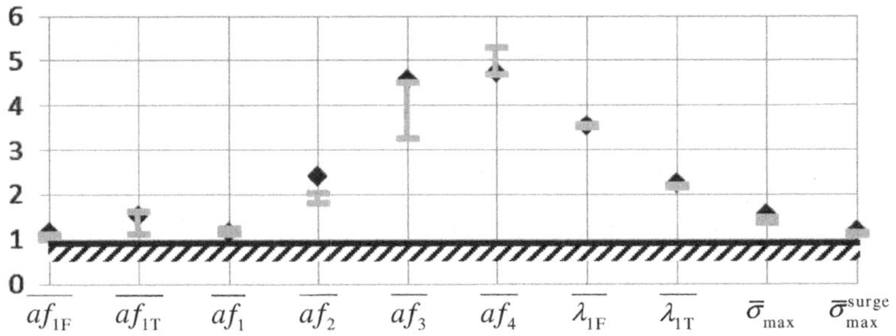

Fig. 16.7 Range of normalised endurance measures for all Pareto-optimal solutions (**I**) compared to the undamaged reference design (◆)

implemented in MATLAB where population size and number of generations are set to 100, respectively. A single solution of the Pareto-optima is chosen such that it has the biggest distance to the already used supporting points. This design point is evaluated according to Fig. 16.5 and used as additional supporting point in the next loop. If it cannot be evaluated, it is assigned to the archive of non-converged designs and the next best Pareto solution is checked. Altogether, 1500 iteration loops are run to update the surrogate models.

The design parameters of the described blending model are limited by confidential geometric bounds $\mathbf{p}^l$, $\mathbf{p}^u$. The use of these parameter limits and the implemented optimisation strategy are able to recover the endurance measures and result in a wide range of different cutout sizes. Figure 16.6 shows the obtained results in the criterion space defined by original objectives (16.6) and (16.7) as non-dominated solutions and some examples of associated blending shapes. All of the Pareto-optima fulfil the endurance constraints and cope with the initial violation of some endurance measures (Fig. 16.7). From these solutions an engineer may choose a specific blending shape for maintenance repair, where further investigations may be done on aerodynamics to account for the impact on compressor efficiency and surge margin.

Acknowledgments The presented work has been performed within the VIT 3 project (Virtual Turbomachinery) partly funded by the Federal State of Brandenburg, Germany, the European Community, and Rolls-Royce Deutschland. Rolls-Royce Deutschland's permission to publish this work is greatly acknowledged.

References

1. ANSYS Release 13.0 (2010) ANSYS mechanical APDL structural analysis guide. ANSYS Inc., Canonsburg, USA
2. Dassault Systèmes Simulia Corp. (2009) Isight 4.0 component guide. Engineous software Inc., Cary
3. Day WD, Fiebiger SW, Patel HN (2012) Parametric evaluation of compressor blade blending, GT2012-68641. In: Proceedings of ASME turbo expo, Copenhagen
4. Giannakoglou KC, Karakasis MK, Kampolis IC (2006) Evolutionary algorithms with surrogate modeling for computationally expensive optimization problems. In: Proceedings ERCOFTAC, Gran Canaria, Spain
5. Gietl T (2012) HPC blade borescope blend repair limits-aerodynamic assessment. Technical Report, Rolls-Royce Deutschland, Dahlewitz
6. Hoenick M (2009) Blisk blending repair-stress assessment. Technical Report, Rolls-Royce Deutschland, Dahlewitz
7. Jones DR (2001) A taxonomy of global optimization methods based on response surfaces. J Glob Optim 21:345–383
8. Karger K, Bestle D (2012) Multi-objective blisk optimisation taking into account environmental influences. In: Proceedings of the 5th IC-SCCE, Athens
9. Lophaven SN, Nielsen HB, Søndergaard J (2002) DACE—A MATLAB Kriging toolbox, Informatics and mathematical modelling, Technical University of Denmark
10. Otto D (2009) Ein Beitrag zur interdisziplinären Prozessintegration und automatischen Mehrzieloptimierung am Beispiel einer Verdichterrotorschaufel. Dissertation, Shaker-Verlag, Aachen
11. Pianka C (2009) MMM21300: Mechanical assessments and criteria for compressor blisk designs regarding the aerofoil integrity. Technical Report, Rolls-Royce Deutschland, Dahlewitz
12. Wuthe F, Pianka C (2012) HPC gaspath boroscope examination. Technical Report, Rolls-Royce Deutschland, Dahlewitz

Chapter 17
Modular Automated Aerodynamic Compressor Design Process

Fiete Poehlmann, Dieter Bestle, Peter Flassig and Michèl Hinz

Abstract Designing complex and challenging machines demands the use of sophisticated methods such as multi-objective optimization. In this paper the aerodynamic design process of a jet engine compressor is used to demonstrate how process automation and optimization may support engineers to find better designs. The design process is divided into four sub-processes starting with a correlation-based 1D meanline code and ending with a 3D CFD analysis. These sub-processes of different fidelity are automated and coupled to enable a cascaded, sequential optimization. This approach allows to start with few basic assumptions and ends with a complete 3D geometry and flow field of an axial jet engine compressor.

Keywords Multi-criterion optimization · Compressor design · Multi-fidelity processes · Process coupling · Meanline · Throughflow · Blading

17.1 Introduction

Increasing air traffic and its negative impact on the environment, e.g. by NO_X or CO_2 emissions, demand for more efficient jet engines. Faster design methods are necessary to enable jet engine manufacturers to cope with these challenges and to stay competitive. Process integration and optimization are powerful tools which may help to account for contradicting requirements on modern jet engines, where

F. Poehlmann (✉) · D. Bestle
Chair of Engineering Mechanics and Vehicle Dynamics, Brandenburg University of Technology, Siemens-Halske-Ring 14, 03046 Cottbus, Germany
e-mail: fiete.poehlmann@tu-cottbus.de

D. Bestle
e-mail: bestle@tu-cottbus.de

P. Flassig · M. Hinz
Rolls-Royce Deutschland Ltd & Co KG, Eschenweg 11, 15827 Blankenfelde-Mahlow, Germany
e-mail: peter.flassig@rolls-royce.com

M. Hinz
e-mail: michel.hinz@rolls-royce.com

D. Greiner et al. (eds.), *Advances in Evolutionary and Deterministic Methods for Design, Optimization and Control in Engineering and Sciences*, Computational Methods in Applied Sciences 36, DOI 10.1007/978-3-319-11541-2_17

the compressor is one of the most complex and challenging components. The flow through a compressor is rather sensitive since it is forced to flow in the direction of increased pressure. These machines, therefore, are designed with great effort using various codes of different fidelity, where from one design step to the next computation time increases. Therefore, it is necessary to exploit the full potential of the fast low-fidelity codes in the beginning and to generate good starting solutions for the subsequent higher fidelity codes.

The first design step is based on a 1D *Meanline* evaluation where the flow field values are determined along the mid-line of the annulus [7]. The next design step uses a 2D streamline curvature method called *Throughflow* resulting in radial distributions for the flow quantities [13]. With the given flow field the next step is to create a proper blade for each stage which can handle the requirements of the flow field. This may be done by section-wise stacking of designed airfoils [8] or by a 3D approach to define the whole blade at once [4]. Finally, stacking and fine tuning of the blade shape is performed to cope with 3D flow effects requiring sophisticated tools such as 3D CFD.

In this paper, a fully automated aerodynamic compressor design process is presented. It is composed of various analysis codes with different fidelity which are integrated by the process integration tool *Isight* [5] and coupled in a sequential overall optimization process. It allows to generate a complete geometry and flow field of a compressor based on few basic assumptions and a rough outline of the compressor. At each step the level of detail is increased until the compressor is finally defined. Recent investigations [4, 7, 8, 13] have focused on automation of the sub-design processes only, whereas feasibility of a design related to the overall aerodynamic design process has been considered only partially. Therefore, this paper focuses on a stronger coupling between the design sub-processes.

17.2 Integrated Optimization Processes for Compressor Design

The first sub-process to be integrated in the overall design process is based on a *Meanline* code determining the outline of a compressor [6]. Typically performance requirements like mass flow or the overall pressure ratio are already defined and a design engineer has more or less design freedom to change parameters within the compressor in order to improve the design. Typical optimization goals are improvement of compressor efficiency η_{ML} and simultaneously of surge margin SM_{ML}, both calculated with *Meanline* (ML).Variations are performed on mid-line values of pressure ratios Π_i, solidities σ_i, axial chord lengths $c_{ax,i}$ and stator exit angles $\alpha^S_{E,i}$ for each stage, as well as annulus shape defined by a superposition of mid-line and compressor height, whereas compressor length and outlet area are maintained. All variations are performed by proper parameterizations of these quantities to keep variations smooth and the number of design parameters small [7]. The total number of parameters $\mathbf{p}_{ML}$ changed during *Meanline* optimization is 43. Constraints are applied to prevent the code from running into unacceptable designs, where limits on de Haller numbers [2], diffusion factors [10], Mach numbers, load coefficients [1], and Koch parameter [9] are summarized as constraints $\mathbf{h}_{ML}\left(\mathbf{p}_{ML}\right) \leq \mathbf{0}$. The optimization problem finally reads as

$$\max_{\mathbf{p}_{ML} \in P_{ML}} \begin{bmatrix} \eta_{ML} \\ SM_{ML} \end{bmatrix}, \quad P_{ML} = \left\{ \mathbf{p}_{ML} \in \mathbb{R}^{43} \mid \mathbf{h}_{ML} \leq \mathbf{0} \right\}. \tag{17.1}$$

The second code used in the design chain is a 2D streamline curvature method called *Throughflow* (TF). In general, the requirements within an optimization are similar to those of *Meanline*. E.g. criteria are the same, i.e., maximization of efficiency η_{TF} and surge margin SM_{TF}, however, now estimated with the higher fidelity *Throughflow* code. Also constraints are almost the same, but now obtained for several streamlines at different radial heights and not only at the mid-line. Therefore, variations during this optimization sub-process regard radial distributions, whereas mid-line values are maintained due to the assumption that they are already an optimal result from the *Meanline* optimization. The varied quantities are the same as for *Meanline* apart from the axial chord length which in *Throughflow* is calculated from other quantities. The number of parameters $\mathbf{p}_{TF}$ used in *Throughflow* optimization is significantly reduced to a total of 27. The problem reads as

$$\max_{\mathbf{p}_{TF} \in P_{TF}} \begin{bmatrix} \eta_{TF} \\ SM_{TF} \end{bmatrix}, \quad P_{TF} = \left\{ \mathbf{p}_{TF} \in \mathbb{R}^{27} \mid \mathbf{h}_{TF} \leq \mathbf{0} \right\}. \tag{17.2}$$

After solving this *Throughflow* optimization problem optimized flow fields are available which need to be realized by filling the annulus with proper blades. For the evaluation of a specific blade the 2D solver *Mises* is used [3] which calculates the 2D flow between blades on the S1 stream surface [15]. This is done for three sections, one near hub, one near casing and one at the blade mid-section. However, the parameterization $\mathbf{p}_B$ describes the complete 3D blade at once and the airfoil sections used for analysis are obtained by blending the 3D blade with the three streamlines of interest coming from *Throughflow*. This quasi-3D approach guarantees smooth blade geometries due to direct 3D blade parameterization and quick results by use of the fast 2D *Mises* solver. The blade is described by a superposition of a dimensionless 3D camber-line angle distribution with a dimensionless 3D thickness distribution, and radial distributions of blade inlet angle, blade outlet angle and thickness to chord ratios which are varied during blade optimization [4]. The optimization criteria are chosen as minimization of the loss coefficients calculated by *Mises* at the design point ω_{DP} and at off-design conditions towards stall ω_{St} and choke ω_{Ch} [4]. Off-design conditions are defined as change of inlet flow angle by $\pm 2°$. The maximum of the two off-design losses is used as second criterion to be minimized. Constraints $\mathbf{h}_B\left(\mathbf{p}_B\right) \leq \mathbf{0}$ for the optimization involve proper flow conditions along the blade, i.e., avoidance of suction sided flow separation and achievement of the flow field predicted by *Throughflow*. This yields the optimization problem

$$\min_{\mathbf{p}_B \in P_B} \left[\frac{\omega_{DP}}{\max\left(\omega_{Ch}, \omega_{St}\right)} \right], \quad P_B = \left\{ \mathbf{p}_B \in \mathbb{R}^{33} \, | \mathbf{h}_B \leq \mathbf{0} \right\}. \tag{17.3}$$

All of the presented stand-alone optimization processes are multi-objective optimization problems which here are solved with the genetic algorithm AMGA [14].

17.3 Coupling of Design Sub-processes

In the presented work these optimization processes are not considered as stand-alone tasks, but a complete process chain is set up which requires a coupling of the codes, respectively the sub-optimization tasks. An efficient way of coupling *Meanline* and *Throughflow* has been investigated in [12] and is shortly recapitulated hereafter. For the *Meanline* optimization problem (17.1) it is necessary to include an immediate *Throughflow* evaluation. However, due to the nature of the *Meanline* optimization design variables $\mathbf{p}_{ML}$ concern only the mid-line but no radial dependence. Therefore, *Throughflow* parameters $\mathbf{p}_{TF}$ are chosen such that the radial distributions are considered to be constant, upper left of Fig. 17.1. This avoids the implementation of $\mathbf{p}_{TF}$ as design variables which would increase the number of optimization variables to 70 and could not be managed by genetic optimization algorithms properly [12]. However, the *Throughflow* evaluation needs to be performed in order to check whether constraints defined for *Throughflow* optimization (17.2) are achievable by a specific design proposed during *Meanline* optimization. It is assumed to be the case if *Throughflow* constraints $\mathbf{h}_{TF}$ are already satisfied within a certain tolerance ε for those preliminary radially constant designs. This prevents *Meanline* from running into critical solutions which cannot be handled by a subsequent *Throughflow* optimization. Since only these relaxed constraints $\mathbf{h}_{TF} \leq \varepsilon$ are transferred from problem (17.2), parameters, objectives and the remaining constraints are still according to formulation (17.1) yielding the *coupled Meanline* optimization problem

$$\max_{\mathbf{p}_{ML} \in \overline{P}_{ML}} \begin{bmatrix} \eta_{ML} \\ SM_{ML} \end{bmatrix}, \quad \overline{P}_{ML} = \left\{ \mathbf{p}_{ML} \in \mathbb{R}^{43} \left| \begin{bmatrix} \mathbf{h}_{ML} \\ \mathbf{h}_{TF} \end{bmatrix} \leq \begin{bmatrix} \mathbf{0} \\ \varepsilon \end{bmatrix} \right. \right\}. \tag{17.4}$$

The result of this *coupled Meanline* optimization is not only a single design but a set of several Pareto optimal design trade-offs. Picking a specific design for further optimization would corrupt the idea of multi-objective optimization, whereas the optimization of every Pareto design would increase computation time enormously. In order to cope with the demands of multi-objective optimization without an increase of computation time, the *Throughflow* optimization problem (17.2) is extended according to an approach proposed by [12]. The idea is not to start from just a single optimal design, but investigate all optimal designs in parallel during one single *Throughflow* optimization sweep, see upper right of Fig. 17.1. This is done by tagging the optimal *Meanline* solutions with an integer parameter $i_{ML} \in \{1, 2, \ldots, K\}$ which is used as an additional design variable and modified during *Throughflow* optimization.

The *Throughflow* optimization problem then handles both, real parameters $\mathbf{p}_{TF}$ and the integer parameter i_{ML} resulting in a challenging, so called mixed integer problem. This kind of problem, when attacked with a genetic algorithm like AMGA, has the tendency of premature convergence and thus lack of diversity. To avoid premature convergence and guarantee diversity over all possible integer values, each representing a specific *Meanline* design as basis for *Throughflow*, a new strategy [12] is applied that extends the problem formulation (17.2) to an unconstrained problem formulation

$$\max_{\mathbf{p}} \begin{bmatrix} \eta_{TF}^* \\ SM_{TF}^* \\ i_{ML.pos}^* \\ i_{ML.neg}^* \\ h^* \end{bmatrix}, \quad \mathbf{p} = \begin{bmatrix} \mathbf{p}_{TF} \\ i_{ML} \end{bmatrix} \tag{17.5}$$

with additional objectives:

$$i_{ML,pos}^* = \begin{cases} i_{ML} & \text{if } j \leq \nu N_E \\ K & \text{else} \end{cases} \tag{17.6}$$

$$i_{ML.neg}^* = \begin{cases} -i_{ML} & \text{if } j \leq \nu N_E \\ 1 & \text{else} \end{cases} \tag{17.7}$$

$$\eta_{TF}^*(\mathbf{p}) = \begin{cases} \eta_{TF}(\mathbf{p}) & \text{if } \mathbf{h}_{TF}(\mathbf{p}) \leq \mathbf{0} \\ \eta_{TF,min} & \text{else} \end{cases} \tag{17.8}$$

$$SM_{TF}^*(\mathbf{p}) = \begin{cases} SM_{TF}(\mathbf{p}) & \text{if } \mathbf{h}_{TF}(\mathbf{p}) \leq \mathbf{0} \\ SM_{TF,min} & \text{else} \end{cases} \tag{17.9}$$

$$h^*(\mathbf{p}) = \begin{cases} 0 & \text{if } \mathbf{h}_{TF}(\mathbf{p}) \leq \mathbf{0} \\ \sum_{j=1}^{N_C} \min\left\{-h_{TF,j}(\mathbf{p}), 0\right\} & \text{else} \end{cases} \tag{17.10}$$

Objectives (17.6) and (17.7) are in contradiction (i.e. all values of i_{ML} are optimal by definition) and therefore guarantee diversity over different *Meanline* solutions as long as the actual number of evaluations j is lower than a part ν of design evaluation limit number N_E. Thus *Meanline* solutions are treated equally and improved according to objectives (17.8)–(17.10) which are a kind of multi-criterion penalty strategy. Only in the second phase the different *Meanline* solutions have to compete and the i_{ML} value is ignored. The penalty strategy chosen here is not the classical one, but violation of constraints is considered as additional objective (17.10). To ensure that feasible solutions are always rated better than infeasible ones, the original objectives η_{TF} and SM_{TF} are allocated with lowest possible numbers in case of constraint violation resulting in modified constraints (17.8) and (17.9). Thus, the optimization algorithm is forced to firstly achieve feasibility and secondly improve designs [12].

Although the *Throughflow* optimization results are a Pareto set like the *Meanline* results, at this stage it is rather difficult to transfer each design to the blading design

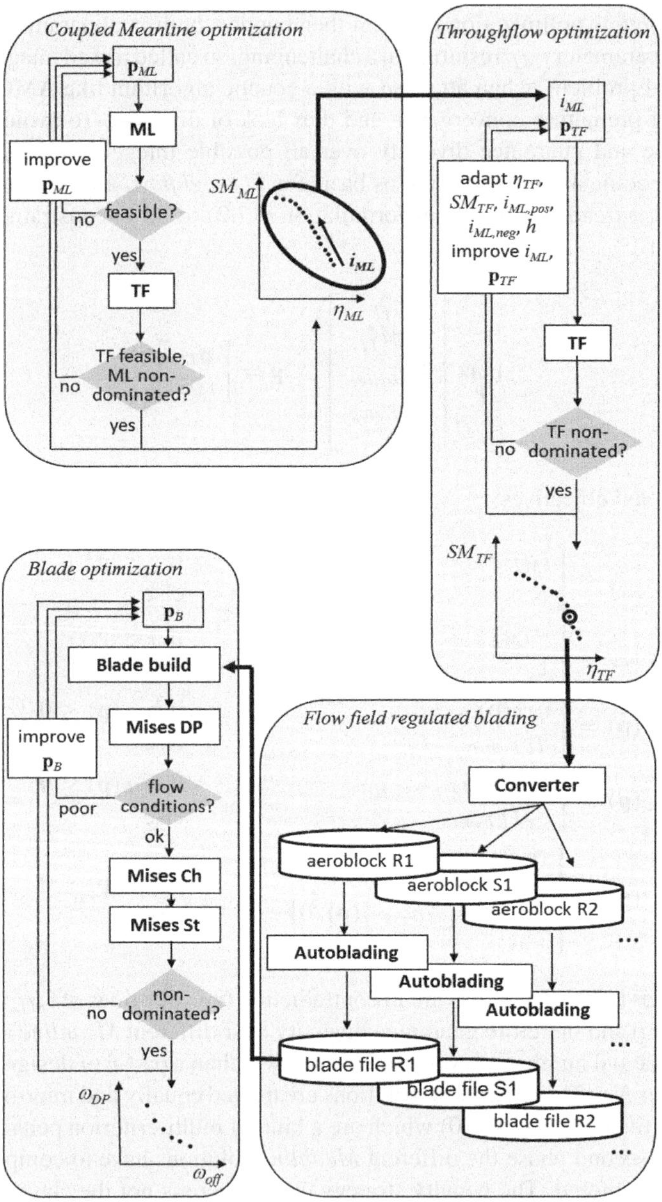

Fig. 17.1 Automated *Isight* modules within the aerodynamic compressor design process

step and optimize it in parallel like the *Throughflow* optimization strategy described above. The necessity of choosing a specific *Throughflow* design for the subsequent blading process may be motivated as follows. Both *Meanline* and *Throughflow* are codes that handle the complete compressor, whereas the blading process is performed for each blade row separately. Therefore, the boundaries between the separate blade rows have to match, i.e., exit flow angles of one blade row have to match with inlet flow angles of the blade row downstream. If multiple *Throughflow* solution were allowed, each blade optimization would prefer another *Throughflow* solution for the specific blade row. This would result in a corrupt annulus where neither flow angles nor hub and casing walls would match. Consequently a specific *Throughflow* solution, i.e., flow field and annulus, has to be selected prior to the blading process.

After selecting a specific optimal flow field from the *Throughflow* solutions, it is converted and split into flow field information for each blade row, called aero block. This defines boundary conditions for the automatic blade generation sub-process shown in the lower right of Fig. 17.1. The blade generation uses knowledge based rules depending on aerodynamic and geometric boundary conditions as e.g. inlet Mach number or solidity, to select blade shapes from an industrial database [11]. Then, these generated blades serve as references for subsequent quasi-3D blade optimizations as can be seen in the lower left of Fig. 17.1. Typically the automatically generated blades show already good aerodynamic performance, but they are unlikely to fulfill the constraints. For the blading process the constraints are split into two groups of different importance. Constraints concerning general flow conditions like separation are considered first-rate, whereas constraints insuring matching with the flow field provided by *Throughflow* are considered second-rate. Further *Mises* calculations for off-design points are calculated only if first-rate flow condition constraints for design point conditions are already satisfied. However, a calculation of off-design points has to be performed before second rate constraints are achieved to avoid running into designs that are feasible at design point, but do not converge at off-design conditions.

Figure 17.1 shows the modular structure of the complete design process. For each design step an *Isight* module was developed which enables automated optimization of designs on a specific design level, but already takes into account coupling effects if necessary. Due to the modular process flow the engineer has the opportunity to verify designs on each level and to adapt the problem formulation if required.

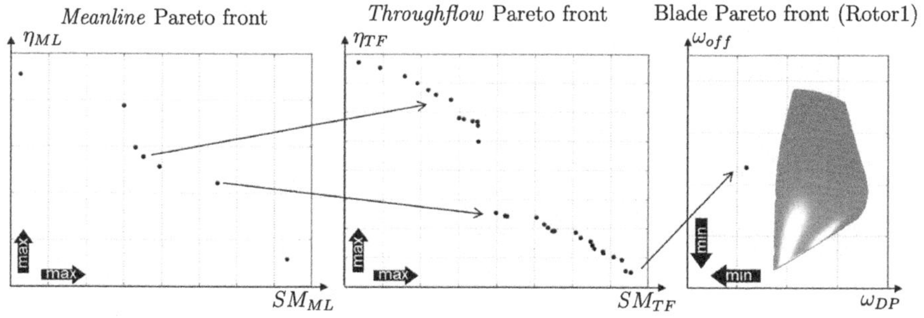

Fig. 17.2 Evolution of optimization results by *Meanline* and *Throughflow* to final blading result shown for first rotor

17.4 Application to an Industrial Compressor

For validation, the described process is applied to a modern 9-stage high pressure research compressor. Just a few performance requirements like mass flow and basic assumptions like tip clearances are provided to start the design process. Seven *Meanline* optimal solutions are proposed by the first design step by solving the optimization problem (17.4), see Fig. 17.2. These *Meanline* results are completely transferred to the extended *Throughflow* optimization problem (17.5) and further optimized in parallel. The resulting Pareto front consists of 32 designs which, however, originate from only two of the seven *Meanline* results. As described above now a selection of a specific design is necessary to continue with the blading step. Here, the design with the highest efficiency is selected from the *Throughfow* Pareto front. The blade optimization is performed for all blade rows on a cluster in parallel. Figure 17.3 shows rotor blades of the first stage for reference design in comparison with the quasi 3D optimized blade. Typical design changes can be associated with the specific sub-process involved. E.g. the final annulus inlet is located at smaller radii $r_j^{opt} < r_j^{ref}$, $j \in \{hub, cas\}$ and the annulus shows a steep contraction which is a result of *Meanline* optimization. The radial distribution of the axial chord length is a result of *Throughflow* optimization and the optimized blade shows a significant difference compared to the reference design. The blade optimization sub-process changed the contour of the blade by changing the dimensionless camber-line angle and thickness distributions. The different contours of both blades can be seen in Fig. 17.3b, where casing near airfoil sections of the blades are displayed. The optimized airfoil is longer, flow angles are different and flow turning has increased. The flow turning of the optimized airfoil is focused on the rear in contrary to the reference design. Finally the stacking of the reference design is applied to the optimized blade and all necessary input files for a 3D CFD are generated. Figure 17.4 shows the meshed blade and annulus, generated for a 3D CFD analysis.

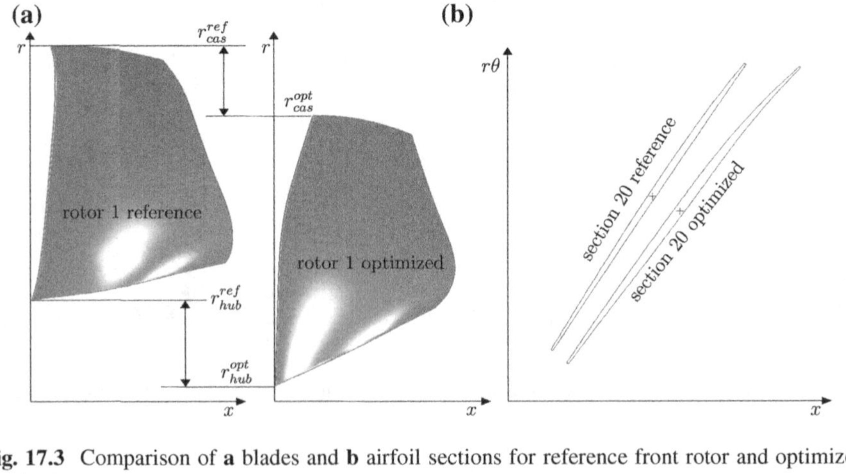

Fig. 17.3 Comparison of **a** blades and **b** airfoil sections for reference front rotor and optimized front rotor

Fig. 17.4 Mesh of optimized blade and annulus for front rotor

17.5 Conclusions

The paper describes a completely automated design process including all steps of an industrial design approach: basic decisions based on *Meanline* and *Throughflow* calculations, blading and final 3D CFD analysis. In order to avoid back-loops, higher fidelity *Throughflow* analysis are already included in the first *Meanline* design step. By transferring all optimized *Meanline* solutions to *Throughflow*, the intermediate results are as good as if all *Meanline* results would have been optimized separately, but computational costs are cut down. The final blading needs to be performed on a single *Throughflow* solution in order to guarantee consistency and results in smooth blades which can be immediately analyzed by 3D CFD. The generation of such optimal blades for the whole compressor can be done on a small-size computer cluster within five days starting almost from scratch.

Acknowledgments The presented work has been performed within the VIT 3 project (Virtual Turbomachinery) in collaboration with Rolls-Royce Deutschland. The project is partly funded by the Federal State of Brandenburg, Germany and the European Community.

References

1. Cumpsty N (2004) Compressor aerodynamics. Krieger Publishing Company, Malabar
2. De Haller P (1953) Das Verhalten von Tragflügeln in Axialverdichtern und im Windkanal. In: Brennstoff-Wärme-Kraft 5, VDI-Verlag, Düsseldorf, pp 333–336
3. Drela M (1986) Two-dimensional transonic aerodynamic design and analysis using the Euler equations. Massachusetts Institute of Technology, Boston
4. Dutta AK, Flassig PM, Bestle D (2008) A non-dimensional quasi-3d blade design approach with respect to aerodynamic criteria. In: Proceedings of ASME Turbo Expo 2008, GT2008-50687, Berlin

5. Engineous Software Inc. (2009) iSIGHT-FD Version 3.5 user's guide. Dassault Systèmes Simulia Corporatio. Cary
6. Gallimore SJ (2001) Axial compressor design. Rolls-Royce plc, Derby
7. Hinz M (2012) Neue Parametrisierungsstrategien und Methoden der Prozessbeschleunigung für die Verdichteroptimierung. Shaker, Aachen
8. Keskin A (2007) Process integration and automated multi-objective optimization supporting aerodynamic compressor design. Shaker, Aachen
9. Koch CC (1981) Stalling pressure rise capability of axial flow compressor stages. J Eng Power 103:645–656
10. Lieblein S, Schwenk FC, Broderick RL (1953) Diffusion factor for estimating losses and limiting blade loadings in axial-flow-compressor blade elements. NACA RM E53D01. Los Angeles
11. Poehlmann F (2009) Knowledge-based automatic 2d compressor blade generation using aspects of aerodynamic robustness. Diploma Thesis, Technical University Berlin
12. Poehlmann F, Bestle D (2012) Multi-objective compressor design optimization using multi-design transfer between codes of different fidelity. In: Proceedings of ASME Turbo Expo 2012, GT2012-68577. Copenhagen
13. Rühle, Bestle D (2010) Ein Verfahren zur optimalen Hochdruckverdichterauslegung auf Basis der Meridianströmungsrechnung. In: Proceedings of Deutscher Luft- und Raumfahrtkongress. Hamburg
14. Tiwari S, Koch P, Fadel G, Deb K (2008) AMGA: An archive-based micro genetic algorithm for multiobjective optimization. In: Proceedings of genetic and evolutionary computation conference, Atlanta, pp 729–736
15. Wu CH (1952) A general theory of three-dimensional flow in subsonic and supersonic turbomachines of axial-, radial- and mixed-flow types. NACA TN-2604. Lewis Flight Propulsion Laboratory, Washington

Chapter 18
Design-Optimization of a Compressor Blading on a GPU Cluster

Konstantinos T. Tsiakas, Xenofon S. Trompoukis, Varvara G. Asouti and Kyriakos C. Giannakoglou

Abstract This paper presents the design/optimization of turbomachinery blades using synchronous and asynchronous metamodel-assisted evolutionary algorithms on a GPU cluster. Asynchronous EAs overcome the synchronization barrier at the end of each generation and exploit better all available computational resources. Radial basis function networks are used as on-line trained surrogate evaluation models (metamodels) according to the inexact pre-evaluation (IPE) concept. With the exception of a few initial evaluations, which are based on the exact evaluation tool, each new candidate solution is approximately evaluated using local metamodels and only the most promising among them are, then, re-evaluated using the exact tool. Suggestions about the number of population members to be re-evaluated on the CFD tool, in the framework of the IPE scheme, are provided. The effect of using more than one GPUs to evaluate each candidate solution in the optimization turnaround time is discussed.

Keywords Design-optimization of turbomachines · Graphics processing units · Evolutionary algorithms · Metamodels

K.T. Tsiakas (✉) · X.S. Trompoukis · V.G. Asouti · K.C. Giannakoglou
Laboratory of Thermal Turbomachines, Parallel CFD & Optimization Unit, School of Mechanical Engineering, National Technical University of Athens (NTUA), Iroon Polytechniou 9, 15780 Athens, Greece
e-mail: tsiakost@gmail.com

X.S. Trompoukis
e-mail: xeftro@gmail.com

V.G. Asouti
e-mail: vasouti@mail.ntua.gr

K.C. Giannakoglou
e-mail: kgianna@central.ntua.gr

D. Greiner et al. (eds.), *Advances in Evolutionary and Deterministic Methods for Design, Optimization and Control in Engineering and Sciences*, Computational Methods in Applied Sciences 36, DOI 10.1007/978-3-319-11541-2_18

18.1 Introduction

Nowadays, evolutionary algorithms (EAs) are successfully applied to many scientific fields including engineering sciences, since they can handle single-or multi-objective, unconstrained or constrained optimization problems by accommodating any evaluation software as a black-box. EAs main disadvantage is related to the great number of evaluations required to reach the optimal solution(s). In engineering problems, based on evaluation software which is computationally demanding (for instance, a CFD code), this noticeably increases the optimization turnaround time.

To decrease the CPU cost and/or the turnaround time of an EA-based optimization, the concurrent evaluation of candidate solutions and/or the implementation of surrogate evaluation models (or metamodels) can be used. Generation-based EAs, to be referred as "synchronous" in this paper, usually implement the master-worker paradigm to concurrently evaluate the generation members. Each population member can optionally be evaluated on many processors, provided that a parallel evaluation software (such as a parallel CFD code, in CFD-based optimization) is available. By overcoming the notion of "generation", the so-called asynchronous EAs (AEAs) have been developed [1, 2]. AEAs may exploit the available computational resources better than a synchronous EA relying upon the master-worker paradigm.

On the other hand, metamodels are tools providing low-cost approximations to the results of costly problem-specific evaluation models. In this work, metamodel-assisted EAs (MAEAs) with on-line trained local metamodels (radial basis function—RBF networks) are used according to the Inexact Pre-Evaluation (IPE) scheme of the candidate solutions. IPE starts after running and archiving a number of individuals using exclusively the problem-specific evaluation model, in order to collect the minimum number of samples to be used to train the metamodels. In the synchronous EA, during the IPE phase [9, 10], all population members of each generation are pre-evaluated on surrogate models built in purpose and only a few top individuals undergo re-evaluation on the exact model. In the asynchronous EAs, [1, 2], instead of generating and evaluating a single new individual every time a CPU becomes idle, a number of trial solutions are generated and pre-evaluated on the metamodel. Then, the best among them, according to the metamodel, is exactly re-evaluated.

In aerodynamic optimization, additional gain is expected from the use of a GPU-enabled Navier-Stokes solver to perform the CFD-based evaluation. Many of the existing CFD solvers running on GPUs use structured grids [4, 11, 15] and thus profit of the aligned access to the GPU memory, leading to high speed-ups. Applications based on unstructured grids are still limited and usually based on cell-centered finite volumes [15]. The GPU-enabled software used in this work solves the Navier-Stokes equations on unstructured grids using the vertex-centered finite volume technique [3, 8, 14]. Though this is the most difficult case regarding GPU memory access, compared to the use of either structured or unstructured grids with cell-centered finite volumes, the optimization of memory access, discussed in [3, 14], along with the use of mixed precision arithmetics [8] make the code running 50$\times$ faster on a single GPU than on a single CPU core.

In this paper, synchronous and asynchronous metamodel-assisted EAs (MAEAs and AMAEAs) are used for the design optimization of a compressor blading on a GPU-cluster which consists of four interconnected server blades, with 3 NVIDIA Tesla M2050 each. In this 12 GPU configuration, various parallelization schemes are investigated in order to minimize the optimization turnaround time. These include MAEAs and asynchronous MAEAs (AMAEAs) allowing up to 12 (as many as the available GPUs) concurrent evaluations. Regarding MAEAs, the impact of the number of individuals to be re-evaluated on the CFD model is investigated.

An additional investigation of the effect of parallelizing the CFD software on many GPUs in the optimization turnaround time is also carried out, for both the MAEA and AMAEA.

18.2 The Navier-Stokes Equations Solver—Implementation on Many GPUs

A GPU-enabled Navier-Stokes solver [3, 8, 14] is used for steady 3D incompressible flows for the evaluation of candidate solutions. The GPU-solver may use GPUs associated with the same or different computational nodes.

18.2.1 The Navier-Stokes Equations

The pseudo-compressibility approach, introduced by Chorin [5], is used to handle incompressible fluid flows. By introducing the artificial compressibility β, the mean flow equations with respect to the rotating frame of reference become

$$\mathbf{R}(\mathbf{W}) = \frac{\partial \mathbf{W}}{\partial t} + \frac{\partial \mathbf{F}_j^{inv}}{\partial x_j} - \frac{\partial \mathbf{F}_j^{vis}}{\partial x_j} - \mathbf{S} = 0 \tag{18.1}$$

where $\mathbf{W} = [p\ w_1\ w_2\ w_3]$ is the vector of the unknowns, $w_i, i = 1, 2, 3$ are the relative velocity components and p is the static pressure. $\mathbf{F^{inv}}$, $\mathbf{F^{vis}}$ are the inviscid and viscous fluxes respectively and $\mathbf{S}$ is the vector of source terms containing the Coriolis and centripetal forces,

$$\mathbf{F}_j^{inv} = \begin{bmatrix} \beta w_j \\ w_j w_1 + p\delta_{1j} \\ w_j w_2 + p\delta_{2j} \\ w_j w_3 + p\delta_{3j} \end{bmatrix}, \quad \mathbf{F}_j^{vis} = \begin{bmatrix} 0 \\ \tau_{1j} \\ \tau_{2j} \\ \tau_{3j} \end{bmatrix}$$

$$\tau_{ij} = (\nu + \nu_t)\left(\frac{\partial w_i}{\partial x_j} + \frac{\partial w_j}{\partial x_i}\right)$$

$$S_i = 2\varepsilon_{kji} w_k \Omega_j + \varepsilon_{jkl}\varepsilon_{hli}\Omega_h \Omega_j x_k$$

$\Omega_i, i = 1, 2, 3$ are the components of the angular velocity vectors. The mean-flow equations are coupled with the one-equation low-Reynolds number Spalart–Allmaras [13] turbulence model. The viscosity coefficient is given by $\nu_t = \tilde{\nu} f_{v_1}$, where $\tilde{\nu}$ is the solution variable in the state turbulence equation, $R_{\tilde{\nu}} = 0$, where

$$R_{\tilde{\nu}} = \frac{\partial (w_i \tilde{\nu})}{\partial x_i} - \frac{\partial}{\partial x_i}\left[\left(\nu + \frac{\tilde{\nu}}{\sigma}\right)\frac{\partial \tilde{\nu}}{\partial x_i}\right] - \frac{c_{b_2}}{\sigma}\left(\frac{\partial \tilde{\nu}}{\partial x_i}\right)^2 - \tilde{\nu} P(\tilde{\nu}) + \tilde{\nu} D(\tilde{\nu}) \quad (18.2)$$

The production $P(\tilde{\nu})$ and destruction $D(\tilde{\nu})$ terms along with f_{v_1}, f_w, $\tilde{S}$, and constants c_{b_1}, c_{b_2}, c_{w_1} and σ are all defined in [13].

18.2.2 Boundary Conditions and Discretization

Concerning the boundary conditions, the no-slip condition is applied along the solid walls. At the inlet, the velocity vector profiles are imposed, while at the outlet a fixed mean pressure value is applied. A zero value of $\tilde{\nu}$ is specified along the solid boundaries.

The discretization of the governing PDEs is based on the time-marching technique and the vertex-centered finite volume method. Thus, in each pseudo-time step, Eqs. 18.1 and 18.2 are integrated over the finite volumes formed around mesh nodes. The CFD-solver used may handle unstructured/hybrid meshes consisting of tetrahedra, pyramids, prisms and hexahedra.

The inviscid numerical fluxes are computed using the Roe's approximate Riemann solver [12] with second-order accuracy. The stresses on the finite volume faces are computed based on the velocity gradients, the computation of which at the mid-point of each edge (PQ) is given by

$$\left.\frac{\partial w_i}{\partial x_j}\right|_{PQ} = \frac{1}{2}\left[\left.\frac{\partial w_i}{\partial x_j}\right|_P + \left.\frac{\partial w_i}{\partial x_j}\right|_Q\right] - \left[\frac{1}{2}\left[\left.\frac{\partial w_i}{\partial x_j}\right|_P + \left.\frac{\partial w_i}{\partial x_j}\right|_Q\right] n_j - \frac{w_i^Q - w_i^P}{(PQ)}\right] n_j \quad (18.3)$$

where n_j is the normal to the finite volume interface. Coriolis and centripetal forces are added as source terms.

18.2.3 Numerical Solution

The discretized Navier-Stokes equations are solved iteratively according to the scheme

$$\frac{\partial \mathbf{R}}{\partial \mathbf{W}} \Delta \mathbf{W} = -\mathbf{R}(\mathbf{W}), \quad \mathbf{W}^{k+1} = \mathbf{W}^{k} + \Delta \mathbf{W} \tag{18.4}$$

with k denoting the pseudo-time iteration. The pseudo-time step is calculated locally at each mesh node, based on stability criteria.

In order to maximize the parallel efficiency of the GPU-enabled solver and reduce the device memory requirements, the solver uses Mixed Precision Arithmetics (MPA) [8], which does not harm the accuracy of the results, due to the delta formulation presented in Eq. 18.4. In the proposed MPA scheme, DPA (Double Precision Arithmetics) is used for computing both the LHS and RHS terms. Then, SPA (Single Precision Arithmetics) is used to store the memory-consuming LHS terms, whereas DPA is used for the RHS (i.e. the residuals) of Eq. 18.4.

18.2.4 Implementation on Many GPUs

The flow solver, written in the CUDA programming environment, uses the MPI protocol for the inter-node communications. Each CPU-process is executed on a different computing node and controls the GPUs associated with this node. Generally, the number of CPU-processes is not equal to the number of mesh partitions since each CPU-process controls many on-node GPUs each of which is associated with a single mesh partition. For the communication between the on-node GPUs, event-stream synchronizations have been employed. Besides, GPUs on the same node use the common CPU (pinned) memory for data interchange. In order to increase the parallel efficiency of the CFD solver, data interchange overlap with computations; thus, GPU cores remain active even while data are transferred through the available devices.

In the beginning, the CPU-process with rank 0 reads and broadcasts the data input which include the flow conditions, the nodal coordinates and the connectivity of mesh elements per subdomain. It also constructs the necessary lists of mesh nodes shared by adjacent subdomains, for which data communications are necessary. Then, each CPU-process performs the computation of topology and mesh related data supporting the finite volume method for the subdomains associated with the GPUs controlled by this CPU-process. The computed data are copied to the GPUs, where the numerical solution of the Navier-Stokes equations takes place through successive GPU kernel launches, data interchange and synchronizations.

As already mentioned, the flow solver uses unstructured grids and the vertex-centered finite volume technique. This is the most difficult case regarding GPU memory access, because (a) the use of unstructured grids leads to "unstructured" memory access due to the random grid element numbering and (b) in the vertex-centered approach the number of neighboring nodes per grid node varies from node to node. On the other hand, in the cell-centered finite volume technique the number of neighboring elements per element is a priori known; for instance, in 3D grids with tetrahedral elements, each internal tetrahedron has 4 adjacent tetrahedra. The optimization of memory access [3, 14] together with the use of mixed precision arithmetics (MPA) [8] make the flow solver running on a single GPU about 50 times faster than its counterpart running on a single CPU core.

18.3 The EA-Based Optimization Platform

The design optimization of the compressor blading presented in this work is based on EAs, using the capabilities of the optimization platform EASY [6] developed by the authors' group. In order to decrease the computational cost of the EA-based optimization, the concurrent evaluation of candidate solutions together with surrogate evaluation models (or metamodels) are implemented. Synchronous and asynchronous metamodel-assisted EAs (MAEAs and AMAEAs) are used to quantify the parallel efficiency from the concurrent evaluation of candidate solutions.

In both MAEA and AMAEA, all evaluated individuals, paired with the corresponding objective function values, are stored in a database (DB). Once a predefined minimum number of DB entries has been archived, the IPE phase (implemented differently in the synchronous and asynchronous mode) starts. On-line trained metamodels are used during the IPE phase. Radial basis function (RBF) networks are trained separately for each new individual on its closest (in terms of Euclidean distances in the normalized design space) DB entries.

In constrained problems, such as the one studied herein, candidate solutions violating one or more constraints are penalized using an exponential penalty function.

In the following sections, the basic features of the MAEA and AMAEA, with emphasis on the parallelization model and the IPE implementation, are described.

18.3.1 Metamodel-Assisted EA

The synchronous or generation-based (μ, λ)EA and (μ, λ)MAEA, [7]) in each generation handle three populations, namely the parent (with μ members), the offspring (with λ members) and the elite (with ε members) ones. The μ parents result from the application of the parent selection operator to the offspring and elite populations of the previous generation. The λ offspring are formed from the parents, via the application of evolution operators, such as crossover, mutation, etc., including elitism.

The first few generations are performed as a conventional EA and the MAEA starts once there are "enough" training patterns in the DB. During the IPE phase all population members are pre-evaluated on surrogate models trained on-the-fly and only a few ($\lambda_{IPE} << \lambda$) top population members, i.e. the most promising based

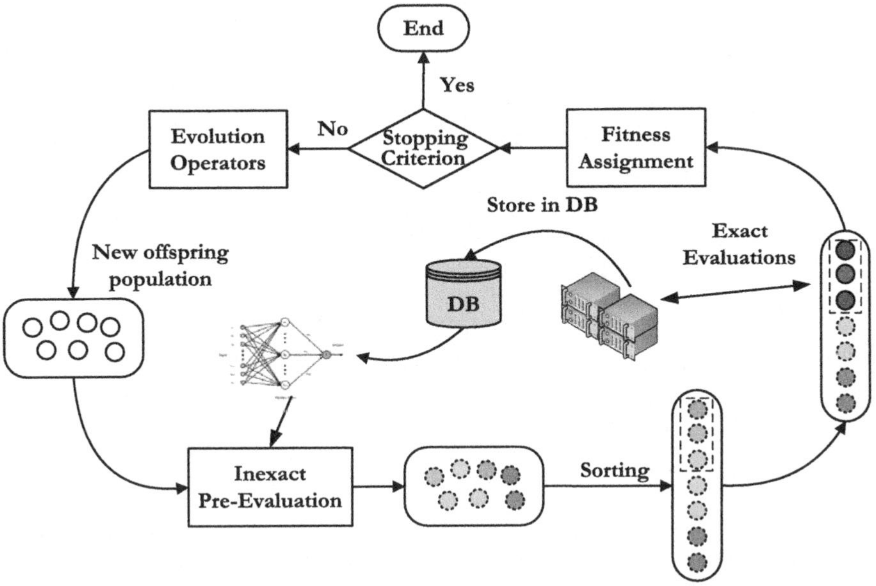

Fig. 18.1 Schematic representation of the IPE phase in a (synchronous) MAEA

on the metamodel, are re-evaluated on the CFD model (Fig. 18.1). The λ_{IPE} value may vary between a lower ($\lambda_{IPE,min}$) and an upper ($\lambda_{IPE,max}$) user-defined bound. Initially, only the $\lambda_{IPE,min}$ top individuals are re-evaluated; then, some more, up to $\lambda_{IPE,max}$ in total, may be re-evaluated too, based on a number of criteria.

The parallelization of EA and MAEA is based on the master-worker paradigm where the master assigns the λ evaluations to the available GPUs (N_{GPU}). The number of CPU processes (N_{CPU}) is equal to N_{GPU} if the evaluation of a candidate solution is assigned to a single GPU or $N_{CPU} < N_{GPU}$ if assigned to more than one GPUs, as described in Sect. 2.4. In the general case where $\lambda < N_{GPU}$, the first N_{GPU} evaluations are assigned to the N_{GPU} devices. The remaining evaluations within the same generation are then assigned to GPUs that become idle anew. The master waits for all GPUs to complete their evaluations before proceeding to the next generation. In the case of a MAEA, the master undertakes the IPE and assigns the λ_{IPE} evaluations to the available GPUs.

18.3.2 Asynchronous Metamodel-Assisted EA (AMAEA)

In the asynchronous EA (AEA) and AMAEA, [1, 2], the population members are associated with the nodes of a supporting mesh which is periodic along its opposite sides. The mesh is subdivided into demes of six nodes each, namely a pole, which acts as the deme's front-end where the best individual of the deme is stored, and

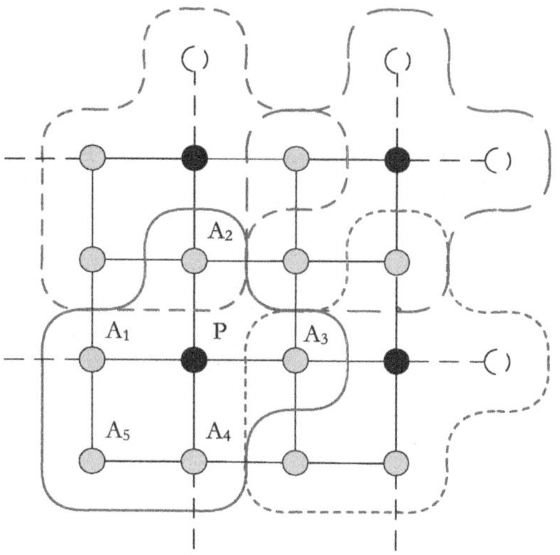

Fig. 18.2 Topology of a 4 × 4 supporting mesh of an asynchronous EA or MAEA. For the deme associated with pole P, agents A_1 to A_4 are shared with its four neighbouring demes whereas agent A_5 is the only non-shared one

five evaluation agents, Fig. 18.2. Demes interact through the shared supporting grid nodes. The application of the evolution operators is restricted within each deme.

Asynchronous EAs overcome the notion of generation and better exploit the available computational resources, in comparison to the master-worker paradigm. In particular, the optimization starts by randomly generating individuals and assigning their

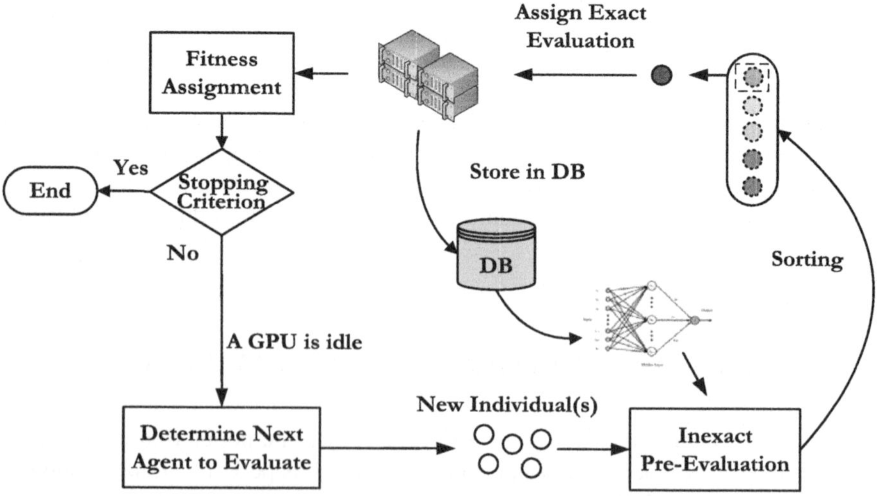

Fig. 18.3 Schematic representation of the IPE phase as implemented by an AMAEA

evaluation to the available GPUs. Upon completion of the evaluation of any individual, the corresponding GPU becomes idle. Instantaneously, a new individual to undergo evaluation is generated through intra- and inter-deme operations. An intra-deme operation, based on dominance criteria, decides whether the just evaluated individual must displace the corresponding pole(s) storing the best so-far computed solution on the deme. Then, an inter-deme operation selects, based on priority criteria, the next agent to undergo evaluation, [1].

In AMAEAs, the metamodels are activated only after completing and archiving a user-defined minimum number of exact evaluations. From this point on (Fig. 18.3), for each idle processor, N_{IPE} trial individuals are instantaneously generated by the evolution operators applied within the corresponding deme. For each one of them, a local metamodel is trained and an approximate ("inexact") fitness value is computed. Then, the "best" among the N_{IPE} individuals, according to the metamodel, is the one to undergo re-evaluation by the problem-specific (CFD) tool.

18.4 Design-Optimization of a Compressor Blade

This section presents the optimization of a peripheral compressor cascade. The existing (reference) compressor comprises 12 blades and operates at 1,300 rpm. Air in axial direction enters at velocity of 15.28 m/s. The blades are mounted on the hub, with a hub-to-tip radius ratio equal to 0.6 and form a 0.005 m clearance with the stationary shroud. In this paper, the blade is redesigned for minimum viscous losses defined as the averaged relative total pressure difference between the rotor inlet and outlet (ΔP_{tR}). A constraint to maintain the operating point of the reference configuration is imposed in terms of the mass averaged difference of the total pressure between the cascade outlet and inlet (ΔP_t).

18.4.1 Parameterization and Grid Generation

The blade shape is defined by superimposing parameterized thickness profiles on a parameterized mean-camber surface. Bezier control polygons are used to specify the distributions of all geometrical quantities. The design variables, 38 in total, are the coordinates of the control points of Bezier curves used to:

1. generate the meridional projection of the leading (LE) and trailing (TE) edges, as well as the hub and shroud generatrices (Fig. 18.4),
2. parameterize the spanwise distributions of (a) the mean-camber surface angles at the LE and TE, (b) the circumferential position of the blade LE and TE and (c) the mean-camber surface curvature,
3. parameterize the non-dimensional thickness profiles at a number of spanwise positions (Fig. 18.5, left) and

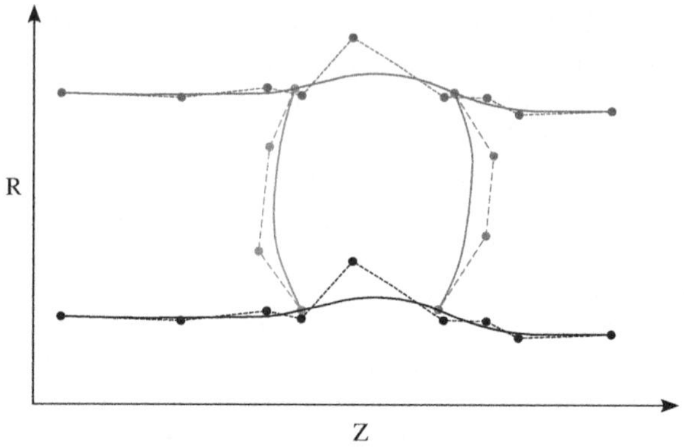

Fig. 18.4 Parameterization of the hub and shroud generatrices and the meridional projection of the leading and trailing edges. *Curves* and their corresponding *Bezier polygons* are shown

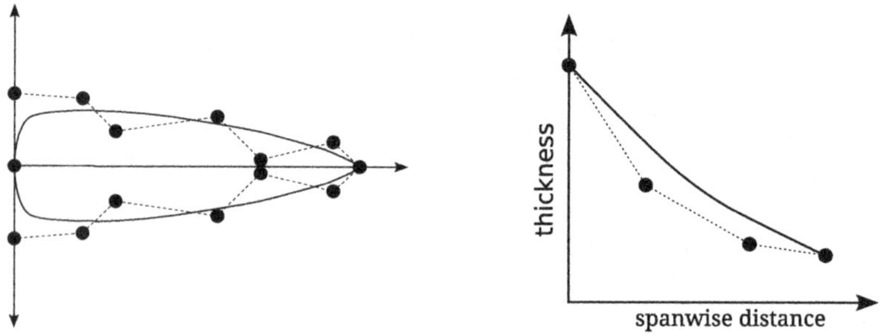

Fig. 18.5 Parameterization of a non-dimensional thickness profile, at certain spanwise positions (*left*) and the spanwise maximum thickness distribution (*right*)

4. dimensionalize the thickness distribution along the spanwise direction (Fig. 18.5, right).

For the evaluation of each candidate solution, a 3D unstructured mesh with about 400,000 nodes is generated using an in-house grid-generation software. The final grid comprises hexahedra and prisms over the blade surface, prisms over the casing and tip region, tetrahedra at the inner part of the domain and a zone of pyramids at the interface between hexahedra and tetrahedra. The CFD evaluation relies on the solver described in Sect. 18.2.

18.4.2 Optimization Results and Discussion

This case is studied using both MAEA and AMAEA with 12 concurrent evaluations, i.e. with $N_{CPU} = N_{GPU} = 12$, i.e. each evaluation is assigned to one GPU. For the

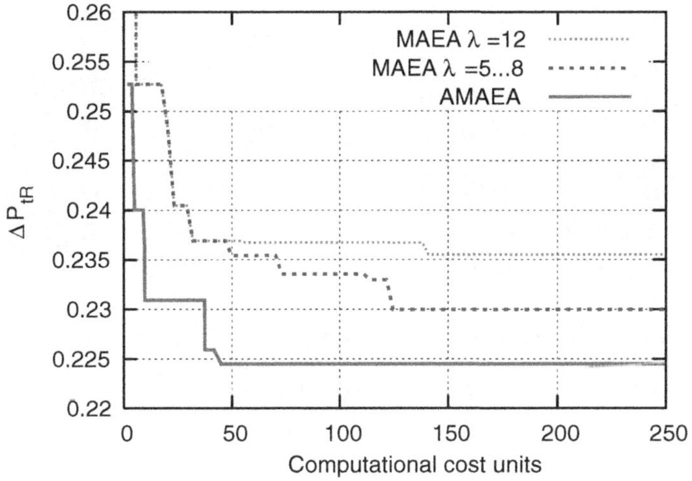

Fig. 18.6 Comparison of the convergence history of MAEA and AMAEA

MAEA $\lambda = 48$, $\mu = 16$ were used. The offspring population size (λ) was selected to be a multiple of N_{GPU} in order to get the "maximum" parallel efficiency from the master-worker model. In the AMAEA, a 8×8 supporting mesh (18.2) is used, which corresponds to 16 poles and 48 evaluation agents.

For both synchronous and asynchronous variants, 27–40 training patterns are used for each RBF network and the IPE started once the first 80 (non-penalized, i.e. feasible) individuals were evaluated and stored in the DB. Using the MAEA, two different optimization runs were carried out. In the first one, $\lambda_{IPE} = 12$ members are selected to be re-evaluated on the CFD code whereas, in the second one, λ_{IPE} value was allowed to vary between $\lambda_{IPE,min} = 5$ and $\lambda_{IPE,min} = 8$. This range was decided so as to have some idle GPUs and measure the effect on the optimization turnaround time. For the AMAEA, $N_{IPE} = 8$ trial individuals are generated and pre-evaluated before assigning the "best" of them to the idle GPU for evaluation on the flow solver.

The convergence histories in terms of computational CPU cost units of the three optimization runs are presented in Fig. 18.6. The AMAEA performs better for the same computational cost. It is interesting to comment on the differences between the two MAEA runs. The use of variable λ_{IPE} value is beneficiary for the optimization algorithm (though some GPUs remain idle during the re-evaluation) compared to the $\lambda_{IPE} = N_{GPU}$ option.

Comparison of the relative total pressure losses between the reference and the optimal blading at two transverse cross-sections at axial positions $z = 0.59\, c_{ax}$ and $z = 1.80\, c_{ax}$ (the z-origin is at the LE and c_{ax} is the axial chord) are shown in Figs. 18.7 and 18.8. It can be seen that, in the optimal blade, the high pressure losses region, corresponding to the tip clearance vortex, is minimized.

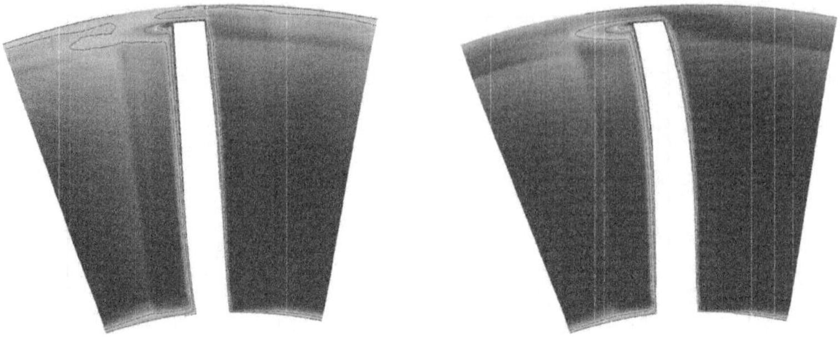

Fig. 18.7 Comparison of the relative total pressure between the reference (*left*) and the optimal (*right*) cascade at a transverse cross-section located at axial position $z = 0.59\, c_{ax}$. The P_{tR} field at $z = 0.59\, c_{ax}$ subtracted from the average inlet P_{tR} value is shown

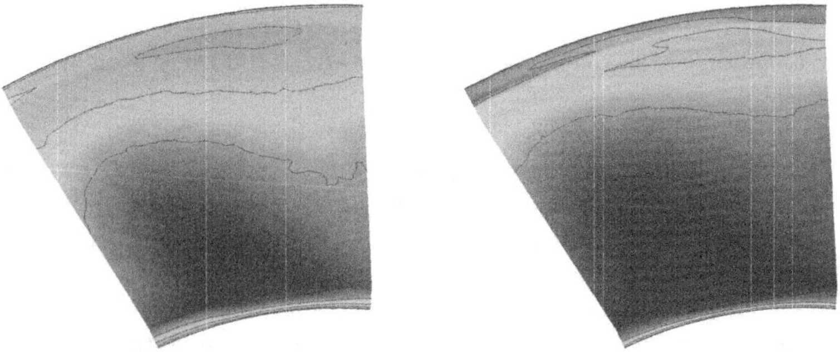

Fig. 18.8 Comparison of the relative total pressure between the reference (*left*) and the optimal (*right*) cascade at a transverse cross-section located at axial position $z = 1.80\, c_{ax}$. The P_{tR} field at $z = 1.80\, c_{ax}$ subtracted from the average inlet P_{tR} value is shown

Table 18.1 Comparison of the objective function and constraint values between the reference and the optimized blade

		Reference blade	Optimized blade
Objective	Losses (ΔP_{tR})	0.329	0.224
Constraint	ΔP_t	0.53	0.54

The corresponding objective function and constraint values are compared in Table 18.1. The pressure distribution on the optimal geometry is shown in Fig. 18.9.

18.4.3 Concurrent Evaluations Using CFD on Many GPUs

Aiming at further minimizing the optimization turnaround time, apart from the "smart" use of metamodels and the concurrent evaluations discussed thus far, the

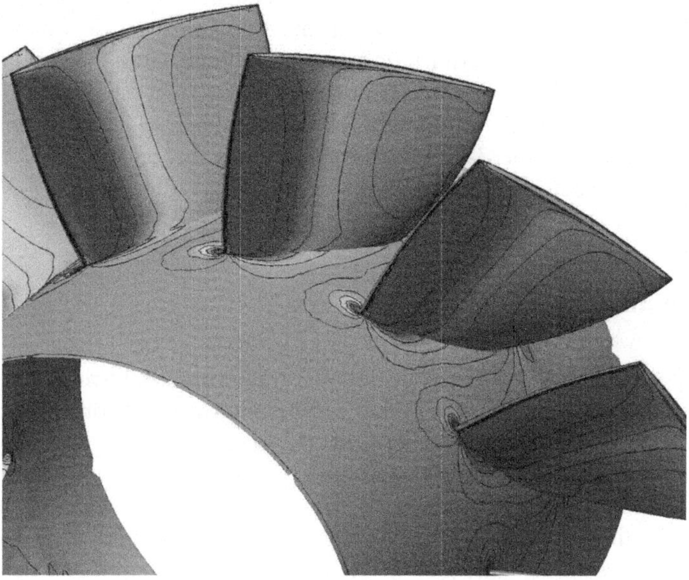

Fig. 18.9 Pressure distribution on the optimal geometry obtained by the AMAEA-based optimization

parallelization of the evaluation software on many GPUs is also considered. In this paper, for parallelizing the evaluation software, one should take into account the search method (synchronous or asynchronous), the possible use of IPE and the computational system in hand (4 computational nodes with 3 Tesla M2050 GPUs on each node).

If an evaluation can be carried out on a single GPU (depending on the computational domain size and the memory of each GPU) the use of asynchronous search with IPE is recommended, as described in the previous section.

Should the computational domain be partitioned and run on many GPUs, things become more complicated. As described in Sect. 18.2.4, running the Navier-Stokes solver on many GPUs can be done using a single CPU thread for the GPUs on the same node, MPI for GPUs on different nodes or a combination of both if more than 3 GPUs are involved. So, one should also select the "best" configuration, i.e. which GPUs (devices) from which node should undertake each evaluation.

Now, assume that a computational domain is partitioned into two subdomains. In such a case, 6 concurrent evaluations should be carried out. The CFD software may run on two GPUs of either the same node or different nodes. Using a single CPU thread to manage more than one GPUs on the same node, the parallel speed-up is greater than that of using MPI for GPUs on different nodes. So, for the asynchronous search, assigning as many evaluations as possible to pairs of GPUs belonging to the same node (4 in the GPU cluster under consideration) is the best practice in terms of the optimization turnaround time. For the synchronous search, if the offspring population is a multiple of 6 (as many as the concurrent evaluations), all combinations will practically lead to the same optimization turnaround time, since the synchronization barrier is determined by the slowest evaluation.

If 3 GPUs are required at minimum for each evaluation, then assigning 4 concurrent evaluations on the 3 GPUs of each node is the optimal choice for both synchronous and asynchronous search on the available GPU cluster.

18.5 Conclusions

In this paper, metamodel-assisted evolutionary algorithms, synchronous (MAEA) and asychronous (AMAEA), are used in combination with a parallel GPU-enabled CFD solver, in order to reduce the overall optimization time of a low speed compressor blading. The overall turnaround time is shown to be greatly reduced by (a) appropriately tuning the IPE scheme, (b) using asynchronous search in order to minimize the idle time, (c) using GPUs for the exact evaluation tool (CFD) instead of CPUs, and (d) "smartly" distributing the concurrent evaluations on the available GPUs of the cluster, especially in cases where the CFD solver is obliged to run on more than one GPUs. Thus, by appropriately combining all the above techniques and features, the use of evolutionary algorithms for aerodynamic optimization problems, can be made very appealing, in terms of optimization turnaround time, even in large scale industrial applications.

References

1. Asouti VG, Giannakoglou KC (2009) Aerodynamic optimization using a parallel asynchronous evolutionary algorithm controlled by strongly interacting demes. Eng Optim 41(3):241–257
2. Asouti VG, Kampolis IC, Giannakoglou KC (2009) A grid-enabled asynchronous metamodel-assisted evolutionary algorithm for aerodynamic optimization. Genet Program Evolvable Mach 10(3):373–389 (SI: Parallel and Distributed Evolutionary Algorithms, Part One)
3. Asouti VG, Trompoukis XS, Kampolis IC, Giannakoglou KC (2011) Unsteady CFD computations using vertex-centered finite volumes for unstructured grids on graphics processing units. Int J Numer Meth Fluids 67(2):232–246
4. Brandvik T, Pullan G (2011) An accelerated 3D Navier-Stokes solver for flows in turbomachines. J Turbomach 133(2):619–629
5. Chorin A (1967) A numerical method for solving incompressible viscous flow problems. J Comput Phys 2(1):12–26
6. Giannakoglou KC (2008) The EASY (Evolutionary Algorithms System) software. http://velos0.ltt.mech.ntua.gr/EASY
7. Kampolis IC, Giannakoglou KC (2008) A multilevel approach to single- and multiobjective aerodynamic optimization. Comput Meth Appl Mech Eng 197(33–40):2963–2975
8. Kampolis IC, Trompoukis XS, Asouti VG, Giannakoglou KC (2010) CFD-based analysis and two-level aerodynamic optimization on graphics processing units. Comput Meth Appl Mech Eng 199(9–12):712–722

9. Karakasis MK, Giannakoglou KC (2006) On the use of metamodel-assisted, multi-objective evolutionary algorithms. Eng Optim 38(8):941–957
10. Karakasis MK, Giotis AP, Giannakoglou KC (2003) Inexact information aided, low-cost, distributed genetic algorithms for aerodynamic shape optimization. Int J Numer Meth Fluids 43(10–11):1149–1166
11. Khajeh-Saeed A, Perot JB (2013) Direct numerical simulation of turbulence using GPU accelerated supercomputers. J Comput Phys 235:241–257
12. Roe P (1981) Approximate Riemann solvers, parameter vectors, and difference schemes. J Comput Phys 43(2):357–372
13. Spalart P, Allmaras S (1994) A one-equation turbulence model for aerodynamic flows. La Recherche Aérospatiale 1:5–21
14. Trompoukis XS, Asouti VG, Kampolis IC, Giannakoglou KC (2011) CUDA implementation of vertex-centered, finite volume CFD methods on unstructured grids with flow control applications, Chapter 17. GPU Computing Gems Jade Edition. Morgan Kaufmann, Burlington
15. Tutkun B, Edis F (2010) A GPU application for high-order compact finite difference scheme. In: 22nd international conference on parallel CFD 2010, Kaohsiung, Taiwan, 17–21 May 2010

Part IV
Engineering Design and Societal Applications: Structures, Materials and Civil Engineering

Chapter 19
Immune and Swarm Optimization of Structures

Tadeusz Burczyński, Arkadiusz Poteralski and Mirosław Szczepanik

Abstract The paper is devoted to applications of two bio-inspired methods: artificial immune systems and particle swarm optimizers to selected shape and topology optimization problems of structures. It contains numerical examples and comparisons of immune and swarm approaches with evolutionary optimization.

Keywords Particle swarm optimizer (PSO) · Artificial immune system (AIS) · Optimization · Finite element method (FEM) · Computational intelligence · Shape and topology optimization

19.1 Introduction

Shape and topology structural optimization is a very active research area. Several competing approaches for topology optimization exist.

Intelligent optimal design techniques based on evolutionary algorithms (EA) have found applications to structural optimization problems. The evolutionary methods are based on the theory of evolution. The main feature of those methods is to simulate biological processes based on heredity principles and the natural selection (the theory of evolution) to create optimal individuals (solutions) presented by single chromosomes.

T. Burczyński
Institute of Fundamental Technological Research,
Polish Academy of Sciences, Warsaw, Poland
e-mail: tburczynski@ippt.pan.pl; tburczyn@pk.edu.pl

T. Burczyński
Cracow University of Technology, Cracow, Poland

A. Poteralski (✉) · M. Szczepanik
Silesian University of Technology, Gliwice, Poland
e-mail: arkadiusz.poteralski@polsl.pl

M. Szczepanik
e-mail: miroslaw.szczepanik@polsl.pl

D. Greiner et al. (eds.), *Advances in Evolutionary and Deterministic Methods for Design, Optimization and Control in Engineering and Sciences*, Computational Methods in Applied Sciences 36, DOI 10.1007/978-3-319-11541-2_19

More recently, other bio-inspired approaches, alternative to EA, as the Artificial Immune Systems (AIS) and the Particle Swarm Optimizers (PSO) have gained popularity.

The paper is devoted to applications of two these approaches to selected shape and topology optimization problems of structures. A short description of biological aspect of natural immune systems and behaviour of swarm living creatures are described in the context of optimization procedures. The clonal selection algorithm which represents one of the main features of the artificial immune system is described. Standard and modified versions of artificial immune systems and swarm algorithms and their applications in different optimization problems of mechanical structures were widely presented by the authors [1–5]. In the present paper applications of these algorithms to topology optimization problems of structures are demonstrated. The paper contains also comparisons of immune and swarm approaches with evolutionary optimization and advantages and drawbacks of both methods are identified.

19.2 Artificial Immune Systems

The artificial immune systems (AIS) are developed on the basis of a mechanism discovered in biological immune systems [6]. An immune system is a complex system which contains distributed groups of specialized cells and organs. The main purpose of the immune system is to recognize and destroy pathogens—funguses, viruses, bacteria and improper functioning cells. The lymphocytes cells play a very important role in the immune system. The lymphocytes are divided into several groups of cells. There are two main groups B and T cells, both contains some subgroups (like B-T dependent or B-T independent). The B cells contain antibodies, which could neutralize pathogens and are also used to recognize pathogens. There is a big diversity between antibodies of the B cells, allowing recognition and neutralization of many different pathogens. The B cells are produced in the bone marrow in long bones. The B cell undergoes a mutation process to achieve big diversity of antibodies. The T cells mature in thymus, only T cells recognizing non self cells are released to the lymphatic and the blood systems. There are also other cells like macrophages with presenting properties, the pathogens are processed by a cell and presented by using MHC (Major Histocompatibility Complex) proteins. The recognition of a pathogen is performed in a few steps. First, the B cells or macrophages present the pathogen to a T cell using MHC, the T cell decides if the presented antigen is a pathogen. The T cell gives a chemical signal to B cells to release antibodies. A part of stimulated B cells goes to a lymph node and proliferate (clone). A part of the B cells changes into memory cells, the rest of them secrete antibodies into blood. The secondary response of the immunology system in the presence of known pathogens is faster because of memory cells. The memory cells created during primary response, proliferate and the antibodies are secreted to blood. The antibodies bind to pathogens and neutralize them. Other cells like macrophages destroy pathogens. The number of lymphocytes in the organism changes, while the presence of pathogens increases, but after attacks a part of the lymphocytes is removed from the organism.

The artificial immune systems [7–10] take only a few elements from the biological immune systems. The most frequently used are the mutation of the B cells, proliferation, memory cells, and recognition by using the B and T cells. The artificial immune systems have been used to optimization problems in classification and also computer viruses recognition. The cloning algorithm presented by von Zuben and de Castro [8, 9] uses some mechanisms similar to biological immune systems to global optimization problems. The unknown global optimum is the searched pathogen. The memory cells contain design variables and proliferate during the optimization process. The B cells created from memory cells undergo mutation. The B cells evaluate and better ones exchange memory cells. In Wierzchoń [10] version of Clonalg the crowding mechanism is used—the diverse between memory cells is forced. A new memory cell is randomly created and substitutes the old one, if two memory cells have similar design variables. The crowding mechanism allows finding not only the global optimum but also other local ones. The presented approach is based on the Wierzchoń algorithm [10], but the mutation operator is changed. The Gaussian mutation is used instead of the nonuniform mutation in the presented approach [22].

The Fig. 19.1 presents the flowchart of an artificial immune system.

The memory cells are created randomly. They proliferate and mutate creating B cells. The number of clones created by each memory cell is determined by the memory cells objective function value.

The objective functions for B cells are evaluated. The selection process exchanges some memory cells for better B cells. The selection is performed on the basis of the geometrical distance between each memory cell and B cells (measured by

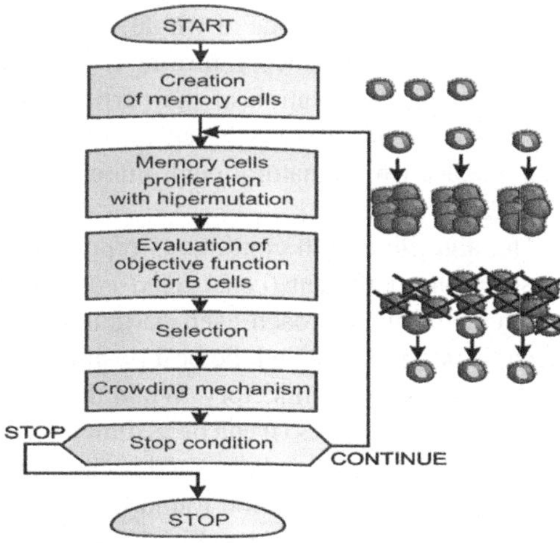

Fig. 19.1 An artificial immune system

using design variables). The crowding mechanism removes similar memory cells. The similarity is also determined as the geometrical distance between memory cells. The process is iteratively repeated until the stop condition is fulfilled. The stop condition can be expressed as the maximum number of iterations.

The unknown global optimum is represented by the searched pathogen. The memory cells contain design variables and proliferate during the optimization process.

19.3 Particle Swarm Optimizers

The particle swarm algorithms [11], similarly to the evolutionary and immune algorithms, are developed on the basis of mechanisms discovered in the nature. The swarm algorithms are based on models of the animals social behaviours: moving and living in the groups. The animals relocate in the three-dimensional space in order to change their stay place, the feeding ground, to find the good place for reproduction or to evading predators. We can distinguish many species of the insects living in swarms, fishes swimming in the shoals, birds flying in flocks or animals living in herds.

A simulation of the bird flocking was published by Reynolds [12]. They assumed that this kind of the coordinated motion is possible only when three basic rules are fulfilled: collision avoidance, velocity matching of the neighbours and flock centering. The computer implementation of these three rules showed very realistic flocking behaviour flaying in the three dimensional space, splitting before obstacle and rejoining again after missing it. The results of this biological examination where used by Kennedy and Eberhart [11], who proposed *Particle Swarm Optimizer*—PSO. This algorithm realizes directed motion of the particles in n-dimensional space to search for solution for n-variable optimization problem. PSO works in an iterative way. The location of one individual (particle) is determined on the basis of its earlier experience and experience of whole group (swarm). Moreover, the ability to memorize and, in consequence, returning to the areas with convenient properties, known earlier, enables adaptation of the particles to the life environment. The optimization process using PSO is based on finding the better and better locations in the search-space (in the natural environment that are for example hatching or feeding grounds).

The algorithm with continuous representation of design variables and constant constriction coefficient (*constricted continuous PSO*) has been used in presented research. In this approach each particle oscillates in the search space between its previous best position and the best position of its neighbours, hopefully finding new best locations on its trajectory. When the swarm is rather small (swarm consists of several or tens particles) it can be assumed that all the particles stay in neighbourhood with currently considered one. In this case we can assume the global neighbourhood version and the best location found by swarm so far is taken into account—current position of the swarm leader (Fig. 19.2b).

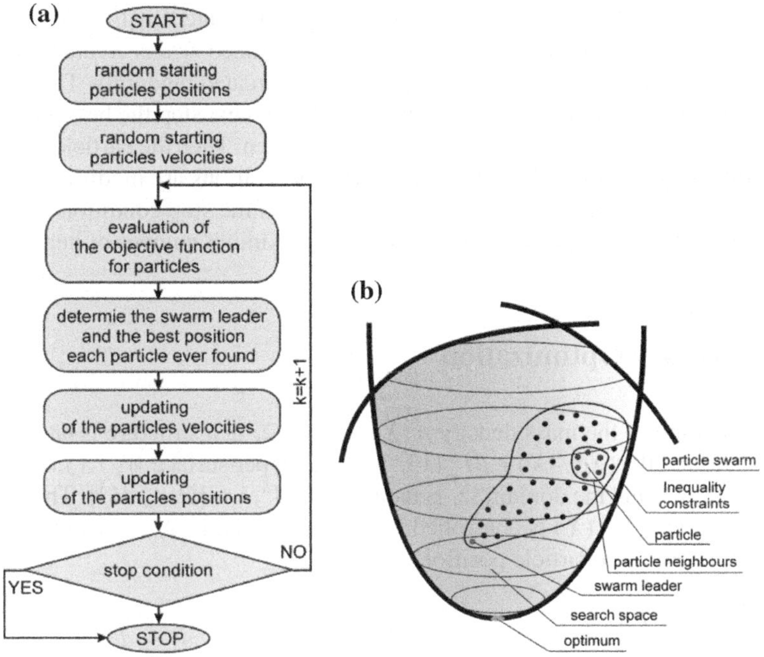

Fig. 19.2 Particle swarm optimizer: **a** blok diagram, **b** the optimization idea

The position of the ith particle is changed by stochastic velocity v_i, which is dependent on the particle distance from its earlier best position and position of the swarm leader. This approach is given by the following equations:

$$v_{ij}(k+1) = wv_{ij}(k)+\phi_{1j}(k)\left[p_{ij}(k) - d_{ij}(k)\right]+\phi_{2j}(k)\left[\hat{p}_{ij}(k) - d_{ij}(k)\right] \quad (19.1)$$

$$d_{ij}(k+1) = d_{ij}(k) + v_{ij}(k+1),\ i = 1, 2, \ldots, m\ ;\ j = 1, 2, \ldots, n \quad (19.2)$$

where:

$$\phi_{1j}(k) = c_1 r_{1j}(k);\ \phi_{2j}(k) = c_2 r_{2j}(k),$$

m	number of the particles,
n	number of design variables (problem dimension),
w	inertia weight,
c_1, c_2	acceleration coefficients,
r_1, r_2	random numbers with uniform distribution [0,1],
$d_i(k)$	position of the ith particle in kth iteration step,
$v_i(k)$	velocity of the ith particle in kth iteration step,
$p_i(k)$	the best found position of the ith particle founf so far,
$\hat{p}_i(k)$	the best position found so far by swarm—the position of the swarm leader,
k	iteration step.

The flowchart of the particle swarm optimiser is presented in Fig. 19.2a. At the beginning of the algorithm the particle swarm of assumed size is created randomly. Starting positions and velocities of the particles are created randomly. The objective function values are evaluated for each particle. In the next step the best positions of the particles are updated and the swarm leader is chosen. Then the particles velocities are modified by means of Eq. (19.1) and particles positions are modified according to Eq. (19.2). The process is iteratively repeated until the stop condition is fulfilled. The stop condition is typically expressed as the maximum number of iterations.

19.4 Topology Optimization

The distribution of the mass density $\rho(X)$, $(X) \in \Omega_t$ in a structure is described by a hyper surface $W_\rho(X), (X) \in H^3$ [13, 14]. The hyper surface $W_\rho(X)$ is stretched under $H^3 \subset E^3$ and the domain Ω_t is included in H^3, i.e. $\left(\Omega_t \subseteq H^3\right)$. The shape of the hyper surface $W_\rho(X)$ is controlled by parameters d_j, j = 1,2,…,G, which create a B-cell receptor or a Particle position:

$$\left.\begin{matrix} B-cell \\ Particle \end{matrix}\right\} = \langle d_1, d_2, \ldots, d_j, \ldots, d_G \rangle_{,} \quad d_j^{\min} \leq d_j \leq d_j^{\max} \tag{19.3}$$

where: $d_j^{\min}$, $d_j^{\max}$—are minimum and maximum values of parameters of B-cell or Particle.

B-cell or Particle parameters are the values of the function $W_\rho(X)$ in the control points $(X)_j$ of the hyper surface, i.e. $d_j = W_\rho\left[(X)_j\right]$, $\mathrm{j} = 0, 1, 2, \ldots, G$. The finite element method[vii] is applied in analysis of the structure. The domain Ω of the structure is discretized using the finite elements, $\Omega = \bigcup_{e=1}^{E} \Omega_e$ [15]. The assignation of the mass density to each finite element $\Omega_e, e = 1, 2, \ldots, E$ is adequately performed by the mappings $\rho_e = W_\rho\left[(X)_e\right], (X)_e \in \Omega_e, e = 1, 2, \ldots, E$. It means that each finite element can have the different mass density. When the value of the mass density for the e-th finite element is included in the interval $0 \leq \rho_e < \rho_{\min}$, the finite element is eliminated and the void is created and when in the interval $\rho_{\min} \leq \rho_e < \rho_{\max}$, the finite element remains. The illustration of this idea[viii] of immune or swarm optimization for a 2-D structure is presented in Fig. 19.3.

In the next step the Young's modulus for the e-th finite element is evaluated using the following equation

$$E_e = E_{\max}\left(\frac{\rho_e}{\rho_{\max}}\right)^r \tag{19.4}$$

where: $E_{\max}$, $\rho_{\max}$—Young's modulus and mass density for the same material, respectively, and r—parameter which can change from 1 to 9.

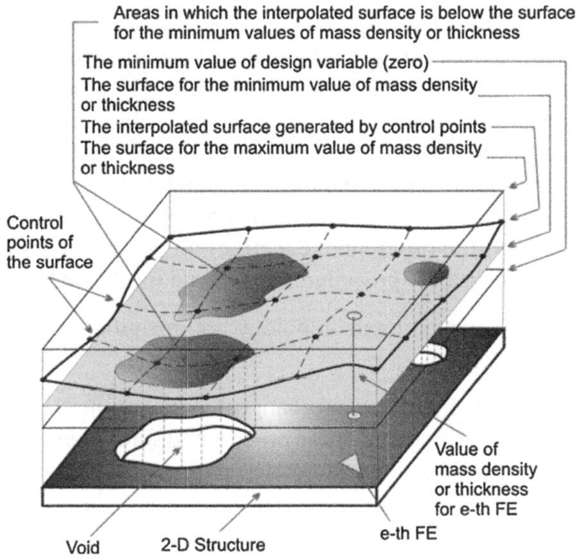

Fig. 19.3 The illustration of the idea of topology optimization for a 2-D structure

The bio-inspired process proceeds in the environment in which the structure fitness is described by the minimization of the mass of the structure

$$J = \int_{\Omega} \rho d\Omega \tag{19.5}$$

with constraints imposed on equivalent stresses σ_{eq} and displacements u of the structure

$$\sigma_{eq}(x, y, z) \le \sigma^{ad}, (x, y, z) \in \Omega \tag{19.6}$$

$$|u(x, y, z)| \le u^{ad}, (x, y, z) \in \Omega \tag{19.7}$$

The proposed approach can be consider as the kind of the level set-based structural optimization method. The idea is to parameterize the level set surface by a set of control points defining a B-cell from the population or Particle of the swarm. In this method the level set function represents a boundary between the material and void domains. The boundary is expressed by means of the level set function $\varphi(X)$:

$$\begin{cases} \varphi(X) > \rho_{\min} & \forall X \in \Omega \backslash \Gamma \\ \varphi(X) = \rho_{\min} & \forall X \in \Gamma \\ \varphi(X) < \rho_{\min} & \forall X \in H^d \backslash \Omega \end{cases} \tag{19.8}$$

In the level set-based methods, the changes of boundaries are defined by solving the Hamilton-Jacobi partial differential equation [16]. Allarie et al. [17] discussed that obtained in this way optimal solutions are strongly dependent on the initial configurations. Our idea, to overcome this problem, is based on the application of the bio-inspired algorithm, which works on the population of individuals (potential solutions). Then changes of the boundaries are realized by an intelligent procedure of the B-cells during the immune or swarm process (without necessity to solve the Hamilton-Jacobi partial differential equation). In order to improve optimization results two different additional procedures have been introduced:

- the additional procedure aiding the topology optimization,
- the smoothing procedure.

Using the described method, one can change material properties of finite elements during the immune or swarm optimization process and some elements are eliminated. As a result optimal shape, topology and material of structure are obtained.

19.5 Immune and Swarm Optimization Examples

The numerical examples present the comparison among the particle swarm optimizer (PSO), the artificial immune system (AIS) and the sequential (SEA) and distributed evolutionary algorithms (DEA). The comparison is performed on the base of optimization of:

- the known mathematical function, i.e. the Rastrigin function with 20 design variables (Example 1),
- a plate in plane stress (Example 2),
- a 3-D solid body (Example 3).

19.5.1 Optimization of Rastrigin Function (Example 1)

The optimization problem of the Rastrigin function (Fig. 19.4a) with 20 design variables is considered. The parameters of AIS, PSO and SEA (Sequential EA) and DEA (Distributed EA) are included in the Table 19.1. The numbers of the objective function evaluations needed to achieve of the value below 0.1 for the Rastrigin function, have been compared. Ten tests have been made with application of each optimization algorithm and the average number of the objective function evaluations have been computed. The result of the comparison are shown in Fig. 19.4b.

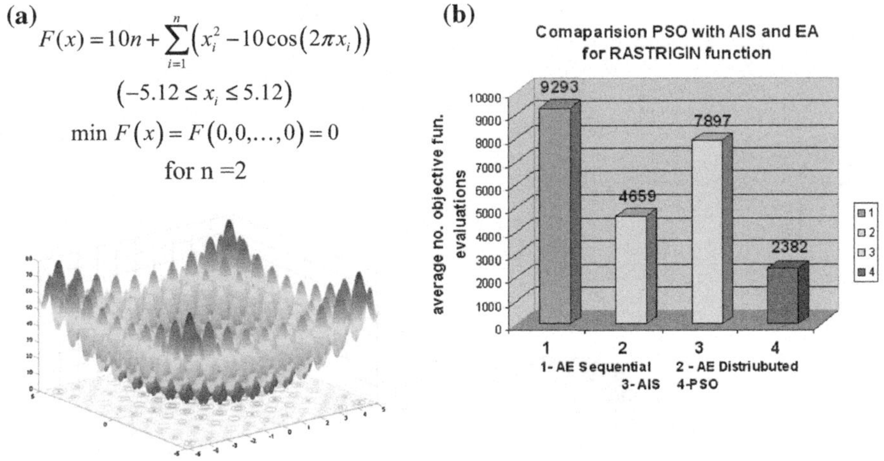

Fig. 19.4 Optimization of multimodal mathematical function: **a** Rastrigin function for 2 design variables, **b** effectiveness comparison among AIS, PSO, SEA and DEA

Table 19.1 Parameters of bioinspired methods (Example 1)

AIS	PSO	SEA	DEA
No. of memory cells = 2,	No. of the particles = 74,	No. of the subpopulations = 2,	No. of the chromosomes = 20,
No. of the clones = 4,	Inertia weight w = 1.0,	No. of the chromosomes in subpopulation = 10,	The probability of simple crossover = 95 %,
Crowding factor = 0.45,	Acceleration coefficients $c_1 = c_2 = 1.9$	The probability of simple crossover = 95 %,	The probability of Gaussian mutation = 50 %
The probability of Gaussian mutation = 40 %		The probability of Gaussian mutation = 50 %	

Table 19.2 The input data to optimization task of a plate in plane stress (Example 2)

a × b (mm)	Thickness (mm)	σ_{min} ; P (MPa)	Q (N)	Range of ρ_e (g/cm^3) existence or elimination of the finite element
200 × 100	4.0	8.0 ; 1.0	4000	$7.3 \le \rho_e < 7.5$ elimination $7.5 \le \rho_e \le 7.86$ existence

19.5.2 Optimization of a Plate in Plane Stress (Example 2)

A rectangular 2-D structure in plane stress loaded with the concentrated force Q in the centre of the lower boundary and fixed on the bottom corners is considered (Fig. 19.5a) [18]. In order to obtain the symmetrical results a half of the structure has been analyzed. The input data to the optimization program are included in Table 19.2.

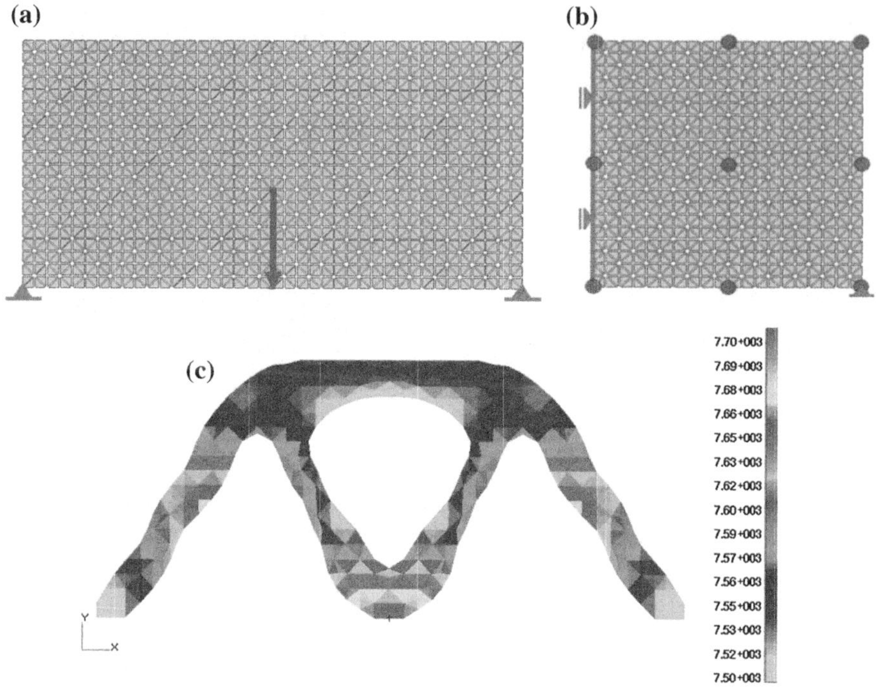

Fig. 19.5 The plate (example 1); **a** the geometry; **b** the distribution of the control points of the interpolation surface; **c** result of the optimization process

Figure 19.5c shows results of the optimization process for the distribution of control points presented in Fig. 19.5b (the symmetry of the structure is taken into account).

To present and compare the effectiveness of the proposed intelligent bio-inspired methods the numbers of the objective function evaluations needed to achieve the mass below 0.14 kg (Fig. 19.5c) [the mass of initial structure equals 0.628 kg (Fig. 19.5a)] were computed (mass reduction 78 %). The results obtained for the sequential and the distributed evolutionary algorithm, artificial immune system and the particle swarm optimizer are presented as the histograms of average number of the objective function evaluation obtained on the basis of 10 tests (Fig. 19.6). The parameters of the algorithms are presented in Table 19.3.

19.5.3 Optimization of 3D L-Solid (Example 3)

Topology optimization of 3D elastic structure like L-solid (Fig. 19.7a) by the minimization of the mass of the structure and with imposed stress or displacement

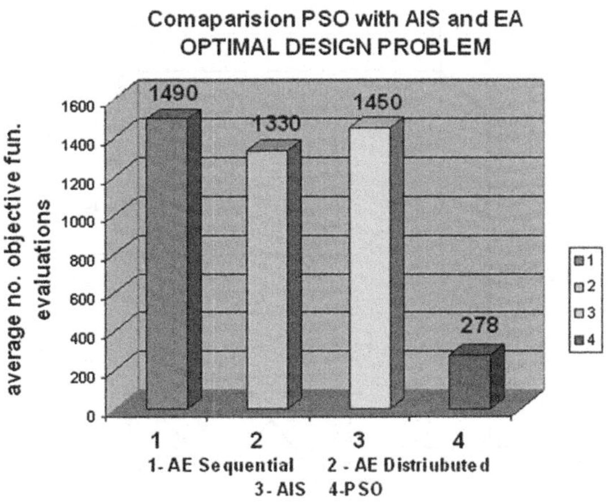

Fig. 19.6 The result of the comparison among PSO, AIS and the SEA and DEA for the design optimization problem

Table 19.3 Parameters of bio-inspired methods (Example 2)

AIS	PSO	SEA	DEA
No. of memory cells = 5,	No .of the particles = 20,	No. of the subpopulations = 2,	No. of the chromosomes = 20,
No. of the clones = 5,	Inertia weight w = 0.73,	No. of the chromosomes in subpopulation = 10,	The probability of simple crossover = 95 %,
Crowding factor = 0.45,	Acceleration coefficients $c_1 = c_2 = 1.47$	The probability of simple crossover = 95 %,	The probability of Gaussian mutation = 50 %
The probability of Gaussian mutation = 50 %		The probability of Gaussian mutation = 50 %	

Table 19.4 The input data to optimization task of a plate in plane stress (Example 3)

a×b×c×d×e (mm)	Maximal displacement (mm)	Maximal stress (MPa)	Q (N)	Range of ρ_e (g/cm^3) existence or elimination of the finite element
48 × 48 × 24 × 24 × 24	0.8	600	8450	$0 \le \rho_e < 3.14$ elimination $3.14 \le \rho_e \le 7.85$ existence

constraints [19]. The input data to the optimization task are included in Table 19.4. Numerical results are presented in Fig. 19.7b. The comparison of the effectiveness among PSO, AIS and the SEA and DEA for this design optimization problem has been also performed. The parameters of the algorithms are presented in Table 19.5. Figure 19.8 shows the comparison of efficiency.

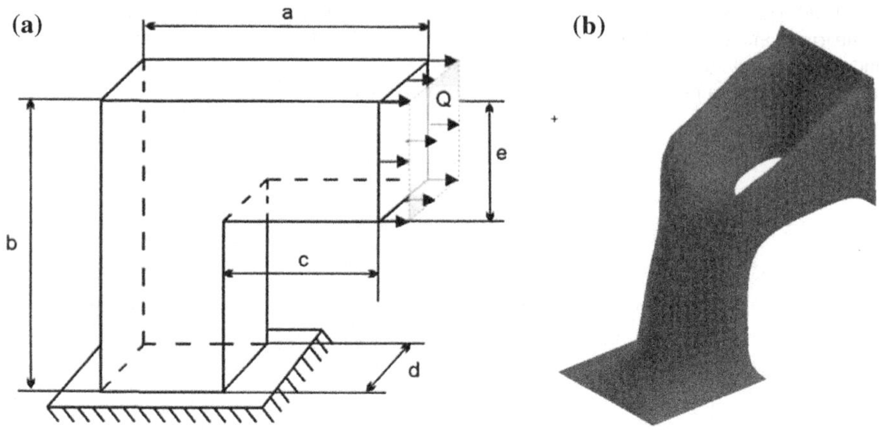

Fig. 19.7 3-D structure like L solid: **a** geometry and scheme of loading of 3-D L solid, **b** optimization result

Table 19.5 Parameters of bioinspired methods (Example 3)

AIS	PSO	SEA	DEA
No. of memory cells = 10,	No. of the particles = 30	No. of the subpopulations = 2,	No. of the chromosomes = 20,
No. of the clones = 10,	Inertia weight w = 0.75,	No. of the chromosomes in subpopulation = 10,	The probability of simple crossover = 95 %,
Crowding factor = 0.5,	Acceleration coefficients $c_1 = c_2 = 1.45$	The probability of simple crossover = 95 %,	The probability of Gaussian mutation = 50 %
The probability of Gaussian mutation = 50 %		The probability of Gaussian mutation = 50 %	

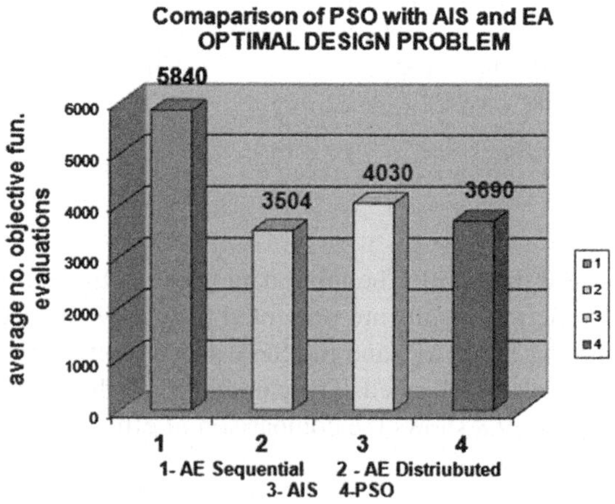

Fig. 19.8 The result of the comparison among PSO, AIS and the SEA and DEA for the design optimization problem

19.6 Conclusions

In the paper, a description of two bio-inspired methods: immune and swarm approaches are presented and applied to optimization of structures.

These bio-inspired approaches can be simply implemented because they need only the values of objective functions. An important feature of these approaches is a strong probability of finding the global optimal solutions. The described approaches are free from limitations of classic gradient optimization methods.

Both approaches belong to methods based on population of solutions and they have some interesting features which can be considered as alternative to evolutionary algorithms.

Described approaches have applied to simultaneous shape, topology and material optimization of 3D structures.

Efficiency of presented approaches and comparison with evolutionary algorithms are performed. One can observe that PSO is generally the most efficient approach.

There are possibilities of further efficiency improvement of the proposed methods by parallelization of computing or hybridization by using sensitivity analysis information [20, 21].

References

1. Burczyński T, Kuś W, Długosz A, Poteralski A, Szczepanik M (2004) Sequential and distributed evolutionary computations in structural optimization ICAISC, In: International conference on artificial intelligence and soft computing, Lecture Notes on Artificial Intelligence, vol 3070. Springer
2. Burczyński T, Dlugosz A, Kuś W, Orantek P, Poteralski A, Szczepanik M (2005) Intelligent computing in evolutionary optimal shaping of solids, In: 3rd international conference on computing, communications and control technologies, vol 3, proceedings, pp 294–298
3. Szczepanik M, Burczyński T (2012) Swarm optimization of stiffeners locations in 2-D structures. Bull Pol Acad Sci Tech Sci 60(2):241–246
4. Szczepanik M, Poteralski A, Ptaszny J, Burczyński T (2012) Hybrid particle swarm optimizer and its application in identification of room acoustic properties. In: Swarm and evolutionary xomputation - international symposia, SIDE 2012 and EC 2012, held in conjunction with ICAISC 2012, Zakopane, Poland, April 29–May 3, 2012. Proceedings. Lecture Notes in Computer Science, vol 7269. Springer. ISBN 978-3-642-29352-8
5. Burczynski T, Górski R, Poteralski A, Szczepanik M (2013) Soft computing in structural dynamics, In: Chapter 24 in computational methods in earthquake engineering. Springer
6. Ptak M, Ptak W (2000) Basics of immunology. Jagiellonian University Press, Cracow (in Polish)
7. Castro LN, Timmis J (2003) Artificial immune systems as a novel soft computing paradigm. Soft Comput 7(8):526–544
8. Castro LN, Von Zuben FJ (2001) Immune and neural network models: theoretical and empirical comparisons. Int J Comput Intell Appl (IJCIA) 1(3):239–257
9. Castro LN, Von Zuben FJ (2002) Learning and optimization using the clonal selection principle. IEEE Trans Evol Comput 6:239–251
10. Wierzchoń ST (2001) Artificial immune systems. Theory and applications, EXIT, Warsaw (in Polish)

11. Kennedy J, Eberhart RC (2001) Swarm Intelligence. Morgamn Kauffman, Burlington
12. Reynolds CW (1987) Flocks, herds, and schools. A distrib behav model. Comput Graph 21:25–34
13. Burczyński T, Poteralski A, Szczepanik M (2003) Genetic generation of 2-D and 3-D structures, In: Second M.I.T. conference on computational fluid and solid mechanics, Massachusetts, Institute of Technology Cambridge, p 02139
14. Burczyński T, Poteralski A, Szczepanik M (2007) Topological evolutionary computing in the optimal design of 2D and 3D structures. Eng Optim Taylor & Francis 39(7):811–830
15. Zienkiewicz OC, Taylor RL (1991) The finite element method. McGraw-Hill, Vol. I., 1989, Vol. II, New York
16. Osher S, Sethian JA (1988) Front propagating with curvature dependent speed: algorithms based on the Hamilton-Jacobi formulations. J Comput Phys 78:12–49
17. Allaire G, Jouve F, Toader AM (2004) Structural optimization using sensitivity analysis and a level-set method. J Comput Phys 194:363–393
18. Burczynski T, Szczepanik M (2013) Intelligent optimal design of spatial structures. Comput Struct 127:102–115
19. Burczyński T, Bereta M, Poteralski A, Szczepanik M (2010) Immune computing: intelligent methodology and its applications in bioengineering and computational mechanics. In: Advanced structured materials, vol 1. Comput Meth Mech, Springer, Berlin
20. Burczyński T, Kane JH, Balakrishna C (1997) Comparison o shape design sensitivity analysis formulations via material derivative-adjoint variable and implicit differentiation techniques for 2-D and 3-D curved boundary element method. Comput Methods Appl Mech Eng 142(1–2):89–108
21. Burczyński T, Bonnet M, Nowakowski M (2002) Sensitivity analysis for shape perturbation of cavity or internal crack using BIE and adjoint variable approach. Int J Solids Struct 39(9):2365–2385
22. Poteralski A, Szczepanik M, Dziatkiewicz G et al (2011) Immune identification of piezoelectric material constants using BEM. Inverse Probl Sci Eng 19(1):103–116

Chapter 20
Investigation of Three Genotypes for Mixed Variable Evolutionary Optimization

Rajan Filomeno Coelho, Manyu Xiao, Aurore Guglielmetti, Manuel Herrera and Weihong Zhang

Abstract While the handling of optimization variables directly expressed by numbers (continuous, discrete, or integer) is abundantly investigated in the literature, the use of nominal variables is generally overlooked, despite its practical interest in plenty of scientific and industrial applications. For example, in civil engineering, the designers of a structure made out of beams might have to select the best cross-section shapes among a list of available geometries (square, circular, rectangular, etc.), which can be modeled by nominal data. Therefore, in the context of single- and multi-objective evolutionary optimization for mixed variables, this study investigates three genetic encodings (binary, real, and real-simplex) for the representation of mixed variables involving both continuous and nominal parameters. The comparison of the genotypes combined with the instances of crossover is performed on six analytical benchmark test functions, as well as on the multi-objective design optimization of a six-storey rigid frame, showing that for mixed variables, real (and to a lesser extent: real-simplex) coding provides the best results, especially when combined with a uniform crossover.

R. Filomeno Coelho (✉) · M. Herrera
ULB–BATir Department, Université Libre de Bruxelles, Avenue F.D. Roosevelt, 50 (CP 194/2), 1050 Brussels, Belgium
e-mail: rfilomen@ulb.ac.be

M. Herrera
e-mail: mherrera@ulb.ac.be

M. Xiao
NPU–Department of Applied Mathematics, Northwestern Polytechnical University, Shaanxi 710072, Xi'an, People's Republic of China
e-mail: manyuxiao@nwpu.edu.cn

A. Guglielmetti · W. Zhang
NPU–School of Mechanical Engineering, Northwestern Polytechnical University, Shaanxi 710072, Xi'an, People's Republic of China
e-mail: wuhong_mathnpu@hotmail.fr

W. Zhang
e-mail: zhangwh@nwpu.edu.cn

D. Greiner et al. (eds.), *Advances in Evolutionary and Deterministic Methods for Design, Optimization and Control in Engineering and Sciences*, Computational Methods in Applied Sciences 36, DOI 10.1007/978-3-319-11541-2_20

Keywords Mixed variables · Evolutionary algorithms · Categorical variables · Genotype

20.1 Introduction

Real-life engineering applications are often characterized by data of versatile natures. Formally, design variables in parameterization/optimization can be divided into five classes:

- *continuous variables* are defined over an interval $I_c \subseteq \mathbb{R}$ (e.g. length, curvature radius);
- *discrete variables* are continuous variables available only within a finite set $I_d = \{d_1, \ldots, d_n\}$ where all $d_i \in \mathbb{R}$ (e.g. cross-section areas from a catalog of beam profiles). It is interesting to note that gradient-based optimizers can be adapted to discrete variables, as developed for instance by [3] through dual formulations and subgradient-based algorithms;
- *integer variables* are defined over an interval $I_i \subseteq \mathbb{N}$ or $\mathbb{Z}$ (e.g. number of holes in a plate). In engineering problems, they differ from discrete variables by the fact that no intermediate values between two integer variables can be defined (e.g. a plate can contain two or three holes, but not 2.5), which has an impact both on the simulation and optimization sides. *Binary* variables are a particular case of integer variables;
- *ordinal categorical variables*, or simply *ordinal variables*, take their values (called *attributes*) among non-numerical values endowed with a ranking, as in the set {'tiny', 'small', 'medium-sized', 'large', 'huge'};
- *nominal categorical variables*, or simply *nominal variables*, are non-numerical variables characterized by no explicit ordering, as the shape of a beam profile: { ○ ; □ ; ■ ; I }, the choice of a material: {'steel', 'aluminum', 'titanium'}, etc.

According to the nature of the variables, distinct fields of optimization have been developed; they can be roughly classified as *continuous* versus *combinatorial* optimization. Interestingly, due to their flexibility in data representation, evolutionary optimization algorithms are ideally suited when both types of data are involved. Additionally, since the categorical variables are characterized by non-numerical entries, they represent a challenging issue for the parameterization—and eventually for the optimization—since they require a careful investigation of the conversion procedure onto a chromosome. The optimization algorithms proposed in the literature for mixed variables have been summarized by the authors in [8], confirming that bio-inspired algorithms are an efficient option for single- and multi-objective optimization, as illustrated for example by [12]. However, a systematic examination of data structures for mixed variables is still missing.

Based on all these considerations, the goal of this paper is to investigate three genotypes (binary, real, and real-simplex) in combination with three types of crossover (line, uniform, and two-site) in order to extract information about the mutual interaction between encoding and genetic operators.

The paper is organized as follows: first, a systematic representation of mixed variables is introduced in Sect. 20.2. Then, the genetic operators used in the evolutionary algorithms are described in Sect. 20.3. Afterwards, six analytical test cases including continuous and nominal variables are thoroughly studied (Sect. 20.4), followed by the multi-objective design optimization of a rigid frame (Sect. 20.5). Finally, the conclusions and future prospects are discussed in Sect. 20.6.

20.2 Data Structures for Mixed Variables

Historically, the initial conversion of design variables to genetic encoding has been done through *binary coding* [4, 9], as illustrated in Fig. 20.1. After conversion, the converted binary strings for all variables can seamlessly be concatenated in a chromosome, regardless of the various natures of the actual variables.

Another popular and straightforward encoding consists in modeling all variables through real values, as shown in Fig. 20.2.

While these operations are straightforward for numerical values (not only continuous, but also discrete and integer), they require an arbitrary mapping for nominal variables, as shown in Table 20.1.

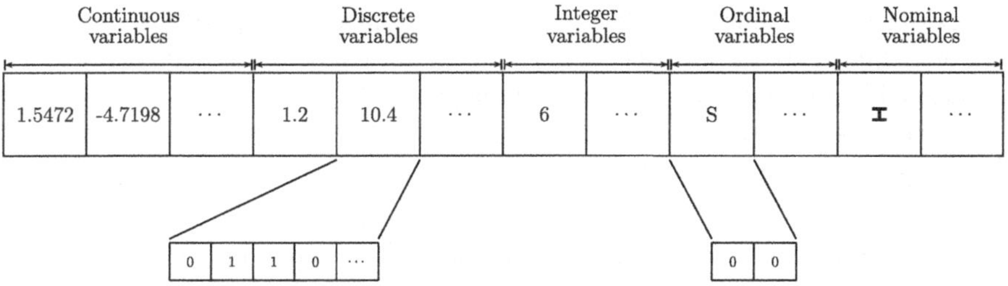

Fig. 20.1 Conversion of a mixed-variable vector to a binary chromosome

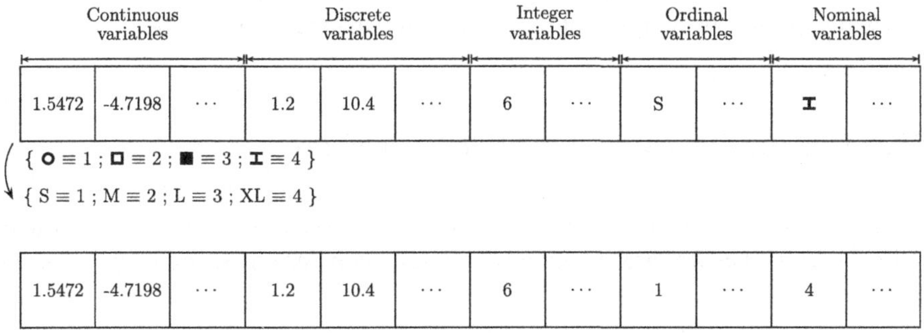

Fig. 20.2 Conversion to a real vector

Table 20.1 Conversion of a nominal variable: two legitimate mappings onto real/binary values

Nominal variable	Mapping 1		Mapping 2	
	Real	Binary	Real	Binary
○	1	(0,0)	2	(0,1)
□	2	(0,1)	4	(1,1)
■	3	(1,0)	1	(0,0)
I	4	(1,1)	3	(1,0)

Fig. 20.3 Mapping of a three-attribute nominal variable onto a regular simplex (the Euclidean distance between each pair of attributes in the regular simplex space is always equal to 1)

Indeed, since there is no intrinsic ordering between attributes, both mappings mentioned in Table 20.1 are legitimate, but they might not exhibit the same behavior in the genetic algorithm. Therefore, another conversion is proposed here for nominal values, constraining the attributes of a nominal variable to be equidistant. This constraint can be ensured by assigning to each of the n attributes the coordinates of the vertex of a regular simplex in a $(n-1)$-space, as depicted in Fig. 20.3. Largely used in learning and classification theory [1] with categorical data, this approach was first introduced by the authors for approximation purposes [7].

To summarize, three encodings are analyzed in this paper:

1. *binary*;
2. *real*;
3. *real-simplex*, i.e. conversion to real numbers for continuous, discrete, integer, and ordered categorical variables, and regular simplex mapping for the nominal parameters.

Since the relation between coding and crossover is critical in genetic algorithms, the next section lists the types of crossovers investigated in this study.

20.3 Genetic Operators for Mixed Variables

In all three genotypes proposed here above, the conversion finally leads to a real-valued vector (the *chromosome*), on which three popular types of crossover can be applied [2]:

1. *line crossover* (LCX): for each pair of parents $(p1, p2)$ selected among the population of the previous generation, the offspring are created as follows:

$$\begin{cases} b_i^{(c1)} = b_i^{(p1)} + \alpha_1 \left(b_i^{(p2)} - b_i^{(p1)} \right) \\ b_i^{(c2)} = b_i^{(p1)} + \alpha_2 \left(b_i^{(p2)} - b_i^{(p1)} \right) \end{cases} \tag{20.1}$$

where b_i are the components of the chromosome (a bit or a real value depending on the coding), $p1$, $p2$ refer to the parents, $c1$, $c2$ to the children, and α_1, α_2 are random numbers uniformly sampled in [0, 1], and constant for the whole chromosome. Geometrically speaking, the children are generated on the line joining the parents (in a space depending on the data structure, viz. the type of coding);

2. *uniform crossover* (UCX): each child c is generated as follows:

$$b_i^{(c)} = \begin{cases} b_i^{(p1)} \text{ if } \alpha < 0.5 \\ b_i^{(p2)} \text{ if } \alpha \geq 0.5 \end{cases} \tag{20.2}$$

where α is sampled independently for each component of the chromosome;

3. *two-site crossover* (TSX): the children are generated by swapping pieces of parental chromosomes between two sites randomly chosen within the chromosome.

After the crossover operation, if values are found that do not correspond to available values (depending on the coding and/or crossover implemented, this problem can happen for discrete, integer, or categorical variables), they are repaired in the chromosome by replacing the wrong value by the closest existing value. The remainder of the paper is devoted to an empirical analysis of the interaction between data structure and crossover on several test cases. Nevertheless, the following conditions can already be devised, and serve eventually as guidelines to analyze the results obtained:

- *Condition of exploration*: the crossover should be able to produce children different from their parents.
- *Condition of invariance*: the children generated from crossover have to be independent from the ranking of the attribute values.

20.4 Analytical Benchmark Test Cases

The first six test functions aim at showing the relation between genotypes and types of crossover for real, nominal, and mixed (real-nominal) variables. Their analytical expression is detailed in Table 20.2. For each test function, three situations are examined: (1) ten real variables, no nominal variables ($nz = 10$, $nc = 0$), (2) ten nominal variables, no real variables ($nz = 0$, $nc = 10$), and (3) five real variables, five nominal variables ($nz = 5$, $nc = 5$).

Table 20.2 Definition of the six analytical benchmarks with mixed variables

Output		
$f_{Ellipsoid,MV}$	$= \sum_{i=1}^{nz} \left(\beta^{\frac{i-1}{nz-1}} z_i\right) + \sum_{i=1}^{nc} \left(\beta^{\frac{i-1}{nc-1}} c_i\right)$	$(\beta = 5)$
$f_{Ackley,MV}$	$= -20e^{-0.2\sqrt{\frac{1}{nz}\sum_{i=1}^{nz} z_i^2}} - e^{\frac{1}{nz}\sum_{i=1} cos(2\pi z_i)}$ $-20e^{0.2\sqrt{\frac{1}{nc}\sum_{i=1}^{nc} c_i^2}} - e^{\frac{1}{nc}\sum_{i=1} cos(2\pi c_i)} + 20 + e$	
$f_{Rastrigin,MV}$	$= 10(nz + nc) + \sum_{i=1}^{nz} \left[z_i^2 - 10cos(2\pi z_i^2)\right]$ $+\sum_{i=1}^{nc} \left[c_i^2 - 10cos(2\pi c_i^2)\right]$	
$f_{Rosenbrock,MV}$	$= \sum_{i=1}^{nz-1} \left[100(z_{i+1} - z_i^2)^2 + (z_i - 1)^2\right]$ $+\sum_{i=1}^{nc-1} \left[100(c_{i+1} - c_i^2)^2 + (c_i - 1)^2\right]$	
$f_{Sphere,MV}$	$= \sum_{i=1}^{nz} z_i^2 + \sum_{i=1}^{nz} c_i^2$	
$f_{Griewank,MV}$	$= \frac{1}{4000} \sum_{i=1}^{nz} z_i^2 - \prod_{i=1}^{nz} cos\left(\frac{z_i}{\sqrt{i}}\right) + \frac{1}{4000} \sum_{i=1}^{nc} c_i^2 - \prod_{i=1}^{nc} cos\left(\frac{c_i}{\sqrt{i}}\right)$	
Input		
Cont. vars.	$z_i = 10^{-3} x_i^{cont}, \; x_i^{cont} \in [-300, 700]$ for $i = 1, \ldots, nz$	
Categ. vars.	$c_i \in o_{permut}^{(i)}([0, 1, \ldots, 10])$ for $i = 1, \ldots, nc$	

Table 20.3 Numerical errors obtained for the six analytical benchmark test functions for 10 *real* variables (the error values, averaged over 10 independent runs, are normalized between 0 and 1)

Coding	Binary			Real			Real-simplex		
Crossover	LCX	UNX	TSX	LCX	UNX	TSX	LCX	UNX	TSX
*TC*1	1	5.45e-01	4.58e-01	0	3.39e-02	6.13e-02	5.17e-02	1.81e-01	2.36e-01
*TC*2	1	3.85e-01	4.87e-01	0	2.97e-02	1.85e-02	1.17e-01	1.79e-01	1.48e-01
*TC*3	1	3.34e-01	3.71e-01	0	5.77e-02	2.42e-02	6.66e-02	1.21e-01	1.37e-01
*TC*4	1	2.20e-01	2.43e-01	2.32e-03	9.24e-03	0	4.22e-02	6.35e-02	6.58e-02
*TC*5	4.59e-01	1	9.24e-01	0	3.43e-02	2.28e-02	4.87e-02	6.39e-02	7.62e-02
*TC*6	1	3.94e-01	4.06e-01	0	3.76e-03	1.13e-03	1.40e-01	1.99e-01	1.05e-01

Since the goal is to compare the interactions between the type of crossover and the data structure (coding), the probability of mutation is set to 0 (no mutation). Moreover, a tournament selection is used; the population size is set to 100, and the number of generations to 50. 10 independent runs are performed for each configuration.

The results are collected in Tables 20.3, 20.4 and 20.5, showing the normalized error between the worst and best solutions (averaged over the ten independent runs). In other words, on each line, 0 corresponds to the best solution found (in average), while 1 refers to the worst.

Table 20.4 Numerical errors obtained for the six analytical benchmark test functions for 10 *nominal* variables (the error values, averaged over 10 independent runs, are normalized between 0 and 1)

Coding	Binary			Real			Real-simplex		
Crossover	LCX	UNX	TSX	LCX	UNX	TSX	LCX	UNX	TSX
$TC1$	1	8.43e-01	8.03e-01	6.86e-02	0	2.47e-02	1.51e-01	4.17e-02	9.79e-02
$TC2$	1	9.79e-01	7.98e-01	9.16e-02	2.95e-03	0	1.23e-01	4.28e-02	6.20e-02
$TC3$	1	9.38e-01	9.02e-01	6.78e-02	0	1.37e-02	5.67e-02	3.87e-02	9.19e-02
$TC4$	9.61e-01	9.35e-01	1	1.36e-01	0	6.78e-02	1.45e-01	1.11e-01	1.19e-01
$TC5$	9.08e-01	1	8.55e-01	1.74e-01	0	5.60e-02	1.66e-01	1.29e-01	1.39e-01
$TC6$	8.85e-01	8.22e-01	1	3.74e-02	0	9.77e-02	3.17e-01	2.73e-01	3.21e-01

Table 20.5 Numerical errors obtained for the six analytical benchmark test functions for 10 *mixed* (viz. 5 real and 5 nominal) variables (the error values, averaged over 10 independent runs, are normalized between 0 and 1)

Coding	Binary			Real			Real-simplex		
Crossover	LCX	UNX	TSX	LCX	UNX	TSX	LCX	UNX	TSX
$TC1$	1	6.05e-01	6.51e-01	7.92e-02	2.90e-03	0	2.46e-01	1.19e-01	1.08e-01
$TC2$	1	7.40e-01	6.80e-01	1.29e-01	0	1.48e-02	1.26e-01	1.16e-01	7.87e-02
$TC3$	1	3.74e-01	4.44e-01	1.81e-02	1.85e-02	0	1.63e-01	3.14e-02	3.68e-02
$TC4$	1	6.44e-01	4.77e-01	9.18e-02	0	2.77e-02	1.44e-01	7.74e-02	1.54e-01
$TC5$	6.25e-01	7.53e-01	1	0	3.39e-02	7.73e-02	4.90e-02	1.86e-01	9.40e-01
$TC6$	1	7.00e-01	8.78e-01	1.57e-03	0	7.27e-02	2.53e-01	1.95e-01	9.43e-02

In the six analytical test functions, it appears that the conversion to real numbers is the most effective for the cases with either continuous or nominal variables only. As intuition would have suggested, the line crossover allows for a better exploration of the design space for continuous variables, while the uniform crossover is more effective with nominal variables, where the swapping of variable values is performed independently and randomly for each variable. The binary coding with line crossover provides the worst results, which could be expected since the linear combination of parental chromosomes is not a meaningful operation for values converted into binary digits. Finally, the real-simplex coding generally provides reasonably good results in comparison with the real coding.

20.5 Application: Structural Design of a Rigid Frame

To analyze the efficiency of the proposed algorithmic instances on a structural design example, a 3D rigid frame is investigated [11]. The quantities of interest are the mass and the compliance, the latter being post-processed from a finite element linear

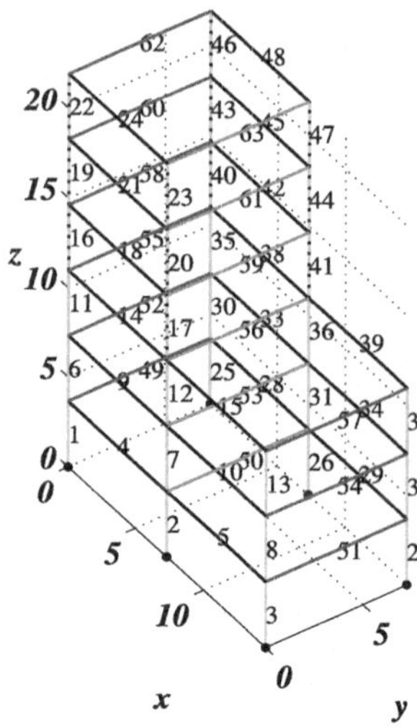

Fig. 20.4 Six-storey rigid frame (with numbering of the beam elements): the five groups of cross-sections are displayed with various shades of *gray* color. Each node connecting two (or more) elements is a rigid connection. The multi-objective optimization consists in finding the best compromise (Pareto) designs with respect to two conflicting objectives, namely the mass and the compliance for the whole structure

analysis with beam elements [6]. The loads are derived from Eurocode 3 [10], and consist in:

- the dead load of the beams and columns;
- the gravity load on the floors (19.16 kPa);
- the lateral load due to the wind (110 kN).

The beams or columns are classified in five groups of common cross-sections (as depicted in Fig. 20.4):

- group 1: {4, 5, 9, 10, 14, 15, 18, 21, 24, 28, 29, 33, 34, 38, 39, 42, 45, 48};
- group 2: {49, 51, 52, 54, 55, 57, 58, 60, 62};
- group 3: {50, 53, 56, 59, 61, 63};
- group 4: {1, 2, 3, 6, 7, 8, 11, 12, 13, 25, 26, 27, 30, 31, 32, 35, 36, 37};
- group 5: {16, 17, 19, 20, 22, 23, 40, 41, 43, 44, 46, 47}.

Ten design variables are necessary to parameterize a given structure:

- for each of the five groups of profiles, a categorical variable c defines the cross-section geometry among seven attributes: { I ; □ ; ○ ; □ ; ■ ; ● ; ■ };
- for all groups of profile, one continuous bounded variable defines the maximum length l of the cross-section (either height or diameter) with $0.09\,\mathrm{m} \leq l\ 0.11\,\mathrm{m}$, and with a fixed thickness (when applicable) set to 0.0025 m. For the rectangular cross-section, the width is defined as half of the height; for the I—section, the width is equal to the height.

The geometry of the cross-section is typically a nominal variable, since no ordering of the available cross-section types can be made a priori. The choice of the cross-section has a direct impact on the calculation of the quantities (area, moments of inertia) necessary to get the normal efforts, shear forces, and bending moments.

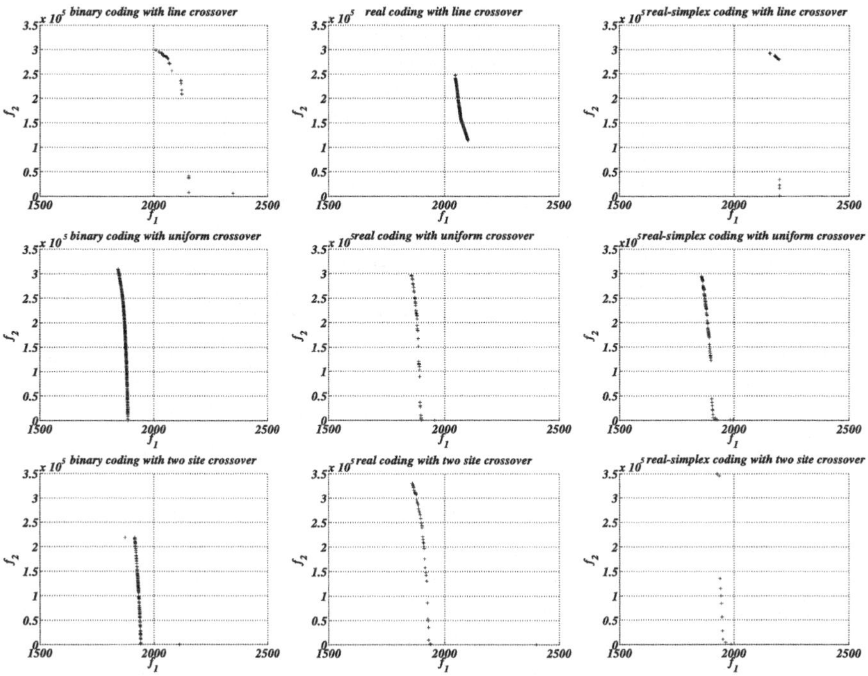

Fig. 20.5 Rigid frame: Pareto fronts obtained for the nine combinations of coding and crossover types

The multi-objective implementation of the genetic algorithm is based on the second version of the *Nondominated Sorting Genetic Algorithm*, or NSGA-II [5]. The corresponding Pareto fronts obtained for the nine possible combinations of coding and type of crossover are depicted in Fig. 20.5.

In this problem, the uniform and two-site crossovers clearly outperfom the line crossover to obtain dense and nondominated Pareto fronts. In particular, the uniform crossover provides comparable results for all codings.

In terms of designs obtained, a close examination of the Pareto set reveals that the optimal cross-sections are mostly circular (**O**) or I-shaped (**I**), and square (**□**) to a lesser extent. The tubular shapes constitute the best compromise between lightness and stiffness.

20.6 Conclusions and Future Prospects

In this paper, three genotypes, along with three types of crossover, are analyzed in order to extract their properties in the modeling of mixed (continuous + nominal) variables. For the analytical test functions, the real coding furnishes the best results in almost all configurations (real, nominal, or mixed variables), while the

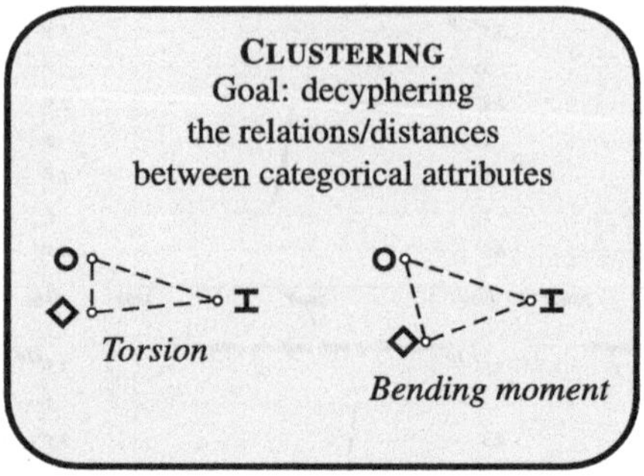

Fig. 20.6 Clustering procedure: considering for example the static structural analysis of a beam with respect to the shape of its cross-section (*square*, *circular*, or *I-shaped*), a priori calculations can build clusters of attributes depending on their impact on output quantities of interest (here: bending moment and torsion effect)

real-simplex performs fairly, and the binary coding poorly. The type of crossover also has a significant influence on the results, and this aspect is mostly visible in the multi-objective design optimization of a rigid frame, where the uniform crossover consisting in randomly swapping genes in the parental chromosome is the most effective one to find dense and widely distributed Pareto fronts.

Future studies are guided by the need for better accounting for the relations between attributes of nominal variables. In the simplex mapping representation of nominal variables, the hypothesis that all attributes are equidistant is acceptable when no information is known a priori about their correlations with the output quantities of interest (objectives and constraints). However, when knowledge about the physics is available, a non-regular simplex might be preferable, with distinct pairwise distances between attributes. For instance, the square and circular cross-section shapes of beam profiles might exhibit a closer behavior than the I-shaped profiles, which might be taken into account directly in the coding. This crucial step (under investigation) requires a *clustering phase* to identify these pairwise relationships to be performed before the optimization process (see Fig. 20.6), by using physical insight from the input-output model.

Acknowledgments This work has been supported by Innoviris (Brussels-Capital Region, Belgium) through a BB2B project entitled "Multicriteria optimization with uncertainty quantification applied to the building industry". The authors also acknowledge support by the Basic Project Foundation of Northwestern Polytechnical University (JC20120241), and by the National Natural Science Fundation of China (Grants No.11302173 and No.51275424).

References

1. Agresti A (1996) An introduction to categorical data analysis. Wiley, New York
2. Bäck T, Fogel DB, Michalewicz Z (1997) Handbook of evolutionary computation. Oxford University Press, New York

3. Beckers M (2000) Dual methods for discrete structural optimization problems. Int J Numer Meth Eng 48:1761–1784
4. De Jong KA (ed) (2006) Evolutionary computation: a unified approach. MIT Press, Massachusetts
5. Deb K, Pratap A, Agarwal S, Meyarivan T (2002) A fast and elitist multiobjective genetic algorithm: NSGA-II. IEEE Trans Evol Comput 6(2):182–197
6. Ferreira AJM (2009) MATLAB codes for finite element analysis: Solids and structures., solid mechanics and its applications Springer Science+Business Media B.V., The Netherlands
7. Filomeno Coelho R (2012) Extending moving least squares to mixed variables for metamodel-assisted optimization. In: 6th European congress on computational methods in applied sciences and engineering (ECCOMAS 2012), Vienna, Austria, 10–14 September
8. Filomeno Coelho R (2013) Metamodels for mixed variables based on moving least squares application to the structural analysis of a rigid frame. Optim Eng
9. Goldberg DE (1989) Genetic algorithms in search, optimization and machine learning. Addison-Wesley Longman, New York
10. Papadrakakis M, Lagaros ND, Plevris V (2005) Design optimization of steel structures considering uncertainties. Eng Struct 27:1408–1418
11. Richard Liew JY, Chen H, Shanmugam NE, Chen WF (2000) Improved nonlinear plastic hinge analysis of space frame structures. Eng Struct 22(10):1324–1338
12. Tang X, Bassir DH, Zhang W (2011) Shape, sizing optimization and material selection based on mixed variables and genetic algorithm. Optim Eng 12:111–128

Chapter 21
A Study of Nash-Evolutionary Algorithms for Reconstruction Inverse Problems in Structural Engineering

D. Greiner, J. Périaux, J.M. Emperador, B. Galván and G. Winter

Abstract In this paper we deal with solving inverse problems in structural engineering (both the reconstruction inverse problem and the fully stressed design problem are considered). We apply a game-theory based Nash-evolutionary algorithm and compare it with the standard panmictic evolutionary algorithm. The procedure performance is analyzed on a ten bar sized test case of discrete real cross-section types structural frame, where a significant increase of performance is achieved using the Nash approach, even achieving super-linear speed-up.

Keywords Structural optimization · Evolutionary algorithms · Nash equilibrium

21.1 Introduction

The use of population based global meta-heuristics as optimizers in real world and complex engineering design problems has been widespread during the last two decades, in single- and multi-objective problems. Among the tools oriented to

D. Greiner (✉) · J.M. Emperador · B. Galván · G. Winter
Institute of Intelligent Systems and Numerical Applications in Engineering (SIANI),
Universidad de Las Palmas de Gran Canaria (ULPGC), 35017 Las Palmas, Spain
e-mail: dgreiner@iusiani.ulpgc.es,david.greiner@ulpgc.es

J.M. Emperador
e-mail: jemperador@iusiani.ulpgc.es

B. Galván
e-mail: blgalvan@gmail.com

G. Winter
e-mail: gabw@step.es

J. Périaux
Mathematical Information Technology Department (MIT), University of Jyvaskyla,
Jyvaskyla, Finland
e-mail: jperiaux@gmail.com

J. Périaux
International Center for Numerical Methods in Engineering (CIMNE),
Universidad Politécnica de Cataluña (UPC), Barcelona, Spain

D. Greiner et al. (eds.), *Advances in Evolutionary and Deterministic Methods for Design, Optimization and Control in Engineering and Sciences*, Computational Methods in Applied Sciences 36, DOI 10.1007/978-3-319-11541-2_21

improve the efficiency of this methods, parallelization and use of game-theory based algorithms have been highlighted in fields like aeronautical engineering, where the computation of fitness is associated frequently with high computational CPU costs (see e.g. [1]). Here we propose to introduce the use of Nash-Evolutionary Algorithm [2–5], to speed up solving inverse problems in structural engineering.

21.2 Nash-Evolutionary Algorithms

Nash-Evolutionary algorithms were introduced in [2] for solving computational fluid dynamics problems. They are based in hybridizing the mathematical concepts of Nash equilibrium [6, 7] (competitive game theory where players maximize their payoffs while taking into account the strategies of their competitors) in the evolutionary search: A set of subpopulations co-evolve simultaneously each of which deals only with a partition of the search variables. These subpopulations interact to evolve towards the equilibrium; when dealing with a single objective problem, a virtual Nash game approach has been applied in inverse shape optimization computational fluid dynamics problems as an improvement technique versus the standard panmictic evolutionary algorithm [4, 8]. This approach has been successfully applied in the case of inverse problems where the fitness function objective is a sum of separable terms (such as the case of many shape optimization problems).

21.3 The Structural Problem

Structural inverse problems are handled here. The objective is to obtain the structure which most fits the maximum stresses of reference. The optimum structural bar design is defined as a design in which some location of every bar member in the structure has a maximum stress value as accurately equal as the maximum stress of reference for that bar.

$$Fitness\ Function = Min \sqrt{\sum_{i=1}^{Nbars} (\sigma_{MAX-i} - \sigma_{MAX-Ri})^2} \tag{21.1}$$

where $\sigma_{\text{MAX-i}}$ is the maximum calculated stress and $\sigma_{\text{MAX-Ri}}$ the maximum stress of reference, both corresponding to bar i.

A value of zero means a perfect fit in maximum stresses between our searched solution and the solution of reference.

In case we define as maximum stresses of reference, the set of a previously known structural design, the problem is a reconstruction inverse (RI) problem. In case we define as maximum stresses of reference, the material admissible stress, the problem is the fully stressed design (FSD) problem.

21.4 Test Case Definition

The purpose of this benchmark is to set a simple test case with frame bar structures based on one which has been applied before to the problem of fully stress design in [9, 10] and following the template of the Finnish design test case database as in http://jucri.jyu.fi. We consider the case of discrete cross-section type variables.

Objective of this benchmarking scenario is to test and compare different optimization approaches for structural design, e.g. see [11]. The computational domain, boundary conditions, loadings and design variable groping set are shown in Fig. 21.1. Boundary conditions consist in three pinned connections at the bottom joining the columns of the structure to the ground.

Table 21.1 show the variable numbering, variable grouping (which is intended to guarantee the structural symmetry), as well as the variable search space of each bar, which is constituted by standard HEB cross-section types. Geometry dimensions of Height H and width W are shown in Table 21.2. Table 21.3 show the material properties, which correspond to those of standard construction steel.

The structural type is a frame bar structure, where the stresses are evaluated by a standard stiffness matrix calculation (rigid nodes: resisting moment capabilities) and elastic behaviour of steel is assumed. The sideways permitted buckling effect case will be investigated in this paper.

Two loading cases are included (see Fig. 21.1):
Case 1 with D = 19674.2 N/m and P = 0 N, own weight is included.
Case 2 with D = 13768.65 N/m and P = 852 kN, own weight is included.

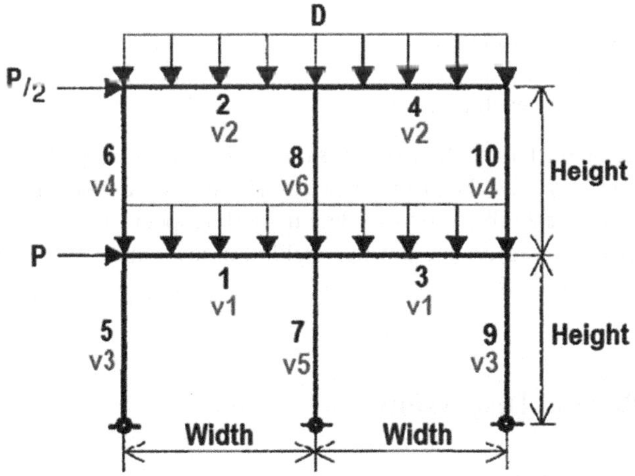

Fig. 21.1 Illustration of computational domain, boundary conditions, loadings and design variable set groupings [10]

Table 21.1 Variable search space

Bar number	Bar variable	Cross-section type set
1	V1	From HEB260 to HEB1000
2	V2	From HEB100 to HEB450
3	V1	From HEB260 to HEB1000
4	V2	From HEB100 to HEB450
5	V3	From HEB260 to HEB1000
6	V4	From HEB100 to HEB450
7	V5	From HEB260 to HEB1000
8	V6	From HEB100 to HEB450
9	V3	From HEB260 to HEB1000
10	V4	From HEB100 to HEB450

Table 21.2 Geometry parameters

Geometry parameters	Value (m)
Column length (Height)—H	3.29
Beam length (Width)—W	3.66

Table 21.3 Material properties (standard steel)

Parameter	Value
Density	7,850 (kg/m^3)
Young modulus	2.1×10^5 (MPa)
Maximum stress	2,600 (kg/cm^2)

The quantities of interest are:

- Values of the fitness function
- Cross-section type sizing for each bar.
- Maximum stress of each bar.

With respect to the RI problem, the cross-section types corresponding to the design shown in Table 21.4 have been taken as reference, where also the correspondent maximum value of each bar stress is shown. With respect to the FSD problem, the maximum stresses of reference are the yield stresses values (2,600 kp/cm^2) in all bars.

21.5 Results and Discussion

21.5.1 Nash Variable Territory Splitting

The considered split territory of variables among Nash players is divided in two subpopulations: v1, v3, v5 (subpopulation 1—Nash player 1, P1) and v2, v4, v6

Table 21.4 Design and stresses (kp/cm^2) of reference in reconstruction inverse problem design

InvProb	Bar 1	Bar 2	Bar 3	Bar 4	Bar 5	Bar 6	Bar 7	Bar 8	Bar 9	Bar 10
Cross section	HEB650	HEB320	HEB650	HEB320	HEB500	HEB360	HEB900	HEB450	HEB500	HEB360
Maximum stress	2427.68	1226.58	2352.79	1148.33	2371.60	1088.43	2393.57	979.21	2243.67	1019.43

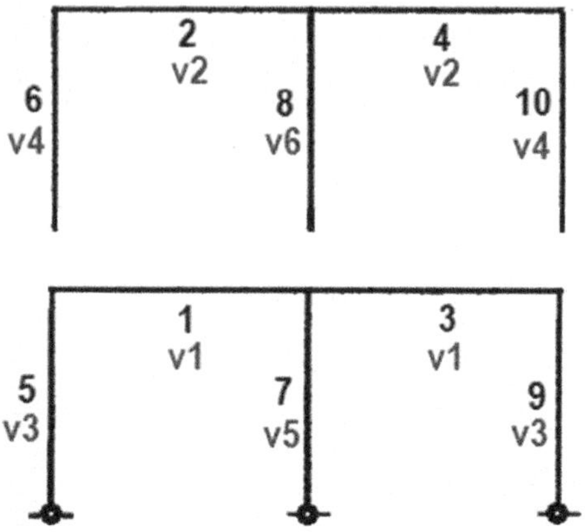

Fig. 21.2 Nash variable territory splitting

(subpopulation 2—Nash player 2, P2), being equally sized both in terms of variable set grouping and number of bars of the structure, as represented in Fig. 21.2. Therefore, the bottom part of the structure is optimized by player—subpopulation P1 and the upper part of the structure is optimized by player—subpopulation P2. This territory splitting does not include boundary conditions in each player.

21.5.2 Experiment Definition: Nash Genetic Algorithm

Statistical metrics obtained from one hundred independent executions for each case will be considered to measure the relative increased performance of the game theory based evolutionary algorithm approach versus the standard approach.

Results corresponding to a population size of 100 individuals, uniform crossover, a mutation rate of 3 % and an elitist generational replacement strategy keeping the two best individuals are considered here. Gray codification is used, in accordance with its good behaviour in structural frame optimum design (e.g., see [11–14]).

The performance of the panmictic GA strategy versus the Nash GA strategy (as described in Sect. 21.5.1) has been compared. In each case, also results from two standard panmictic populations (Panmictic P1 and Panmictic P2) evolving in parallel without interaction are shown for comparison, each of them having been executed 100 independent times to perform the following figures and statistics.

21.5.3 Results: Nash Genetic Algorithm

In relation with the reconstruction inverse problem, Fig. 21.3 shows the evolution of the fitness function average, Fig. 21.4 shows the evolution of the fitness function

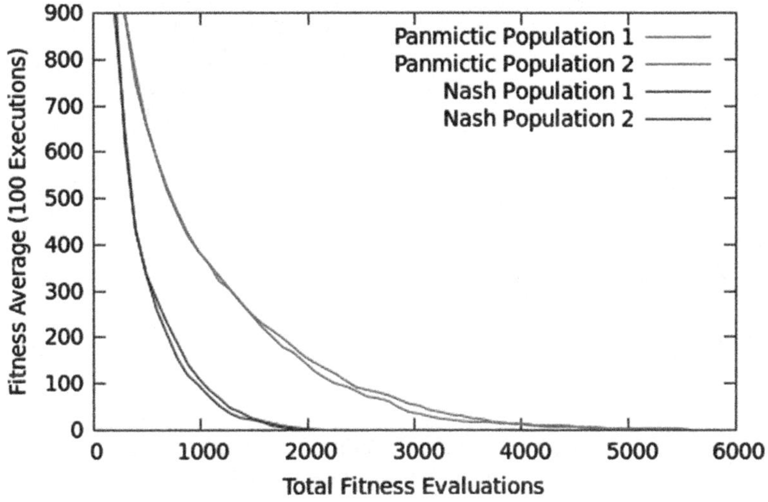

Fig. 21.3 100 population size, 3 % mutation rate, elitist generational strategy; Average over 100 executions (Reconstruction Inverse Problem)

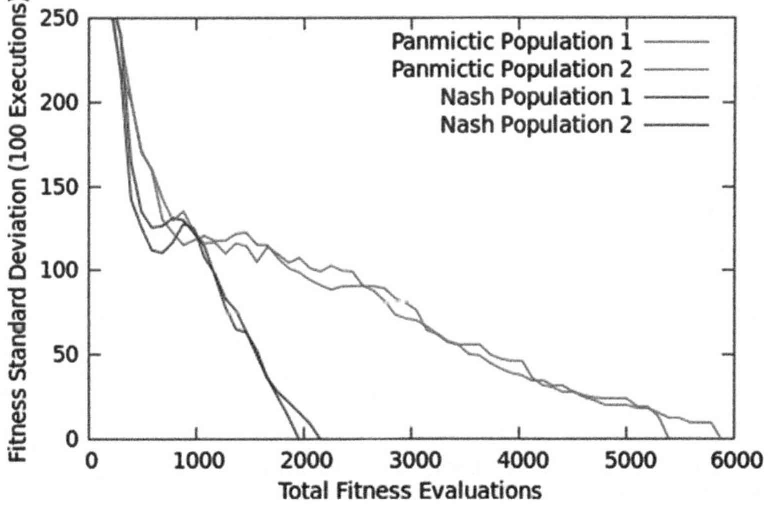

Fig. 21.4 100 population size, 3 % mutation rate, elitist generational strategy; Standard Deviation over 100 executions (Reconstruction Inverse Problem)

standard deviation and Fig. 21.5 shows the evolution of the fitness function best, all of them over 100 executions. In relation with the fully stressed design problem, Figs. 21.6, 21.7 and 21.8 show the evolution of fitness function average, standard deviation and best, respectively, all of them over 100 executions also.

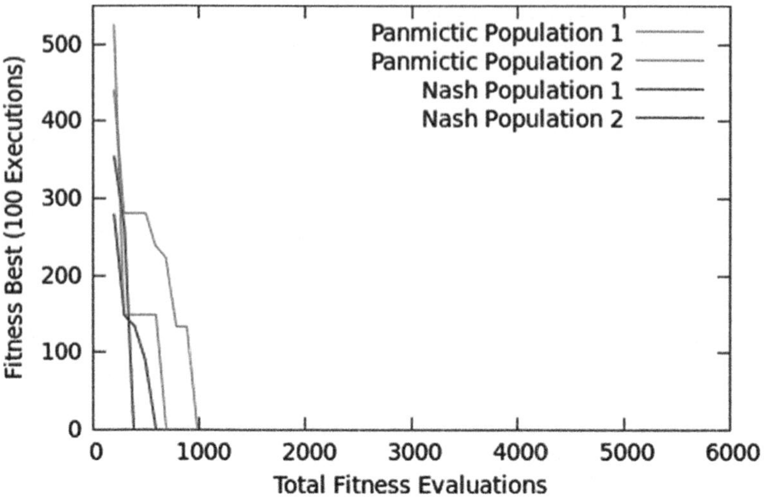

Fig. 21.5 100 population size, 3 % mutation rate, elitist generational strategy; Best over 100 executions (Reconstruction Inverse Problem)

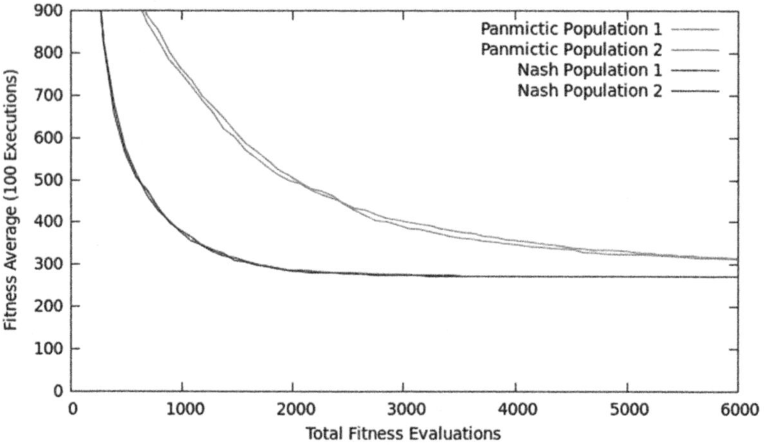

Fig. 21.6 100 population size, 3 % mutation rate, elitist generational strategy; Average over 100 executions (Fully Stressed Design Problem)

Tables 21.5 and 21.6 show N, the number of times out of 100 independent runs where the optimum solution of each problem (RI in Table 21.5 and FSD in Table 21.6) was achieved after a maximum stopping criterion of 100,000 fitness function evaluations; in addition, the average, standard deviation and best values of the number of fitness evaluations required to reach the optimum solution are included in each case.

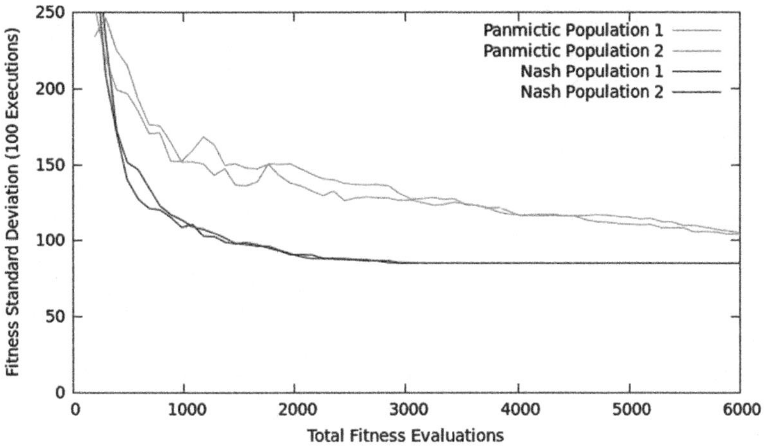

Fig. 21.7 100 population size, 3 % mutation rate, elitist generational strategy; Standard Deviation over 100 executions (Fully Stressed Design Problem)

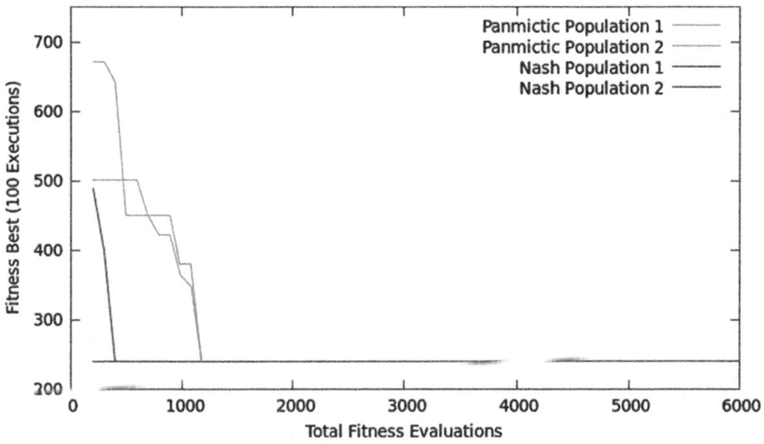

Fig. 21.8 100 population size, 3 % mutation rate, elitist generational strategy; Best over 100 executions (Fully Stressed Design Problem)

21.5.4 Discussion: Nash Genetic Algorithm

In case of the reconstruction inverse problem, seen Table 21.5, the Nash-GA strategy requires much less number of average and best evaluations, in a factor of 2.50 in case of the average and of 1.17 in case of the best value. The lower average value corresponds to the case Nash P2, which requires an average of 1059.4 fitness evaluations to reach the optimum (null value). Both algorithms were able to achieve the optimum solution in 100 out of 100 of the independent runs.

Table 21.5 Fitness Evaluations statistics of number of times out of 100 independent executions where optimum solution is achieved (N); reconstruction inverse problem

Algorithm	N	Aver.	St. dev.	Best
Panmictic P1	100	2647.0	833.7	688
Panmictic P2	100	2884.2	930.1	982
Nash P1	100	1177.0	390.6	394
Nash P2	100	1059.4	358.5	590

Table 21.6 Fitness Evaluations statistics of number of times out of 100 independent executions where optimum solution is achieved (N); fully stressed design problem

Algorithm	N	Aver.	St. Dev.	Best
Panmictic P1	97	7083.3	6705.6	1178
Panmictic P2	96	7957.4	8157.0	1178
Nash P1	86	1354.6	579.4	198
Nash P2	86	1399.1	603.8	394

In case of the fully stressed design problem, seen Table 21.6, the Nash-GA strategy requires much less number of average and best evaluations, in a factor of 5.06 in case of the average and of 11.12 in case of the best value. The lower average value corresponds to the case Nash P1, which requires an average of 1354.6 fitness evaluations to reach the optimum (a fitness function of 239.946 stress value, see detailed design solution in Table 21.7). A super-linear speed-up is achieved when compared the best fitness evaluations required to achieve the best fitness value (2.99 speed-up); this speed-up is even enhanced if the comparison is based in the fitness evaluations required to achieve the best average/standard deviation value, which evidences an increased robust behaviour of the Nash approach in this problem. The standard panmictic algorithm was able to find the optimum in 97 and 96 times out of 100, while the Nash algorithm found the optimum both in 86 times out of 100 executions (slightly lower than panmictic algorithm).

In summary, the Nash GA approach allows a super-linear improvement in terms of convergence speed up of the algorithm, requiring much less time to reach the optimum design.

21.6 Conclusions

The performance of Nash genetic algorithms in inverse problems in structural engineering (both the reconstruction inverse problem and the fully stressed design problem) has been tested in a ten bar sized frame test case showing a remarkable increased speed-up, even achieving super-linear gains in terms of fitness function

Table 21.7 Design and stresses (kp/cm^2) of best solution in Fully Stressed Design problem (FSD); Reference Stress = 2600.0

InvProb	Bar 1	Bar 2	Bar 3	Bar 4	Bar 5	Bar 6	Bar 7	Bar 8	Bar 9	Bar 10
Cross section	HEB600	HEB220	HEB600	HEB220	HEB500	HEB220	HEB800	HEB300	HEB500	HEB220
Maximum stress	2695.83	2511.46	2661.69	2585.86	2648.65	2685.72	2657.21	2664.32	2521.43	2484.90

evaluations in certain circumstances when compared with the standard panmictic genetic algorithm. In addition, it should be considered the benefits coming from enhanced parallel implementation capability of this type of algorithms.

Extension of experiments to other test cases and other related structural engineering problems will be performed in future works to explore the potential of this game theory based evolutionary algorithms in the field of structural optimum design, as well as other kinds of hybridized games like Stackelberg or Pareto games for multiobjective optimization problems and their implementation in parallel computing environments.

Acknowledgments This research work is funded through contract CAS12/00400 by Ministerio de Educación, Cultura y Deporte of the Government of Spain, through the Programa Nacional de Movilidad de Recursos Humanos del Plan Nacional de I+D+I 2008–2011, extended by agreement of Consejo de Ministros of 7th October 2011. The first author gratefully acknowledges support given at the Mathematical Information Technology Department, University of Jyväskylä (Finland), and in particular to Prof. Pekka Neittaanmäki.

References

1. Periaux J, Gonzalez F, Lee DC (2014) Evolutionary optimization and game strategies for advanced design: application to aeronautics. Intelligent systems, control and automation (ISCA) Series. Springer
2. Sefrioui M, Periaux J (2000) Nash genetic algorithms: examples and applications. In: Proceedings of the 2000 congress on evolutionary computation CEC-2000, IEEE Press, La Jolla Marriott Hotel La Jolla, California, pp 509–516, ISBN: 0-7803-6375-2
3. Lee DS, Gonzalez LF, Srinivas K, Periaux J (2009) Multifidelity Nash-Game strategies for reconstruction design in aerospace engineering problems. In: Proceedings of the 13th Australian international aerospace congress (AIAC)
4. Leskinen J, Périaux J (2012) Distributed evolutionary optimization using Nash games and GPUs—applications to CFD design problems. Comput Fluids. http://dx.doi.org/10.1016/j.compfluid.2012.03.017
5. Leskinen J, Wang H, Périaux J (2013) Increasing parallelism of evolutionary algorithms by Nash games in design inverse flow problems. Eng Comput 30(4):581–600
6. Nash JF (1950) Equilibrium points in N-person games. Proc Natl Acad Sci 36:46–49
7. Nash JF (1951) Non-cooperative games. Ann Math 54(2):286–295
8. Lee DS, Periaux J, Gonzalez LF, Srinivas K, Oñate E (2010) Active flow control bump design using hybrid Nash-game coupled to evolutionary algorithms. In: Pereira JCF, Sequeira A, Pereira JMC (eds) Proceedings of V European conference on computational fluid dynamics ECCOMAS CFD 2010, ECCOMAS CFD, Lisbon, Portugal, pp 1–14
9. Mueller KM (2000) Sizing of members in the fully stressed design of frame structures, Ph.D thesis, Deptartment of Civil and Environmental Engineering, University of Illinois at Urbana-Champaign
10. Mueller KM, Liu M, Burns SA (2002) Fully stresses design of frame structures and multiple load paths. J Struct Eng ASCE 128(6):806–814
11. Greiner D, Winter G, Emperador JM (2001) Optimising frame structures by different strategies of genetic algorithms. Finite Elem Anal Des 37–5:381–402
12. Greiner D, Winter G, Emperador JM (2004) Single- and multi-objective frame optimization by evolutionary algorithms and the auto-adaptive rebirth operator. Comput Methods Appl Mech Eng 37–35:3711–3743

13. Greiner D, Winter G, Emperador JM, Galván B (2005) Gray coding in evolutionary multicriteria optimization: application in frame structural optimum design. Lecture Notes in Computer Science, Evolutionary Multi-Criterion Optimization, vol 3410, pp 576–591
14. Greiner D, Diaz N, Emperador JM, Galván B, Winter G (2013) A comparative study of the influence of codification on discrete optimum design of frame structures. In: Proceedings of the third international conference on soft computing technology in civil, structural and environmental engineering, Paper 7, Italy

Chapter 22
A Comparative Study on Design Optimization of Polygonal and Bézier Curve-Shaped Thin Noise Barriers Using Dual BEM Formulation

Rayco Toledo, Juan J. Aznárez, Orlando Maeso and David Greiner

Abstract The inclusion of sound barriers for abating road traffic noise is a broadly used strategy that is often constrained by the requirements associated with its effective height. Due to this fact, the searching process has to deal with compromise solutions between the effective height and the acoustic efficiency of the barrier, assessed by the insertion loss (IL) in this paper. Two different barrier designs are studied herein for two different receivers configurations and for three clearly distinguishable regions in terms of closeness to the barrier. These models are based on the optimization of the IL of thin-cross section profiles proposed by an Evolutionary Algorithm. The special nature of these sorts of barriers makes necessary the implementation of a dual BEM formulation in the optimization process. Results obtained show the usefulness of representing complex thin-cross section barrier configurations as null boundary thickness-like models.

Keywords Thin noise barriers · Shape optimization · Genetic algorithms · Dual boundary element formulation

22.1 Introduction

The inclusion of sound barriers for abating the negative effects of road traffic noise near residential areas is a broadly used strategy. Considerable research work and studies focused on sound diffraction around barriers have been carried out in the past

---

R. Toledo (✉) · J.J. Aznárez · O. Maeso · D. Greiner
Instituto de Sistemas Inteligentes y Aplicaciones Numéricas en Ingeniería (SIANI), Edificio Central Del Parque Científico y Tecnológico del Campus Universitario de Tafira, 35017 Las Palmas de G.C., Spain
e-mail: rtoledo@iusiani.ulpgc.es

J.J. Aznárez
e-mail: jaznarez@iusiani.ulpgc.es

O. Maeso
e-mail: omaeso@iusiani.ulpgc.es

D. Greiner
e-mail: dgreiner@iusiani.ulpgc.es

D. Greiner et al. (eds.), *Advances in Evolutionary and Deterministic Methods for Design, Optimization and Control in Engineering and Sciences*, Computational Methods in Applied Sciences 36, DOI 10.1007/978-3-319-11541-2_22

two decades, specifically in the prediction of the performance and the development of more efficient designs. Amongst all of the different theoretical methods proposed concerning the issue, the Boundary Element Method (BEM hereinafter) has been previously used by the authors of this work [6] in the analysis of complex barrier configurations.

Evolutionary Algorithms (EA) have been widely used for Shape Design Optimization problems in numerous Engineering fields. The combined use of optimization problems using EA with a BEM code has been implemented in sound barriers design problems within the institute where this work is developed (see [2–4]).

The Insertion Loss coefficient (IL henceforward) is a valuable estimator to assess the acoustic efficiency of the barrier. This parameter represents the sound pressure level difference in the situation with and without the presence of the barrier at a particular point (receiver). Whilst the parameters involving the efficiency are numerous, the effective height of the barrier (h_{eff}) is the factor with greatest influence. Constraints related to this factor force the searching process to find compromise solutions between the effective height and the acoustic efficiency of the barrier. These profiles generally feature complex configurations and its implementation in optimization processes is often found to be difficult in terms of validating its topological feasibility. Two different barrier designs are studied in this work. These models are based on the optimization of the acoustic efficiency of thin-cross section profiles, idealized as null boundary thickness, proposed by an EA. The special nature of these sorts of barriers makes necessary the implementation of a dual BEM formulation in the optimization process.

22.2 Modelling and Discretization by Implementing a Dual BEM Formulation

The next lines are devoted to the implementation of a dual BEM formulation in thin noise barriers idealized as null boundary thickness profile (see Fig. 22.1). The special nature of these sorts of barriers makes necessary the addition of a complementary formulation (hyper-singular) that coupled with the conventional BEM formulation yields a compatible system of equations.

22.2.1 Singular BEM Formulation

The integral equation for the i boundary point to be solved by the singular BEM formulation can be expressed as follows:

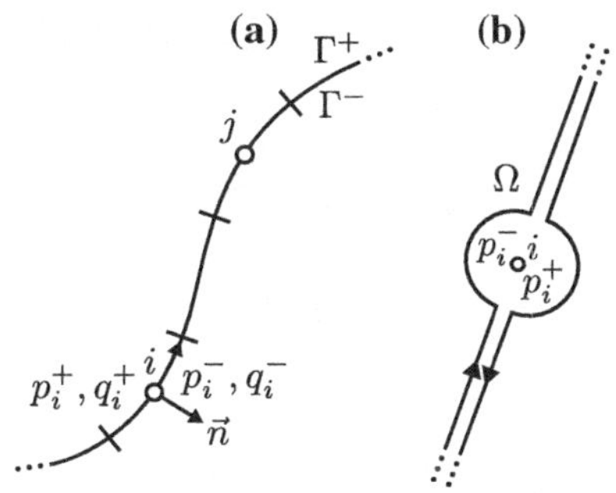

Fig. 22.1 **a** Idealization of a generic thin-cross section noise barrier profile as null thickness boundaries. **b** Strategy used to avoid the singularity around the collocation point in BEM formulation

$$c_i p_i + \fint_{\Gamma_b} p \frac{\partial p^*}{\partial \mathbf{n}_j} d\Gamma = p_0^* + \int_{\Gamma_b} \frac{\partial p}{\partial \mathbf{n}_j} p^* \, d\Gamma \tag{22.1}$$

This integral equality just involves the boundary of the barrier under investigation. The $\fint$ symbol represents the integral along the boundary to be understood in the Cauchy principal value sense, once the singularity around the collocation point i has been extracted (c_i). In Eq. (22.1), p is the acoustic pressure field over the barrier surface and p^* is the half-space fundamental solution [the acoustic pressure field when the source is placed at the collocation point i over a plane with admittance β_g (ground admittance)] and c_i is the free term. On the whole: $c_i = \theta/2\pi$, where θ represents the inner angle to the boundary measured in radians. It is easily shown that $c_i = 0.5$ for smooth boundaries.

The expressions of the fundamental solution and its derivative for a perfectly reflective ground for bi-dimensional, harmonious problems are:

$$\begin{aligned} p^*(k, r) &= \frac{1}{2\pi} \left[K_0(ikr) + K_0(ik\overline{r}) \right] \\ \frac{\partial p^*}{\partial n} &= -\frac{ik}{2\pi} \left[K_1(ikr) \frac{\partial r}{\partial \mathbf{n}_j} + K_1(ik\overline{r}) \frac{\partial \overline{r}}{\partial \mathbf{n}_j} \right] \end{aligned} \tag{22.2}$$

being i the imaginary unit, k the wave number, r and $\overline{r}$ the distances from the source and the image point to the observation point respectively, and K_0 and K_1 the Bessel modified functions of order 0 and 1 respectively.

By discretizing the boundary, the integral kernels of the fundamental solution of the singular BEM formulation are yielded:

$$h_k^{ij} = \int_{\Gamma_j} \frac{\partial p^*}{\partial \mathbf{n}_j} \phi_k d\Gamma_j; \quad g_k^{ij} = \int_{\Gamma_j} p^* \, \phi_k \, d\Gamma_j \tag{22.3}$$

A system of equations is obtained from this process and leads to values of acoustic pressure on the barrier boundary.

22.2.2 Hyper-Singular BEM Formulation

The integral equation for the i boundary point to be solved by the hyper-singular BEM formulation can be written as follows:

$$c_i\left(\frac{\partial p_i}{\partial \mathbf{n}_i}\right)+\neq\!\!\!\!\!\!\int_\Gamma p\frac{\partial^2 p^*}{\partial \mathbf{n}_i \partial \mathbf{n}_j}d\Gamma=-\!\!\!\!\!\!\int_\Gamma \frac{\partial p^*}{\partial \mathbf{n}_i}\frac{\partial p}{\partial \mathbf{n}_j}d\Gamma+\frac{\partial p_0^*}{\partial \mathbf{n}_i} \tag{22.4}$$

where the $\neq\!\!\!\!\!\int$ and $-\!\!\!\!\!\int$ symbols represent the integral along the boundary to be understood in the Hadamard finite part integral and in the Cauchy principal value sense, respectively. The hyper-singular formulation of the method demands that the source placement (collocation point i) to be inside the element (non-nodal collocation point) (see [8]). Thus, in (22.4) it is satisfied that $c_i = 0.5$.

Expression (22.5) shows the values of the fundamental solution and its derivative for the hyper-singular formulation:

$$\begin{aligned}\frac{\partial p^*}{\partial \mathbf{n}_i}&=-\frac{ik}{2\pi}\left[K_1(ikr)\frac{\partial r}{\partial \mathbf{n}_i}+K_1(ik\overline{r})\frac{\partial \overline{r}}{\partial \mathbf{n}_I}\right]\\ \frac{\partial^2 p^*}{\partial \mathbf{n}_i\partial \mathbf{n}_j}&=\frac{(ik)^2}{2\pi}\left[\left(K_2(ikr)\frac{\partial r}{\partial \mathbf{n}_i}\frac{\partial r}{\partial \mathbf{n}_j}+\frac{K_1(ikr)}{r}\mathbf{n}_i\cdot\mathbf{n}_j\right)\right.\\ &\quad\left.+\left(K_2(ik\overline{r})\frac{\partial \overline{r}}{\partial \mathbf{n}_I}\frac{\partial \overline{r}}{\partial \mathbf{n}_j}+\frac{K_1(ik\overline{r})}{\overline{r}}\mathbf{n}_I\cdot\mathbf{n}_j\right)\right]\end{aligned} \tag{22.5}$$

Similarly to Eq. (22.2), i is the imaginary unit, k the wave number and r, $\overline{r}$ the distances to the observation point from the collocation point and its symmetric point with respect to the ground plane, respectively. It is worth making a distinction here regarding the normal vectors involved in the expressions above. $\mathbf{n}_j$ is the normal to the boundary at the integration point and $\mathbf{n}_i$ (n_x^i, n_y^i), $\mathbf{n}_I$ $(n_x^i, -n_y^i)$ represent the normal vectors to the real boundary at the collocation point (i) and at its symmetric point (I) placed on a fictitious, symmetric boundary with respect to the ground plane, respectively. K_1 and K_2 represent the Bessel modified functions of order 1 and 2, respectively.

After a discretization process along the boundary, expression (22.4) yields a numerical solution from which the integral kernels of the hyper-singular BEM formulation are obtained, for i collocation point when integrated over j element:

$$m_k^{ij}=\int_{\Gamma_j}\frac{\partial^2 p^*}{\partial \mathbf{n}_i\,\partial \mathbf{n}_j}\phi_k\,d\Gamma_j;\quad l_k^{ij}=\int_{\Gamma_j}\frac{\partial p^*}{\partial \mathbf{n}_i}\phi_k\,d\Gamma_j \tag{22.6}$$

The numerical resolution of these integrals deserve a thorough treatment and can be consulted in [7, 8].

22.2.3 Dual BEM Formulation

Figure 22.1 represents a generic thin-cross section noise barrier to be solved by dual BEM formulation. After a discretization process, each node holds the values of pressure and flux with respect to the boundary normal, i.e., both at the left and at the right according to the direction of travel on the boundary (p^+, q^+, p^-, q^- hereinafter).

Figure 22.1b represents the strategy used to isolate the singularity of the method in this sort of domains. Thus, the matrix equality of the singular BEM formulation for thin-cross section noise barriers can be expressed as follows:

$$c_i\left(p_i^+ + p_i^-\right) + \sum_{j=1}^{N}\left(\mathbf{H}_j^+ p_j^+ + \mathbf{H}_j^- p_j^-\right) = \sum_{j=1}^{N}\left(\mathbf{G}_j^+ q_j^+ + \mathbf{G}_j^- q_j^-\right) \tag{22.7}$$

being N the overall nodes number of the discretization over the boundary. Considering that $\mathbf{n}^+ = \mathbf{n}^-$ at the collocation point j, it is easily shown that:

$$\mathbf{H}_j^+ = -\mathbf{H}_j^-; \quad \mathbf{G}_j^+ = \mathbf{G}_j^- \tag{22.8}$$

For internal noise sources and smooth boundaries, the final expression can be written as follows (see [1]):

$$\left(\frac{1}{2}\right)\Sigma p_i + \sum_{j=1}^{N}\mathbf{H}_j^+ \Delta p_j = \sum_{j=1}^{N}\mathbf{G}_j^+ \Sigma q_j + p_0^* \tag{22.9}$$

where:

$$\Sigma p_i = p_i^+ + p_i^-; \quad \Delta p_j = p_j^+ - p_j^-; \quad \Sigma q_j = q_j^+ + q_j^- \tag{22.10}$$

Deriving (22.7) with respect to $\mathbf{n}_i^+$ an integral equality of the hyper-singular BEM formulation is obtained:

$$c_i\left(\frac{\partial p_i^+}{\partial \mathbf{n}_i^+} + \frac{\partial p_i^-}{\partial \mathbf{n}_i^+}\right) + \sum_{j=1}^{N}\left(\mathbf{M}_j^+ p_j^+ + \mathbf{M}_j^- p_j^-\right) = \sum_{j=1}^{N}\left(\mathbf{L}_j^+ q_j^+ + \mathbf{L}_j^- q_j^-\right) \tag{22.11}$$

where:

$$\frac{\partial p_i^-}{\partial \mathbf{n}_i^+} = -q_i; \quad \mathbf{M}_j^+ = -\mathbf{M}_j^-; \quad \mathbf{L}_j^+ = \mathbf{L}_j^- \tag{22.12}$$

The hyper-singular formulation of the method requires that the collocation point i to be inside the element (see [8]) what assurances that the inner angle to the boundary at that point is always $\theta = \pi$. In this way, the final dual BEM expressions for internal noise sources for both the singular and hyper-singular formulation are:

$$\left(\frac{1}{2}\right)\Sigma p_i + \sum_{j=1}^{N} \mathbf{H}_j^+ \Delta p_j = \sum_{j=1}^{N} \mathbf{G}_j^+ (A^+ \Sigma p_j + A^- \Delta p_j)$$

$$+ p_0^* \left(\frac{1}{2}\right)(A^- \Sigma p_i + A^+ \Delta p_i) + \sum_{j=1}^{N} \mathbf{M}_j^+ \Delta p_j$$

$$= \sum_{j=1}^{N} \mathbf{L}_j^+ (A^+ \Sigma p_j + A^- \Delta p_j) + \frac{\partial p_0^*}{\partial \mathbf{n}_i} \quad (22.13)$$

being:

$$\Delta q_j = A^- \Sigma p_j + A^+ \Delta p_j \;\; ; \;\; \Sigma q_j = A^+ \Sigma p_j + A^- \Delta p_j$$
$$A^+ = -(1/2)\, ik\, (\beta^+ + \beta^-) \;\; ; \;\; A^- = -(1/2)\, ik\, (\beta^+ - \beta^-) \quad (22.14)$$

Finally, expression (22.13) can be expressed matricially as:

$$\begin{bmatrix} \frac{\mathbf{I}}{2} - \mathbf{G}^+ A^+ & \mathbf{H}^+ - \mathbf{G}^+ A^- \\ \frac{A^-}{2}\mathbf{I} - \mathbf{L}^+ A^+ & \frac{A^+}{2}\mathbf{I} + \mathbf{M}^+ - \mathbf{L}^+ - A^- \end{bmatrix} \begin{bmatrix} \Sigma p \\ \Delta p \end{bmatrix} = \begin{bmatrix} p_o^* \\ \frac{\partial p_o^*}{\partial \mathbf{n}_i^+} \end{bmatrix} \quad (22.15)$$

22.3 Problem Definition

Figure 22.2 represents the general configuration of the model under study. It deals with a two-dimensional model concerning an infinite, coherent mono-frequency source of sound, placed parallel to an infinite noise barrier of thin cross-section that

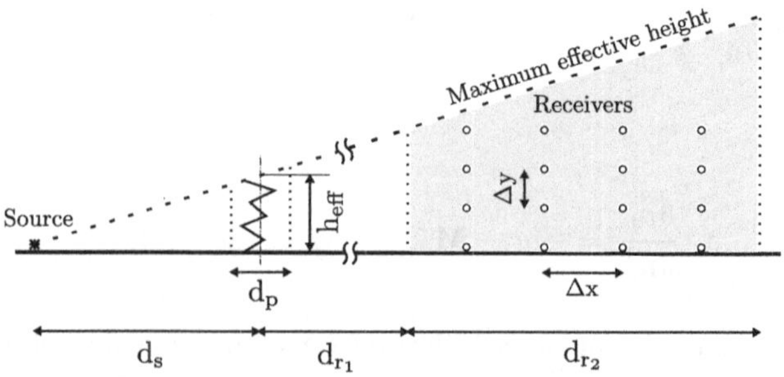

Fig. 22.2 Two-dimensional configuration for thin cross-section acoustic barriers

Table 22.1 Data concerning regions under study

Region	d_s(m)	d_p(m)	d_{r_1} (m)	d_{r_2} (m)	Δx(m)	Δy(m)
1			0.5	10.0	2.0	1.0
2	10.0	1.0	10.5	40.0	8.0	2.0
3			50.5	50.0	10.0	5.0

stands on a flat plane (ground) of uniform admittance at $d_s = 10$ m. Both the ground and the barrier feature a perfectly reflective surface in this article ($\beta_g = \beta_b = 0$). A trapezoidal section holds the area for feasible profiles, defined by the barrier projection to the ground, that is constant and $d_p = 1$ m, and the maximum effective height to be achieved, that is $h_{eff} = 3$ m at the median of the rectangle trapezium.

Two different receiver configurations are studied. In one configuration (Ca) a group of four receivers placed on the ground and separated Δx from one another is considered. In the other configuration (Cb), four groups of four receivers are studied. The first group is laid on the ground and the remaining ones are placed at different heights, separated among them by a distance of Δy. In accordance with the former configuration, the horizontal distance among the receivers of a group is Δx.

In addition to this, three clearly distinguishable regions in terms of closeness to the median of the feasible region (d_{r_1}) are proposed for both receiver configurations. Table 22.1 holds the data concerning these regions.

The results achieved are given in terms of *insertion loss* (IL), defined as follows:

$$IL = -20 \log_{10} \left(\frac{P_B}{P_{HS}} \right) \text{(dB)} \tag{22.16}$$

on every frequency of the band spectra, and represents the sound pressure level at the receiver points as a difference between the situation with (P_B) and without (P_{HS}) considering the barrier. This parameter is a widely used estimator to assess the acoustic efficiency of sound barriers.

22.4 Methodology

This section provides an overview of the proposed methodology for the optimization of thin-cross section noise barriers idealized as null boundary thickness-like models.

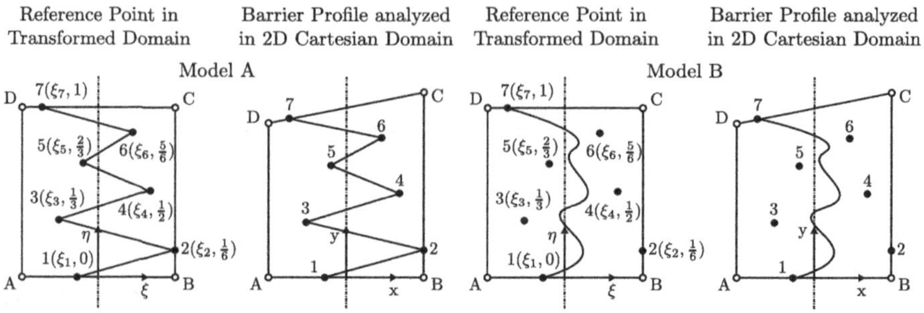

Fig. 22.3 Design variables and models under study

22.4.1 Shape Optimization

Shape design optimization is carried out by the combined use of an EA and a code that implements a dual BEM formulation. The EA software used in this work applies the GAlib package [9]. This library is a collection of C++ genetic algorithm (GA) components from which it is possible to quickly construct GA's to attack a wide variety of problems.

In this paper, a steady-state genetic algorithm is used replacing the two worst individuals (in terms of their fitness function) at every generation, with a population size of 100 individuals. A single-point crossover operator is used in this study, with a crossover rate of 0.9. The considered mutation rate is $1/\mathrm{n_{ch}}$, where $\mathrm{n_{ch}}$ is the chromosome length ($\mathrm{n_{ch}} = 8 \times \mathrm{n}$, being n the overall number of the design variables—of 8 bits precision each). Five independent executions of the optimization process are considered for each model and configuration. The stop-criterion condition is met for 20,000 evaluations of the fitness function (FF).

A transformed domain is considered (see [3]). This domain holds the set of design variables of the model under study, denoted by (ξ_i, η_i), and represents the rectangular search space for the GA (see left part of Fig. 22.3). Every (ξ_i, η_i) point in the transformed domain has its image $(\mathrm{x_i}, \mathrm{y_i})$ in the Cartesian space, that is the real domain where the barrier operates. In this paper $\mathrm{h_{eff}} = 3\,\mathrm{m}$ is proposed. This generates a trapezoidal search space in the Cartesian barrier domain (see right part of Fig. 22.3).

Two acoustic barrier designs are studied along this paper (Fig. 22.3). The horizontal projection ($\mathrm{d_p}$) and the effective height ($\mathrm{h_{eff}}$) are identical for each design. Both models are built from seven points, being the first and the last one on the ground and on the effective height line respectively. The vertical distance among the points is $\mathrm{d_i} = 1/6$ in the search space (transformed domain) and they are just allowed to feature horizontal movements. Model A is a polygonal curve-shaped barrier built from points through which straight slopes pass. Model B is a 6th degree Bézier curve-shaped barrier built from seven control points of which only the first (0) and the last (7) belong to it.

22.4.2 Assessment of the Insertion Loss

Taking into account the overall value of the IL of the frequency band spectrum analyzed for each receiver seems to be a more realistic estimator to evaluate the efficiency of a sound barrier. Consequently, (22.17) represents the average IL value for each frequency and receiver when using the ISO 717.2 normalized traffic noise spectra for third-octave band center frequencies [5], ranging from 100 to 2,000 Hz.

$$\overline{\mathrm{IL}} = -10 \cdot \log_{10}\left(\frac{\sum_{i=1}^{NF} 10^{(A_i - IL_i)/10}}{\sum_{i=1}^{NF} 10^{A_i/10}}\right) \text{(dBA)} \tag{22.17}$$

being NF the studied spectra number of frequencies, i.e. NF = 14, A_i the spectra A-weighted noise level and IL_i the insertion loss value for sources pulsing at every frequency of the spectra.

22.4.3 Definition of the Fitness Function

Shape optimization is carried out entirely based on the overall IL mean value of all receiver points.

$$FF = \sum_{j=1}^{NR} \overline{IL}_j / NR \tag{22.18}$$

being $\overline{IL}_j$ the IL mean value for each receiver (see (22.18)) and NR the total number of receivers. This value corresponds to the fitness function (FF) to be maximized, so the higher its value the higher the acoustic efficiency of the sound barrier.

22.5 Results and Discussion

Tables 22.2 and 22.3 collect the acoustic efficiency and the coordinates of the design variables (see Fig. 22.3) of the best individuals respectively, for each receiver configuration, region and model.

Figures 22.4 and 22.5 show the barrier profile of the best individuals of the models under study, in terms of its acoustic efficiency as well as the average frequential IL evolution of the receivers for each region and model.

Figures 22.6 and 22.7 show the evolution of the average of the fitness function ($FF_{Average}$), the best individual (FF_{Best}) and the average of the standard deviation for both models under study and every receiver configuration.

Table 22.2 Acoustic efficiency of the models under study

RC*	Region	Model	L_c(m)	ΔL_c(m)	FF_{best}(dBA)	ΔFF_{best}(dBA)
Ca	1	A	4.08177	+1.08177	17.92628	+4.32966
		B	3.60547	+1.60547	16.77495	+3.17833
	2	A	3.97839	+0.97839	12.93384	+1.02711
		B	3.51863	+0.51863	14.08611	+2.17938
	3	A	3.52785	+0.52785	12.46634	+1.04716
		B	3.84417	+0.84417	13.57048	+2.15130
Cb	1	A	4.10065	+1.10065	16.95941	+2.41822
		B	3.41333	+0.41333	16.83553	+2.29434
	2	A	3.63842	+0.63842	14.36767	+0.92615
		B	3.71865	+0.71865	14.87088	+1.42936
	3	A	3.60034	+0.60034	13.64344	+0.88584
		B	3.68994	+0.68994	13.75215	+0.99450

RC Receiver configuration

Table 22.3 Design variables of the best individuals

			Design variables						
RC*	Region	Model	ξ_1	ξ_2	ξ_3	ξ_4	ξ_5	ξ_6	ξ_7
Ca	1	A	−0.48824	0.06078	0.01765	0.35098	0.02941	0.49608	−0.31961
		B	−0.01373	0.05882	−0.48235	−0.57647	0.17647	−1.23529	0.50000
	2	A	−0.39412	−0.02549	0.39020	0.08824	0.04902	0.48039	0.34314
		B	−0.50000	0.36471	−1.44706	−0.01176	0.38823	−1.02353	0.40980
	3	A	−0.44902	−0.08039	−0.22549	0.35490	0.44510	0.26078	0.32353
		B	0.43726	−0.95294	0.15294	−1.30588	0.38824	−1.23529	0.31177
Cb	1	A	0.50000	0.22941	−0.15882	0.32745	−0.03333	0.37843	−0.44118
		B	−0.06078	−0.17647	0.10588	−0.92941	0.74118	−0.90588	0.50000
	2	A	−0.22156	−0.50000	0.29216	0.37451	0.50000	0.50000	0.31569
		B	−0.46470	−0.67059	2.05882	−0.95294	1.37647	−0.90588	0.50000
	3	A	−0.25294	−0.50000	0.24902	0.37843	0.50000	0.50000	0.29608
		B	−0.48039	−0.62353	2.01176	−1.04706	1.49412	−0.88235	0.50000

RC Receiver configuration

In the light of the results the following analysis is carried out:

- The polygonal-shaped barrier outperforms the acoustic efficiency of the 6th degree Bézier curve-shaped model for the near region when the receivers are placed on the ground (Ca configuration). However, the latter model performs a better acoustic behaviour for non-near regions (over 1 dBA).
- Both models under study display similar acoustic performances when a grid of receivers is considered in the shadow region of the barrier, with the exception of the intermediate region in which the Bézier model outperforms the polygonal design in half a decibel.

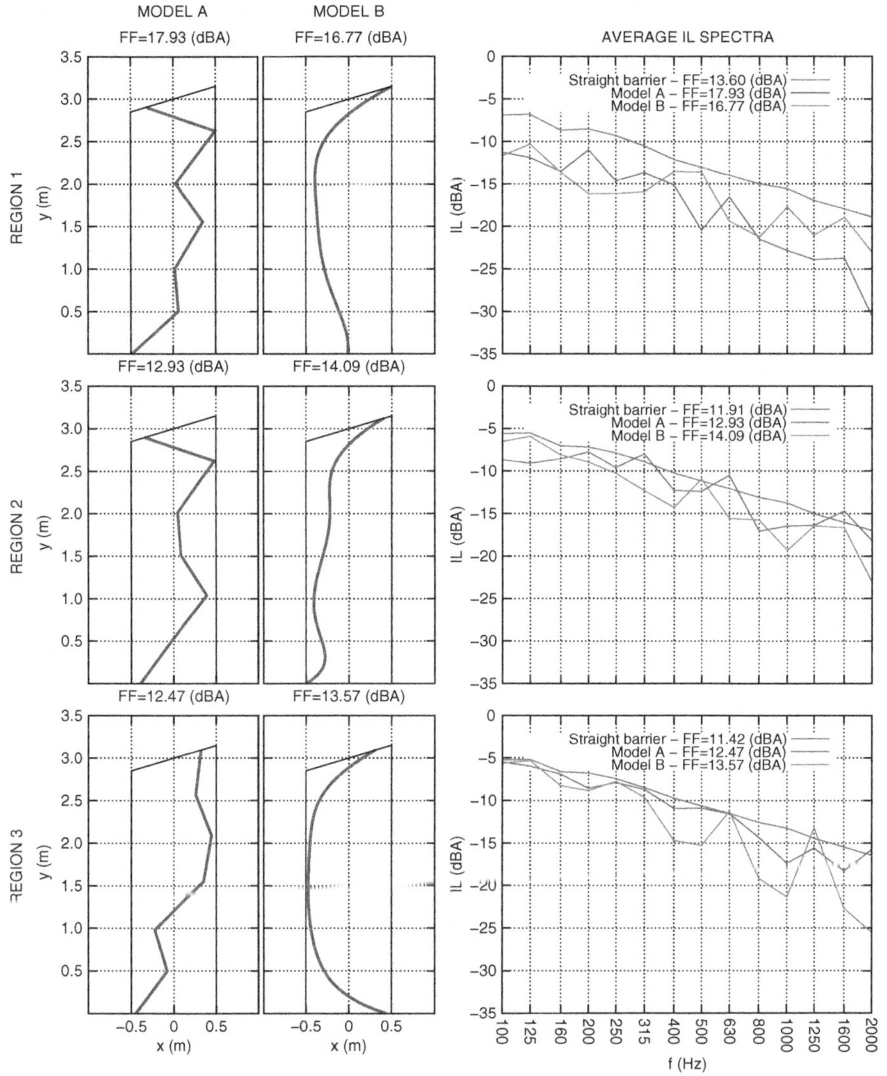

Fig. 22.4 Ca receiver configuration. *Left* barrier profile of the best individuals for each region and model. *Right* average frequential IL evolution for models A and B and for the 3 m height straight barrier

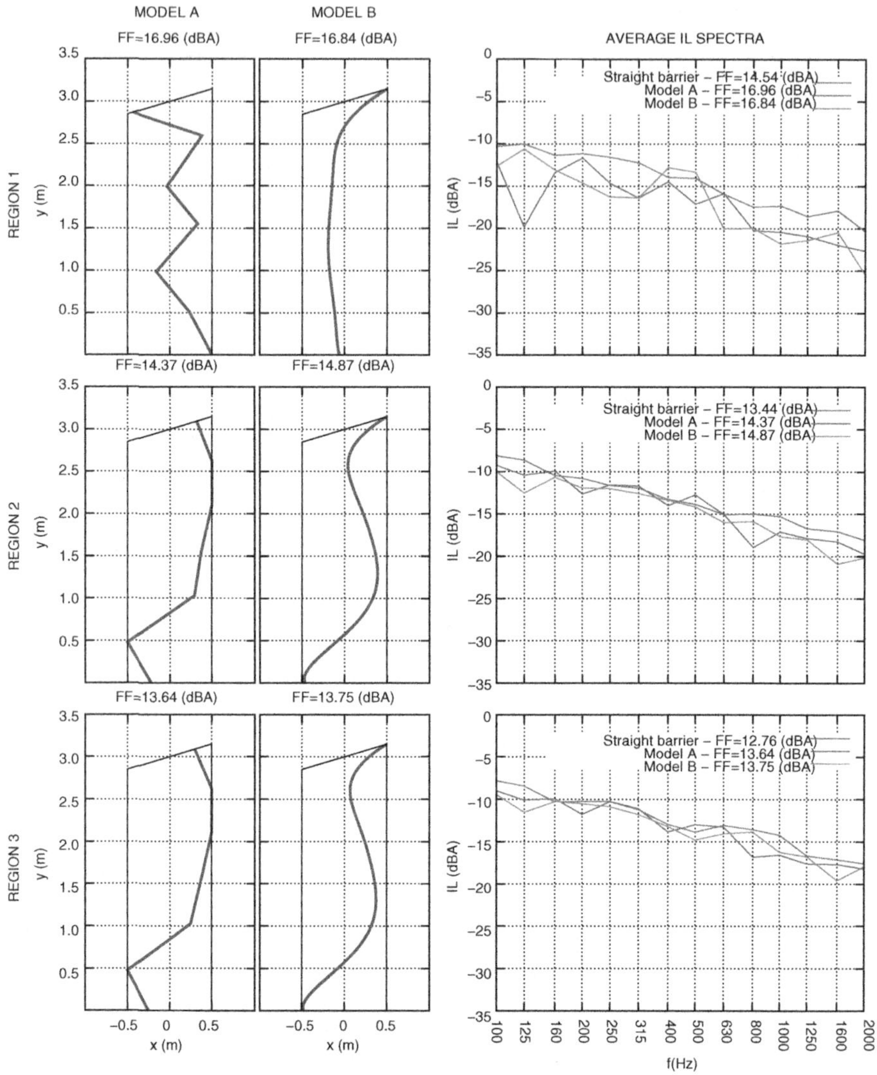

Fig. 22.5 Cb receiver configuration. *Left* barrier profile of the best individuals for each region and model. *Right* average frequential IL evolution for models A and B and for the 3 m height straight barrier

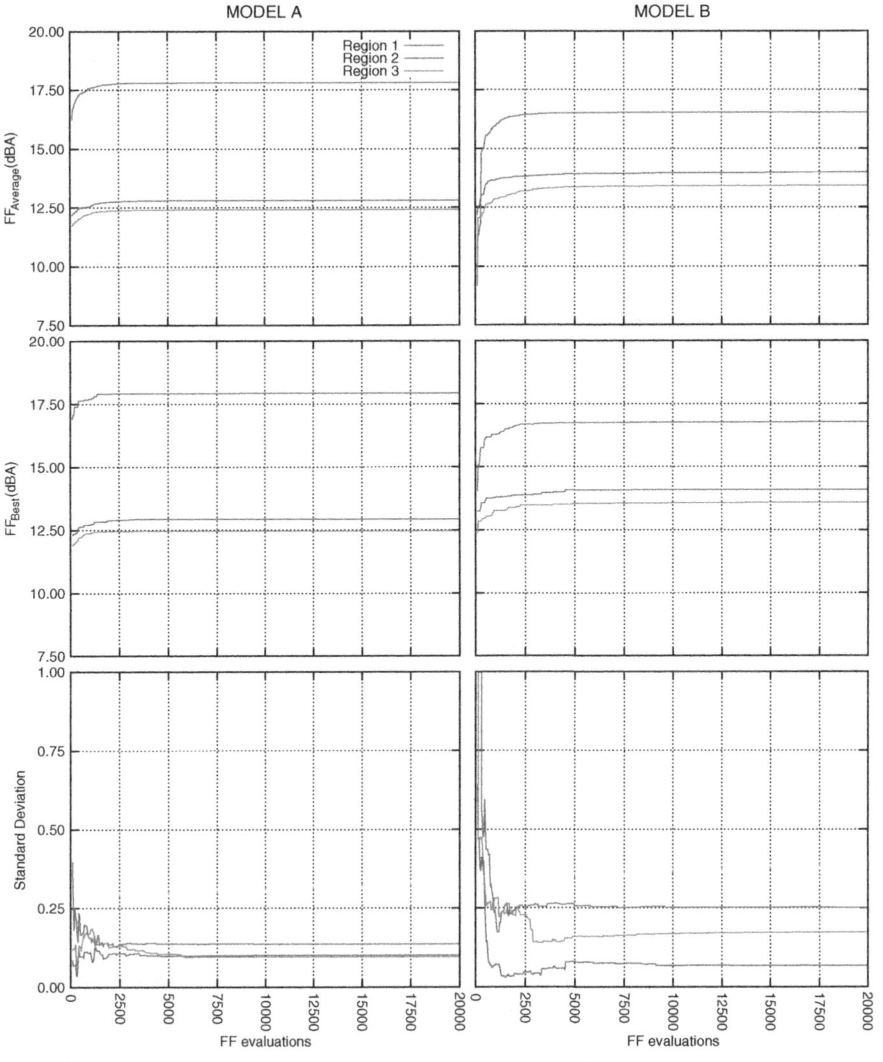

Fig. 22.6 Ca receiver configuration. *Left* model A. *Right* model B

- According to the comparative analysis between the optimized models and the straight barrier the need to study designs alternative to the latter is suggested, even for far regions.
- Model B displays a wider variety among best individuals of the population than in the case for Model A according to the evolution of the highest standard deviation (Figs. 22.6 and 22.7), meaning that the convergence of the optimization process turns out to be more cumbersome in Model B (optimization process is easier in Model A).

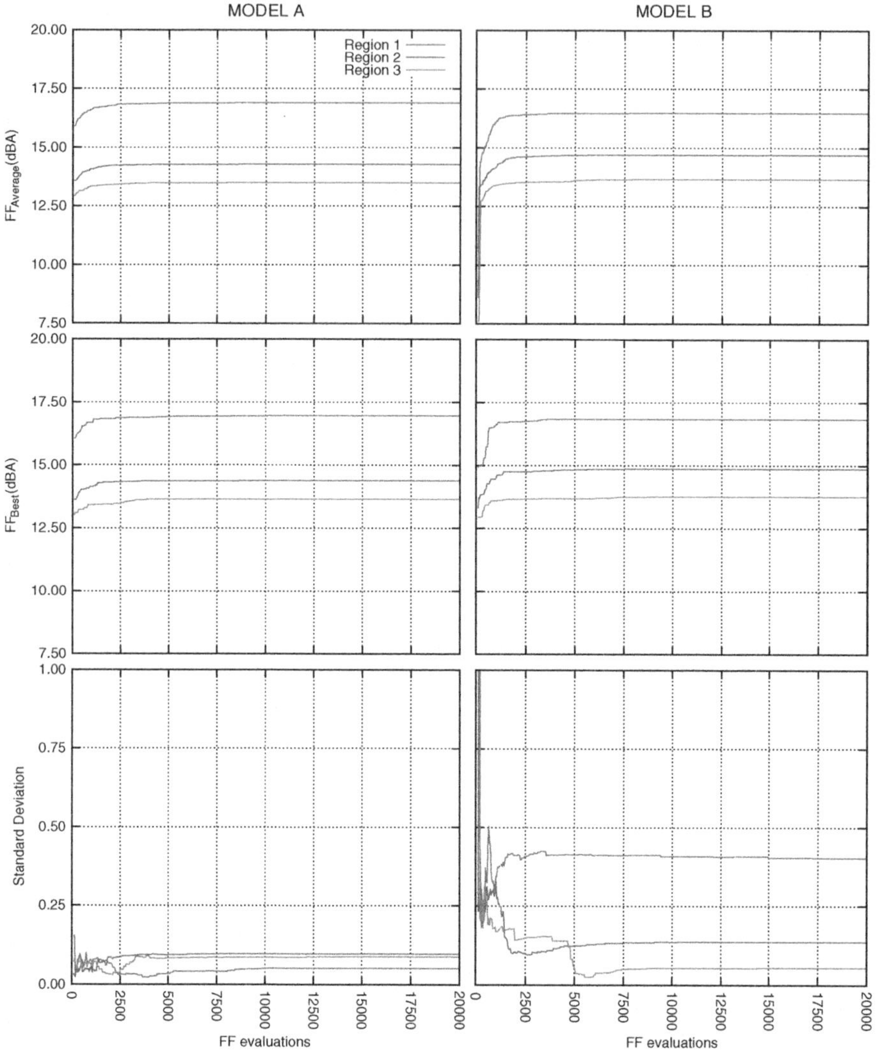

Fig. 22.7 Cb receiver configuration. *Left* model A. *Right* model B

22.6 Conclusions

A methodology to successfully optimize thin cross-section noise barriers by idealizing their profiles as null cross-section boundaries has been presented. This procedure has been applied to two specific noise barrier models although its applicability covers a wide designs spectra, ranging from complex straight boundary configurations to curve-shaped profiles like those built from Spline expressions, amongst others.

The versatility of the algorithm responsible for the geometry generation of the barrier makes the building of the profile to be easily accomplished. This is a significant advantage over the case when dealing with geometries of real barrier profiles, as the evaluation process for the feasibility of the design proposed by the EA is often complex and difficult to establish.

The procedure here presented is a useful method to assess the behaviour of complex noise barriers configurations and yields conclusions that might have been hardly drawn without its implementation.

Acknowledgments This work was supported by the Ministerio de Economía y Competitividad (MINECO) of Spain and FEDER through research project BIA2010-21399-C02-01 and also by the Agencia Canaria de Inv., Inn. y Soc.de la Inf. (ACIISI) of the Government of the Canary Islands and FEDER through research project ProID20100224. R. Toledo is a recipient of a fellowship from the Subprogram of Predoctoral Fellowships of Research Personnel in Trainning (FPI-MICINN), granted by Ministerio de Ciencia e Innovación of Spain. The authors are grateful for this support.

References

1. de Lacerda L, Wrobel L, Mansur W (1997) A dual boundary element formulation for sound porpagation around barriers over an impedance plane. JSV 202(2):235–247
2. Greiner D, Galván B, Aznárez J, Maeso O, Winter G (2009) Robust design of noise attenuation barriers with evolutionary multiobjective algorithms and the boundary element method. Evol Multi-Criterion Optim 5467:261–274
3. Greiner D, Aznárez JJ, Maeso O, Winter G (2010) Single- and multi-objective shape design of Y-noise barriers using evolutionary computation and boundary elements. Adv Eng Softw 41(2):368–378
4. Greiner D, Aznárez J, Maeso O, Winter G (2006) Shape design of noise barriers using evolutionary optimization and boundary elements. In: The fifth international conference on engineering computacional technology, civil-comp-press, vol 43. Stirlingshire
5. ISO 717.2 Rating of sound insulation in buildings and of building elements—Part 2: Impact sound insulation
6. Maeso O, Aznárez JJ (2005) Estrategias para la reducción del impacto acústico en el entorno de carreteras. Una aplicación del Método de los Elementos de Contorno. Universidad de Las Palmas de Gran Canaria. ISBN: 84-689-0340-X. DOI 846890340X
7. Sáez A, Gallego R, Domínguez J (1995) Hypersingular quarter-point boundary elements for crack problems. Int J Numer Meth Eng 38:1681–1701
8. Toledo R, Aznárez JJ, Maeso O, Greiner D (2013) Optimización de diseños de pantallas acústicas de pequeño espesor mediante la implementación de la formulación dual del mec. In: Congress on numerial method in engineering, SEMNI
9. Wall M (ed) (1996) GAlib: a C++ library of genetic algorithm components (http://lancet.mit.edu/ga/). Mechanical engineering department, MIT

Chapter 23
A Discrete Adjoint Approach for Trailing-Edge Noise Minimization Using Porous Material

Beckett Y. Zhou, Nicolas R. Gauger, Seong R. Koh and Wolfgang Schröder

Abstract In this paper, we present a discrete adjoint-based optimization framework to obtain the optimal distribution of the porous material over the trailing edge of a 3-D flat plate. The near-body strength of the noise source generated by the unsteady turbulent flow field is computed using a high-fidelity large-eddy simulation (LES). By optimally controlling the material porosity and permeability, it is possible to minimize the turbulence intensity responsible for noise generation at the trailing edge and thus significantly reduce the radiated noise. We demonstrate, using a simple geometry as a first step, the efficacy of the discrete adjoint method in achieving minimum-noise design via optimal distribution of porous media, with future applications to aircraft high-lift devices.

Keywords Discrete adjoint method · Noise reduction · Automatic differentiation · Large-eddy simulation · Turbulence

23.1 Introduction

In its 2012 annual report, the US Federal Aviation Administration (FAA) predicted the air travel to nearly double over the next 20 years for US carriers alone [1]. However, the ability of the aviation industry to meet such growing demands, is limited by the increasingly stringent noise emission requirements imposed by the FAA and

B.Y. Zhou (✉) · N.R. Gauger
Computational Mathematics Group, Department of Mathematics and CCES, RWTH Aachen University, Schinkelstraße 2, 52062 Aachen, Germany
e-mail: zhou@mathcces.rwth-aachen.de

N.R. Gauger
e-mail: gauger@mathcces.rwth-aachen.de

S.R. Koh · W. Schröder
Institute of Aerodynamics, RWTH Aachen University, Wüllnerstraße 5a, 52062 Aachen, Germany
e-mail: s.koh@aia.rwth-aachen.de

W. Schröder
e-mail: office@aia.rwth-aachen.de

D. Greiner et al. (eds.), *Advances in Evolutionary and Deterministic Methods for Design, Optimization and Control in Engineering and Sciences*, Computational Methods in Applied Sciences 36, DOI 10.1007/978-3-319-11541-2_23

European regulatory bodies, due to high population densities around major airports. Consequently, accurate prediction and subsequently minimization of noise are recognized as an important issue in aircraft design and have garnered much research interests in the recent years.

Two major noise sources of modern aircraft are the engine and airframe noises, with the former playing the dominant role during the turbojet era. The introduction of high bypass ratio turbofan engines has led to a drastic reduction of engine noises. As a result, the airframe noise is now generally on-par with the engine noise, even surpassing it under certain operating conditions. This is particularly true during the landing phase of the flight in which the aircraft is operating at reduced thrust with the engines throttled down and the high-lift devices such as slats and flaps deployed. Note that the noise produced by the aircraft is extremely hazardous during this particular phase of the flight given its proximity to the ground population. Therefore in this work, we focus on airframe noise minimization in low speed operating conditions.

The airframe noises are generated primarily due to the interaction of unsteady turbulent flows with solid boundaries such as aerodynamic surfaces and landing gears. Due to the 3-D, unsteady and turbulent nature of noise generation, high-fidelity simulations are required to sufficiently resolve the turbulent structures for accurate predictions. In addition, to obtain physically meaningful far-field noise prediction the computational mesh must extend on the order of one hundred chord-length in at least the normal direction from the body. Other numerical challenges related to artificial dissipation and boundary conditions are discussed in a review by Colonius and Lele [2]. Consequently, high-fidelity aeroacoustic simulation remains a computational daunting task. This perhaps serves to explain partially why examples involving aeroacoustic optimizations are rare in literature while in comparison, significant progresses have been made on aerodynamic design optimization. It has been shown however, that while the aerodynamic and aeroacoustic design objectives are not mutually conflicting (i.e. a design optimized for aerodynamic efficiency typically also results in noise reduction from its baseline configuration and vise versa), the two optimizations do not lead to the same optimal designs [3, 4]. In landing situation, it is sensible to judiciously adopt a design biased towards the aeroacoustic objective for airframe noise minimization, provided that aerodynamic performance constraints are minimally satisfied. Therefore, it is by incorporating aeroacoustic considerations in the initial design stage that the aircraft industry can meet the increasingly stringent noise reduction requirements. Three predominant approaches exist for airframe noise minimization—via shape optimizations [4–6], injection of gas mixtures [7, 8], and modification of the trailing edge with porous materials [9, 10]. In this work, we adopt the last approach.

Porous media modifies the surface in such a way so as to reduce the discontinuity that the fluid convecting over a trailing-edge experiences as it transitions from a wall-bounded flow to a free-shear flow. In essence, the porous media makes the trailing-edge 'acoustically soft', thereby reducing the near-body turbulence intensity and the strength of vortex-shedding responsible for noise generation. However, applying porous media to the trailing-edge in practice is not a trivial matter, in that no clear design guidelines exist for the ideal placement of the porous media. Existing experimental investigations indicate that the spatial distribution of the porous media strongly influences the extent of noise reduction [11]. An uninformed porous trailing-edge design may even lead to an amplification of far-field noise. To that end, numerical optimization algorithms can be used to search for the *optimal* distribution of the porous material in noise reduction.

In this paper, we present a discrete adjoint-based optimization framework to obtain the optimal distribution of the porous material over the trailing edge of a 3-D flat plate. The near-body strength of the noise source generated by the unsteady turbulent flow field is computed using a high-fidelity large-eddy simulation (LES). By optimally controlling the material porosity and permeability, it is possible to minimize the turbulent intensity responsible for noise generation at the trailing edge and thus significantly reduce the radiated noise signal. The objective of this work is to demonstrate, using a simple geometry as a first step, the efficacy of the discrete adjoint method in achieving minimum-noise design via optimal distribution of porous media, with future applications to aircraft high-lift configurations.

The remainder of the paper is organized as follows. In Sect. 23.2, the LES aeroacoustic solvers for flows over a flat plate with porous trailing edge as well as the optimization framework based on discrete adjoint and automatic differentiation (AD) are presented. Results are presented in Sect. 23.3 while conclusion and future work are outlined in Sect. 23.4.

23.2 Aeroacoustic Optimization Methodology

23.2.1 Flow Through Porous Media

To simulate the flow field containing porous media, we employ the Brinkman penalization method in which the flow inside the permeable material as well as the external flow are modelled in a monolithic fashion as a continuum. Therefore, a single set of equations governs the entire computational domain containing multi-phases with fluid, solid and homogeneous porous medium, eliminating the cumbersome task to specify interface conditions directly. Developed on the basis of the Brinkman equation [12], Liu and Vasilyev [13] extended the method by modelling the influence of material permeability as penalizations to both the momentum and energy equations, as well as accounting for the effect of porosity in the continuity equation. The Navier-Stokes equations for flows through porous media can be written as follows:

$$\frac{\partial \rho}{\partial t} = -[1 + (\frac{1}{\varepsilon} - 1)\chi]\frac{\partial \rho u_j}{\partial x_j} \tag{23.1}$$

$$\frac{\partial \rho u_i}{\partial t} = -\frac{\partial}{\partial x_j}(\rho u_i u_j) - \frac{\partial p}{\partial x_i} + \frac{1}{Re_a}\frac{\partial \tau_{ij}}{\partial x_j} - \frac{\chi}{K_v}(u_i - U_{oi}) \tag{23.2}$$

$$\frac{\partial e}{\partial t} = -\frac{\partial[(e+p)u_j]}{\partial x_j} + \frac{1}{Re_a}\frac{\partial u_i \tau_{ij}}{\partial x_j} + \frac{1}{Re_a Pr(\gamma - 1)}\frac{\partial}{\partial x_j}(\mu\frac{\partial T}{\partial x_j}) - \frac{\chi}{K_t}(T - T_o) \tag{23.3}$$

where ρ is the fluid density, u_i is the fluid velocity, p is the pressure, τ_{ij} is the shear stress, μ is the dynamic viscosity, e is the total energy and T is the temperature. The two nondimensional numbers $Re_a = \frac{\rho_o c_o L}{\mu}$ and $Pr = \frac{\mu c_p}{k}$ are the acoustic Reynolds number and Prandtl number respectively. U_{oi} and T_i are the velocity and the temperature of the body respectively. χ is a step function that takes the value of 1 within the permeable media and 0 outside.

The influence of the permeable material is modeled using three parameters ε, K_v and K_t. ε is the material porosity defined as the ratio of the volume occupied by fluid to the total material volume. $\varepsilon = 1$ represents void space where only fluid and no solid material is present. On the other hand as $\varepsilon \to 0$, the material volume approaches that of a pure solid with no passages for fluid flow. Note that $\varepsilon = 0$ leads to a singularity in the continuity equation, therefore it is constrained in the range $\varepsilon \in (0, 1]$. K_v is the viscous permeability and K_t is the thermal permeability. They provide a measure of flow conductance indicating how easy it is for fluid to flow through the material at a given porosity ratio. It is clear from the momentum equations that the term involving the viscous permeability acts as a retarding force to the flow in the porous media. Analogous effect by the thermal permeability can also be seen in the energy equation. As K_v, $K_t \to 0$, the material approaches an impenetrable solid. Note that when $\varepsilon \to 1$, $K_v \to \infty$ and $K_t \to \infty$, the above equations reduce to the classical Navier-Stokes equations describing pure fluid flow.

23.2.2 LES Solvers

Two LES solvers developed by the Aachen Institute of Aerodynamics (AIA) are used in this study. The first is an unsteady incompressible solver with a dynamic subgrid scale model proposed by Germano et al. [14]. The inviscid fluxes are discretized by a high-order scheme based on the summation-by-parts operator for a sixth-order dispersion preserving relation (DRP) scheme by Johansson [15]. An explicit four-stage Runge-Kutta method is used for time integration. The second and more elaborate LES solver solves the unsteady compressible Navier-Stokes equations. It is based on a finite-volume method, in which the inviscid fluxes are spatially discretized by the AUSM scheme with the MUSCL approach and a centered approximation for the pressure term. The viscous terms are approximated by a centered discretization. For the temporal integration an explicit five-step Runge-Kutta formulation is used. The numerical details are described in Meinke et al. [16]. To eliminate undesirable wave reflections on the boundaries, a sponge layer is imposed.

23.2.3 Optimization Framework

As discussed in Sect. 23.1, near-body turbulence is the main source responsible for the far-field noise generation. It is therefore of practical importance to control the turbulence intensity via optimal distribution of permeable material in the plate trailing edge. The objective function of this study is defined as the temporally and spatially averaged normal velocity components at N_p user-defined locations over all N_t time steps:

$$J = \frac{1}{N_t N_p} \sum_j^{N_t} \sum_i^{N_p} |v_i^j| \tag{23.4}$$

J thus defined provides an indication of the level of the turbulence intensity in the near-field source region.

The design variables are the spatially-varying material properties of the surface, namely ε, K_v and K_t. The evaluation of the gradient of the objective function with respect to these design variables is based on the discrete adjoint method, making its computational cost independent of the number of design variables.

Furthermore, the implementation of the discrete adjoint formulation in this work is eased by the use of automatic differentiation (AD),[1] eliminating the error-prone hand-differentiation of the discretized equations. AD was developed based on the observation that any simulation code, regardless of its complexity is merely a sequence of elementary operations whose differentiation rules are well known. Therefore, by successive applications of the chain-rule through the computer program, it is possible to compute both the simulation output and its derivative with respect to prescribed design variables simultaneously. A remarkable feature of AD, owing to its construction, is that it does not incur any truncation errors compared to the traditional finite difference method. In particular, the derivatives are accurate to machine accuracy. The AD can be performed in the forward and reverse mode. The forward mode, albeit exact, requires one evaluation for each component of the gradient vector. In contrast, the reverse mode is capable of computing the entire gradient vector in one stroke, at the expense of high memory requirement due to the need to save all intermediate variables. A good discussion of these two AD modes can be found in [18].

A quasi-Newton optimizer in which an estimate of the inverse Hessian based on the BFGS (Broyden-Fletcher-Goldfarb-Shanno) rank-two update formula is used to compute a search direction [19]. The step size is determined using a line search, which enforces the strong Wolfe conditions [19]. To enforce physically meaningful solutions, box constraints are imposed on the design variables during the line search.

23.3 Results

23.3.1 *Minimization of Trailing Edge Vortex Shedding at Startup*

Turbulent flow over a flat plate with thickness h, a free-stream Mach number of 0.06 and a Reynolds number of 13,500 is considered in this work. Furthermore, porous material is used in the last 12 % chord ($d = 0.12c$) of the flat plate as

[1] Performed using AD tool TAPENADE [17].

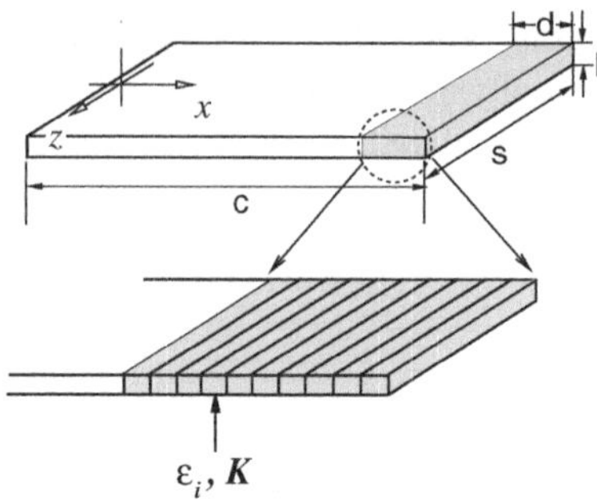

Fig. 23.1 Flat plate with porous trailing edge consisting of porous strips laid in the spanwise direction

shown on Fig. 23.1. Within this area, ten porous strips are laid along the spanwise direction. While the porosities (ε) are allowed to vary from strip to strip, they are considered uniform within each strip. The material permeability is also held constant over the entire porous domain. Along the streamwise direction, the computational mesh extends $330h$ in front of the leading-edge and $300h$ behind the trailing-edge. In the normal direction it extends $330h$ and $37h$ in the spanwise direction. A total of 2.5 million nodes are used. The computational domain is divided into 16 blocks and parallelized with the Message Passing Interface (MPI). The first, incompressible LES solver discussed in Sect. 23.2.2 is used for this test case.

The objective function is defined by Eq. 23.4 where the $N_p = 11$ user-defined observation points are equispaced along a line in the spanwise direction at a distance of 1 % chord behind the upper corner of the trailing-edge ($x = 1.01c$, $y = h/2$). This objective function gives a direct indication of the vorticity at the trailing-edge. For such blunt trailing edge, periodic vortext shedding contributes to strong tonal noises.

The simulation is started from free-stream at a step size of $\Delta t = 10^{-5}$ for $N_t = 80$ time steps. The first 30 time steps in each optimization iterate are omitted in the objective function and gradient computations to bypass the unphysical transient due to the change of design parameters. The next 50 time steps are optimized (see Fig. 23.3). The design gradient is computed using the forward and reverse modes of AD and compared with the second-order finite difference (FD) results. It can be seen on Table 23.1 that FD gradient incurs significant error after merely 100 time steps. The highlighted digits show the discrepancy of the gradient values as compared to the forward AD gradient which is known to be exact. It should be stressed that the reverse AD, aside from being more accurate than FD is also (and more importantly) far more efficient, capable of obtaining all 10 components of the gradient vector in one stroke. The second-order FD on the other hand, requires 20 unsteady evaluations. Therefore, reverse AD gradients is used to perform efficient unsteady optimization in this case.

The optimization is started with a baseline design with 10 strips of equal porosity: $\varepsilon_i = 0.1, i = 1, \ldots, 10$. (i.e. ‘nearly’ hard plate). Quasi-Newton optimization using the BFGS algorithm is performed for 10 iterations which achieves a 92 % reduction

Table 23.1 Comparison between the gradients computed using 2nd order finite difference ($\delta = 10^{-5}$), forward-mode and reverse-mode of AD, over 100 time steps

i	Finite difference	Forward AD	Reverse AD
1	1.599252978309873E-6	1.599242842229384E-6	1.599242842229349E-6
2	1.175208229342583E-5	1.175208354564064E-5	1.175208354564077E-5
3	1.612330616240776E-5	1.612332568496750E-5	1.612332568496753E-5
4	3.118079878496050E-5	3.118079812541050E-5	3.118079812541059E-5
5	5.342393276895618E-5	5.342392467928700E-5	5.342392467928717E-5
6	9.228763434877828E-5	9.228763273798646E-5	9.228763273798641E-5
7	1.940379114329094E-4	1.940379225337220E-4	1.940379225337226E-4
8	4.980055826921126E-4	4.980055913482977E-4	4.980055913482987E-4
9	1.302111566139907E-3	1.302111561659751E-3	1.302111561659750E-3
10	4.517035093404634E-3	4.517035119303520E-3	4.517035119303535E-3

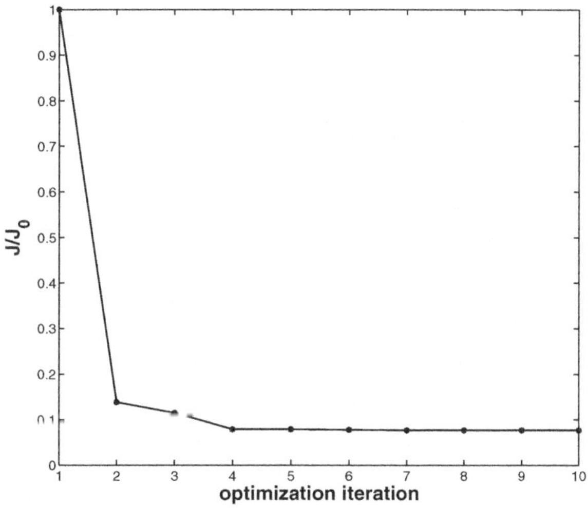

Fig. 23.2 Convergence history over 10 optimization iterations. Note that the objective function is scaled by its initial value

in objective function compared to the baseline design (see Fig. 23.2). This is also reflected on Fig. 23.3 which shows that the instantaneous turbulence intensity measure of the optimized design is drastically reduced from its baseline values over the entire optimization window.

The effect of the optimization can be visualized by comparing the normal velocity fields of the baseline and optimal designs near the trailing edge at the end of the optimization window as shown on Fig. 23.4. Note that the noise-generating trailing edge vortex pair clearly visible in the baseline design has been removed completely in the optimal design.

A comparison on the strip porosities between the baseline and optimized designs is show on Fig. 23.5. Note that the results indicate that highly porous material with as

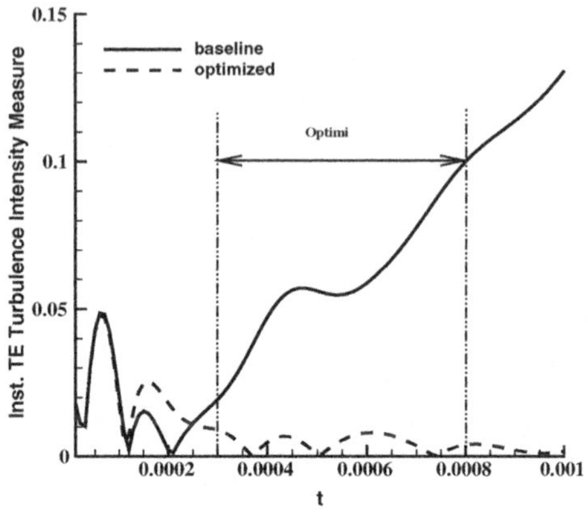

Fig. 23.3 Instantaneous trailing-edge turbulence intensity measure at each time step

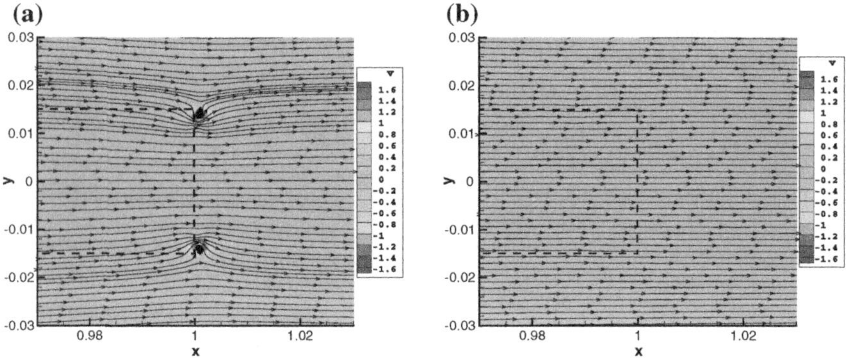

Fig. 23.4 Comparison of trailing edge normal velocity field between baseline and optimal design. Note that *dashed lines* outline the boundaries of the porous trailing edge. **a** Baseline. **b** Optimized

much as 85 % porosity should be used in the immediate vicinity of the trailing-edge while the porosities of the 3 strips further upstream largely remain unchanged from the baseline design. This promotes a 'smoother' transition from the wall-bounded flow to free-shear flow and hence reducing the noise generation. Although the result appears to be intuitively obvious, it serves as a validation for our optimization methodology.

Since the incompressible flow solution is used, the acoustic pressure fluctuation cannot be distinguished from the hydrodynamic pressure. Therefore we defer the computations of the acoustic pressure fluctuation and the sound pressure level to the next test case which employes a compressible LES solver.

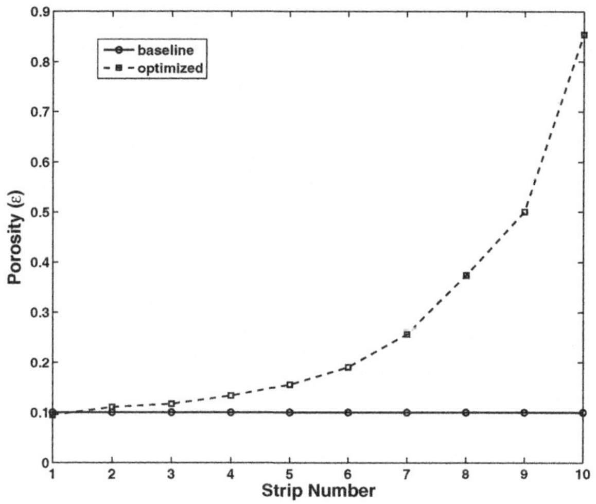

Fig. 23.5 Comparison of strip porosities between baseline and optimal design

23.3.2 Minimization of Trailing Edge Turbulence Intensity

Having validated our methodology using a simple test case in the previous section, we now proceed to a noise minimization problem over a longer simulation time. The same geometry and flow conditions discussed in Sect. 23.3.1 are used. Within the porous domain, five porous strips are laid along the spanwise direction. In this study we fix the porosity (ε) and thermal permeability (K_t) and allow the viscous permeability (K_v) to vary from strip to strip. A total of 3.5 million grid points are used. The computational domain is divided into 32 blocks and parallelized with the Message Passing Interface (MPI).

The objective function is defined by Eq. 23.4 where the $N_p = 5$ measurement points are equi-spaced at a height h along the mid-span over the last 12 % chord where the porous material is used. The aim is to enhance the overall noise performance via the optimization of a near-body objective function—the turbulence intensity measure over the modified porous trailing edge.

In each optimization cycle, 1,000 time steps of the flow solution are computed at a constant CFL number of $CFL = 1$. In this case, a prohibitively large memory overhead arises in the reverse mode AD due to its need to store all the many millions of intermediate solutions and transitional variables over such large number of time steps. This challenge will be addressed in our future work by employing more advanced techniques such as reverse accumulation and checkpointing. Instead, in this work we opt to use the forward AD for the gradient computations which still affords us the exact derivatives. The reduction in computational efficiency from the one-stroke reverse AD is negligible since only 5 design variables are used. A comparison between the forward AD gradient and finite difference gradient are shown on Table. 23.2.

Table 23.2 Comparison between the gradients computed using 2nd order finite difference ($\delta = 10^{-6}$) and forward-mode of AD, over 100 time steps

i	Finite difference	Forward AD
1	1.167141014946083	1.167141550726590
2	0.979045005067292	0.979047994868520
3	−0.037741010316950	−0.037741196023650
4	−0.172801009057366	−0.172800017556677
5	2.172569004699199	2.172574472509500

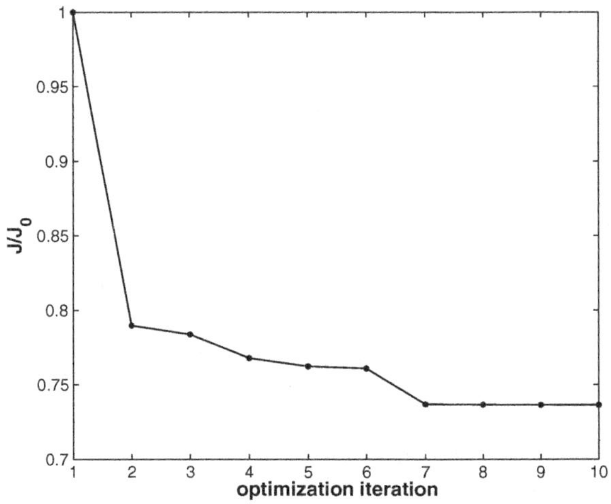

Fig. 23.6 Convergence history over 10 optimization iterations. Note that the objective function is scaled by its initial value

The optimization is started with a baseline design with 5 strips of equal viscous permeability: $K_v^i = 0.1, i = 1, \ldots, 5$ as control variables while the porosity and thermal permeability are held constant throughout the optimization at $\varepsilon = 0.2$ and $K_t = 0.05$ respectively. The BFGS algorithm is performed for 10 iterations which achieves a 27 % reduction in trailing edge noise measure compared to the baseline design (see Fig. 23.6). This is also clearly seen on Fig. 23.7 that the instantaneous trailing edge turbulence intensity measure of the optimized design is drastically reduced from its baseline values at almost every time step of the simulation horizon.

Figure 23.8 compares the vorticity fields between the baseline and optimized design at a later time $t = 4\frac{c}{U_\infty}$. The baseline design exhibits a strong trailing edge roller with a strong spanwise correlation. Since the plate thickness is much larger than the boundary layer thickness in this test case, such roller is known to generate very strong tonal noise. In contrast, the coherent turbulent structure in the wake region is broken up in the optimized design, indicating a weaker noise source. This is clearly demonstrated by Fig. 23.9 which shows that the pressure fluctuation is noticeably reduced in the optimal design over the entire domain. Consequently, the overall sound pressure level (OASPL) computed at a radius $R = 1.5c$ around the

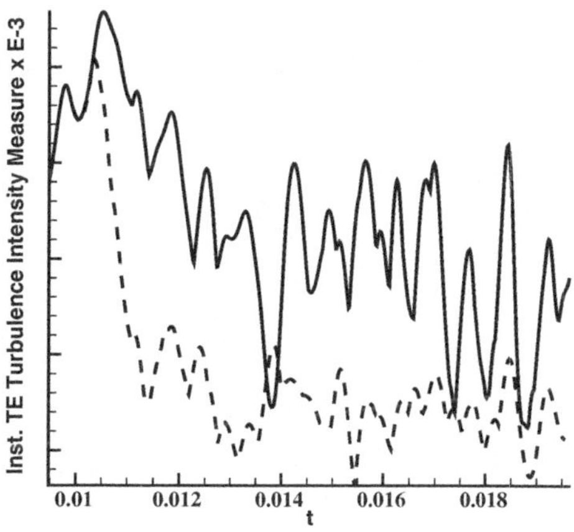

Fig. 23.7 Instantaneous trailing-edge turbulence intensity measure at each time step

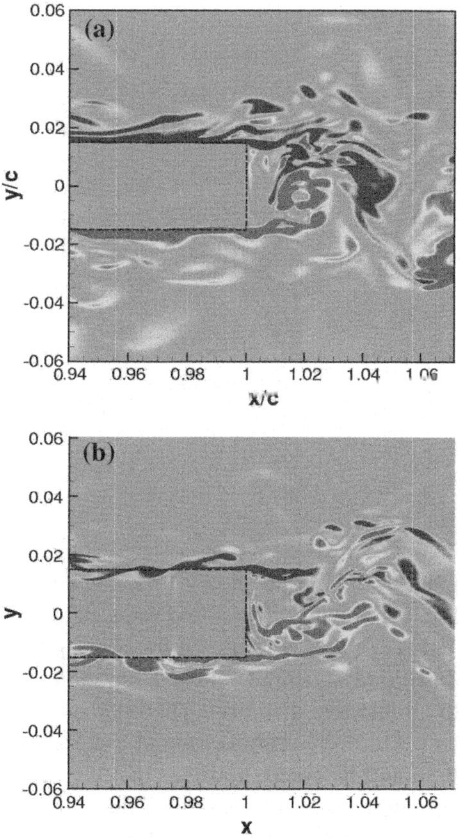

Fig. 23.8 Vorticity field of the baseline (**a**) and optimized design (**b**) at $t = 4\frac{c}{U_\infty}$

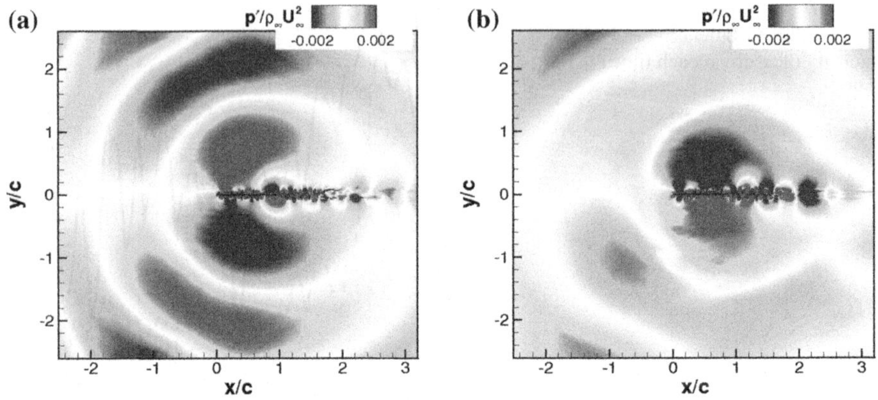

Fig. 23.9 Pressure fluctuation field of the baseline (**a**) and optimized design (**b**) at $t = 4\frac{c}{U_\infty}$

Fig. 23.10 Overall sound pressure level of three designs

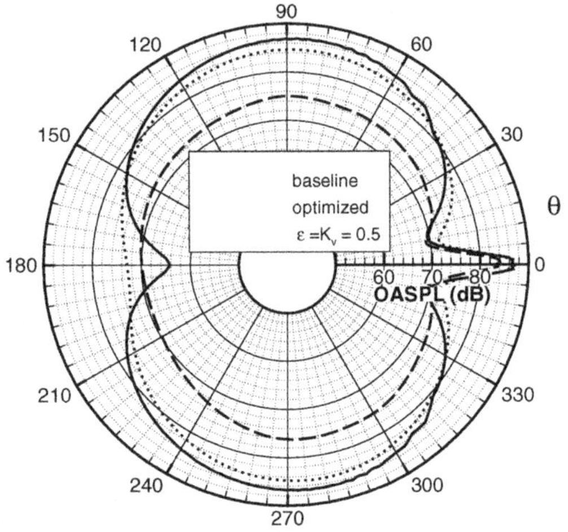

plate trailing edge shows a 2 dB reduction in the transverse direction, as seen on Fig. 23.10. Note that sound pressure level of a third design with uniform $\varepsilon = 0.5$ and $K_v = 0.5$ over the entire porous domain is also shown on the same figure (in dash) and appears to have a far more impressive noise reduction of 10–12 dB in the transverse direction. This indicates that one may obtain more superior designs by performing optimization at a higher ε value or even controlling both ε and K_v. It should be noted, however that the degradation of aerodynamic performance and structural integrity at higher values of porosity must be also be carefully taken into account for practical high-lift applications envisaged beyond current work.

A remarkable feature of the optimized design, as shown on Fig. 23.11 is that except for the 4-th strip, all other strips have their respective K_v reduced significantly

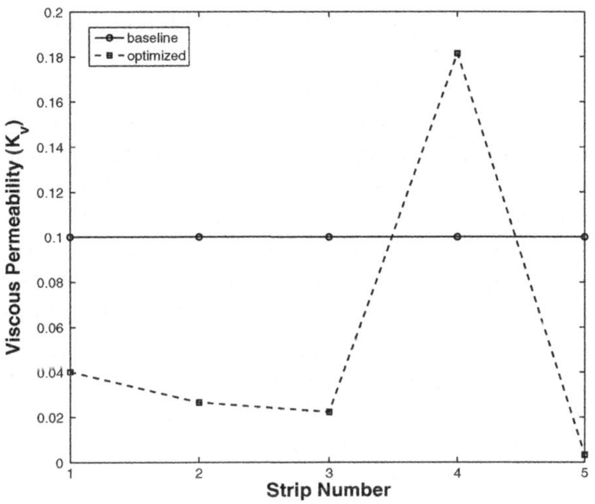

Fig. 23.11 Comparison of strip viscous permeabilities between baseline and optimal design

below the baseline value of 0.1. In particular, the K_v of the last strip of the trailing edge is decreased to almost zero. In fact, a box constraint is imposed so that $K_v^i > 0.003, i = 1, \ldots, 5$ to prevent singularity in the flow solution. The optimizer pushes K_v^5 right up to this bound over the 10 design iterations. This indicates that it is advantageous from the noise-reduction standpoint to use non-permeable material at the very end of the trailing edge. From the physical point of view, it is unclear *a priori* why such distribution would be favorable—one would assume that the optimal distribution would adopt a trend as exhibited in the porosity case in the previous test case (see Fig. 23.5) which increase towards trailing edge where wall-bounded flow transitions into free-shear flow. This serves to highlight the power of combining high-fidelity flow simulations with numerical optimization algorithms—non-intuitive and unconventional designs may be explored, particularly in areas such as this where no clear design guidelines exist.

23.4 Conclusion

We present in this paper the development of a discrete adjoint-based optimization framework to obtain the optimal distribution of the porous material over the trailing edge of a flat plate. The near-body strength of the noise source generated by the unsteady turbulent flow field is computed using a high-fidelity LES solver. By optimally controlling the material porosity and permeability, it is possible to minimize the turbulence intensity responsible for noise generation at the trailing edge and therefore significantly reduce the radiated noise. In this work, exact design gradients are computed by applying AD to the LES code. The results suggest that the proposed framework is effective in achieving minimum-noise designs via optimal distribution of porous media.

To the best of our knowledge, this is the first work in the direction of noise-minimization that couples the discrete adjoint approach with a high-fidelity three-dimensional unsteady LES solver. This new venture presents unique challenges. Ideally, to ensure optimal performance over long times, one must perform the optimization over a few cycles of fundamental frequency of the trailing edge noise. However, due to the low speed nature of the problem and the plate thickness, such frequency is very low which calls for many thousands of time steps just to capture one cycle. For a high-fidelity setup with millions of grid nodes explored in current work, this presents an impossible task for the reverse mode AD due to the memory overhead and can only be accomplished using advanced AD techniques such as reverse accumulation and checkpointing, which we intend to implement next.

Furthermore, we intend to perform far-field noise minimization at observation locations many tens of chord lengths away from the plate. This will involve applying our optimization framework to an efficient hybrid noise prediction approach developed by Ewert and Schröder [20, 21] in which the near-body turbulent flow field is computed using a high-fidelity LES solver to obtain the noise source which is then propagated to the far-field using the acoustic perturbation equations (APE).

Finally, from the applications standpoint, as the turbulence responsible for the noise generation is typically 3-D by nature, we plan to apply porous tiles in both streamwise and spanwise directions with the aim of further reducing the far-field noise from the current porous strips.

Acknowledgments The funding of all authors by the German Research Foundation (DFG) is gratefully acknowledged. In addition, the first author would like to acknowledge the partial funding by the Natural Science and Engineering Research Council of Canada.

References

1. Federal Aviation Administration (2012) FAA Aerospace Forecast
2. Colonius T, Lele SK (2004) Computational aeroacoustics: progress on nonlienar problems of sound generation. Prog Aerosp Sci 40:345–416
3. Jones BR, Crossley WA, Lyrintzis AS (2000) Aerodynamic and aeroacoustic optimization of rotorcraft airfoils via a parallel genetic algorithm. J Aircr 37(6):1088–1096
4. Rumpfkeil MP, Zingg DW (2010) A hybrid algorithm for far-field noise minimization. Comput Fluids 39(9):1516–1528
5. Economon TD, Palacios F, Alonso JJ (2012) A coupled-adjoint method for aerodynamic and aeroacoustic optimization. AIAA paper 2012-5598
6. Marsden AL, Wang M, Dennis JE, Moin P (2007) Trailing-edge noise reduction using derivative-free optimization and large-eddy simulation. J Fluid Mech 572:13–36
7. Koh SR, Schröder W, Meinke M (2011) Airframe-noise reduction by suppressing near-wall turbulent structures. AIAA paper 2011-2904
8. Koh SR, Schröder W, Meinke M (2009) Sound generation control by fluid bleeding. AIAA paper 2009-3225

9. Choudhari M, Khorrami MR (2003) Computational study of porous treatments for altering flap side-edge flow field. AIAA paper 2003-3113
10. Schulze J, Sesterhenn J (2013) Optimal distribution of porous media to reduce trailing edge noise. Comput Fluids 78(20):41–53
11. Geyer T, Sarradj E, Fritzsche C (2010) Measurement of the noise generation at the trailing edge of porous airfoils. Exp Fluids 48:291–308
12. Brinkman HC (1949) A calculation of the viscous force exerted by a flowing fluid on a dense swarm of particles. Appl Sci Res 1(1):27–34
13. Liu Q, Vasilyev O (2007) A Brinkman penalization method for compressible flows in complex geometries. J Comput Phys 227:946–966
14. Germano M, Piomelli U, Moin P, Cabot WH (1991) A dynamic subgrid-scale eddy viscosity model. Phys Fluids 3(7):1760–1765
15. Johansson S (2004) High order finite difference operators with the summation by part property based on DRP schemes. Technical report 2004-035, Uppsala University
16. Meinke M, Schröder W, Krause E, Rister T (2002) A comparison of second- and sixth-order methods for large-eddy simulations. J Comput Phys 31:695–718
17. Hascoet L, Pascual V (2004) Tapenade 2.1 users guide. Technical report 0300, INRIA
18. Nemili A, Özkaya E, Gauger N, Carnarius A, ThieleF (2011) Optimal control of unsteady flows using discrete adjoints. AIAA paper 2011-3720
19. Nocedal J, Wright S (2006) Numerical optimization, 2nd edn. Springer, New York
20. Ewert R, Schröder W (2003) Acoustic perturbation equations based on flow decomposition via source filtering. J Comput Phys 188:365–398
21. Ewert R, Schröder W (2004) On the simulation of trailing edge noise with a hybrid LES/APE method. J Sound Vib 270:509–524

Part V
Engineering Design and Societal Applications: Aeronautics and Astronautics

Chapter 24
Conceptual Design of Single-Stage Launch Vehicle with Hybrid Rocket Engine Using Design Informatics

Kazuhisa Chiba, Masahiro Kanazaki, Koki Kitagawa and Toru Shimada

Abstract A single-stage launch vehicle with hybrid rocket engine has been conceptually designed by using design informatics, which has three points of view as problem definition, optimization, and data mining. The primary objective of the design in the present study is that the sufficient down range and the duration time in the lower thermosphere are achieved for aurora scientific observation whereas the initial gross weight is held down. Multidisciplinary design optimization and data mining were performed by using evolutionary hybrid computation under the conditions that polypropylene as solid fuel and liquid oxygen as liquid oxidizer were adopted and that single-time ignition is implemented in sequence. Consequently, the design information regarding the tradeoffs and the behaviors of the design variables in the design space was obtained in order to quantitatively differentiate the advantage of hybrid rocket engine.

Keywords Design informatics · Evolutionary optimization · Data mining · Single-stage launch vehicle for scientific observation · Hybrid rocket engine using solid fuel and liquid oxidizer

K. Chiba (✉)
Hokkaido University of Science, 7-15-4-1, Maeda, Teine, Sapporo 006-8585, Japan
e-mail: kazchiba@hus.ac.jp

M. Kanazaki
Tokyo Metropolitan University, 6-6, Asahigaoka, Hino, Tokyo 191-0065, Japan
e-mail: kana@tmu.ac.jp

K. Kitagawa · T. Shimada
Institute of Space and Astronautical Science, Japan Aerospace Exploration Agency, 3-1-1, Yoshinodai, Chuo, Sagamihara 252-5210, Japan
e-mail: kitagawa.koki@jaxa.jp

T. Shimada
e-mail: shimada.toru@jaxa.jp

D. Greiner et al. (eds.), *Advances in Evolutionary and Deterministic Methods for Design, Optimization and Control in Engineering and Sciences*, Computational Methods in Applied Sciences 36, DOI 10.1007/978-3-319-11541-2_24

24.1 Introduction

Single-stage rockets have been being researched and developed for the scientific observations and the experiments of high-altitude zero-gravity condition, whereas multi-stage rockets have been being also studied for the orbit injection of payload. The Institute of Space and Astronautical Science (ISAS), Japan Aerospace Exploration Agency (JAXA) has been operating K, L, and M series rockets as the representatives of solid rocket in order to contribute to the space scientific research. A next-generation single-stage rocket as well as multi-stage rocket is necessary due to the retirement of M-V in 2008 and in order to promote space scientific research. In fact, E rocket will begin to be operated from August 2013. On the other hand, the launch vehicle with hybrid rocket engine using solid fuel and liquid oxidizer has been being researched and developed as an innovative technology in mainly Europe and United States [12, 19]. The present study will investigate the conceptual design in order to develop a next-generation single-stage launch vehicle with hybrid rocket engine. Since the technologies of hybrid rocket engine for single-stage and multi-stage are not independent, the solution of the fundamental physics regarding single-stage hybrid rocket is also the knowledge for multi-stage. Although three-stage hybrid rocket is parallel studied, design requirements are different because of the difference of the operation objectives. A hybrid rocket offers the several advantages as higher safety, lower cost, and environmental friendly. In fact, the SpaceShipOne successfully uses a hybrid rocket engine for a private manned space flight.[1] The multi-time ignition is the especial advantage of hybrid rocket engine [17]. On the other hand, the disadvantage of a hybrid rocket engine is in its combustion. As a hybrid rocket engine has low regression rate of solid fuel, the thrust of hybrid rocket engine is less than that of pure solid and pure liquid engines. Multidisciplinary design requirements should be considered in order to surmount the disadvantage of hybrid rocket engine. Moreover, design information will be obtained in order to exhaustively grasp the design space.

Design informatics is essential for practical design problems. Although solving design optimization problems is important under the consideration of many disciplines of engineering [1], the most significant part of the process is the extraction of useful knowledge of the design space from results of optimization runs. The results produced by multiobjective optimization (MOO) are not an individual optimal solution but rather an entire set of optimal solutions due to tradeoffs. That is, the result of a MOO is not sufficient from the practical point of view as designers need a conclusive shape and not the entire selection of possible optimal shapes. On the other hand, this set of optimal solutions produced by an evolutionary MOO algorithm can be considered a hypothetical design database for design space. Then, data mining techniques can be applied to this hypothetical database in order to acquire not only useful design knowledge but also the structurization and visualization of design space for the conception support of basic design. This approach was suggested as

[1] "Scaled Composites, LLC" available online at http://www.scaled.com/projects/tierone/ [retrieved 24 May 2013].

the design informatics [5]. The goal of this approach is the conception support for designers in order to materialize innovation. This methodology is constructed by the three essences as (1) problem definition, (2) efficient optimization, and (3) structurization and visualization of design space by data mining. A design problem including objective function, design variable, and constraint, is strictly defined in view of the background physics (it takes several months because this process determines the quality of the design space), then optimization is implemented in order to acquire nondominated solutions (quasi-Pareto solutions) as hypothetical database. Data mining is performed for this database in order to obtain design information. Mining has the role of a postprocess for optimization. Mining result is the significant observations for next design phase and also becomes the material to redefine a design problem.

In the present study, a single-stage launch vehicle with hybrid rocket engine of solid fuel and liquid oxidizer for the scientific observation of aurora will be conceptually designed by using design informatics approach in order to quantitatively reveal the ascendancy and in order to discover the fundamental physics regarding hybrid rocket engine. As a first step, an optimization problem on single-time ignition, which is the identical condition of the current solid rocket, is defined under the present studying constructions, and then the design information, which is the correlation among objective functions (design requirements) and the influence of design variables, is obtained. As a second step, the implication of solid fuels in performance of hybrid rocket will be revealed because the regression rate is one of the key elements for the performance of hybrid rocket. The validity of the problem definition will be considered by using the design information from the two-step results. Finally, the sequence using multi-time ignition, which is the great advantage of a hybrid rocket, will be investigated in order to reveal the ascendancy of hybrid rocket and also practically contribute to space science. The standing point of the present research is on the first step as the milestone to observe the quantitative difference of performance between conventional solid rocket and the present hybrid rocket. Moreover, the present research investigates the role as the reference result to quantitatively show the ascendancy of multi-time ignition.

24.2 Design Informatics

Design informatics after the definition of detailed problem is constructed by two phases as optimization and data mining. Evolutionary computation is used for optimization. Although a surrogate model [11] like as the Kriging model [16], which is a response surface model developed in the field of spatial statistics and geostatistics, can be employed as optimization method, it will not be selected because it is difficult to deal with a large number of design variables. In addition, since the designers require to present many exact optimum solutions for the decision of a compromise one, an evolutionary-based Pareto approach as an efficient multi-thread algorithm, which the plural individuals are parallel conducted, is employed instead of gradient-based methods. The optimizer used in the present study is the evolutionary hybrid method between the differential evolution (DE) and the genetic algorithm (GA) [4]. Moreover, global design information is primarily essential in order to determine a compromise solution. Therefore, a self-organizing map (SOM) [13] is used as a data mining technique in the present study because SOM extracts the global information in design space [6].

24.2.1 Hybrid Optimization Method

The view of hybridization is inspired by the evolutionary developmental biology [2]. When there is the evolution which the Darwinism cannot explain in the identical species, each individual might have a different evolutionary methodology. When the practical evolution is imitated for the evolutionary computation, the different evolutionary algorithms might ultimately be applied to each individual in population. The making performance of next generation for each methodology depends on not only their algorithms but also the quality of candidate of parent in the archive of non-dominated solutions. The present hybridization is intended to improve the quality of candidate of parent by sharing the nondominated solutions in the archive among each methodology. In the present study, the evolutionary hybrid optimization methodology between DE and GA is employed. It was confirmed that this methodology had the high performance regarding the convergence and diversity, as well as the strength for noise [4]. Note that noise imitates the error on computational analyses and experiments and is described as the perturbation on objective functions. It is an important factor when the optimization for practical engineering problem is considered.

The flowchart of the present hybrid methodology is shown in Fig. 24.1. First, multiple individuals are generated randomly as an initial population. Then, objective functions are evaluated for each individual. The population size is equally divided into

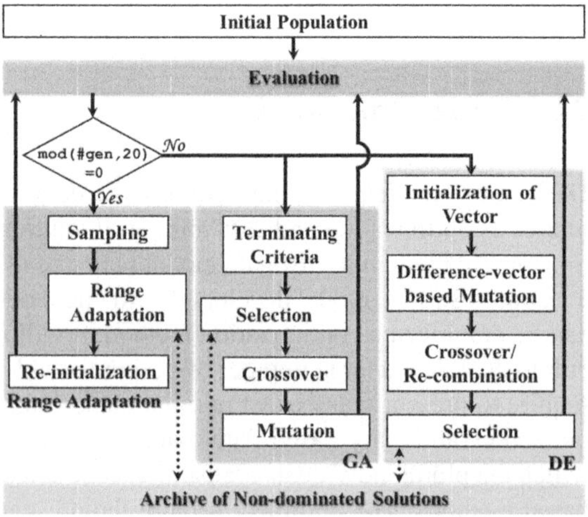

Fig. 24.1 Conceptual flowchart of the present evolutionary hybrid methodology

sub-populations between DE and GA (although sub-population size can be changed at every generations on the optimizer, the determined initial sub-populations are fixed at all generations in the present study). New individuals generated by each operation are combined in next generation. The nondominated solutions in the combined population are archived in common. It is notable that only the archive data is in common between DE and GA. The respective optimization methods are independently performed in the present hybrid methodology.

24.2.2 Configurations of Operators for Optimizer

The present optimization methodology is a real-coded optimizer [14]. Although GA is based on the real-coded NSGA-II (the elitist nondominated sorting genetic algorithm) [7], it is made several improvements on in order to be progressed with the diversity of solutions. Fonseca's Pareto ranking [8] and the crowding distance [7] are used for the fitness value of each individual. The stochastic universal sampling [3] is employed for parents selection. The crossover rate is 100 %. The principal component analysis blended crossover-α (PCABLX) [20] and the confidence interval based crossover using L_2 norm (CIX) [10] are used because of the high performance for the convergence and the diversity as well as the strength for noise [4]. The subpopulation size served by GA is equally divided for these two crossovers. The mutation rate is set to be constant as the reciprocal of the number of design variables. For alternation of generations, the Best-N selection [7] is used. DE is used as the revised scheme [15] for multiobjective optimization from DE/rand/1/bin scheme. The scaling factor F is set to be 0.5. The present optimizer has the function of range adaptation [18], which changes the search region according to the statistics of better solutions, for all design variables. In the present study, the range adaptation is implemented at every 20th generations.

24.2.3 Data-Mining Technique

In the present study, SOM is selected as a data-mining technique because the primary objective of data mining is the acquisition of global design information in order to implement the structuring of design space. The previous study [6] indicated that SOM extracted the global design information for whole design space. The distinguishing feature of SOM is the generation of a qualitative description. The advantage of this method includes the intuitive visualization of two-dimensional colored maps of design space using bird's-eye-views. As a result, SOM reveals the tradeoffs among objective functions. Moreover, SOM addresses the effective design variables and also reveals how a specific design variable gives effects on objective functions and other design characteristics. One SOM is colored for one variable of objective function, design variable, and other characteristic value so that the coloring pattern is compared each other. In the present study, SOMs are generated by using commercial software

Viscovery® SOMine 4.0 plus produced by Eudaptics, GmbH.[2] The uniqueness of the map generated by SOMine is assured due to Kohonen's Batch SOM algorithm and search of the best-matching unit for all input data and adjustment of weight vector near the best-matching unit.

24.3 Problem Definition

The conceptual design for a single-stage hybrid rocket, simply composed of a payload chamber, an oxidizer tank, a combustion chamber, and a nozzle is considered in the present study. A single-stage hybrid rocket for aurora scientific observation will be focused because the rocket for more efficient scientific observation is desired for successfully obtaining new scientific knowledge on the aurora observation by ISAS in 2009. In addition, a single-stage hybrid rocket problem fits for the resolution of the fundamental physics regarding hybrid rocket engine and for the improvement of the present design problem due to its simplification.

24.3.1 Objective Functions

Three objective functions are defined in the present study. First objective is the maximization of the down range in the lower thermosphere (altitude of 90 to 150 km) R_d (km) (obj1). Second is the maximization of the duration time in the lower thermosphere T_d (s) (obj2). It recently turns out that atmosphere has furious and intricate motion in the lower thermosphere due to the energy injection, which leads aurora, from high altitude. The view of these objective functions is to secure the horizontal distance and time for the competent observation of atmospheric temperature and the wind for the elucidation of atmospheric dynamics and the balance of thermal energy. Note that the priority between the down range and the duration time depends on scientific mission requirement. The present study investigates the correlation between them on the definition of the exhaustive design space because of no concrete mission requirement. Third objective is the minimization of the initial gross weight of launch vehicle $M_{\mathrm{tot}}(0)$ (kg) (obj3), which is generally the primary proposition for space transportation system.

24.3.2 Design Variables

Seven design variables are used as the initial mass flow of oxidizer $\dot{m}_{\mathrm{oxi}}(0)$ (kg/s) (dv1), fuel length L_{fuel} (m) (dv2), the initial radius of port $r_{\mathrm{port}}(0)$ (m) (dv3), combustion time t_{burn} (s) (dv4), the initial pressure in combustion chamber

[2] "Eudaptics" available online at http://www.eudaptics.com [retrieved 19 October 2012].

Table 24.1 Limitation of upper/lower values of each design variable

Serial number	Design variable	Design space		
dv1	Initial mass flow of oxidizer	1.0≤	$\dot{m}_{\text{oxi}}(0)$ (kg/s)	≤30.0
dv2	Fuel length	1.0≤	L_{fuel} (m)	≤10.0
dv3	Initial radius of port	0.01≤	$r_{\text{port}}(0)$ (m)	≤0.30
dv4	Combustion time	10.0≤	t_{burn} (s)	≤40.0
dv5	Initial pressure in combustion chamber	3.0≤	$P_{\text{cc}}(0)$ (MPa)	≤6.0
dv6	Aperture ratio of nozzle	5.0≤	ε (−)	≤8.0
dv7	Elevation at launch time	50.0≤	$\phi(0)$ (deg)	≤90.0

$P_{\text{cc}}(0)$ (MPa) (dv5), the aperture ratio of nozzle ε (−) (dv6), and the elevation at launch time ϕ (deg) (dv7). Note that there is no constraint except the limitations of upper/lower values of each design variable summarized in Table 24.1. These upper/lower values are exhaustively covering the region of the design space which is physically admitted. When there is a sweet spot (the region that all objective functions proceed optimum directions) in the objective-function space, the exploration space would intentionally become narrow due to the operation of the range adaptation on the evolutionary computation.

24.3.3 Evaluation Method

First of all, the mixture ratio between liquid oxidizer and solid fuel $O/F(t)$ is computed by the following equation.

$$\begin{aligned} \frac{O}{F}(t) &= \frac{\dot{m}_{\text{oxi}}(t)}{\dot{m}_{\text{fuel}}(t)} \\ \dot{m}_{\text{fuel}}(t) &= 2\pi r_{\text{port}}(t) L_{\text{fuel}} \rho_{\text{fuel}} \dot{\bar{r}}_{\text{port}}(t) \\ r_{\text{port}}(t) &= r_{\text{port}}(0) + \int \dot{r}_{\text{port}}(t)\, dt \end{aligned} \tag{24.1}$$

$\dot{m}_{\text{oxi}}(t)$ and $\dot{m}_{\text{fuel}}(t)$ are the mass flow of oxidizer (kg/s) and the mass flow of fuel (kg/s) at time t, respectively. $r_{\text{port}}(t)$ is the radius of port (m) at t, L_{fuel} describes fuel length, and ρ_{fuel} is the density of fuel (kg/m^3). $\dot{r}_{\text{port}}(t)$ describes the regression rate of fuel. After that, an analysis of chemical equilibrium is performed by using NASA-CEA (chemical equilibrium with applications)[3] [9], then trajectory, thrust, aerodynamic, and structural analyses are respectively implemented. The present rocket is assumed as a point mass. As the time step is set to be 0.5 s in the present study, it takes roughly 10 s for the evaluation of an individual using a general desktop computer. The contents of each analysis are briefly summarized as follows.

[3] "Chemical Equilibrium with Applications" available online at http://www.grc.nasa.gov/www/ceaweb/ceahome.htm [retrieved 8 November 2012].

24.3.3.1 Trajectory/Thrust Analysis

The following equation of motion, which ignores the influence of atmosphere, described by using $T(t)$ (N) and drag $D(t)$ (N) is computed for rocket motion.

$$M_{\text{tot}}(t)(a(t) - g) = (T(t) - D(t)) \sin\phi(t) \tag{24.2}$$

$M_{\text{tot}}(t)$ is the gross weight (kg) at t, $a(t)$ describes acceleration (m/s^2) at t, and g is gravity (m/s^2). $T(t)$ is evaluated by using the following equation.

$$T(t) = \eta_T \left\{\lambda \dot{m}_{\text{prop}}(t) \cdot u_e + (P_e - P_a) \cdot A_e\right\} \tag{24.3}$$

where, η_T is total thrust loss coefficient, λ is momentum loss coefficient at nozzle exit by friction, $\dot{m}_{\text{prop}}(t)$ is the mass flow of propellant, u_e is the velocity at nozzle exit, P_e is the pressure at nozzle exit, P_a is the pressure of atmosphere at flight altitude, and A_e describes the area of nozzle exit. $\phi(t)$ describes the Euler angle at t, that is, the inclination against the ground. $\phi(0)$ describes the elevation at launch time.

$$\dot{m}_{\text{prop}}(t) = -(\dot{m}_{\text{oxi}}(t) + \dot{m}_{\text{fuel}}(t)) \tag{24.4}$$

A combustion chamber is filled with solid fuel with a single port at the center to supply oxidizer. As the regression rate to the radial direction of the fuel $\dot{r}_{\text{port}}(t)$ (m/s) generally governs the thrust power of hybrid rocket engine, it is a significant parameter. The following empirical model [21] is used in the present study.

$$\begin{aligned}\dot{r}_{\text{port}}(t) &= a_{\text{fuel}} \times G_{\text{oxi}}^{n_{\text{fuel}}}(t) \\ &= a_{\text{fuel}} \times \left(\frac{\dot{m}_{\text{oxi}}(t)}{\pi r_{\text{port}}^2(t)}\right)^{n_{\text{fuel}}}\end{aligned} \tag{24.5}$$

where, $G_{\text{oxi}}(t)$ is oxidizer mass flux [kg/(m^2 s)] and it has upper limitation of 1,000 in the present study. a_{fuel} (m/s) and n_{fuel} (−) are the constant values empirically determined by fuels. In the present study, liquid oxygen as liquid oxidizer and polypropylene as thermoplastic resin for solid fuel in order to adopt swirling flow for the supply mode of oxidizer. Therefore, a_{fuel} and n_{fuel} are respectively set to be 8.26×10^{-5} m/s and 0.5500.

24.3.3.2 Structural Analysis

Body is divided into the components as combustion chamber, oxidizer tank, and nozzle in order to decide weight and shape. First, total length L_{tot} is defined by using the length of combustion chamber L_{cc}, the length of oxidizer tank L_{oxi}, and the length of nozzle L_{noz} as follows;

$$L_{\text{tot}} = 1.5 \times (L_{\text{cc}} + L_{\text{oxi}} + L_{\text{noz}}) \tag{24.6}$$

It is assumed that the outside radius of fuel r_{fuel} is equal to the inside radius of combustion chamber. The outside radius of rocket R_{tot} is also defined as the outside radius of oxidizer tank by using the radius of fuel r_{fuel} and the thickness of oxidizer tank t_{oxi}.

$$R_{\mathrm{tot}} = r_{\mathrm{fuel}} + t_{\mathrm{oxi}} \tag{24.7}$$

where, oxidizer tank and combustion chamber are assumed as thin cylindrical/spherical structure. The thickness of oxidizer tank t_{oxi} is defined as the following equation.

$$t_{\mathrm{oxi}} = f_s \cdot \frac{P_{\mathrm{oxi}} \cdot r_{\mathrm{fuel}}}{\sigma_{\mathrm{oxi}}} \tag{24.8}$$

f_s is the safety factor (in the present study, the constant value of 1.25 is set), P_{oxi} is the internal pressure of oxidizer tank, σ_{oxi} is the allowable stress for oxidizer tank. The internal pressure of combustion chamber $P_{\mathrm{cc}}(t)$ is described as the following equation.

$$P_{\mathrm{cc}}(t) = \frac{(\dot{m}_{\mathrm{oxi}}(t) + \dot{m}_{\mathrm{fuel}}(t)) \cdot c^*(t)}{A_{\mathrm{throat}}} \tag{24.9}$$

where, $c^*(t)$ and A_{throat} respectively represents characteristic exhaust velocity and the area at nozzle throat. $c^*(t)$ is estimated by NASA-CEA. Note that there is no constraint regarding the structural requirements for strength and vibration due to the simplification of the present problem. Initial gross weight $M_{\mathrm{tot}}(0)$ is evaluated by the following equation.

$$\begin{aligned} M_{\mathrm{tot}}(0) &= \frac{M_{\mathrm{prop}}(0)}{0.65} + M_{\mathrm{pay}} \\ &= \frac{M_{\mathrm{oxi}}(0) + M_{\mathrm{fuel}}(0)}{0.65} + M_{\mathrm{pay}} \\ M_{\mathrm{oxi}}(0) &= \int_0^{t_{\mathrm{burn}}} \dot{m}_{\mathrm{oxi}}(t)\,dt \\ M_{\mathrm{fuel}}(0) &= \int_0^{t_{\mathrm{burn}}} \dot{m}_{\mathrm{fuel}}(t)\,dt \end{aligned} \tag{24.10}$$

$M_{\mathrm{prop}}(0)$, $M_{\mathrm{oxi}}(0)$, and $M_{\mathrm{fuel}}(0)$ are the mass of propellant, the mass of oxidizer, and the mass of fuel, respectively. M_{pay} describes the mass of payload. The present M_{pay} is the constant value of 40 kg. The constant value of 0.65 in Eq. (24.9) represents that the mass of propellant assumes 65 % of the initial gross weight $M_{\mathrm{tot}}(0)$. Total weight is defined as the summation of all components. The weight of each component is calculated by the product of volume and density.

24.3.3.3 Aerodynamic Analysis

$D(t)$ is described by using pressure drag $D_p(t)$ and friction drag $D_f(t)$, which are respectively estimated by using the flight data of S-520 as the solid rocket in ISAS.

$$\begin{aligned} D(t) &= D_p(t) + D_f(t) \\ D_p(t) &= \frac{1}{2}\rho V^2 S_{\text{ref}} C_{D_p}^{(\text{S}-520)} \\ D_f(t) &= \frac{1}{2}\rho V^2 S_{\text{tot}} C_{D_f} \end{aligned} \tag{24.11}$$

where, S_{ref} is reference area and S_{tot} is total surface area. Pressure drag coefficient C_{D_p} and friction drag coefficient C_{D_f} are calculated as follows;

$$\begin{aligned} C_{D_p}^{(\text{S}-520)} &= C_D^{(\text{S}-520)} - C_{D_f}^{(\text{S}-520)} \cdot \frac{S_{\text{tot}}^{(\text{S}-520)}}{S_{\text{ref}}^{(\text{S}-520)}} \\ C_{D_f}^{(\text{S}-520)} &= \frac{0.455}{(\log_{10} Re)^{2.58}} \cdot \frac{1}{(1+0.144M^2)^{0.655}} \\ Re &= \frac{V L_{\text{tot}}^{(\text{S}-520)}}{\nu} \\ M &= \frac{V}{\sqrt{\gamma R T}} \end{aligned} \tag{24.12}$$

Re, M, and V respectively describe Reynolds number, Mach number, and velocity. Total length $L_{\text{tot}}^{(\text{S}-520)} = 8.715\,\text{m}$, specific heat ratio $\gamma = 1.4$, and gas constant $R = 287\,\text{J/(kg}\cdot\text{K)}$.

$$\begin{aligned} C_{D_f} &= \frac{0.455}{(\log_{10} Re)^{2.58}} \cdot \frac{1}{(1+0.144M^2)^{0.655}} \\ Re &= \frac{V L_{\text{tot}}}{\nu} \end{aligned} \tag{24.13}$$

Kinematic viscosity coefficient ν (m^2/s) and atmospheric temperature T (K) are variables for altitude, referring International Standard Atmosphere.

24.4 Results

24.4.1 Optimization Result

The population size is set to be 18 and evolutionary computation is performed until 3,000 generations when the evolution is roughly converged. The plots of acquired nondominated solutions are shown in Fig. 24.2, which reveals that there generates no multimodal and clean convex curved surface.

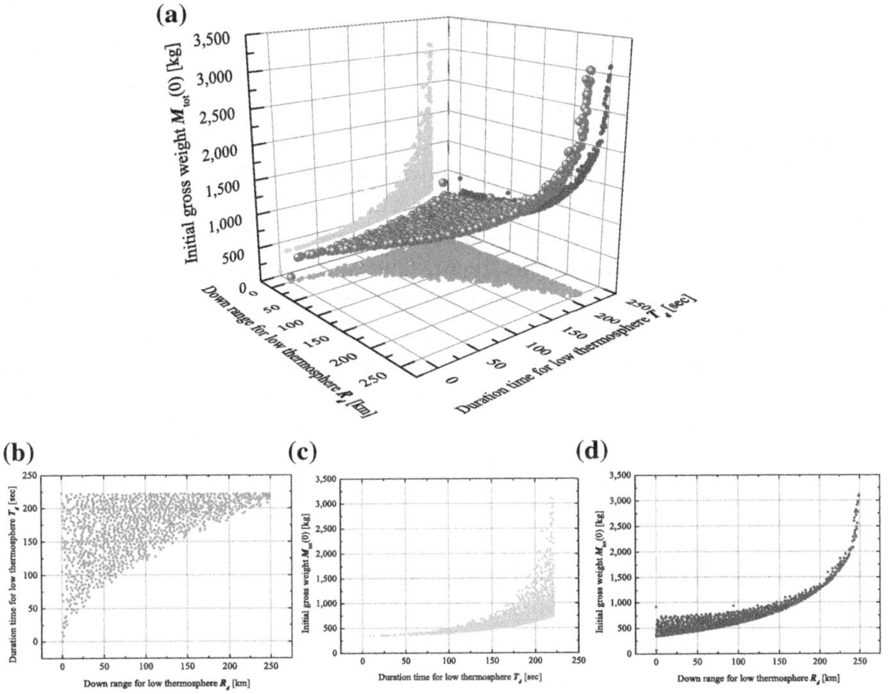

Fig. 24.2 Plots of nondominated solutions derived by optimization, **a** plotted in three dimensional objective-function space, **b** projected plot onto two dimension as down range R_d(obj1) versus duration time T_d(obj2), **c** projected plot onto two dimension as duration time T_d(obj2) versus initial gross weight $M_{tot}(0)$(obj3), and (d) projected plot onto two dimension as down range R_d(obj1) versus initial gross weight $M_{tot}(0)$(obj3)

There is no tradeoff between the down range R_d and the duration time T_d in the lower thermosphere shown in Fig. 24.2b. This figure also shows that there are upper limitations of roughly 250 km for the down range R_d and of roughly 220 s for the duration time T_d. Therefore, the projection plots onto two dimension between the down range R_d and the duration time T_d do not converge in one point. In the present study, the initial mass flow of oxidizer $\dot{m}_{oxi}(0)$ (dv1) has the limitation of upper/lower values. Since the regression rate of fuel $\dot{r}_{port}(t)$ as an empirical model uses the mass flow of oxidizer $\dot{m}_{oxi}(t)$, $\dot{r}_{port}(t)$ has constraints. As a result, the limitations are generated for the down range R_d and the duration time T_d.

There is an incomplete tradeoff between the duration time T_d and the initial gross weight $M_{tot}(0)$ shown in Fig. 24.2c. The convex nondominated surface to optimum direction with incompleteness is generated due to the limitation of the duration time T_d. As the inclination $\Delta M_{tot}(0)/\Delta T_d$ is small on the convex curve, the duration time T_d can be substantially improved when trifling initial gross weight $M_{tot}(0)$ would be sacrificed. In addition, Fig. 24.2c shows that the minimum initial gross weight to reach the limitation of the duration time (roughly 220 s) is approximately 700 kg.

And also, the minimum initial gross weight to attain to the lower thermosphere (altitude of 90 km) is approximately 350 kg. As these values are better than those of the solid rockets which are operated at present for scientific observation, it suggests that hybrid rocket has an advantage even when hybrid rocket does not have a sequence of multi-time ignition.

There is a severe tradeoff between the down range R_d and the initial gross weight $M_{tot}(0)$ shown in Fig. 24.2d (although the down range strictly has the upper limitation, it seems that the clean convex curve is generated because the limitation is on the edge of the nondominated surface). This figure shows that the maximum down range is roughly 130 km when the minimum initial gross weight to reach the limitation of the duration time T_d (roughly 700 kg) is adopted. The initial gross weight $M_{tot}(0)$ should be absolutely increased in order to have more down range R_d (greater than 130 km) despite no increase of the duration time T_d (remaining roughly 220 s). This fact suggests that the design strategies for the maximizations of the down range R_d and the duration time T_d are different.

24.4.2 Data-Mining Result Using SOM

Figure 24.3 shows SOMs colored by the objective functions and the design variables. As this SOM learning is implicated based on the values of the objective functions as the indicator for the similarity on the neural network, SOMs colored by the objective functions have absolutely gradation shown in Fig. 24.3a. Figure 24.3b is the SOMs colored by the design variables. The upper/lower values of coloring range are set to be upper/lower values defined in the problem summarized in Table 24.1.

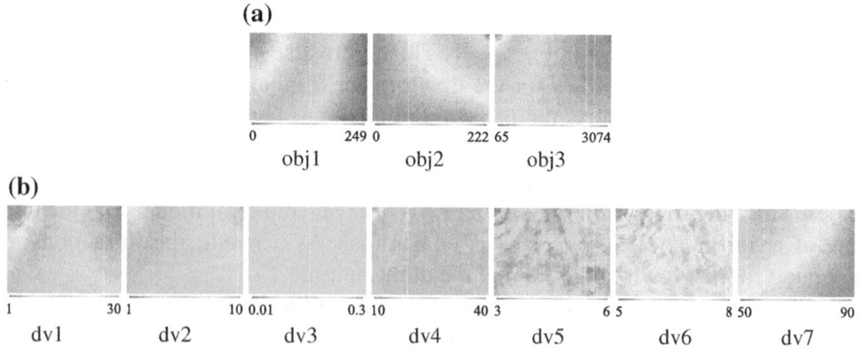

Fig. 24.3 SOMs generated by the objective-function values **a** colored by the objective functions, **b** colored by the design variables, upper/lower values are set to be those in problem definition in Table 24.1

24.4.2.1 Acquired Information Regarding Relationship Among Objective Functions

The comparison of the coloring pattern in Fig. 24.3a reveals the tradeoffs among the objective functions. When obj1 is high value (red region), obj2 absolutely becomes high. However, as obj1 does not always become high whenever obj2 is high, this relationship is irreversible. This is because not only the down range R_d (obj1) but also the attained maximum altitude gives the effect on the duration time T_d (obj2). In contrast, when obj2 is low value (blue region), obj1 absolutely becomes low. However, as obj2 does not always become low whenever obj1 is low, this relationship is similarly irreversible. Although there is no tradeoff in the global space of the objective functions, there is locally tradeoff at the right bottom of Fig. 24.3. This local tradeoff is caused by the limitation of the duration time T_d(obj2), that is, reaching the maximum value of the duration time T_d(obj2) is easier than achieving the maximum value of the down range R_d(obj1).

Since the initial gross weight $M_{tot}(0)$(obj3) is the minimization function, there are severe tradeoffs among obj3 and the others. It especially reveals the severe problem that the optimum direction of the down range R_d(obj1) and the pessimum direction of the initial gross weight $M_{tot}(0)$(obj3) accord (observing the upper left on SOM regarding obj1 and obj2 reveals that the better obj1 and the worse obj3 accord). The structural constraints and the combustion mode should be reconsidered in order to avoid this problem. On the other hand, the optimum region of the duration time T_d(obj2) and the pessimum direction of the initial gross weight $M_{tot}(0)$(obj3) overlap only in part. Therefore, the initial gross weight $M_{tot}(0)$ can become low when the duration time T_d is the primary objective. On the other hand, the minimum initial gross weight is decided by the expected down range when the down range R_d is the primary objective, that is, the minimum initial gross weight depends on the mission requirement as the necessary down range.

24.4.2.2 Acquired Information Regarding Behavior of Design Variables

Figure 24.3b reveals the behavior of each design variable in the design space with the defined wide range which is physically available in order to become the nondominated solution. All nondominated solutions have higher dv1 than the lower bound of dv1 defined in Table 24.1. This fact suggests that the minimum initial mass flow of oxidizer is necessary in order to attain to the lower thermosphere (altitude of 90 km). The mass flow of oxidizer $\dot{m}_{oxi}(t)$ affects the structural weight because of the increase of the filling pressure of oxidizer tank and the pressure of the combustion chamber $P_{cc}(t)$. Threfore, the initial mass flow of oxidizer $\dot{m}_{oxi}(0)$(dv1) is essential in order to improve the initial gross weight $M_{tot}(0)$(obj3).

The value of dv2 does not have both high and low. As the minimum fuel length is necessary in order to attain to the lower thermosphere (altitude of 90 km), dv2 does not have low. On the other hand, as it is considerable that the fuel length L_{fuel}(dv2) does not affect strongly on the maximization of the down range R_d and the duration time T_d rather than the initial mass flow of oxidizer $\dot{m}_{oxi}(0)$(dv1), the fuel length L_{fuel}(dv2) does not have high value.

The value of dv3 is roughly constant. This fact indicates that there is the optimum initial radius of port $r_{port}(0)$(dv3), which might be determined by the combustion mode and the swirl intensity. The other combustion mode should be investigated in order to confirm it.

The value of dv4 is in the narrow region of the design space. But, the combustion time t_{burn}(dv4) becomes high when the initial gross weight $M_{tot}(0)$(obj3) is high. That is, the combustion time t_{burn}(dv4) has identical behavior of the initial radius of port $r_{port}(0)$(dv3) in the design space except the direct affection on the initial gross weight $M_{tot}(0)$(obj3). Threfore, the implication of other fuels should be investigated.

The value of dv5 has low value. The high value of the initial pressure in combustion chamber $P_{cc}(0)$(dv5) fundamentally gives high thrust. As it is expected that the structural requirement is not fulfilled due to high pressure, the structural fulfillment should be confirmed by the parametric study regarding the structural safety factor. In addition, the time fluctuation regarding the pressure in combustion chamber $P_{cc}(t)$ should be observed in the next-step design problem.

The value of dv6 has the coloring pattern in a muddle which is similar to that of dv5, although there is the difference of color. Since the aperture ratio of nozzle ε(dv6) becomes high in order to keep the high thrust, ε(dv6) indirectly give the effect on the objective functions. In fact, the high value of dv6 is on the upper left of SOM, which is the region to become high objective functions.

The coloring pattern of SOM by dv7 is similar to that by obj1. As the vertical launch would be implemented when the elevation at launch time ϕ(dv7) becomes high, it is easily understandable that the down range R_d(obj1) is low. The definition of the next-step design problem will be discussed by all designers using the above design information.

24.5 Conclusions

The next-generation single-stage launch vehicle with hybrid rocket engine of solid fuel and liquid oxidizer on behalf of the present pure solid-fuel rockets has been conceptually designed by using design informatics in order to contribute to the low cost launch vehicle system and efficient space scientific observation. The objective functions as the design requirements in the design problem is the maximization of the down range and the duration time in the lower thermosphere as well as the minimization of the initial gross weight. The evolutionary hybrid computation between the differential evolution and the genetic algorithm is employed for the efficient exploration in the design space. A self-organizing map is used in order to structurize and visualize the design space.

As a result, the design information has been revealed regarding the tradeoffs among the objective functions and the behavior of the design variable in the design space. Consequently, the design strategy for the maximizations of the down range and the duration time is different because the duration time can easily attain the limitation rather than the down range. The characteristics as the regression rate of fuel and structure coefficients shound be investigated as a next design phase in order to reveal the performance limitation of single-time ignition on hybrid rocket eingine. And also, these results indicate that the ascendancy of multi-time ignition as the advantage of hybrid rocket will be quantitatively shown. The results show the quantitative data to compare the performances of solid-fuel rocket in present and hybrid rocket with multi-time ignition.

References

1. Arias-Montano A, Coello CAC, Mezura-Montes E (2012) Multiobjective evolutionary algorithms in aeronautical and aerospace engineering. IEEE Trans Evol Comput 16(5):662–694
2. Arthur W (2002) The emerging conceptual framework of evolutionary developmental biology. Nature 415(6873):757–764
3. Baker JE (1985) Adaptive selection methods for genetic algorithms. In: Proceedings of the international conference on genetic algorithms and their applications, Lawrence Erlbaum Associates, Hillsdale, pp 101–111
4. Chiba K (2013) Evolutionary hybrid computation in view of design information by data mining. In: Proceedings on IEEE congress on evolutionary computation. IEEE, pp 3387–3394
5. Chiba K, Makino Y, Takatoya T (2012) Design-informatics approach for intimate configuration of silent supersonic technology demonstrator. J Aircr 49(5):1200–1211
6. Chiba K, Obayashi S (2008) Knowledge discovery in aerodynamic design space for flyback-booster wing using data mining. J Spacecr Rocket 45(5):975–987
7. Deb K, Pratap A, Agarwal S, Meyarivan T (2002) A fast and elitist multiobjective genetic algorithm: Nsga-ii. IEEE Trans Evol Comput 6(2):182–197
8. Fonseca CM, Fleming PJ (1993) Genetic algorithms for multiobjective optimization: formulation, discussion and generalization. In: Proceedings of the fifth international conference on genetic algorithms, Morgan Kaufmann, pp 416–423
9. Gordon S, McBride BJ (1994) Computer program for calculation of complex chemical equilibrium compositions and applications i. analysis. In: NASA Reference Publication RP-1311, NASA
10. Hervas-Martinez C, Ortiz-Bayer D, Garcia-Pedrajas N (2002) Theoretical analysis of the confidence interval based crossover for real-coded genetic algorithms. In: The 7th international conference on parallel problem solving from nature, LNCS, vol 2439. Springer, Berlin, pp 153–161
11. Jones DR, Schonlau M, Welch WJ (1998) Efficient global optimization of expensive black-box functions. J Glob Optim 13(4):455–492
12. Karabeyoglu MA (2013) Advanced hybrid rockets for future space launch. In: Proceedings on 5th European conference for aeronautics and space sciences, EUCASS
13. Kohonen T (1995) Self-organizing maps. Springer, Berlin
14. Oyama A, Obayashi S, Nakamura T (2001) Real-coded adaptive range genetic algorithm applied to transonic wing optimization. Appl Soft Comput 1(3):179–187
15. Robic T, Filipic B (2005) Demo: differential evolution for multiobjective optimization. In: The 3rd international conference on evolutionary multi-criterion optimization, LNCS, vol 3410. Springer, Guanajuato, Mexico, pp 520–533
16. Sacks J, Welch W, Mitchell T, Wynn H (1989) Design and analysis of computer experiments. Stat Sci 4(4):409–423
17. Saraniero MA, Caveny LH, Summerfield M (1973) Restart transients of hybrid rocket engines. J Spacecr Rocket 10(3):215–217

18. Sasaki D, Obayashi S (2005) Efficient search for trade-offs by adaptive range multi-objective genetic algorithms. J Aerosp Comput Inf Commun 2(1):44–64
19. Simurda L, Zilliac G, Zaseck C (2013) High performance hybrid propulsion system for small satellites. In: AIAA paper 2013-3635. AIAA
20. Takahashi M, Kita H (2001) A crossover operator using independent component analysis for real-coded genetic algorithms. In: Proceedings of IEEE congress on evolutionary computation 2001, IEEE, pp 643–649
21. Yuasa S, Shiraishi N, Hirata K (2012) Controlling parameters for fuel regression rate of swirling-oxidizer-flow-type hybrid rocket engine. In: AIAA paper 2012-4106, AIAA

Chapter 25
Robust Optimization of a Helicopter Rotor Airfoil Using Multi-fidelity Approach

F. Fusi, P.M. Congedo, A. Guardone and G. Quaranta

Abstract A robust optimization technique is developed for the aerodynamic shape optimization of a helicopter rotor airfoil considering uncertain operating conditions. Both a CFD model and a coupled panel/integral boundary layer model of the aerodynamics are coupled with an optimization code based on Genetic Algorithms. In order to reduce the computational cost of the robust optimization, a multi-fidelity strategy is developed which employs both aerodynamic models inside the optimization loop.

Keywords Helicopter rotor airfoil design · Multi-fidelity optimization · Robust Optimization

25.1 Introduction

Shape optimization is a very powerful tool for aerodynamic design: it is well-suited to indicate possible directions of improvement of actual design, explore unusual design and deal with multiple operating conditions and objectives [9]. However, the employment of a theoretical model of the actual system is based on assumptions and simplifications that introduce errors in the prediction of the performance. In addition, a large portion of the data necessary to build the numerical model is affected by numerous sources of uncertainties. A very active research field is devoted to the

F. Fusi (✉) · A. Guardone · G. Quaranta
Department of Aerospace Science and Technology, Politecnico di Milano,
Via La Masa 34, 20156 Milan, Italy
e-mail: francesca.fusi@polimi.it

A. Guardone
e-mail: alberto.guardone@polimi.it

G. Quaranta
e-mail: giuseppe.quaranta@polimi.it

P.M. Congedo
INRIA Bordeaux Sud-Ouest, 200 Avenue de la Vieille Tour, 33405 Talence Cedex, France
e-mail: pietro.congedo@inria.fr

D. Greiner et al. (eds.), *Advances in Evolutionary and Deterministic Methods for Design, Optimization and Control in Engineering and Sciences*, Computational Methods in Applied Sciences 36, DOI 10.1007/978-3-319-11541-2_25

development of robust optimization techniques, namely algorithms that seek an optimal design capable of satisfactory performances when considering the variability of system parameters [1]. The application of uncertainty-based analysis is very attractive in the field of aerospace vehicle design where reliability and performance are important issues; a crucial point is to design stable configurations without loss of performance in off-design conditions [20]. In particular, one challenging area for the application of robust optimization techniques is the aerodynamic design of helicopter rotor blades. As a matter of fact, rotors feature a very complex aerodynamics with unsteady effects due to the blade motion, as well as the presence and interaction of tip vortices and wakes trailed from each blade [12] (see Fig. 25.1). These phenomena require an accurate prediction, which could be achieved by means of high fidelity analysis tools, i.e. Computational Fluid Dynamics (CFD) codes. Unfortunately, CFD codes can be very expensive from a computational point of view, especially when employed in the design stage where many computations are needed. In addition, rotor blades should deliver an adequate level of performance both in hover and forward flight, two operating conditions which are characterized by extremely different flows: in the former the velocity variation along the blade is azimuthally axisymetric and radially almost linear, so that the flow on each section of the blade can be considered as steady; in the latter, in addition to the radial variation the flow varies with respect to blade azimuth angle and the freestream velocity adds to or subtracts from the rotational velocity of the blade, yielding an unsteady periodically changing flow. In most cases, good performance in hover is unlikely to be carried over to forward flight and this problem can only be tackled by a multi-objective approach in the design of the blade. Even in the case of a single operating conditions optimization, the complex flowfield around the helicopter rotor puts each section of the blades under largely variable operating conditions; this variability makes the blade aerodynamic optimization a very interesting case for the development and application of robust approaches. All these elements undoubtedly aggravate the numerical cost of the optimization procedure to the extent that comprehensive algorithms with the computational efficiency and robustness of those developed for fixed-wing design optimization have not emerged yet [3]. In such a context, taking into account the variability of the model parameters in the optimization procedure has a great impact on the computational cost, because of the higher number of evaluations of the

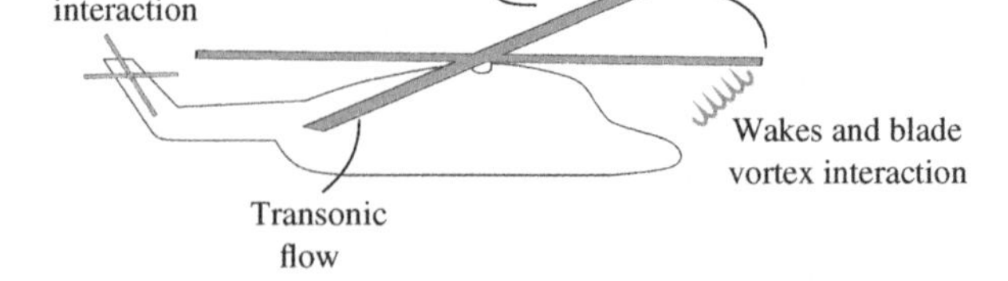

Fig. 25.1 Characteristics of the complex flowfield around a helicopter rotor

aerodynamic performance required to compute the optimal design. It is then required to develop optimization strategies capable to reduce the computational burden without jeopardizing the accuracy of the results. A possibility is then replacing the high-fidelity model of the aerodynamics provided by the accurate, yet expensive CFD codes with low-fidelity models; examples of these models are potential flow models, which have been employed in the past both in fixed wing [8] and rotor airfoil design [6, 18], as well as approximate models of the performance, such as response surfaces [13, 16]. These models require small amount of CPU time, but can have a small range of applicability. Another option is given by the development of a procedure that employs models of different fidelity and level of approximation in the attempt to balance pros and cons of each model; in the literature this solution is usually referred to as multi-fidelity optimization and it has been applied to achieve accurate results with lower computational cost [4, 15].

The present work is devoted to the assessment of the numerical efficiency of the proposed robust optimization techniques with application to the problem of rotor airfoil design. A genetic algorithm is coupled with a non-intrusive uncertainty propagation technique to solve the robust optimization problem of finding the optimal airfoil which maximizes the aerodynamic efficiency under uncertain operating condition. Two different aerodynamic models with different fidelity are employed in order to assess the effect of the choice of the model on the optimal design as well as the computational cost: (i) a high-fidelity CFD code and (ii) an aerodynamic code coupling a panel method and integral boundary layer method. A controlled multi-fidelity strategy is then implemented in order to reduce the computational cost of the optimization procedure. Finally, a comparison among the design obtained with the various approaches is performed.

25.2 Optimization Problem

The robust optimization technique is applied to the optimal design of helicopter rotor airfoil in hovering condition. Only one flight condition is considered in this paper. Instead of considering a deterministic, known-a-priori operating condition for hover, it is assumed that the freestream condition is affected by some uncertainties, such as the uncertainty arising from modeling errors (e.g. the blade inflow model), as well as by other physical sources of uncertainty like blade structural flexibility or variation in the environmental condition. In particular, since the airfoil operating condition in design is typically given by the blade pitch angle θ and the induced velocity V_i, uncertainties on these two system variables are considered in the present analysis.

The objective of the aerodynamic optimization is to maximize the lift-to-drag ratio E, which is the measure of the aerodynamic efficiency of the airfoil, in order to ensure greater aerodynamic lift while reducing the aerodynamic drag. Thus, the objective of this analysis is to find an optimal airfoil shape that could be robust: this means that the design parameters $\mathbf{x}$ defining the shape of the airfoil simultaneously permit to optimize the mean value μ_E and to minimize the variance

σ_E^2 of the objective function, computed when taking into account the uncertainties of the operating conditions. Unlike a single-objective optimization where there is only one global optimum, in multi-objective optimization there is a set of optimal solutions (the so-called Pareto front) corresponding to various trade-offs between a high mean value and a low variance.

Furthermore, constraints $\mathbf{g}$ need to be set in order to obtain reasonable design; in this sense, the pitching moment coefficient computed at quarter-chord location is restrained to prevent excessive loads on the blade structure, whereas the geometry of the airfoil is constrained in such a way that only airfoils having one inflection points per surface are allowed, so that geometries unfeasible from a structural point of view are discarded.

In mathematical terms, the resulting optimization problem can be stated as:

$$\begin{aligned} \text{optimize:} \quad & \left(\mu_E\left(\mathbf{x}, \boldsymbol{\xi}\right), \sigma_E^2\left(\mathbf{x}, \boldsymbol{\xi}\right)\right) \\ \text{subject to:} \quad & \mathbf{g}\left(\mathbf{x}, \boldsymbol{\xi}\right) \leq \mathbf{0} \end{aligned} \tag{25.1}$$

with respect to the design variables $\mathbf{x}$ varying in the solution space σ^2 and with uncertain system parameters $\boldsymbol{\xi} = \{\theta, V_i\}$ varying in the stochastic space Ω.

25.2.1 Shape Parameterization

It is necessary to parameterize the airfoil shape by using a finite set of design variables. In this work, the airfoil is described by means of a parameterization called Class/Shape function Transformation (CST) [11]. This strategy is very efficient for representing rounded-nose airfoils with a small number of parameters. The parameterization is well-defined by specifying a geometry class function (which determines a group or a class of geometries) and a shape function that defines the actual shape of the geometry. The former is defined as

$$C(\psi) = \psi^{N1}(1-\psi)^{N2}, \tag{25.2}$$

where ψ is the chordwise coordinate divided by chord length ($\psi = x/c$) and $N1$ and $N2$ are appropriate coefficient that determine the class; in the case of an airfoil with a rounded nose and a sharp trailing edge: $N1 = 0.5$ and $N2 = 1$ (see Ref. [11] for further details). The shape function $S(\psi)$ is given by a Bernstein polynomial of order n. The CST is applied to the camber mean-line ζ_c and to the thickness distribution ζ_t of the airfoil and it reads

$$\begin{aligned} \zeta_c(\psi) &= C(\psi) \cdot S(\psi) + \psi \zeta_{T_c} \\ &= C(\psi) \cdot \sum_{i=0}^{n} A_{c_i} S_{n,i}(\psi) + \psi \zeta_{T_c} \end{aligned}$$

$$\zeta_t(\psi) = C(\psi) \cdot S(\psi) = C(\psi) \cdot \sum_{i=0}^{n} A_{t_i} S_{n,i}(\psi) \tag{25.3}$$

where the coefficients A_{c_i}, A_{t_i} and the coordinate of the trailing edge of the mean camberline ζ_{T_c} represent the design variables of the optimization $\mathbf{x}$ (please note that the coordinate of the trailing edge of the thickness distribution is set to zero to consider closed trailing edge airfoil). The order of the polynomial is set to seven after an analysis of the ability of the CST to represent a group of typical helicopter rotor airfoils; the resulting design variables are 17. Finally, from the meanline and camber distribution the upper surface ζ_u and lower surface ζ_l are then computed as follows

$$\zeta_u = \zeta_c + \frac{1}{2}\zeta_t \qquad \zeta_l = \zeta_c - \frac{1}{2}\zeta_t. \tag{25.4}$$

25.2.2 Aerodynamic Models

The evaluation of the objective function is performed by means of two aerodynamic models. The high-fidelity model is the ROtorcraft Software ITAlia (ROSITA) flow solver [2], which is a numerical solver of the Reynolds-Averaged Navier Stokes equations coupled with the turbulence model of Spalart-Allmaras. The solver provides space discretization schemes, such as cell-centered finite-volume implementation of either Roe's or Jameson's scheme, with scalar and matricial numerical viscosity; second order accuracy is obtained through the use of MUSCL extrapolation supplemented with a modified version of the Van Albada limiter introduced by Venkatakrishnan [17]. Time advancement is carried out with a dual-time stepping scheme, employing a second-order backward differentiation scheme to approximate the time derivative and an implicit scheme in pseudo-time. The generalized conjugate gradient, in conjunction with a block preconditioner, is used to solve the resulting system. The solver is integrated in the optimization loop and each computation of the flowfield of a new design is obtained starting from a baseline computation and applying a suited displacement to the airfoil boundary of the mesh; the solver provides a mesh deformation tool based on Inverse Distance Weighting [21] that propagates the displacement of the boundary to the mesh volumes. Computations with the ROSITA solver are performed on a C-mesh built around the wing section which is smoothly refined in radial sense from the far-field boundary to the body; the mesh extends ±20 chords both chordwise and in the airfoil thickness direction with 41,961 volumes.

The low-fidelity model is XFOIL, which is an aerodynamic code with coupled panel and integral boundary layer methods developed for the analysis of subsonic, isolated airfoils [8]. In particular, an inviscid linear-vorticity panel method is coupled to the viscous layers represented by the two-equation lagged dissipation integral method in order to correct the potential flow with the viscous boundary layer. Transition of the boundary layer from laminar to turbulent flow is predicted with the e^n criterion, while compressible effects for low Mach numbers are take into account with the Karman-Tsien correction. Computations with XFOIL are performed on a unit-chord airfoil with maximum number of panel and a refinement of the trailing and leading edge; at each iteration loop the optimization algorithms passes the new design to the solver, which computes the aerodynamic loads.

25.2.3 Operating Conditions

Modelling uncertain variables is very challenging in robust or uncertainty-based optimization. This is especially true in the application of helicopter rotor blade sections with uncertain operating conditions, because small uncertainties can yield strong variations in the flowfield characteristics. In fact, critical sections of rotor blades operate in flow conditions where the combination of Reynolds number, Mach number and angle of attack may lead to instability of the boundary layer, large separation of the flow and complex compressibility effects. Therefore, a careful analysis of the uncertain operating conditions is needed, which must also take into account the applicability of the chosen fluid dynamics model.

In this work a nominal operating condition is first chosen, which represents a typical condition for a section of a blade in hover; to this end data from Ref. [19] are employed, considering the blade section at 40 % of the span (see Table 25.1). From the data, the induced velocity and blade pitch angle are computed with the Blade Element Momentum theory [12] using a uniform inflow model. Then, an estimate of the uncertainty on each variable is added to and subtracted to the nominal value. In particular, for the blade pitch angle a variation of $\pm 2°$ is estimated due to longitudinal cyclic and lateral cyclic pitch. A value of $\pm 0.5°$ is added in order to model the effect of blade flexibility with an uncertainty in the torsional stiffness. This level of uncertainty has also been found on the pitch measures taken during the HART II experimental campaign [19]. Concerning the induced velocity, a variation of $\pm 15\,\%$ of the nominal value is considered in order to approximate the effects of the losses due to the wakes trailed from the blades and other tridimensional effects of the rotor which are not yet known in the design stage of the blade.

The uncertainties of the induced velocity V_i and the blade pitch angle θ affect both the freestream speed V_∞ and the angle of attack α of the airfoil. Even though the change in freestream speed slightly changes the Reynolds number, its influence on the Mach number is the most significant to the extent that it jeopardizes the applicability

Table 25.1 Nominal operating condition from Ref. [19]

V_i (m/s)	θ (deg)	α (deg)	M_∞ (−)	Re (−)
10.8	16.76	9.7	0.2582	1.6e6

of the low-fidelity model. Likewise, both the uncertainty on the induced velocity and pitch angle may increase the angle of attack up to values slightly below the stall angle. Note that higher angles of attack are encountered for higher value of the blade pitch angle but lower downwash, which implies lower values of induced velocity. As a final remark, the extreme cases in terms of flow physics are not those with highest or lowest value of induced velocity and pitch angle, but those which maximize and minimize freestream Mach number and angle of attack.

25.3 Robust Optimization Method

In the present work, the optimization strategy is based on the Nondominated Sorting Genetic Algorithm (NSGA) [5, 8]. Main tuning parameters of the algorithm are the population size, the number of generations, the crossover and mutation probabilities pc, pm and the so-called sharing parameter r used to take into account the relative isolation of an individual along a dominance front. Typical values for pc, pm are, respectively, 0.9 and 0.1; values of r are retained following a formula given in Ref. [7] that takes into account the population size and the number of objectives.

25.3.1 Uncertainty Quantification

Let us consider a stochastic differential equation of the form:

$$L(\mathbf{x}, \boldsymbol{\xi}, \phi) = f(\mathbf{x}, \boldsymbol{\xi}) \tag{25.5}$$

where L is a non-linear spatial differential operator (for instance, L is the steady Navier-Stokes operator) depending on a set of uncertainties, defined with a random vector $\boldsymbol{\xi}$ (whose dimension depends on the number of uncertain parameters in the problem) and $f(\mathbf{x}, \boldsymbol{\xi})$ is a source term depending on the design parameters $\mathbf{x}$ and on the uncertain parameters $\boldsymbol{\xi}$. The solution of the stochastic equation (25.5) is the unknown dependent variable $\phi(\mathbf{x}, \boldsymbol{\xi})$. In this paper, we assume a uniform probability density function for each uncertainty. The complete stochastic problem is solved using a quasi-Monte Carlo method (QMC). QMC is similar to the usual Monte Carlo simulation but uses quasi-random sequences instead of (pseudo) random numbers; these sequences are deterministic and fill the space more uniformly than random sequences (known as low discrepancy sequences, i.e. the Sobol sequences). Main advantage of this method is that the coupling between GA and the stochastic strategy is very simple, since a non-intrusive method is used. Mean and variance are then computed for each individual (i.e. for each geometry) by using the classical formulation based on a summation over the samples. These values are then directly exploited by the GA algorithm.

25.3.2 Evolution Control Strategy

The coupling of the uncertainty quantification analysis with the optimization algorithm yield an increase of the computational cost of the robust optmization. In the case of the low-fidelity model, the global cost is feasible, but it becomes huge when more accurate models are used. In order to reduce this cost, a hybrid optimization algorithm is developed, which is based on Evolution Control [10]. The underlying idea is that the employment of evolutionary algorithms in industrial application may be effective in the design stage to find new unusual design or to explore the design variable space; once a set of new solutions is found, it is possible to refine that design or examine its performance by means of a solver with higher accuracy. A possibility is based on the use of a lower fidelity model in the optimization process in order to speed up the initial procedure, where many solutions have to be assessed. During the final steps of the optimization, then only a small set of design can be validated with the higher fidelity model. However, the aerodynamic models inevitably affect the optimization results and may yield false optima; to avoid this problem, a technique is developed which employes the low-fidelity model but drives the evolution of the generations inside the optimization loop by leveraging the estimate of the performance computed with the high-fidelity model. This technique is called Evolution Control and it has been applied in aerospace application with CFD solvers and surrogate models [10, 14]. Usually, either a certain number of individuals within a generation or a certain number of generations in an evolution is evaluated with the high-order fitness function. In the application to robust optimization, we propose to control each individual of every generation by computing the fitness function with the high-fidelity model for a subset of samples of the operating conditions. This subset includes the most significant and critical operating conditions, which are the nominal operating condition and the conditions with minimum/maximum Mach number and angle of attack. So, for each individual in a generation, the following steps are performed:

- compute the performance for each operating conditions in the complete set of samples S in the stochastic space by means of the low-fidelity algorithm and obtain the mean and variance to get an estimate of the objectives $\left(\tilde{J}_1, \tilde{J}_2\right)$;
- employ the high-fidelity model to compute the performance for the operating conditions belonging to the subset of critical operating conditions S_c;
- compare the ranking of the individuals obtained with the two approaches and penalize the objective function $\left(\tilde{J}_1, \tilde{J}_2\right)$ of the ith individual depending on the ranking of the individual based on the high-fidelity computations. Since a two-objectives optimization is considered, the penalization is carried out as follows

$$
\begin{aligned}
J_{1,i} &= \tilde{J}_{1,i} + c_1 R_{1,i}/N_{ind} \\
J_{2,i} &= \tilde{J}_{2,i} + c_2 N_{ind}/R_{2,i},
\end{aligned}
\qquad (25.6)
$$

where R_i is the ranking of the ith individual (e.g. $R_{1,i} = 1$ for the individual having highest mean value in the generation), N_{ind} is the number of individual in the current generation and c_1, c_2 are scale factors chosen as $c_1 = \left(\max_i(\tilde{J}_{1,i}) - \min_i(\tilde{J}_{1,i})\right)$ and $c_2 = \left(\max_i(\tilde{J}_{2,i}) - \min_i(\tilde{J}_{2,i})\right)$.

25.4 Results

Three robust optimization loops are performed: one with the hybrid approach, and the other loops by using only XFOIL and ROSITA, respectively. For each loop, the genetic algorithm is run first imposing just the geometrical constraint for five generations of 100 individuals. Then, one generation of 100 individuals is computed evaluating the aerodynamic performance of the individuals that satisfy the geometrical constraint; this step is performed to explore the design parameters space. Finally, five generation of 40 individuals are computed.

Results obtained from robust optimization using the XFOIL solver are presented in Fig. 25.2 where the red color is used for the individual with higher mean value

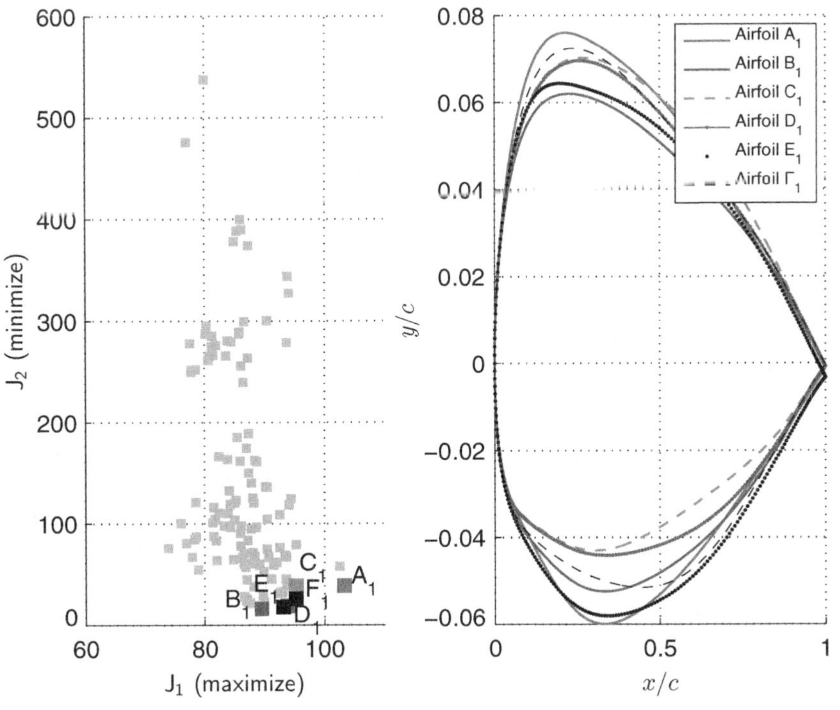

Fig. 25.2 Results of robust optimization with low-fidelity model: objective functions of global population (*left*) and airfoils belonging to Pareto front (*right*)

Table 25.2 Comparison of optimization results with deterministic approach and robust approach coupled with the low-fidelity model

	Airfoil A_1	Airfoil B_1	Airfoil C_1	Airfoil D_1	Airfoil E_1	Airfoil F_1	Deterministic
μ_E	103.28	89.57	95.34	93.89	93.18	95.31	103.95
σ_E^2	38.89	16.26	38.79	18.89	18.05	26.31	85.73

and higher variance, the blue color is used for the airfoil with lowest variance and a scale of dark grey color is used for trade-off solutions. Pareto front is flattened and characterised by low values of the variance of the performance. To assess the validity of the robust approach, a deterministic optimization is performed, considering only the nominal operating condition; after the optimization, the mean and variance of the lift-to-drag ratio of the deterministic optimal airfoil are computed on the same set of uncertain operating conditions considered in the robust approach and the comparison presented in Table 25.2 shows the efficiency of the robust approach.

The robust optimization is then performed with the ROSITA solver and results are presented in Fig. 25.3; in this case, the Pareto front is more spread and the variance of the performance appears greater with respect to the results obtained of the

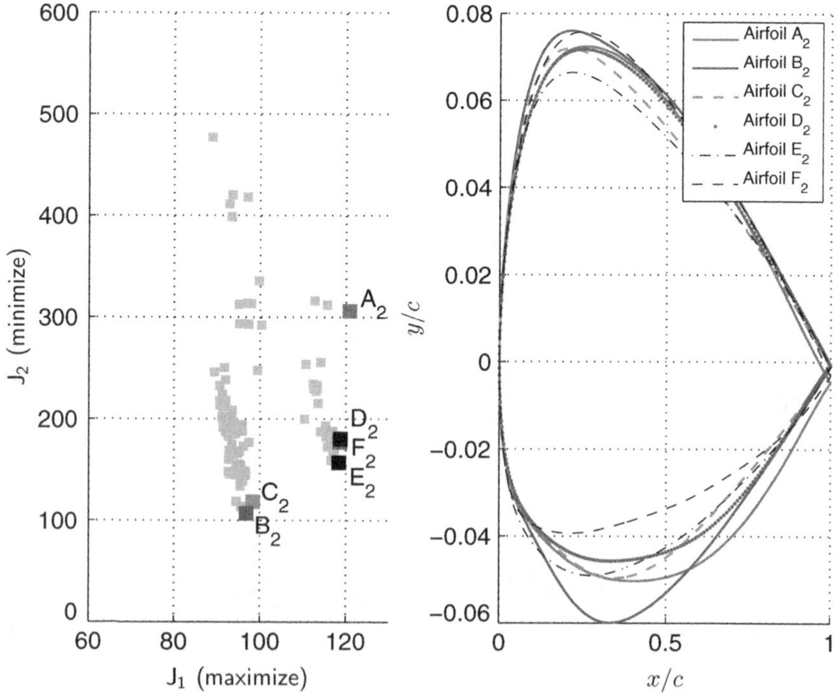

Fig. 25.3 Results of robust optimization with high-fidelity model: objective functions of global population (*left*) and airfoils belonging to Pareto front (*right*)

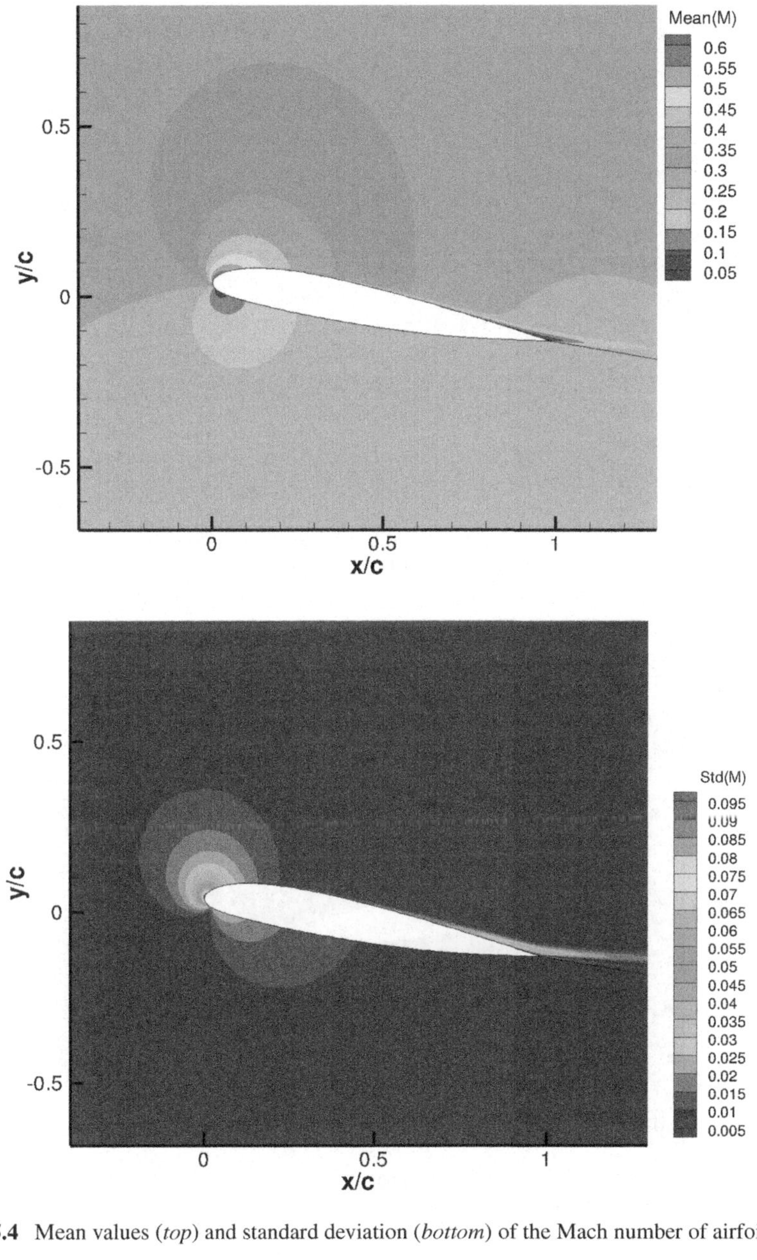

Fig. 25.4 Mean values (*top*) and standard deviation (*bottom*) of the Mach number of airfoil A_2

low-fidelity model. Such difference is due to the fact that ROSITA solver is more accurate in representing the flowfield at high angles of attack whereas the XFOIL code may be less sensitive to the change in operating conditions at high angle of attack due to the boundary layer treatment. In terms of computational cost, the

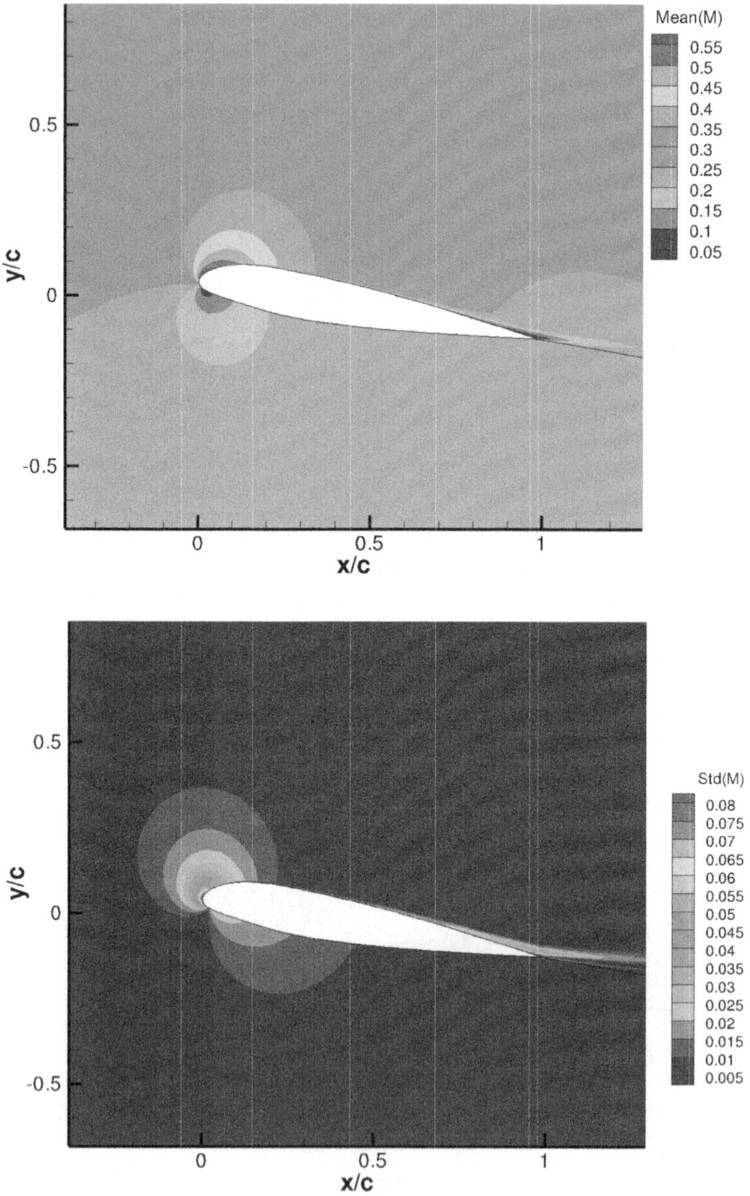

Fig. 25.5 Mean values (*top*) and standard deviation (*bottom*) of the Mach number of airfoil B_2

coupling of the optimization procedure with the ROSITA solver requires about 23 h using 192 cores to evaluate one generation, whereas the XFOIL code takes about 15 min on 8 cores. In Figs. 25.4 and 25.5 the mean value and standard deviation of the local Mach number are presented for airfoil A_2 which has highest variance and

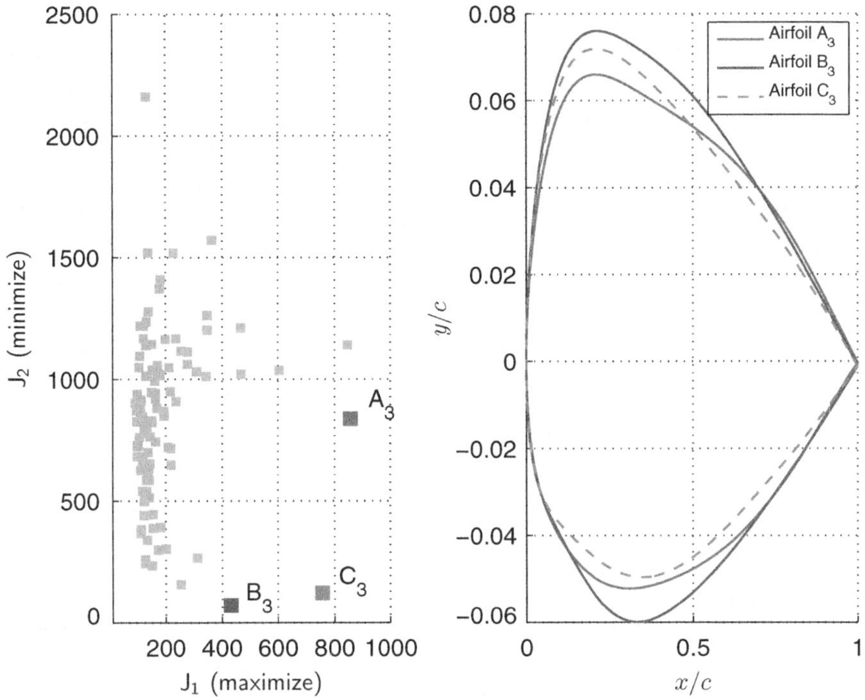

Fig. 25.6 Results of robust optimization with the hybrid approach: objective functions of global population (*left*) and airfoils belonging to Pareto front (*right*)

mean efficiency and airfoil B_2 which has lowest variance and mean value of E. This feature translates in the mean value and deviation of the local Mach number which has higher standard deviation and maximum negative value in the case of airfoil A_2. It is also interesting to note that the deviation is very low everywhere in the flowfield except for the area close to the leading edge of the airfoil and in the wake; in fact, a change in the operating conditions affect the stagnation point location, the peak of the pressure coefficient next to the leading edge and the thickness of the boundary layer close to the trailing edge.

The results of the hybrid approach are presented in Fig. 25.6, where the controlled fitness functions are shown. In this case, the values of J_1, J_2 are very different due to the application of the control strategy in Eq. (25.6); also, it appears that the controlled strategy requires more generation to build a dense Pareto front and more iterations are still needed to get reliable results. It is worth noting that the Pareto front of the controlled strategy shares two airfoils (namely B_3 and C_3) with the Pareto front obtained with the ROSITA solver (airfoils B_2 and C_2), although these airfoils belong to the first generation and are mantained in the Pareto front due to the low variability of the performance. To assess the effectiveness of the controlled approach, a computation of the mean value and variance of the lift-to-drag ratio based on the ROSITA solver is performed which actually proves that the airfoil A_3 belong to the Pareto front obtained with the high-fidelity model. In particular the values of the objectives are $\mu_E = 99.24$ and $\sigma_E^2 = 141.01$, so that airfoil A_3 would belong to the Pareto front obtained with the high-fidelity solver (Fig. 25.3); computation of further generations could explore individuals with higher mean efficiency that belong to the Pareto front. The computational cost of the multi-fidelity approach is a trade-off value between the previous two methods: the computation of one generation takes about 4 h using 192 cores.

25.5 Conclusion

In conclusion, this work is the first step in the development of a low-cost hybrid robust optimization tecnhique for the aerodynamic shape optimization of helicopter rotor blade airfoil. Two different models of the aerodynamics are considered, and used in an hybrid optimization strategy aiming at reducing the global computational cost when uncertainties are taken into account. Results of this strategy are compared with the reference results obtained by performing a classical robust optimization with the high-fidelity model. Moreover, they are compared with the results coming from the fast low-fidelity based optimization. Results show the effect of the aerodynamic model on the set of robust optima in the Pareto front. Preliminary results display the good performances of the hybrid approach, that is capable to capture some elements of the real Pareto front, obtained by means of the high-fidelity model optimization. Further developments are demanded for improving the control criterion, and for improving the computational cost associated to the uncertainty quantification.

References

1. Beyer HG, Sendhoff B (2007) Robust optimization—a comprehensive survey. Comput Methods Appl Mech Eng 196:3190–3218
2. Biava M (2007) RANS computations of rotor/fuselage unsteady interactional aerodynamics. PhD Thesis, Politecnico di Milano
3. Celi R (1999) Recent applications of design optimization to rotorcraft—a survey. J Aircr 36:176–189
4. Choi S, Alonso JA, Kroo I (2005) Two-level multi-fidelity design optimization studies for supersonic jets. In: 43rd AIAA aerospace sciences meeting and exhibit
5. Congedo PM, Corre C, Martinez JM (2011) Shape optimization of an airfoil in a BZT flow with multiple-source uncertainties. Comput Methods Appl Mech Eng 200:216–232
6. Jones BR, Crossley WA, Lyrintzis AS (2000) Aerodynamic and aeroacoustic optimization of rotorcraft airfoils via a parallel genetic algorithm. J Aircr 37:1088–1096
7. Srinivas N, Deb K (1995) Multiobjective function optimization using nondominated sorting genetic algorithms. Evol Comput 2:221–248
8. Drela M (1989) XFOIL: An analysis and design system for low reynolds number airfoils. In: Conference on low Reynolds number airfoil aerodynamics, University of Notre Dame
9. Drela M (1998) Pros and cons of airfoil optimization. In: Caughey DA, Hafez MM (eds) Frontiers of computational fluid dynamics. World Scientific, Singapore, pp 363–382

10. Jin Y, Olhofer M, Sendhoff B (2002) A framework for evolutionary optimization with approximate fitness functions. IEEE Trans Evol Comput 5:481–494
11. Kulfan BM, Bussoletti JE (2006) Fundamental parametric geometry representation for aircraft component shapes. In: 11th AIAA/ISSMO multidisciplinary analysis and optimization conference
12. Leishman JG (2002) Principles of helicopter aerodynamics. Cambridge University Press, Cambridge
13. Massaro A, Bennini E (2012) Multi-objective optimization of helicopter airfoils using surrogate-assisted memetic algorithm. J Aircr 49:375–383
14. Minisci E, Vasile M (2013) Robust design of a re-entry unmanned space vehicle by multi-fidelity evolution control. AIAA J 51:1284–1295
15. Robinson TD, Willcox KE, Eldred MS, Haimes R (2006) Multifidelity optimization for variable-complexity design. In: 11th AIAA/ISSMO multidisciplinary analysis and optimization conference
16. Sun H, Lee S (2005) Response surface approach to aerodynamic optimization design of helicopter rotor blade. Int J Numer Meth Eng 64:125–142
17. Venkatakrishnan V, Mavriplis DJ (1996) Implicit method for the computation of unsteady flows on unstructured grids. J Comput Phys 127:380–397
18. Vu NA, Lee JW, Byun YH, Kim S (2010) Aerodynamic design optimization of helicopter rotor blades including airfoil shapes. In: 66th annual forum of the American helicopter society
19. van der Wall BG (2003) 2nd HHC aeroacoustic rotor test (HART II)—Part I: test documentation. Braunschweig, Germany
20. Yao W, Chen X, Luo W, van Tooren M, Guo J (2011) Review of uncertainty-based multidisciplinary design optimization methods for aerospace vehicles. Prog Aerosp Sci 47:450–479
21. Witteveen J (2009) Explicit mesh deformation using inverse distance weighting interpolation. In: 47th AIAA aerospace sciences meeting

Chapter 26
Computational Results for Flight Test Points Distribution in the Flight Envelope

Lina Mallozzi, Pierluigi De Paolis, Gabriele Di Francesco and Alessandro d'Argenio

Abstract In this paper we present a computational methodology to solve the problem of the proper design of the test matrix for an envelope expansion test campaign, where both flutter and systems testing are required (i.e. a new store integration). There are two different stakeholders involved: Structural Engineers (StE), who want to verify their predictions about the flutter free area, and the Systems Engineers (SyE), who want to investigate environmental aspects in the entire operational flight envelope. The test matrix, representing the test points distribution in the flight envelope, can be found solving an optimization problem with hard constraints (flight envelope boundaries) and different objective functions for the two stakeholders StE and SyE. Given the goals of the two stakeholders, the problem is formulated as a noncooperative game, where StE control M distribution and SyE control H distribution, according to their respective strategies. The two players make their decision about test points location simultaneously, playing a spatial competition game and a genetic algorithm is adopted to estimate the Nash equilibrium solutions to the multiple test points location problem. Results for a multiple test points location problem are shown.

Keywords Experimental design · Nash equilibrium · Potential games

---

L. Mallozzi (✉)
Dipartimento di Matematica Ed Applicazioni, Università Degli Studi di Napoli Federico II, Via Claudio 21, 80125 Napoli, Italy
e-mail: mallozzi@unina.it

P. De Paolis · A. d'Argenio
Dipartimento di Ingegneria Aerospaziale, Università Degli Studi di Napoli Federico II, Via Claudio 21, 80125 Napoli, Italy
e-mail: de_paolis.pierluigi@libero.it

A. d'Argenio
e-mail: ale.dargenio@gmail.com

G. Di Francesco
Dipartimento di Ingegneria Industriale e dellInformazione, Seconda Università Degli Studi di Napoli, Via Roma 29, 81031 Aversa, CE, Italy
e-mail: gabriele.difrancesco@libero.it

D. Greiner et al. (eds.), *Advances in Evolutionary and Deterministic Methods for Design, Optimization and Control in Engineering and Sciences*, Computational Methods in Applied Sciences 36, DOI 10.1007/978-3-319-11541-2_26

26.1 Introduction

In this paper we describe the design of a test matrix in an envelope expansion flight test activity, called flight envelope. The Flight Envelope is a region defined by aircraft limits. It represents the area where A/C flying is allowed. Outside the flight envelope flight could be conducted only under a restricted and controlled test environment; for example in order to define and certify the flight envelope flight test pilots and engineers have to fly also outside its limits, that actually during the first flight still do not exist if not by analysis or modeling and simulation: flight envelope is always a result of interpolation of the flown test points, never an extrapolation. Figure 26.1 shows an example of flight envelope typical of a fighter type aircraft and the limiting factors per each area where flying the A/C is forbidden.

- A: Stall limit/High AoA (angle of attack)
- B: Engine performance (service altitude)
- C: Compressibility effects (Mach effect)
- D: Dynamic pressure effects (structural limits)

Classical methods used in this kind of test are the so called Economy Method, which consists on a choice of a subset of flight conditions in accordance with the build-up approach principle in dynamic pressure, and the Extensive Method, which basically attempts to cover the most part of the flight envelope, resulting very expensive and time consuming. We considered that the main driving factors to be considered in this kind of test are the requirements to be demonstrated by two categories of engineers: Structural Engineers and Systems Engineers. Test matrix is designed in order to give the opportunity to flight test engineers to gather all relevant data, necessary for the new store certification process.

Section 26.2 describes the specific test points location problem in the flight test envelope for a store integration certification program. The competition between SyE and StE test requirements is presented according to the Game Theory as a

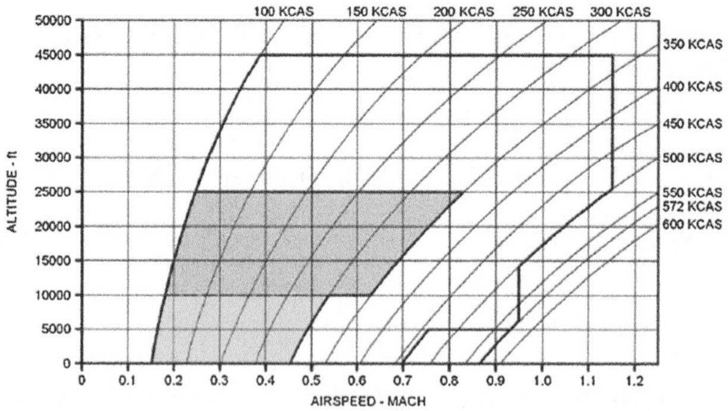

Fig. 26.1 The flight envelope

non-cooperative two players facility location problem. Section 26.3 is dedicated to the analytic model that represents the basis of the case study solution. The two players objective functions and a detailed demonstration about how to face the location problem as a potential game are presented in this section. Sections 26.4 and 26.5 present the results and conclusions of the genetic algorithm used to solve the proposed case study and leave an open eye on future possible follow on activities and way-ahead to face the same problem using different tools and modeling.

26.2 A Location Problem: Test Points Identification

The design of a test matrix in an envelope expansion flight test activity is an experimental design problem [5]. We present the problem from a different point of view, namely we look for the design of a test matrix as a facility location problem. The classical facility location problem deals with determining the site of a number of centers that serve the market places or communities in a given region of the plane where the facilities are considered as dimensionless points [7]. A typical situation is to minimize the distance between facilities and their assigned customers (minsum problem). Sometimes a one-dimensional space is used as in the case that includes competition between firms: recall the pioneering Hotelling problem where two ice-cream seller on a beach compete in prices [2, 8, 10]. In [12], a minsum facility location problem in a bidimensional region has been studied by using game theoretical equilibrium solutions.

In this paper we consider that the requirements of the different engineers categories can be formalized with two different objective functions. The StE, being more interested on combined true airspeed, load factor and compressibility effects on the structures, want to optimize distribution in Mach range, while SyE, being more focused on environmental effects on the aircraft and store systems, want to optimize distribution in Pressure Altitude range.

However, not all the test points in the flight envelope have the same importance. Flight test engineers need to investigate and report the effects at different flight conditions in the entire flight envelope, but they also know that at high dynamic pressure, both structure and systems face their worst environmental and aerodynamic conditions (Fig. 26.2). That is why an additional requirement is to test at the conditions corresponding to the maximum equivalent airspeed, which is a function of both M and H. The dichotomy of the requirements can be interpreted as a particular non-cooperative game, a spatial competition that is also known as Hotelling competition where the two groups of engineers represent the two players. The facility is identified with each single test point and the spatial domain corresponds to the flight envelope.

We want to decide the design variables $V1$, $V2$, whose values are in suitable real intervals, for n experiments $E1, \ldots, En$. Two players StE and SyE decide for each experiment i the values of $V1_i$ and $V2_i$ in the spatial domain optimizing a payoff function.

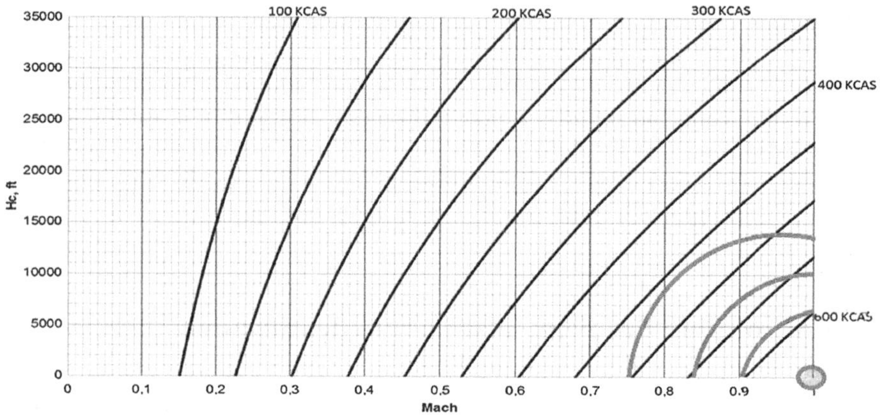

Fig. 26.2 Flying in the hot spot

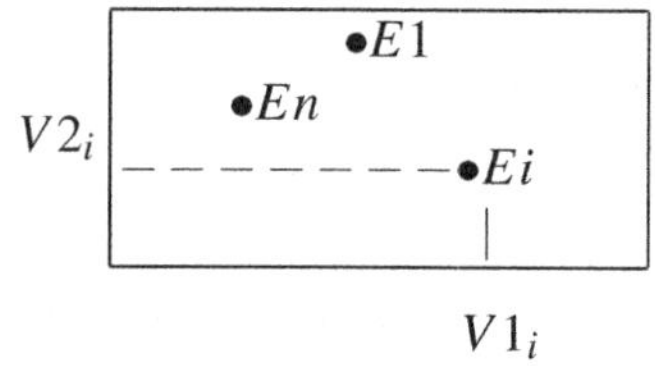

In our contest, the aim of the StE is to optimize distribution in Mach (M) range; the aim of SyE is to optimize distribution in Pressure Altitude (H) range; both of them want to maximize test points density near maximum equivalent airspeed (V_E) area. Because of their own aim, players StE and SyE are in competition, so we will define in the next section the Flight Test Location Game.

26.3 The Model

Let n be a fixed natural number ($n \geq 1$) that is the number of the prescribed flight tests. We consider the two-player normal form game [2, 8] $\Gamma^{FT} = \langle 2; X_1^n, X_2^n; f_1, f_2 \rangle$ where player 1 is the StE team and player 2 is the SyE team. The sets X_1, X_2 are real intervals and represent the variable ranges: for each $i \in [1, n]$ (we denote $[1, n] = \{1, \ldots, n\}$), player 1 choses the Mach number M_i in the set $X_1 = [M_L, M_U]$ and player 2 the pressure altitude H_i in $X_2 = [H_L, H_U]$. The ith flight test point has coordinates (M_i, H_i) and $(\mathbf{M}, \mathbf{H})$ is the $2n$-dimensional vector $(M_1, \ldots, M_n, H_1, \ldots, H_n)$. Player 1 (resp. player 2) has to choose a n-dimensional vector $\mathbf{M} \in X_1^n$ (resp. $\mathbf{H} \in X_2^n$). The objective functions are real valued functions defined on $X_1^n \times X_2^n$ and defined by

$$f_1(\mathbf{M},\mathbf{H}) = \sum_{i=1}^{n}\left[\sqrt{\alpha_1(M_i - M_L)^2 + \alpha_2(M_i - M_U)^2} + \delta(V_U - V_i) - \min_{j\in[1,n],j\neq i}\sqrt{\frac{(M_i - M_j)^2}{M_U^2} + \frac{(H_i - H_j)^2}{H_U^2}}\right] \tag{26.1}$$

and

$$f_2(\mathbf{M},\mathbf{H}) = \sum_{i=1}^{n}\left[\sqrt{\beta_1(H_i - H_L)^2 + \beta_2(H_i - H_U)^2} + \delta(V_U - V_i) - \min_{j\in[1,n],j\neq i}\sqrt{\frac{(M_i - M_j)^2}{M_U^2} + \frac{(H_i - H_j)^2}{H_U^2}}\right] \tag{26.2}$$

where $\alpha_1, \alpha_2, \beta_1, \beta_2, \delta$ are positive real numbers, $V_U = 400$ KCAS and $V_i = V_E(M_i, H_i)$ is the equivalent airspeed that is a function of M_i and H_i under the assumption of International Standard Atmosphere [1]. Here the equivalent airspeed is given by $V_E(M_i, H_i) = aM_i\sqrt{(1 - bH_i)^c}$ with a, b, c positive constants.

The first term of each objective function represents the position of the points with respect to the lower bound and the upper bound of the variable range, the second term is the distance in terms of equivalent airspeed and the last one considers the opposite distance from the closest test point. So that each player asks to minimize his own objective function in order to obtain an optimal points distribution: the task is to distribute the points maximizing their dispersion in the flight envelope and in the same time to be close as possible to the right lower corner of the envelope.

In our model the optimal flight test distribution will be a Nash equilibrium solution of the game Γ^{FT}, i.e. a vector $(\bar{\mathbf{M}}, \bar{\mathbf{H}}) \in X_1^n \times X_2^n$ such that

$$f_1(\bar{\mathbf{M}}, \bar{\mathbf{H}}) \le f_1(\mathbf{M}, \bar{\mathbf{H}}), \quad \forall\, \mathbf{M} \in X_1^n$$
$$f_2(\bar{\mathbf{M}}, \bar{\mathbf{H}}) \le f_2(\bar{\mathbf{M}}, \mathbf{H}), \quad \forall\, \mathbf{H} \in X_2^n$$

In terms of facility location problems, the payoff functions of the flight test location game present a minsum part as well a minmax one [7, 9].

In order to prove the existence of Nash equilibria of the game Γ^{FT}, we recall the definition of potential game, a class of games that have pure Nash equilibrium strategies under suitable assumptions on the data.

Let $\langle A, B, K, L\rangle$ be a *two-person game* with strategy space A for player 1, strategy space B for player 2, and $K : A \times B \mapsto \mathcal{R}$, $L : A \times B \mapsto \mathcal{R}$ the payoff real valued function of player 1, 2 respectively. If the players 1 and 2 choose $a \in A$ and $b \in B$ respectively, then player 1 obtains a payoff $K(a, b)$ and player 2 obtains $L(a, b)$.

Such a game is called a *potential game* [13, 14] if there is a (*potential*) *function* $P : A \times B \mapsto \mathcal{R}$ such that

$$K(a_2, b) - K(a_1, b) = P(a_2, b) - P(a_1, b), \quad \text{for all } a_1, a_2 \in A \text{ and for each } b \in B,$$
$$L(a, b_1) - L(a, b_2) = P(a, b_1) - P(a, b_2), \quad \text{for each } a \in A \text{ and for all } b_1, b_2 \in B.$$

Clearly, elements of $argmax(P)$ are Nash equilibria of the game.

The next lemma will be useful [3]. It states that for a two-person potential game the payoff function of player 1 (player 2) can be written as the sum of a potential and a function on the Cartesian product of the strategy spaces, which only depends on the strategy choice of player 2 (player 1).

Lemma 26.1 *Let $\langle A, B, K, L\rangle$ be a potential game with potential P. Then there exist functions $h\colon A \mapsto \mathscr{R}$ and $g\colon B \mapsto \mathscr{R}$ such that*

$$K(a, b) = P(a, b) - 2g(b),$$
$$L(a, b) = P(a, b) - 2h(a)$$

for each $a \in A$ and $b \in B$.

In the next section we present the existence of Nash equilibria of the flight test location game together with computational results.

26.4 The Results

The following results guarantees that the flight test location game Γ^{FT} admits at least a Nash equilibrium thanks to the potential structure of the considered game.

Theorem 26.1 *$\Gamma^{FT} = \langle 2; X^n, Y^n; f_1, f_2\rangle$, where X, Y are real intervals and f_1, f_2 given as before, is a potential game with potential function*

$$P(\mathbf{M}, \mathbf{H}) = \sum_{i=1}^{n} \Bigg[\sqrt{\alpha_1(M_i - M_L)^2 + \alpha_2(M_i - M_U)^2}$$
$$+ \sqrt{\beta_1(H_i - H_L)^2 + \beta_2(H_i - H_U)^2}$$
$$+ \delta(V_U - V_i) - \min_{j\in[1,n], j\neq i} \sqrt{\frac{(M_i - M_j)^2}{M_U^2} + \frac{(H_i - H_j)^2}{H_U^2}} \Bigg].$$

Then, Γ^{FT} admits at least a Nash equilibrium solution.

Proof By using Lemma 26.1, the function P is a potential function since the function

$$f_1(\mathbf{M}, \mathbf{H}) - P(\mathbf{M}, \mathbf{H})$$

does not depend on $\mathbf{M}$ and

$$f_2(\mathbf{M}, \mathbf{H}) - P(\mathbf{M}, \mathbf{H})$$

does not depend on $\mathbf{H}$. Moreover P admits a minimum point.

The problem of the test matrix design is now reduced to the following optimization one: finding a pair $(\bar{\mathbf{M}}, \bar{\mathbf{H}})$ s.t.

$$P(\bar{\mathbf{M}}, \bar{\mathbf{H}}) = \min_{(\mathbf{M},\mathbf{H}) \in X^n \times Y^n} P(\mathbf{M}, \mathbf{H}).$$

From the analysis of the two objective functions representing the pillars of the problem, it turns out that we are facing a potential game; this reduces its resolution to the determination of the minimum of the potential function, which represents a Nash Equilibrium (NE) solution [11, 13, 14].

A Genetic Algorithm, consisting in a heuristic search technique modeled on the principle of evolution with natural selection, by reproduction of the best elements with possible crossover and mutation, was used in order to find the minimum values of the potential function. The following Table 26.1 shows the parameters setup used in order to achieve a fast convergence; this setup refers to "ga" function setting in MatLab MathWorks software application. After generating the solution of the proposed problem the output was analyzed to evaluate the goodness of the result and robustness of the solution iterating the process applying step by step minor changes to the setup configuration. Furthermore, results validation was accomplished comparing the test cases results with the test matrix structure given by other standard empirical testing method, as the Economy method already mentioned in this study.

A typical store integration test campaign requires a test matrix dimension ranging from 10 to 30 test points. Figure 26.3 shows the test matrix plot in the flight envelope for a 30-test points distribution. From a quick analysis of Fig. 26.3 is possible to verify that the initial requirements of the test team have been met by the solution in terms of test points distribution and compliance to the buildup approach philosophy. At higher altitude the test points are more spaced than at low altitude where is possible to appreciate that test conditions range from low to high dynamic pressure portions of the flight envelope.

Table 26.1 GA details

Parameter	Setup value
Population size	200
Crossover fraction	0.80
Mutation fraction	0.20
Fitness scaling	Rank
Selection function	Tournament
Crossover mode	Scattered
Mutation mode	Adaptive feasible

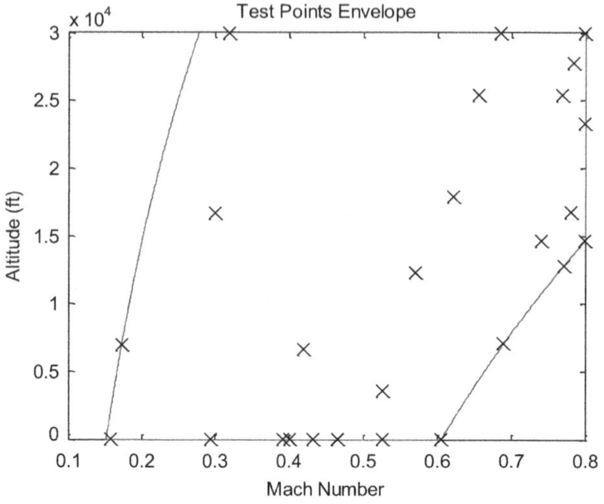

Fig. 26.3 The optimal distribution for 30 flight test points

26.5 Conclusions and Future Works

A solution is presented for a multiple test points location problem, based on a GA algorithm. By means of this tool is possible to approach the optimal test points distribution for a test campaign of a new store integration, where optimality is assessed in terms of prescribed objective functions [6]. Flutter, which represents one of the most dangerous aero-elastic instability (divergent induced oscillations), and systems test can be performed simultaneously in an effective way. All procedures implemented in the algorithm presented in this paper and relative results were deemed to be reliable, also compared with studies of previous literature and test philosophy [4, 12, 15]. One of the subject that could be the focus for future studies is the definition of new methodologies to translate in math the attractiveness of the portion of the flight envelope with higher dynamic pressure; in this paper the proposed way is to assign weights to the reference corner points, but other modeling tools are currently being developed.

A way-ahead could be represented by a computational methodology for an experimental design problem based on the concept of potential and repulsive fields. Each test point could be assumed to be the source of different fields which expose all other points to repulsive forces acting in different directions. The result of the mutual repulsive forces would be a dynamic evolution of the configuration of test points in the two-dimensional domain (pursuant to hard constraints—permitted boundaries of the domain, and soft constraints—minimization of potential), which eventually converges to a condition of minimum potential, where forces are balanced. Furthermore, it is not unusual during a test campaign that the total amount of flight test point needs to be cut drastically, due to time and cost constraints. Therefore, it is paramount, when a change of the number of test points (either an increment or reduction) is needed, to create a method to dynamically relocate the remaining test points, after an initial subset has been performed. A method and an algorithm showing effectiveness and computationally efficiency are required in order to allow an easy tuning of the process to match flexibility and generalization of the case study treated in this document.

Acknowledgments This work has been partially supported by F.A.R.O. 2012 "Metodi Matematici per la modellizzazione di fenomeni naturali", University of Naples Federico II, Italy.

References

1. Anderson JD (2007) Fundamentals of aerodynamics, 4th edn. McGraw-Hill, New York
2. Başar T, Olsder GJ (1999) Dynamic noncooperative game theory. Reprint of the second (1995) edition. Classics in applied mathematics, 23. Society for industrial and applied mathematics (SIAM), Philadelphia, PA
3. Branzei R, Mallozzi L, Tijs SH (2003) Supermodular games and potential games. J Math Econ 39:39–49
4. Clarich A, Periaux J, Poloni C (2003) Combining game strategies and evolutionary algorithms for CAD parametrization and multi-point optimization of complex aeronautic systems, Barcelona, EUROGEN 2003
5. Dean AM, Voss D (2000) Design and analysis of experiments. Series: Springer texts in statistics, Springer, New York, p XIX
6. Deb K (2001) Multi-objective optimization using evolutionary algorithms. Wiley, Chichester
7. Eiselt HA, Marianov V (2011) Foundations of location analysis, International series in operations research & management science, vol 115. Springer, New York
8. Fudenberg D, Tirole J (1993) Game theory. The MIT Press, Cambridge
9. Hansen P, Peeters D, Richard D, Thisse J-F (1985) The minisum and minimax location problems revisited. Oper Res 33:1251–1265
10. Hotelling H (1929) Stability in competition. Econ J 39:41–57
11. Mallozzi L (2007) Noncooperative facility location games. Oper Res Lett 35:151–154
12. Mallozzi L, D'Amato E, Daniele E, Petrone G (2011) Evolutionary and deterministic methods for design, optimization and control. In: Poloni C, Quagliarella D, Periaux J, Gauger N, Giannakoglou K (eds) N leader—M follower coalition games with genetic algorithms and applications. CIRA, Capua
13. Mallozzi L (2013) An application of optimization theory to the study of equilibria for games: a survey. Centr Eur J Oper Res 21:523–539
14. Monderer D, Shapley LS (1996) Potential games. Games Econ Beh 14:124–143
15. Periaux J, Chen HQ, Mantel B, Sefrioui M, Sui HT (2001) Combining game theory and genetic algorithms with application to DDM-nozzle optimization problems. Finite El Anal Des 37:417–429

Chapter 27
Optimal Separation Control on the Flap of a 2D High-Lift Configuration

Anil Nemili, Emre Özkaya, Nicolas R. Gauger, Felix Kramer and Frank Thiele

Abstract Flow separation on the flap of a high-lift device degrades the overall aerodynamic performance and hence results in a drop in the lift coefficient. However, by employing the active flow control techniques, separation can be delayed and thus the lift can be enhanced. In these methods, the flow is controlled by varying the parameters of actuation. In the present work, the optimal set of actuation parameters is found using the gradient-based optimisation algorithms combined with an accurate and robust discrete adjoint method for unsteady RANS. Numerical results are presented for the optimal separation control on the flap of a high-lift configuration over a large time interval.

Keywords Unsteady RANS · Optimal flow control · Discrete adjoints · Algorithmic differentiation · High-Lift configuration

---

A. Nemili (✉) · E. Özkaya · N.R. Gauger
Computational Mathematics Group, CCES, RWTH Aachen University, Schinkelstr. 2, 52062 Aachen, Germany
e-mail: nemili@mathcces.rwth-aachen.de

E. Özkaya
e-mail: ozkaya@mathcces.rwth-aachen.de

N.R. Gauger
e-mail: gauger@mathcces.rwth-aachen.de

F. Kramer
Institute of Fluid Dynamics and Technical Acoustics, Technische Universität Berlin, Müller-Breslau-Str. 8, 10623 Berlin, Germany
e-mail: felix.kramer@cfd.tu-berlin.de

F. Thiele
CFD Software Entwicklungs- und Forschungsgesellschaft mbH, Marie-Elisabeth-Lüders-Straße 1, 10625 Berlin, Germany
e-mail: frank.thiele@cfd-berlin.com

D. Greiner et al. (eds.), *Advances in Evolutionary and Deterministic Methods for Design, Optimization and Control in Engineering and Sciences*, Computational Methods in Applied Sciences 36, DOI 10.1007/978-3-319-11541-2_27

27.1 Introduction

It is well-known that the trailing edge flap of a high-lift device is prone to turbulent flow separation on the suction side, which results in a rapid fall in the lift while the drag increases enormously. In order to delay the flow separation and to enhance the lift coefficient, blowing and suction type of active flow control techniques can be used [2, 12]. In these techniques, the flow control is realised by actuators that apply blowing and/or suction on the suction side of the flap surface. The additional momentum due to actuation delays the separation further downstream and hence increases the lift. Typically, the separation is controlled by varying the parameters of actuation like the amplitude, the direction and the phase shift.

In previous studies [5, 7], the strategies employed to find these parameters are often heuristic in nature and are limited to either simple cases, or involved only small quantities of actuators. With the increase in the number of actuation parameters, an efficient way of finding the optimal set of parameters is by employing gradient-based optimisation algorithms combined with discrete adjoint methods.

In general, a discrete adjoint method for optimal active flow control can be developed either by using hand-discrete approach [11] or by employing Algorithmic Differentiation (AD) [9, 10]. In the hand-discrete approach, the adjoint equations are first hand-derived and a computer code is then written to solve the unsteady adjoint equations and to compute the sensitivity gradients. In AD based approach, the adjoint code is generated by applying AD techniques [4] to the corresponding primal CFD code. Accurate computation of sensitivity gradients require the exact differentiation of all discrete residual terms. However, the hand differentiation of higher order scheme for convective terms, limiters and residuals due to turbulence models is quite complex and often prone to errors. Any approximation made by neglecting the differentiation of these terms will result in inaccurate computation of sensitivity gradients. In fact, in unsteady flows, the effect of these approximations on the accuracy of sensitivities will be much more significant as the errors generated in the adjoint solution accumulate while solving the adjoint equations in the backward-in-time integration. On the other hand, AD performs the exact differentiation of all residual terms with much ease and thus computes the functional gradients that are always accurate and consistent with primal solutions. Since all the terms in the primal solver are differentiated exactly, the adjoint solver inherits the robustness and asymptotic convergence rate of the primal solver.

In this paper, an attempt has been made to find the optimal actuation parameters for a three-element $2D$ high-lift configuration by employing an accurate unsteady discrete adjoint Reynolds-averaged Navier Stokes (RANS) solver based on AD.

This paper is organised as follows. In Sect. 27.2, the governing unsteady incompressible RANS equations are presented. Section 27.3 presents the development of an accurate discrete adjoint approach for optimal active flow control. In Sect. 27.4, the three-element high-lift configuration and the actuation boundary condition are described. Details concerning the primal and adjoint solvers are also presented. In Sect. 27.5, numerical results are shown for the un-actuated and optimal actuated flows. Finally, conclusions are drawn in Sect. 27.6.

27.2 Governing Equations

The unsteady Reynolds-averaged Navier Stokes (RANS) equations that govern the incompressible turbulent flows are given by

$$\frac{\partial \boldsymbol{u}}{\partial t} + \boldsymbol{u} \cdot \nabla \boldsymbol{u} = -\nabla p + \nabla \cdot \left((\nu + \nu_t) \left[\nabla \boldsymbol{u} + \nabla \boldsymbol{u}^T \right] \right) \tag{27.1}$$
$$\nabla \cdot \boldsymbol{u} = 0$$

Here $\boldsymbol{u}$ is the time-averaged velocity vector, p is the time-averaged modified pressure divided by the constant density ρ, ν is the molecular kinematic viscosity and ν_t is the kinematic turbulent or eddy viscosity. In the present work, the unknown eddy viscosity is approximated by the SST $k - \omega$ turbulence model [8].

27.3 Unsteady Discrete Adjoint Method for Optimal Active Flow Control

Consider the optimisation problem of finding the actuation parameters that result in effective separation control on the suction side of the flap surface and maximum lift. The objective function is defined as the maximisation of the time-averaged (mean) lift coefficient over the interval $[0, T]$, which in the discrete form is given by

$$\overline{C_l} = \frac{1}{N} \sum_{n=1}^{N} I^n \left(\boldsymbol{U}^n, \boldsymbol{\alpha} \right), \quad I^n = C_l^n \tag{27.2}$$

where $\boldsymbol{U}^n = (\boldsymbol{u}^n, p^n, k^n, \omega^n)^T$ and I^n are respectively the state vector and lift coefficient at time iteration n. The state variables k and ω are the turbulent kinetic energy and its specific dissipation rate respectively. $\boldsymbol{\alpha}$ is the control vector consisting of the actuation parameters. N is the number of time iterations that span the given time interval, which can be obtained from the relation $T = N\Delta t$. Here, Δt is the step size of the time discretisation. At each time iteration n, the state vector $\boldsymbol{U}^n$ has to satisfy the discretised unsteady incompressible RANS equations with SST $k - \omega$ turbulence model as constraints. In the semi-discrete form, the constraints based on a pressure velocity coupling scheme can be written as

$$\begin{aligned}\frac{d\boldsymbol{u}}{dt} + R_{\boldsymbol{u}}\left(\boldsymbol{U}^n, \boldsymbol{\alpha}\right) &= 0 \\ R_p\left(\boldsymbol{U}^n, \boldsymbol{\alpha}\right) &= 0 \\ \frac{dk}{dt} + R_k\left(\boldsymbol{U}^n, \boldsymbol{\alpha}\right) &= 0 \\ \frac{d\omega}{dt} + R_\omega\left(\boldsymbol{U}^n, \boldsymbol{\alpha}\right) &= 0\end{aligned} \tag{27.3}$$

Here $R_{\boldsymbol{u}}$ is the discrete spatial residual vector corresponding to momentum equations, while R_p, R_k and R_ω are the spatial residuals due to pressure Poisson, turbulent kinetic energy and specific dissipation rate equations respectively. Approximating the transient terms in the above equations using a second-order implicit backward difference formula, the constraints can be rewritten as

$$\begin{aligned}R^n_{\boldsymbol{u}}\left(\boldsymbol{U}^n, \boldsymbol{U}^{n-1}, \boldsymbol{U}^{n-2}, \boldsymbol{\alpha}\right) &:= \frac{3\boldsymbol{u}^n - 4\boldsymbol{u}^{n-1} + \boldsymbol{u}^{n-2}}{\Delta t} + R_{\boldsymbol{u}}\left(\boldsymbol{U}^n, \boldsymbol{\alpha}\right) = 0 \\ & \qquad R_p\left(\boldsymbol{U}^n, \boldsymbol{\alpha}\right) = 0 \\ R^n_k\left(\boldsymbol{U}^n, \boldsymbol{U}^{n-1}, \boldsymbol{U}^{n-2}, \boldsymbol{\alpha}\right) &:= \frac{3k^n - 4k^{n-1} + k^{n-2}}{\Delta t} + R_k\left(\boldsymbol{U}^n, \boldsymbol{\alpha}\right) = 0 \\ R^n_\omega\left(\boldsymbol{U}^n, \boldsymbol{U}^{n-1}, \boldsymbol{U}^{n-2}, \boldsymbol{\alpha}\right) &:= \frac{3\omega^n - 4\omega^{n-1} + \omega^{n-2}}{\Delta t} + R_\omega\left(\boldsymbol{U}^n, \boldsymbol{\alpha}\right) = 0\end{aligned} \tag{27.4}$$

where $R^n_{\boldsymbol{u}}$, R^n_k and R^n_ω are the discrete residuals of the unsteady momentum, turbulent kinetic energy and specific dissipation rate equations respectively. At each time iteration n, the residual equations are solved for the solution of the state vector $\boldsymbol{U}^n$ by a contractive fixed point iterative scheme of the form

$$\boldsymbol{U}^n_{i+1} = \boldsymbol{G}^n\left(\boldsymbol{U}^n_i, \boldsymbol{U}^{n-1}, \boldsymbol{U}^{n-2}, \boldsymbol{\alpha}\right), \quad n = 1, \ldots, N. \tag{27.5}$$

Here, $\boldsymbol{G}^n$ represents an iteration of the pressure-velocity coupling scheme based on the SIMPLE algorithm [13]. $\boldsymbol{U}^{n-1}$ and $\boldsymbol{U}^{n-2}$ are the converged state vectors at time iterations $n-1$ and $n-2$ respectively. The fixed point iteration converges to the numerical solution $\boldsymbol{U}^n$, given by

$$\boldsymbol{U}^n = \boldsymbol{G}^n\left(\boldsymbol{U}^n, \boldsymbol{U}^{n-1}, \boldsymbol{U}^{n-2}, \boldsymbol{\alpha}\right), \quad n = 1, \ldots, N. \tag{27.6}$$

The discrete optimisation problem defined by Eqs. (27.2) and (27.4) can then be posed as

$$\begin{aligned}\max_{\boldsymbol{\alpha}} \qquad & \overline{C}_l = \frac{1}{N}\sum_{n=1}^{N} I^n\left(\boldsymbol{U}^n, \boldsymbol{\alpha}\right) \\ \text{subject to} \quad & \boldsymbol{U}^n = \boldsymbol{G}^n\left(\boldsymbol{U}^n, \boldsymbol{U}^{n-1}, \boldsymbol{U}^{n-2}, \boldsymbol{\alpha}\right), \quad n = 1, \ldots, N.\end{aligned} \tag{27.7}$$

The Lagrangian associated with the above constrained optimisation problem is given by

$$L = \frac{1}{N}\sum_{n=1}^{N}\{I^n(U^n,\alpha)\} - \sum_{n=1}^{N}\left\{\left(\overline{U}^n\right)^T\left(U^n - G^n\left(U^n, U^{n-1}, U^{n-2}, \alpha\right)\right)\right\} \tag{27.8}$$

where $\overline{U}^n$ is the vector of Lagrangian multipliers or the adjoint state vector at time iteration n. The first order necessary conditions for optimality (KKT) are given by

$$\frac{\partial L}{\partial \overline{U}^n} = 0, \quad n = 1, \ldots, N. \qquad \text{(State equations)} \tag{27.9a}$$

$$\frac{\partial L}{\partial U^n} = 0, \quad n = 1, \ldots, N. \qquad \text{(Adjoint equations)} \tag{27.9b}$$

$$\frac{\partial L}{\partial \alpha} = 0 \qquad \text{(Control equation)} \tag{27.9c}$$

From Eq. (27.9b), the unsteady discrete adjoint equations can be derived in the fixed point form as

$$\overline{U}^n_{i+1} = \left(\frac{\partial G^n}{\partial U^n}\right)^T \overline{U}^n_i + \left(\frac{\partial G^{n+1}}{\partial U^n}\right)^T \overline{U}^{n+1} + \left(\frac{\partial G^{n+2}}{\partial U^n}\right)^T \overline{U}^{n+2} + \frac{1}{N}\left(\frac{\partial I^n}{\partial U^n}\right)^T, \quad \text{for} \quad n = N, \ldots, 1 \tag{27.10}$$

where $\overline{U}^{n+1}$ and $\overline{U}^{n+2}$ are the converged adjoint state vectors at time iterations $n+1$ and $n+2$ respectively. It can be observed that the adjoint solution $\overline{U}^n$ requires the flow solution U^n to compute $\frac{\partial G^n}{\partial U^n}$ and $\frac{\partial I^n}{\partial U^n}$. This clearly shows that the primal solution U^n must be available at all time iterations a priori as the adjoint equations are solved in backward-in-time integration from $n = N, \ldots, 1$. Note that $\frac{\partial G^{n+1}}{\partial U^n}$ and $\frac{\partial G^{n+2}}{\partial U^n}$ result in constant Jacobians as the spatial residuals inside G^{n+1} and G^{n+2} do not depend on U^n. The above fixed point iteration converges to the adjoint solution $\overline{U}^n$, given by

$$\overline{U}^n = \left(\frac{\partial G^n}{\partial U^n}\right)^T \overline{U}^n + \left(\frac{\partial G^{n+1}}{\partial U^n}\right)^T \overline{U}^{n+1} + \left(\frac{\partial G^{n+2}}{\partial U^n}\right)^T \overline{U}^{n+2} + \frac{1}{N}\left(\frac{\partial I^n}{\partial U^n}\right)^T, \quad \text{for} \quad n = N, \ldots, 1 \tag{27.11}$$

The solution of the adjoint equations is then substituted in Eq. (27.9c) to compute the sensitivity gradients as

$$\frac{dL}{d\boldsymbol{\alpha}} = \sum_{n=1}^{N} \left\{ \frac{1}{N} \frac{\partial I^n}{\partial \boldsymbol{\alpha}} + \left(\frac{\partial \boldsymbol{G}^n}{\partial \boldsymbol{\alpha}} \right)^T \overline{\boldsymbol{U}}^n \right\} \tag{27.12}$$

In the present work, at the first step, AD is applied in a black-box fashion to generate the adjoint code by differentiating the unsteady incompressible RANS code with respect to the control vector $\boldsymbol{\alpha}$. An advantage of the black-box AD approach is that all terms in the pressure-velocity coupling scheme and time discretisation are differentiated exactly and therefore the adjoint code computes the sensitivity gradients that are always accurate and consistent to the primal solutions. Further, the adjoint solver inherits the robustness and asymptotic convergence of the primal solver. However, a major disadvantage of the black-box AD approach is that the adjoint code demands prohibitively expensive memory and computational time, thus making it unfeasible to use in practical applications. In order to circumvent the excessive memory and run-time costs, various advanced AD strategies like binomial checkpointing [3] for adjoining the unsteady time iterations and reverse accumulation technique [1] for adjoining the fixed point schemes in the incompressible solver are employed. The resulting adjoint code exactly performs the fixed point scheme in (27.10) and the sensitivity evaluation in (27.12). Note that the implementation of these techniques does not effect the accuracy and robustness of the adjoint solver, as the exact differentiation of the fixed point iterator $\boldsymbol{G}$ is still maintained. Numerical investigations have shown that the run-time of the unsteady discrete adjoint RANS code is around a factor of 6 compared to the primal code. Although this AD based adjoint code may not be as computationally efficient as the hand differentiated code, the lack of efficiency is well compensated by accuracy, robustness and easy maintenance.

27.4 Computational Setup

27.4.1 Description of the High-Lift Configuration

The geometry of interest is the three-element Swept Constant Chord Half (SCCH) model high-lift configuration, as shown in Fig. 27.1. The flap angle and the angle of attack are chosen such that the un-actuated flow separates distinctively at the flap shoulder. More details on this configuration are described in [5, 7], where extensive

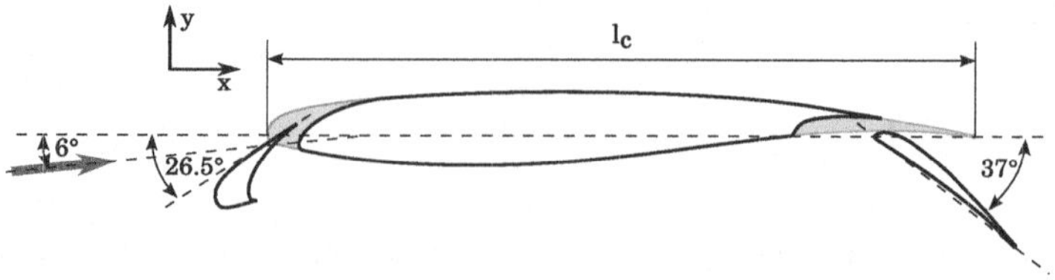

Fig. 27.1 Sketch of the SCCH high-lift configuration

numerical and experimental investigations are performed to control the flow separation.

27.4.2 Actuation

In the present work, the flow control is realised by applying synthetic jet actuation boundary condition at 27 slots, that are distributed starting right in front of the shoulder and passing nearly the whole of the flap surface, as shown in Fig. 27.2. The actuation boundary condition at a slot face is defined by

$$\begin{pmatrix} u \\ v \end{pmatrix} = A \cdot u_\infty \begin{pmatrix} \frac{\cos\theta}{\tan\beta} - \sin\theta \\ \frac{\sin\theta}{\tan\beta} + \cos\theta \end{pmatrix} \sin[2\pi F \cdot (t - t_0)] \qquad (27.13)$$

Here u and v are components of the actuation velocity vector, A is the non-dimensional amplitude of actuation, u_∞ is the free-stream velocity, β is the blowing angle, F is the non-dimensional frequency, t is the non-dimensional physical time and t_0 is the non-dimensional phase shift. The angle of the slit face, θ is fixed by the geometry of the airfoil. The named quantities are appropriately non-dimensionalised by u_∞. The flap length $l_{\text{flap}} = 0.254 l_c$, where l_c is the chord length.

27.4.3 Primal and Adjoint Solvers

In the present work, the unsteady RANS calculations are performed using the multi-purpose incompressible finite volume code ELAN [5, 7]. The mean flow equations

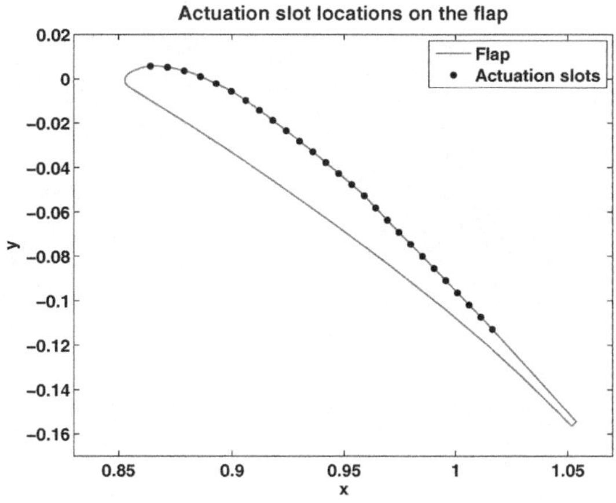

Fig. 27.2 Location of the actuation slots on the surface of the flap of SCCH high-lift configuration

are solved with the pressure correction scheme SIMPLE coupled with the Rhie-Chow interpolation method on a collocated grid. The code is fully implicit and is of second order accuracy both in space and time. The convective fluxes are approximated by a total variation diminishing (TVD) scheme. In addition, the code offers various options for RANS/LES turbulence models. It is parallelised by domain decomposition and uses the MPI library for communication.

The unsteady discrete adjoint RANS code [9, 10] is generated by differentiating the ELAN code with the AD tool Tapenade [6]. The adjoint code retains all the features of the primal code ELAN.

27.5 Results and Discussion

27.5.1 Un-actuated Flow

We first consider the base flow without actuation around the two-dimensional SCCH high-lift configuration. The unsteady RANS simulations are performed over the time interval $[0, 118]$ with the SST $k-\omega$ turbulence model at Reynolds number $Re = l_c u_\infty/\nu = 10^6$ and angle of attack $AoA = 6°$. The step size in the time discretisation scheme is chosen as $\Delta t = 7.874 \times 10^{-3} l_{flap}/u_\infty$. The computational domain consists of a block-structured grid with 43,098 finite volumes. Figure 27.3 shows the grid around the high-lift configuration and the flap region. Figures 27.4 and 27.5 show the corresponding vorticity contours. It can be observed that the high flap deflection angle has caused massive flow separation on the suction side, starting from the flap's leading edge. The early separation reduces the circulation around the flap and hence the lift contribution from the flap decreases.

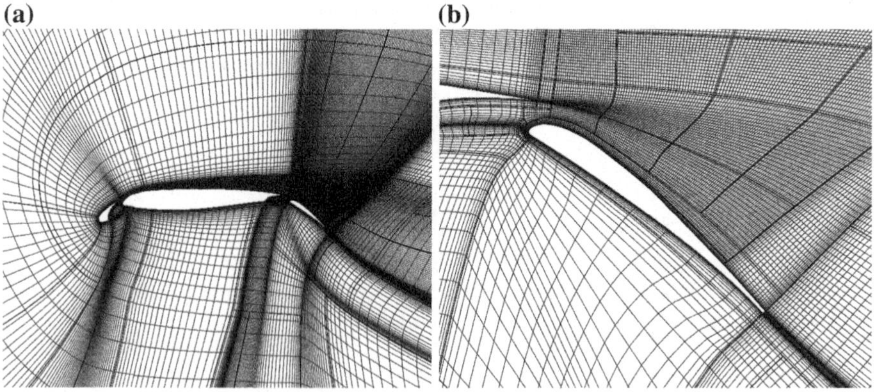

Fig. 27.3 Computational mesh for the RANS simulations. **a** Entire high-lift configuration. **b** Flap region

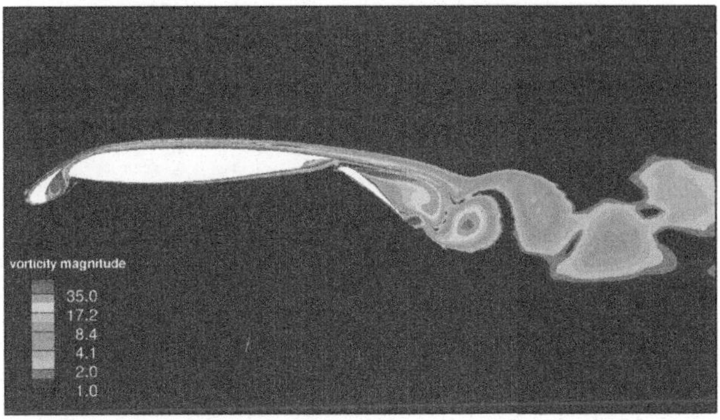

Fig. 27.4 Un-actuated base flow. Vorticity contours on the entire high-lift configuration

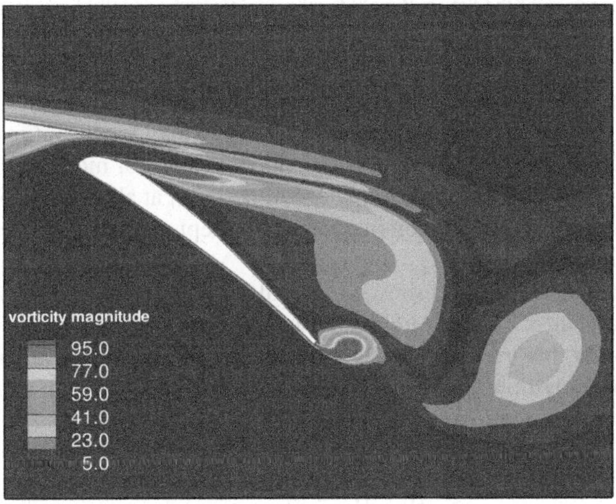

Fig. 27.5 Un-actuated base flow. Vorticity contours on the flap

27.5.2 Initial Actuation and Sensitivity Validation

In order to delay the flow separation and to enhance the mean lift coefficient, sinusoidal blowing and suction with zero net mass flux is applied at 27 actuation faces along the suction side of the flap, as shown in Fig. 27.2. In the computational domain, each slot is resolved by a single cell face. Initially, the values of actuation parameters are chosen as amplitude $A = 0.3$, angle $\beta = 90°$, frequency $F = 0.508$ and phase shift $t_0/T = -0.25$. Here T is the period of actuation. Note that the chosen frequency matches the natural separation frequency of the un-actuated flow. From Fig. 27.6, it can be observed that the actuation has significantly improved the mean

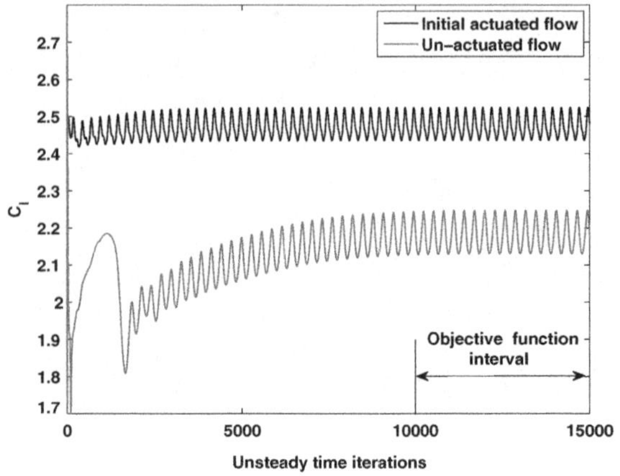

Fig. 27.6 Comparison of the lift coefficient for the un-actuated and initial actuated flows

lift coefficient over the un-actuated flow. However, the mean lift thus achieved need not be maximum as the actuation control parameters may not be optimal. To find the optimal actuation parameters, we pose the maximisation of mean lift as an optimal control problem. The actuation parameters A, β and t_0 at all the 27 slots are chosen as control variables. The actuation frequency F is kept constant. We then have a total of 81 control variables.

The lift profile in Fig. 27.6 shows a typical initial transient phase, after which a periodic behaviour of the flow field is observed. For the objective function evaluation, we exclude this transient phase and consider the contributions from the fully developed unsteady flow. The time-averaged lift coefficient is then defined as

$$\overline{C}_l = \frac{1}{N - N*} \sum_{n=N*+1}^{N} C_l^n \left(\boldsymbol{U}^n, \boldsymbol{\alpha}\right) \tag{27.14}$$

while the Lagrangian reduces to

$$L = \frac{1}{N - N*} \sum_{n=N*+1}^{N} C_l^n - \sum_{n=1}^{N} \left\{ \left(\overline{\boldsymbol{U}}^n\right)^T \left(\boldsymbol{U}^n - \boldsymbol{G}^n \left(\boldsymbol{U}^n, \boldsymbol{U}^{n-1}, \boldsymbol{U}^{n-2}, \boldsymbol{\alpha}\right)\right)\right\} \tag{27.15}$$

Here N^* is the number of time iterations for the transient phase. By choosing $N^* = 10{,}000$ and $N = 15{,}000$, the mean lift is evaluated from time iteration 10,001–15,000.

We first validate the sensitivity gradients computed using the unsteady discrete adjoint RANS solver. Figure 27.7 shows a comparison of the sensitivity gradients of the mean lift with respect to amplitude at eight randomly selected slot faces on the flap. It can be observed that the sensitivities based on the adjoint code are in excellent

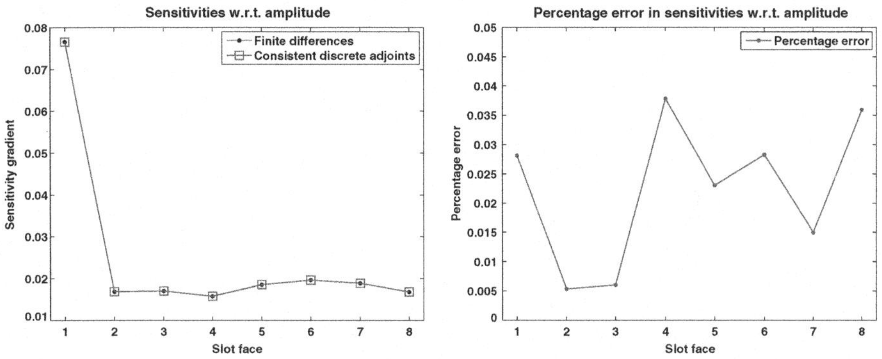

Fig. 27.7 Comparison of the amplitude sensitivity gradients based on consistent discrete adjoint method and second order accurate finite differences

agreement with second order accurate finite differences, as the errors in sensitivities are found to be within 0.04 %.

27.5.3 Optimal Actuated Flow

To find a maximum of the mean lift coefficient, the unsteady discrete adjoint RANS solver is combined with the BFGS quasi-Newton optimisation algorithm. Figure 27.8 shows the variation of the lift coefficient after 10 cycles of optimisation. The enhancement in the mean lift with respect to the optimisation cycles is shown in Fig. 27.9.

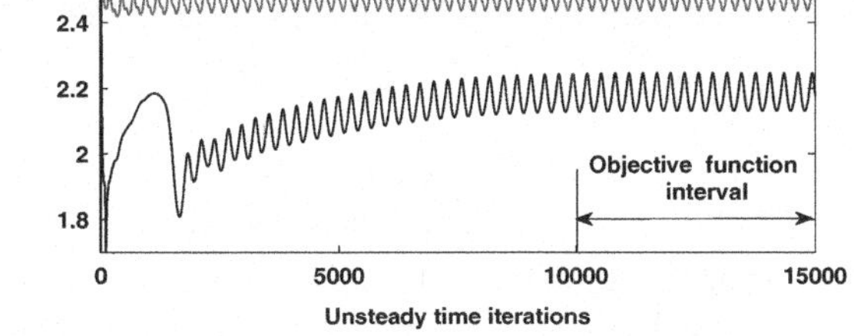

Fig. 27.8 Comparison of the lift coefficient for the un-actuated, initial actuated and optimal actuated flows

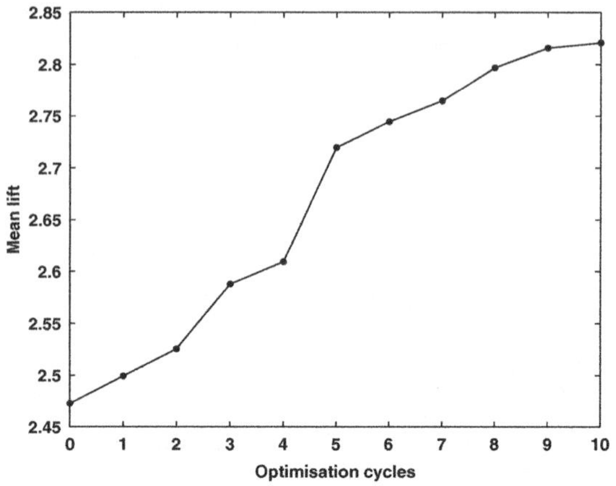

Fig. 27.9 Variation of the mean lift coefficient with the number of optimisation cycles

Table 27.1 Comparison of the mean lift coefficient for un-actuated, initial actuated and optimal actuated flows

Flow	Mean lift coefficient
Un-actuated	2.1802
Initial actuated	2.4728
Optimal actuated	2.8207

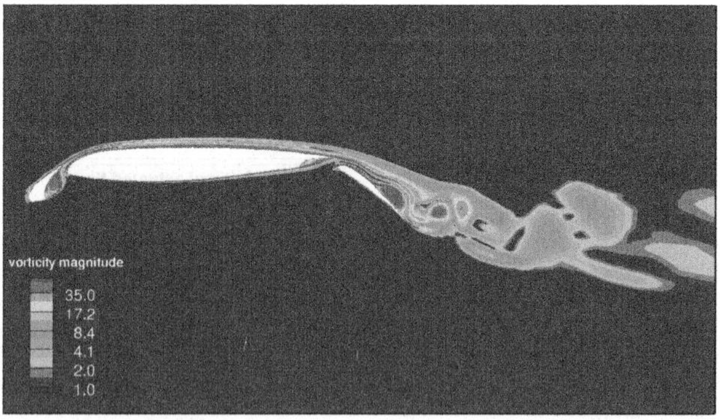

Fig. 27.10 Optimal actuated flow. Vorticity contours on the entire high-lift configuration

From Table 27.1, it can be observed that the optimal actuation has resulted in 14 % improvement in the mean lift compared to the initial actuation and 29 % enhancement over the un-actuated flow. Figures 27.10 and 27.11 show the vorticity contours for the optimal actuated flow. It is clearly evident that the optimal actuation has delayed the separation point further downstream.

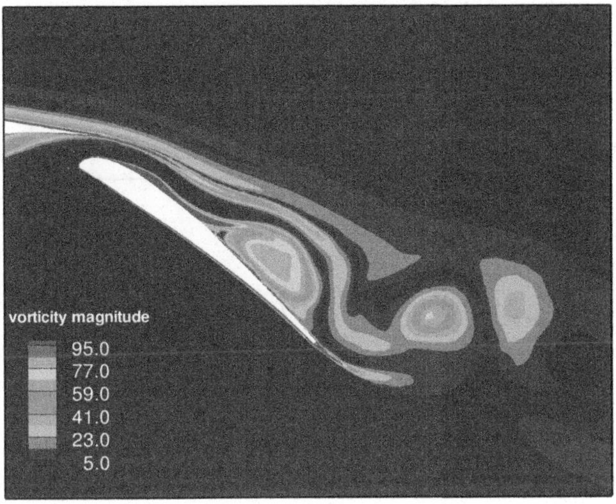

Fig. 27.11 Optimal actuated flow. Vorticity contours on the flap

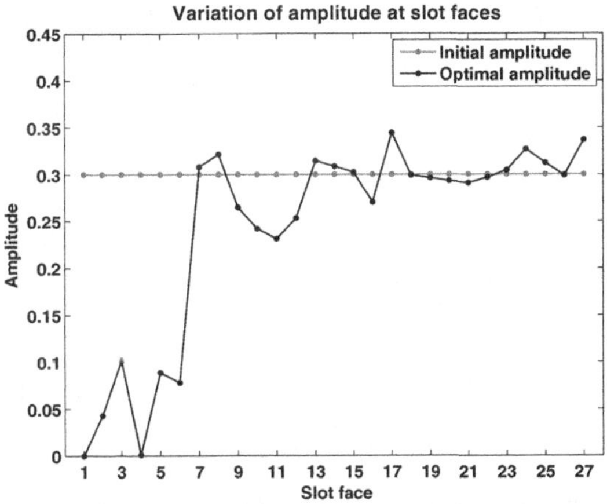

Fig. 27.12 Comparison of the initial and optimal distribution of amplitude at actuation slot faces on the flap surface

Figures 27.12, 27.13 and 27.14 respectively show the optimal distribution of amplitude, angle and phase shift at the actuation faces. From these plots, it can be observed that the actuation slots from index 1–6 are within the region, where the flow stays attached. Actuation at these slots is unfavourable as it triggers the separation. Therefore, the amplitudes at these slots are tending towards zero. At slot 7, the amplitude increases rapidly and causes the flow to counteract the periodic separation present around this region, which spans the large downstream. For slots with indices

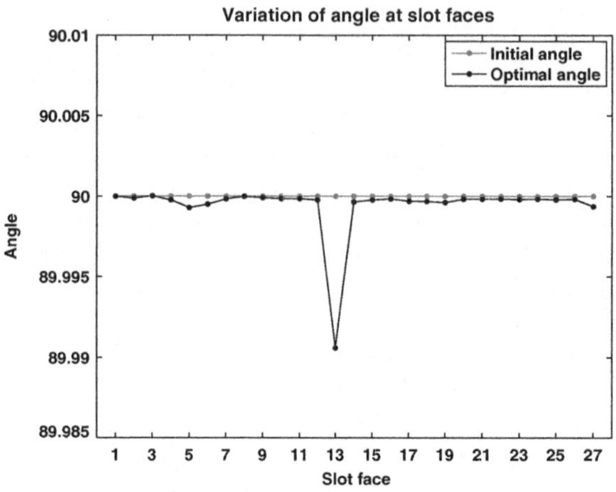

Fig. 27.13 Comparison of the initial and optimal distribution of angle at actuation slot faces on the flap surface

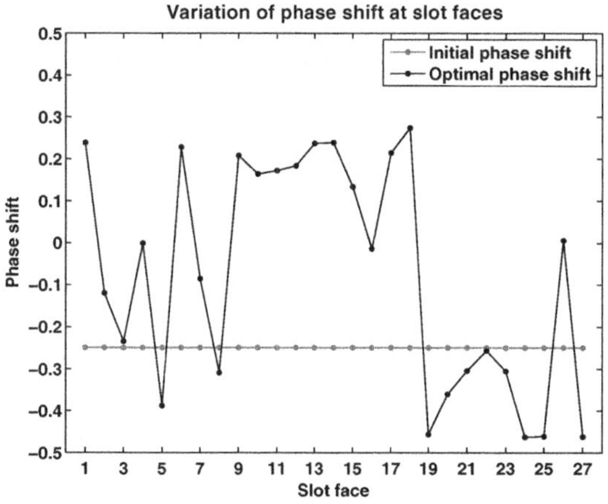

Fig. 27.14 Comparison of the initial and optimal distribution of phase shift normalised by the period of actuation at slot faces on the flap surface

8–13, the optimal amplitudes are observed to be smaller. For many slots with indices above 13, where the flow is separated for large time frames, the sensitivities are found to be very small compared to other actuation slots. It is not clear whether the flow is insensitive to the actuation at these slots or it is a required part of the optimal actuation. In future, this will be investigated by introducing a suitable cost function that accounts for the saving of energy at slots where the actuation has no effect.

27.6 Conclusions

In this paper, we presented the optimal active separation control on the flap of a three-element high-lift configuration. The optimal set of actuation parameters is obtained by combining an accurate unsteady discrete adjoint RANS solver with the BFGS quasi-Newton optimisation algorithm. Numerical results have shown that the optimal actuation has delayed the separation point further downstream and thus enhanced the lift coefficient significantly. Overall, the optimisation procedure has yielded 14 % enhancement in the mean lift coefficient compared to the non-optimised initial actuation and 27 % gain over the un-actuated base flow.

Acknowledgments This research is funded by the German Science Foundation (DFG) under the project numbers $GA\ 857/5 - 1$ and $TH - 288/37 - 1$.

References

1. Christianson B (1994) Reverse accumulation of attractive fixed points. Opt Meth Soft 3:311–326
2. Galbraith MC (2006) Numerical simulations of a high-lift airfoil employing active flow control. In: AIAA paper 2006-0147
3. Griewank A, Walther A (2000) Revolve: an implementation of checkpointing for the reverse or adjoint mode of computational differentiation. ACM Trans Math Soft 26:19–45
4. Griewank A, Walther A (2008) Evaluating derivatives: principles and techniques of algorithmic differentiation. SIAM, Philadelphia
5. Günther B, Thiele F, Petz W, Nitsche R, Sahner J, Weinkauf T, Hege HC (2007) Control of seperation on the flap of a three-element high-lift configuration. In: AIAA paper 2007-0265
6. Hascoet L, Pascual V (2004) Tapenade 2.1 users guide. Technical Report 0300, INRIA
7. Höll T, Job AKV, Giacopinelli P, Thiele F (2012) Numerical study of active flow control on a high-lift configuration. J Aircr 49(5):1406–1422
8. Menter FR (1994) Two-equation eddy-viscosity turbulence models for engineering applications. AIAA J 32:269–289
9. Nemili A, Özkaya E, Gauger N, Carnarius A, Thiele F (2011) Optimal control of unsteady flows using discrete adjoints. In: AIAA paper 2011-3720
10. Nemili A, Özkaya E, Gauger N, Kramer F, Thiele F (2013) Discrete adjoint based sensitivity analysis for optimal flow control of a $3d$ high-lift configuration. In: AIAA paper 2013-2585
11. Nielsen EJ, Jones WT (2011) Integrated design of an active flow control system using a time-dependent adjoint method. Math Model Nat Phenom 6(3):141–165
12. Melton LGP, Yao CS, Seifert A (2006) Active control of flow separation from the flap of a supercritical airfoil. AIAA J 44(1):34–41
13. Patankar SV, Spalding DB (1972) A calculation procedure for heat, mass and momentum transfer in three-dimensional parabolic flows. Int J Heat Mass Transfer 15(10):1787–1806

Part VI
Engineering Design and Societal Applications: Societal Applications

Chapter 28
New Challenges and Opportunities in Reliability and Risk Based Optimization

Sebastian Martorell, Maryory Villamizar, Isabel Martón, Ana Sánchez and Sofia Carlos

Abstract Safety (S) improvement of industrial installations leans on the optimal allocation of designs that use equipment that is more reliable and testing and maintenance activities to assure a high level of reliability, availability and maintainability (RAM) for their safety-related systems. However, this also requires assigning a certain amount of resources (C) that are usually limited. Therefore, the decision-maker in this context faces in general a multiple-objective optimization problem (MOP) based on RAMS+C criteria where the parameters of design, testing and maintenance act as decision variables. A general framework for such MOP based on RAMS+C criteria was proposed in [1]. There, a number of alternatives were proposed based on the use of a combination of RAMS+S formulation and Genetic Algorithms (GAs) based optimization to solve the problem of testing and maintenance optimization based only on system unavailablity and cost criteria. The results showed the capabilities and limitations of alternatives. Based on them, challenges were identified in this field and guidelines were provided for further research. In [2], a full scope application of RAMS+S based optimization using GAs was reported. Since then, the reliability and risk based optimization of design and operation of equipment and facilities has evolved into a set of technical documents, conference contributions and technical papers published elsewhere. Many of them have already addressed to some extent the effect of both random and epistemic uncertainties within this reliability and risk informed decision-making framework. This paper discusses the importance of appropriate formulation, treatment and analysis of model and parameter uncertainties in reliability and risk informed decision-making. It faces on how treatment and analysis of uncertainties should be integrated within an approach for evaluation

S. Martorell (✉) · M. Villamizar · I. Martón · S. Carlos
Department of Chemical and Nuclear Engineering, Universitat Politècnica de València, Camino de Vera 14, 46022 Valencia, Spain
e-mail: smartore@iqn.upv.es

A. Sánchez
Department of Statistics and Operational Research, Universitat Politècnica de València, Camino de Vera 14, 46022 Valencia, Spain

D. Greiner et al. (eds.), *Advances in Evolutionary and Deterministic Methods for Design, Optimization and Control in Engineering and Sciences*, Computational Methods in Applied Sciences 36, DOI 10.1007/978-3-319-11541-2_28

of reliability and risk impact of safety issues, i.e. equipment design, operational requirements, etc. The approach would consist of modeling, assessment and analysis of the safety concern, which is intended to be used within an optimization context to support the decision-making on the most effective safety requirements. The paper focuses on Reliability and Risk of Nuclear Installations, where particular attention is paid to address the effect of uncertainties in the reliability and risk informed optimization of testing and maintenance of safety related equipment. Similar challenges can be observed for many other complex installations, such as energy generation and distributions, process industry, aeronautics, etc.

Keywords Reliability · Risk informed · Optimization · Nuclear power plant

28.1 Introduction

At the last decades many studies have been developed aimed at improving safety systems, with the main focus on implementing an appropriate surveillance and maintenance policy to assure that an acceptable standard of reliability, availability and maintainability (RAM) of the safety systems is kept during all the plant operational life (e.g. testing and preventive maintenance optimization).

Probabilistic Risk Assessment (PRA) is a standard method for assessing, maintaining, assuring and improving plant safety, which integrates RAM models and data. It is the most effective and efficient tool for safety and risk management in Nuclear Power Plants (NPPs). PRA studies not only evaluate risk/safety of systems but also their results are very useful in safe, economical and effective design and operation of NPPs. The latter application is known as Risk-Informed Decision Making.

The risk informed approaches are intended to make design, requirements and activities more risk effective and at the same time utilizing fewer resources by making use of PRA results to focus better on what is critical to safety. The optimization of the test surveillance and maintenance intervals is one of the main issues in risk-informed applications.

28.2 Risk-Informed Decision-Making Regulation

The Nuclear Community has been encouraging the use of PRA to support a risk-informed decision-making framework. In this context, the NRC issued the first draft of Regulatory Guide RG 1.174 in 1998 [3], which remains a major milestone in the NRC initiative to risk-inform the regulations on changes to licensing basis for operation of NPPs.

Since then, the risk-informed process introduced in RG 1.174 has evolved into a suite of regulatory guides and NUREG reports that define an integrated approach to risk-informed regulation [3–9]. Nowadays, there are draft versions of Revision 3 to

RG 1.174 (DG-1285) and Revision 2 to RG 1.177 (DG-1287). RG 1.174 [3] presents a framework umbrella for using PRA in risk-informed decision-making on specific changes to licensing basis, while RG 1.177 [7] proposes a more specific approach that focuses, in particular, on plant specific changes to Technical Specifications (TS), e.g. Limiting Conditions for Operation (LCO) and Surveillance Requirements (SR), which are parts of the licensing basis.

28.3 Overview of Previous Work

The original US NRC policy statement in 1995 and the first drafts of RG 1.174 and RG 1.177 in 1998 already established that all sources of uncertainty must be identified and analysed such that their impacts are understood.

Prior work in this field has already faced the problem of addressing uncertainties in reliability and risk based decision making on changes to licensing basis and particularly to TS [10–16]. However, comprehensive guidance on the systematic treatment of epistemic uncertainties associated with the specific use of the PRA in risk-informed decision making of changes to licensing basis has expanded mainly in the last years [4, 17, 18]. Moreover, no specific guidance has been proposed yet for the treatment and analysis of epistemic uncertainties particularly in evaluating the risk impact of changes to TS based on the use of the PRA; therefore, there was a need of adapting the generic guidance to this particular PRA based application. This was the aim of the work published in Refs. [19–21], which show the origins of a methodology that has evolved into the integrated RIDM approach proposed in this paper.

28.4 An Approach for Evaluation of Reliability and Risk Impact of TS Changes Addressing Uncertainties

RG 1.174 [3] establishes that licensee's risk assessment may be used to address the fourth principle of the integrated risk informed decision-making on plant-specific changes to the licensing basis, where the necessary sophistication of the evaluation of the risk impact of the change depends on the magnitude of the potential risk impact. It discusses the use of PRA results in decision-making consisting of three parts:

1. A fundamental element is a PRA of sufficient scope, level of detail, and technical acceptability for the intended application.
2. PRA results are to be used in two ways, i.e. to assess the overall baseline CDF/LERF and to assess the CDF/LERF impact of the proposed change as compared to the acceptance guidelines.
3. One of the strength of the PRA framework is its ability to characterize the impact of the uncertainty in the analysis, and it is essential that these uncertainties be

recognized when assessing whether the fourth principle is being met, so that guidelines on how the uncertainty has to be addressed must be provided.

What concerns the third part, RG 1.174 [3] introduces the following sources of uncertainty linked to PRA. Uncertainties can be categorized as either random or epistemic. Random uncertainty reflects our inability to predict random observable events. Epistemic uncertainty represents our confidence in the model and in the numerical values of its parameters. This type of uncertainty is also called 'state-of-knowledge' uncertainty or just 'uncertainty'. Epistemic uncertainties can be split into three categories [3]:

1. Parameter uncertainty is that which relates to the parameters of the PRA, given a choice of model. Even with a known model, the parameter values may still be unknown. Examples of parameter uncertainties include equipment failure rates, initiating-event frequencies, and human error probabilities.
2. Model uncertainty relates to the uncertainty in the assumptions made in the analysis and the models used. In many cases, there is limited knowledge and some disagreement on the proper model to represent a system. The result is that for a particular process, there are multiple competing models, each of which necessarily produces a different approximation of the same real-world system. Because the correct model is unknown, there is additional uncertainty in the output of any model, representing the uncertainty in the model's itself.
3. Completeness uncertainty represents the uncertainty due to the portion of risk that is not explicitly included or modeled in the PRA.

General guidance on addressing uncertainties from modeled and non-modeled risk contributors in the context of RG 1.174, i.e. identification of sources key to decision, treatment and analysis of uncertainties, are specifically addressed in NUREG-1855 [4] and EPRI-1026511 [17]. While the analysis of parametric uncertainty and, to some extent, model uncertainty is fairly mature, the analysis of completeness uncertainty cannot be handled in a similar formal manner. So that, only parameter and model uncertainties will be addressed in detail in the approach proposed in this paper.

Figure 28.1 presents the three steps based approach proposed for the evaluation of reliability and risk impact of changes to Technical Specifications, i.e. risk modelling, risk assessment and risk analysis, which is based on the use of the PRA and includes identification, treatment and analysis of uncertainties in an integrated manner. The approach proposed is coherent with the general principles of risk-informed decision-making framework on plant-specific changes to the licensing basis as introduced above.

What concerns reliability and risk modelling, it emphasizes first the identification of not only the usually addressed sources of uncertainty linked to PRA models and data but also the sources of model and parameter uncertainties associated with the assumptions in Surveillance Requirement change evaluation [7].

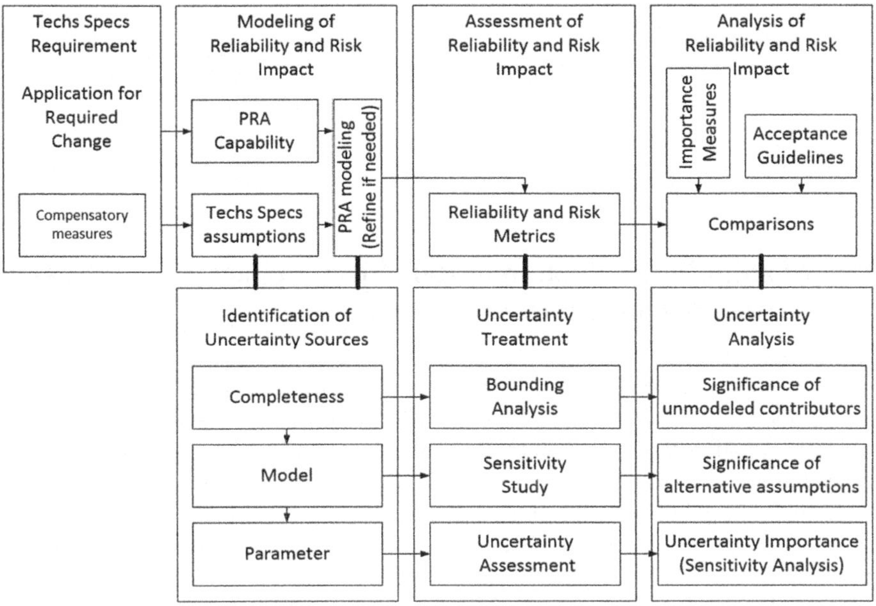

Fig. 28.1 Schematic view of the PRA based approach to analyzing changes to Technical Specifications

Reliability and risk assessment is proposed by formulating the usual reliability and risk metrics for analyzing Surveillance Requirement changes in the literature [5, 22, 23]. It is proposed treatment of model and parameter uncertainties based on traditional sensitivity studies and uncertainty assessment respectively, the latter based of the probabilistic approach for uncertainty formulation and propagation by standard Monte Carlo Sampling technique [21].

Reliability and risk analysis is based on the well established comparison of the assessment of reliability and risk impact of the change including treatment of uncertainties with acceptance guidelines [3]. In addition, it is proposed also the use of both traditional importance measures and uncertainty importance measures (sensitivity analysis) in order to analyze respectively how basic events and parameter uncertainties influence the reliability and risk impact of the change proposed and its uncertainty [24, 25]. The former helps to identify the main risk contributors and eventually the chance for adopting compensatory measures in order to the make the proposed change acceptable. Uncertainty importance measures help to identify which of uncertain parameters are most significant contributors to reliability and risk impact uncertainty and eventually the need of limiting the impact of the effect of particular uncertainties or of developing additional treatment of uncertainties.

Table 28.1 SIL for low demand mode of operation

SIL	PFD_{avg}
4	$\geq 10^{-5}$ to $< 10^{-4}$
3	$\geq 10^{-4}$ to $< 10^{-3}$
2	$\geq 10^{-3}$ to $< 10^{-2}$
1	$\geq 10^{-2}$ to $< 10^{-1}$

28.5 Example of Evaluation of Reliability Impact of a TS Change for the Reactor Protection System

The Reactor Protection System (RPS) is a very important system for NPP safety, as it is responsible for shutting down the NPP by means of reactor trip in case of accidental conditions. So that, current regulation requires a very high reliability of the system in performing its vital and required safety function, i.e. reactor trip.

The RPS is a sort of Safety Instrumented System (SIS) [26]. A SIS is defined as an "instrumented system used to implement one or more safety instrumented control functions. A SIS is composed of any combination of sensors, logic solver and final elements". The standard IEC 61508 requires every safety function to achieve a determined Safety Integrity Level (SIL). For low demand operating systems the SIL levels are defined in terms of average probability of failure on demand (PFD_{avg}, see Table 28.1) [27, 28].

Surveillance Requirements (SR) are part of Technical Specifications that are included into the Licensing Basis (LB) for operation of Nuclear Power Plants. Surveillance tests aim at limiting risk of undetected downtimes of safety related equipment by imposing equipment operability checks, which consist of testing of equipment operational parameters with established Surveillance Frequency (SF) and Test Strategy.

Surveillance testing is executed periodically, and can be implemented using several strategies. The strategy establishes how the tests of the redundant components are scheduled with respect to one another. IEC 61508 (1998–2005) [27, 28] defines proof test as a "periodic test performed to detect failures in a safety-related systems so that the system can be restored to an 'as new' condition or as close as practical to this condition". This standard establishes the need of routine maintenance action in order to detect unrevealed failures, being proof test one of these activities. It thus has an important role in the achievement of safety integrity.

Surveillance requirements were established taking into account mainly deterministic criteria. A number of problems have been identified connected to SR that can jeopardize plant safety. The development of Probabilistic Risk Assessment and its application since the early 80s to analyze SR changes has brought the opportunity to review SR consistency from a reliability-risk viewpoint, i.e. addressing the impact of the changes on plant safety on the basis of the reliability-risk information provided by the PRA, with particular attention to the role of the Surveillance Frequency included within Surveillance Requirements.

The case study presents the results of the evaluation of the risk impact of a Surveillance Frequency change of circuit breakers of the Reactor Protection System of a PWR Nuclear Power Plant, which could be used to obtain a refined model for the MC-GA based SF optimization.

28.5.1 Problem Description

Figure 28.2 shows a schematic view of the RPS, which consists of two redundant and identical channels, A and B. The RPS can be actuated manually from the control room or automatically after a signal requiring reactor trip. This signal opens circuit breakers 52RTA and 52RTB. Manual or automatic actuation of the RPS deenergizes control rods, which are inserted in the reactor core by gravity. The by-pass breakers 52BYA and 52BYB allows testing of the main circuit breakers above using manual trip/rep.

Surveillance Requirements of the RPS establish a functional test of the redundant channels with a Surveillance Test Interval (TI), i.e. 1/SF, of two months and sequential testing. This means the first circuit breaker, e.g. 52RTA, is tested the first month and the second one, 52RTB, is tested the next month. The TI change consist of extending the current TI from 2–3 months; i.e. 1,440–2,160 h.

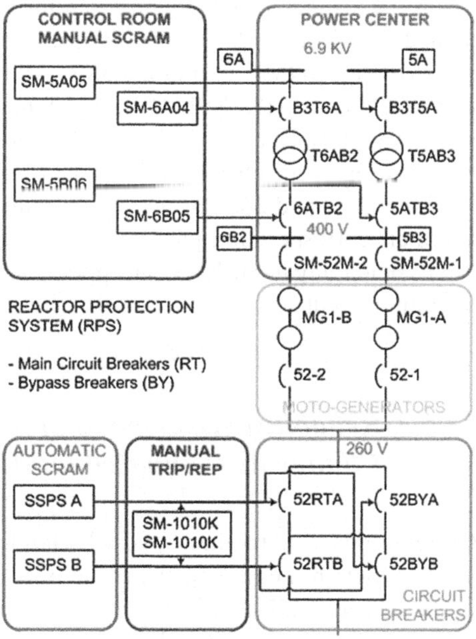

Fig. 28.2 Schematic view of the RPS

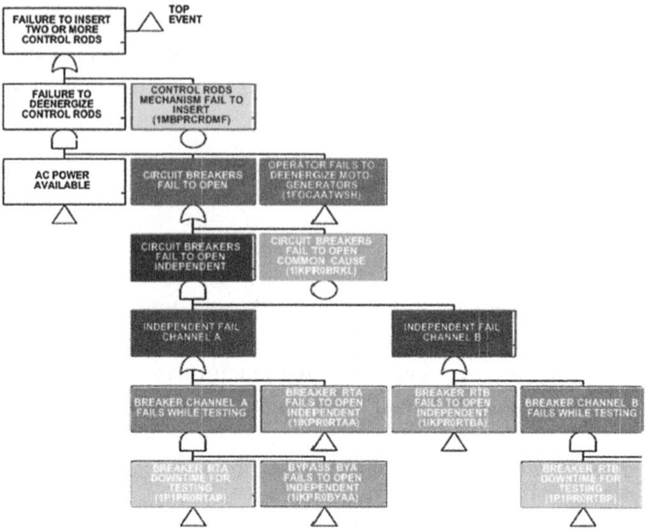

Fig. 28.3 Fault tree for RPS failure

28.5.2 Plant Specific PRA

Figure 28.3 shows a fault tree representing the failure of the RPS as top event included in the Level 1 PRA. Basic events representing control rods mechanism fail to insert, i.e. 1MBPRCRDMF, common cause failure (CCF) of the circuit breakers 52RTA and 52RTB, i.e. 1IKPR0BRKL, and human error to disconnect motor generators MG1-A y MG1-B, i.e. 1FOCAATWSH, are highlighted in addition to the AND gate representing occurrence of independent failures of RPS channels A and B.

Tables 28.2 and 28.3 present models and data used for modeling the basic events and their corresponding parameters in current PRA only for most important basic events belonging to the RPS that appear in the final Boolean equation after generating the minimal cut sets using a cut-off criteria 10^{-12}.

28.5.3 Reliability Assessment of the RPS Before and After the SF Change as Compared to SIL Levels

Current regulation requires a very high reliability of the RPS in performing its vital safety function, i.e. reactor trip. Table 28.4 shows there is no significant reduction of the RPS reliability after the SF extension (base case), which remains very high. See also the results in Figs. 28.4 and 28.5. Table 28.4 summarizes also the results of a number of sensitivity studies performed in order to estimate the impact of model uncertainties on the assessment of risk impact of the SF change including parameter

Table 28.2 More important basic events

Basic event	Description	BE model	Unavailability formula (u)
1MBPRCRDMF	Control rods mechanisms fail to insert	Standby	$1/2^*\lambda_R{}^*MTBRT$
1FOCAATWSH	Operator fails to deenergize generators	Demand	ρ_D
1IKPR0BRKL	Circuit breakers common cause fail to open (CCF)	Standby	$1/2^*\lambda_C{}^*TI$ (sequential tests)
1IKPR0RTAA 1IKPR0RTBA 1IKPR0BYAA 1IKPR0BYBA	Circuit breaker independent fail to open	Standby	$1/2^*\lambda_I{}^*TI$
1FOPRATWSH	Operator fails to manually scram reactor	Demand	ρ_S
1P1PR0RTAP 1P1PR0RTBP	Downtime for testing	Test	τ/TI

Table 28.3 Data and parameters of basic events

Parameter	Description	Parameter type	Parameter/value
λ_R (h^{-1})	Control rods mechanisms failure rate	PDF	Gamma (4.32×10^{-9}; 0.49; 1.13×10^{8})
ρ_D (−)	Human error probability to deenergize generators	PDF	Log-normal (1.05×10^{-1}; 5)
λ_C (h^{-1})	Circuit breakers common cause failure rate	PDF	$\lambda_C = \beta * \lambda_I$ Log-normal (1.21×10^{-7}; 3)
β (−)	β-factor for Circuit breakers common cause failure rate	Constant	0,1277 (sequential tests)
λ_I (h^{-1})	Circuit breaker independent failure rate	PDF	Gamma (9.47×10^{-7}; 1.49; 1.57×10^{6})
ρ_S (−)	Human error probability to manually scam reactor	PDF	Log-normal (1.66×10^{-1}; 5)
MTBRT (h)	Mean time between reactor trips	Constant	2,160 (true trips) [Max TBRT = 13,140]
TI (h)	Test Interval (1/SF)	Constant	1,440 (current TI in TS)
τ (h)	Test time	Constant	1.37

uncertainty as well (see also Fig. 28.4). In addition, comparing results in Table 28.4 (Fig. 28.4) with SIL requirements in Table 28.1, the highest SIL level 4 is achieved any case.

Table 28.4 Results of quantification of unavailability for the base case and sensitivity studies

	Unavailability RPS—2 months			Unavailability RPS—3 months		
	Mean	5 %	95 %	Mean	5 %	95 %
Base case	1.38E-5	1.64E-6	3.97E-5	1.83E-5	2.19E-6	5.78E-5
Sensitivity						
MTBRT	3.74E-5	2.83E-6	1.18E-4	4.20E-5	3.66E-6	1.36E-4
FFNCT	1.38E-5	1.66E-6	3.56E-5	1.61E-5	1.92E-6	4.21E-5
HEP “rookie”	2.29E-5	2.15E-6	7.53E-5	3.20E-5	3.63E-6	1.122E-4
Test strategy staggered	7.11E-6	5.31E-7	2.21E-5	8.33E-6	7.71E-7	2.60E-5

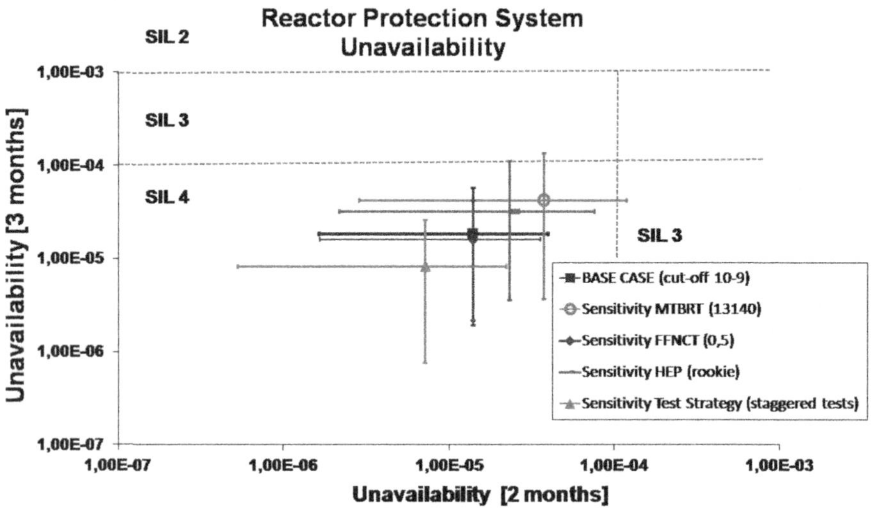

Fig. 28.4 Unavailability of RPS before and after SF change

28.5.4 Measures of Importance of Basic Events and Parameters

Table 28.5 summarizes the results derived for the traditional importance measures for RPS Unavailability before the SF change (2 months) and ranked according to RAW.

Reference [29] proposes also the use of threshold values to determine the value of RAW below which we deem the basic event to be not risk-important. Threshold values of RAW are given for RPS Unavailability as compared to SIL levels using the following equations:

$$\mathrm{RAW}_{\mathrm{U,\,SIL\,x}} = \frac{\mathrm{U}_{\mathrm{SIL\,x}}}{\mathrm{U}} \tag{28.1}$$

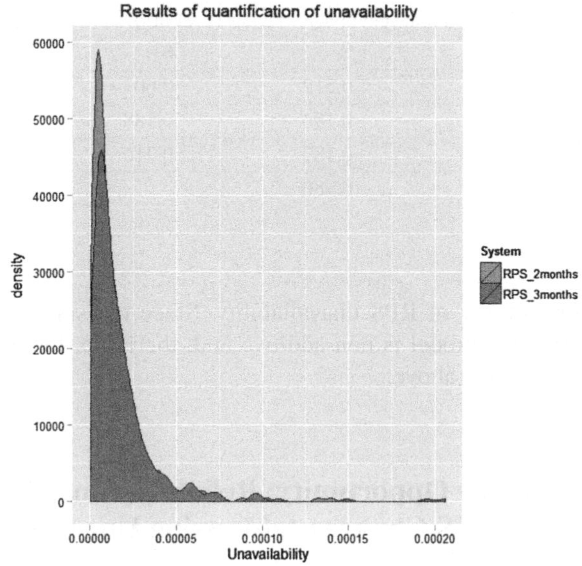

Fig. 28.5 PDF of unavailability RPS for the base case

Table 28.5 Traditional importance measures of basic events (2 months)

Basic event	Unavailability	FV	RAW	RRW
1MBPRCRDMF	4.67×10^{-6}	3.38×10^{-1}	72330	1.51
1IKPR0BRKL	8.70×10^{-5}	6.61×10^{-1}	7596	2.95
1BLBC0G1CF	2.24×10^{-6}	1.41×10^{-5}	7.3	1
1FOCAATWSH	1.05×10^{-1}	6.63×10^{-1}	6.65	2.96

The following set of threshold RAW values for U is obtained using Eq. (28.1) and corresponding SIL levels:

$$RAWU, SIL\,x = \{7246;\ 724.6;\ 72.46;\ 7.26\} \tag{28.2}$$

Comparing RAW of basic events in Table 28.5 with threshold values in Eq. (28.2) it is found that uncertainty of basic events 1MBPRCRDMF and 1IKPR0BRKL is very significant, e.g. RAW beyond 7300, which means RPS reliability could move from current SIL 4 to SIL 1.

Table 28.6 summarizes the results derived for the uncertainty importance measures for RPS Unavailability before the SF change (2 months). In particular, Sobol and Spearman indices are presented. Both Spearmen and Sobol indices rank parameters in the same position.

Table 28.6 results show, based on Sobol indices, the probability of fail to insert control rods mechanisms is the parameter uncertainty, which influence the most

Table 28.6 Global sensitivity analysis (2 month)

Parameter	S_i	S_{Ti}	$S_{Ti} - S_i$	Spearman
ρ_D	0.612	0.601	−0.01	0.576
λ_R	0.186	0.188	0.002	0.566
λ_C	0.187	0.216	0.029	0.422
Total	0.980	–	–	–

uncertainty of RPS unavailability. Since the sum of all indices S_i is less than 1, then, the model is non-additive and, therefore, there are interactions among input parameters above.

28.6 The Opportunity: Reliability and Risk Based Optimization of TS Changes Addressing Uncertainties

Accounting for the state-of-art, there is an opportunity to make significant progress in this field by means of the integration of the previous approach for evaluation of reliability and risk impact into a multi-objective optimization problem of TS changes addressing uncertainties.

For sake of clarity in the sequel only a simplified A[U]+C problem is considered because of the generalization is straightforward. The general multi-objective optimization problem (MOP) can be formulated for the A[U]+C problem based on the tolerance interval method to minimize the vector of multi-objective uncertain functions, **f**(**x**), subject to the vector of uncertain constraints, **g**(**x**):

$$\mathbf{f}(\mathbf{x}) = \{U\{\mathrm{U}(\mathbf{x})\}_{\gamma/\beta}, U\{\mathrm{C}(\mathbf{x})\}_{\gamma/\beta}\}$$
$$\mathbf{g}(\mathbf{x}) = \{U\{\mathrm{U}(\mathbf{x})\}_{\gamma/\beta} \leq \mathrm{U_L}, U\{\mathrm{C}(\mathbf{x})\}_{\gamma/\beta} \leq \mathrm{C_L}\}$$

Figure 28.6 shows the U-C plot of the possible solutions to the MOP considering herein uncertain objectives and constraints.

Figure 28.7 shows the framework adopted in previous works (see for example Ref. [30]) for the reliability and risk based optimization of Technical Specification changes addressing uncertainties using the principles depicted in Fig. 28.6.

In Fig. 28.7 it is shown that only comparison of the reliability and risk assessment results against safety goals, which is a small part of the reliability and risk analysis, third step, in Fig. 28.1, is considered by means of the formulation of the constraint function. Thus, only steps 1 and 2, of the three steps based approach proposed in this paper, were fully accounted for in previous work.

Figure 28.8 shows a new framework proposed in this paper for the reliability and risk based optimization of Technical Specification changes addressing treatment and analysis of uncertainties in an integrated manner.

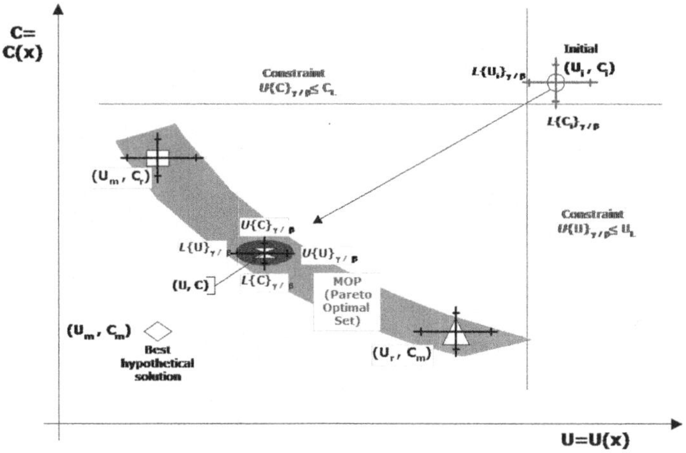

Fig. 28.6 Objective space of feasible solutions for MOP

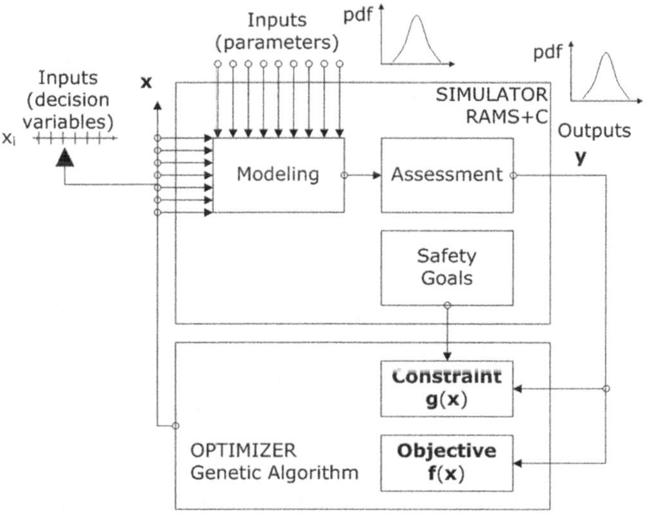

Fig. 28.7 Basic optimization approach

28.7 Future Challenges

The framework depicted in Fig. 28.8 introduces consideration of step 3, i.e. reliability and risk analysis, in full.

Thus, major challenge involves integration of the three steps based approach within the reliability and risk based optimization of TS changes addressing uncertainties. In particular, the following topics bring new challenges requiring further research.

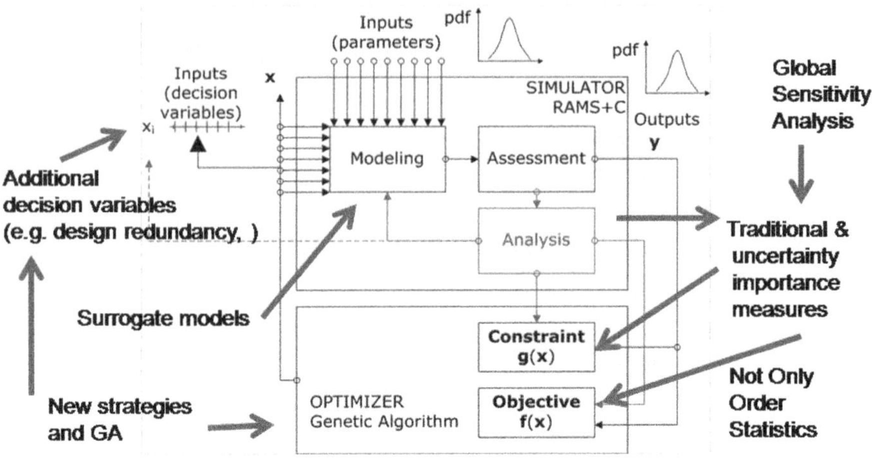

Fig. 28.8 Extended optimization approach

- Integration of the above approach within a reliability and risk based optimization framework can benefit of previous results presented on the combined use of MC-GA.
- Decision variables: New (additional) variables may be introduced into the decision-making because of the analysis of the importance of uncertainties, e.g. adoption of design changes like a sort of compensatory measure.
- Modeling: Move from model to parameter uncertainty as much as possible. In addition, use of surrogate models may be necessary to reduce computational cost based on the results of classical and uncertainty importance measures computation.
- Assessment: Account for uncertainty introduced by meta-models as compared with PRA based models. Adoption of uncertainty reduction methods.
- Analysis: Use of traditional versus uncertainty importance measures. Link to global sensitivity analysis with/without meta-modeling.
- Optimization: Introduce additional variables in the decision-making. Introduce the results of not only the reliability and risk assessment but also the results of reliability and risk analysis in the decision making, for example, include classical and uncertainty importance measures as part of the objective and/or constraint functions.
- Optimizer: Develop efficient EA to cope with previous MOP formulation.

Acknowledgments Authors are grateful to the Spanish Ministry of Science and Innovation for the financial support of this work (Research Project ENE2010-17449).

References

1. Martorell S, Sánchez A, Carlos S, Serradell V (2004) Alternatives and challenges in optimizing industrial safety using genetic algorithms. Reliab Eng Syst Saf 86(1):25–38
2. Martorell S, Villanueva JF, Carlos S, Nebot Y, Sánchez A, Pitarch JL, Serradell V (2005) RAMS+C informed decision-making with application to multi-objective optimization of technical specifications and maintenance using genetic algorithms. Reliab Eng Syst Saf 87(1):65–75
3. RG 1.174 (2011) An approach for using probabilistic risk assessment in risk-informed decisions on plant-specific changes to the licensing basis, Revision 2, USNRC
4. NUREG 1855 (2009) Guidance on the Treatment of Uncertainties Associated with PRAs in Risk-Informed Decisionmaking, Vol 1, USNRC (see also NUREG 1855 Draft Report, Revision 1, for Comment March 2013)
5. NUREG/CR-6141 (1995) Handbook of methods for risk-based analyses of technical specifications, USNRC
6. NUREG-0800 (2007) Review of risk information used to support permanent plant-specific changes to the licensing basis: general guidance, Section 19.2, USNRC
7. RG 1.177 (2011) An approach for plant-specific, risk-informed decision making: technical specifications, Revision 1, USNRC
8. RG 1.200 (2009) An approach for determining the technical adequacy of probabilistic risk assessment results for risk-informed activities, Revision 2, USNRC
9. RG 1.201 (2005) Guidance for categorizing structures, systems, and components in nuclear power plants according to their safety significance, Revision 1, USNRC
10. Cepin M (2005) Analysis of truncation limit in probabilistic safety assessment. Reliab Eng Syst Saf 87:395–403
11. Martorell S, Sánchez A, Carlos S (2007) A tolerance interval based approach to address uncertainty for RAMS+C optimization. Reliab Eng Syst Saf 92:408–422
12. Martorell S, Sánchez A, Villanueva JF, Carlos S, Serradell V (2008) A multi-objective genetic algorithm for RAMS+C optimization with uncertain decision variables. In: Proceedings of IMechE, Vol 222 Part O: J Risk Reliab. JRR140. IMechE 2008, pp 153–160
13. Martorell S, Villanueva JF, Nebot Y, Carlos S, Serradell V (2004) Current trends in risk-informed changes to limiting conditions for operation. In: Proceedings of 15th European safety and reliability conference (ESREL 2004), Berlin
14. Sánchez A, Carlos S, Martorell S, Villanueva JF (2009) Addressing imperfect maintenance modelling uncertainty in unavailability and cost based optimization. Reliab Eng Syst Saf 94:22–32
15. Smith CL, Knudsen JK, Calley MB (1999) Calculating and addressing uncertainty for risk-based allowable outage times. Reliab Eng Syst Saf 66:41–47
16. Volkanovski A, Cepin M (2011) Implication of PSA uncertainties on risk-informed decision making. Nucl Eng Des 241:1108–1113
17. EPRI Report (2012) Practical guidance on the use of pra in risk-informed applications with a focus on the treatment of uncertainty. EPRI-1026511, Draft report, August 2012
18. EPRI-1016737 (2008) Treatment of parameter and model uncertainty for probabilistic risk assessments. EPRI, Palo Alto
19. Martorell S, Villamizar M, Villanueva JF, Carlos S, Sanchez AI (2010) Risk-informed decision-making on changes to allowed outage times addressing uncertainties. In: Proceedings of European safety and reliability conference (ESREL 2010), pp 2278–2286
20. Martorell S, Villamizar M, Villanueva JF, Carlos S, Sanchez AI (2011) Importance analysis in risk-informed decision-making of changes to allowed outage times addressing uncertainties. In: Proceedings of European safety and reliability conference (ESREL 2011), pp 2143–2151
21. Martorell S, Villamizar M, Villanueva JF, Serradell V, Sanchez AI (2009) Addressing uncertainties in risk-informed decision-making of changes to nuclear power plant technical specifications. In: Proceedings of European safety and reliability conference (ESREL 2009), pp 353–359

22. Kim IS, Martorell S, Vesely WE, Samanta PK (1992) Quantitative evaluation of surveillance test intervals including test-caused risks. NUREG/CR-5775, BNLNUREG-52296, Feburary 1992
23. Kim IS, Martorell SA, Vesely WE, Samanta PK (1994) Risk analysis of surveillance requirements including their adverse effects. Reliab Eng Syst Saf 45(3):225–234
24. Aven T (2010) On the use of uncertainty importance measures in reliability and risk analysis. Reliab Eng Syst Saf 95(2010):127–133
25. Borgonovo E, Apostolakis GE, Tarantola S, Saltelli A (2003) Comparison of global sensitivity analysis techniques and importance measures in probabilistic safety assessment. Reliab Eng Syst Saf 79(2):175–185
26. Torres-Echeverria AC, Martorell S, Thompson HA (2009) Modelling and optimization of proof testing policies for safety instrumented systems. Reliab Eng Syst Saf 94(4):838–854
27. International Electrotechnical Commission (1998–2005) IEC 61508. Functional safety of electri-cal/electronic/programmable electronic safety-related systems. Parts 1–7. Geneva, Switzerland
28. International Electrotechnical Commission (2003) IEC 61511 Functional safety–safety instrumented systems for the process industry sector. Parts 1–3. Switzerland
29. Reinert JM, Apostolakis GE (2006) Including model uncertainty in risk-informed decision making. Ann Nucl Energy 33:354–369
30. Villanueva JF, Sanchez AI, Carlos S, Martorell S (2008) Genetic algorithm-based optimization of testing and maintenance under uncertain unavailability and cost estimation: a survey of strategies for harmonizing evolution and accuracy. Reliab Eng Syst Saf 93(12):1830–1841

Chapter 29
Bi-objective Discrete PSO for Service-Oriented VRPTW

Julio Brito, Airam Expósito and José A. Moreno-Pérez

Abstract In this paper we deal with a variant of the VRPTW that is oriented to the quality of service to customers. In this model, we incorporate a measure of quality associated with the time the vehicles reach customers within their time window as an objective. We apply a bi-objective discrete PSO to deal with the problem. The procedure performance is analyzed on classical and real data based instances.

Keywords Multicriteria optimization · Vehicle routing · Particle swarm optimization

29.1 Introduction

Distribution route planning plays an important role in supply chain management, among other reasons, because it improves quality of service and customer satisfaction while reducing costs. Vehicle Routing Problems (VRPs) is a well-known class of problems consisting in finding the best set of routes for a set of vehicles in order to serve the demand for goods of a set of geographical scattered customers [39]. The usual objective of these problems is to determine the set of feasible routes for the available vehicles that minimizes the total cost of operations, generally expressed in term of distances or times. However, in order to guarantee a minimum quality of service, in the variant Vehicle Routing Problem with Time Windows (VRPTW), each customer fixes a time window within which the service has to be completed. However, with present-day competitive markets, companies strategically increase their interest in the quality of service by increasing customer satisfaction. Thus,

J. Brito (✉) · A. Expósito · J.A. Moreno-Pérez
I.U.D.R., University of La Laguna, 38271 La Laguna, Spain
e-mail: jbrito@ull.es

A. Expósito
e-mail: aexposito@ull.es

J.A. Moreno-Pérez
e-mail: jamoreno@ull.es

D. Greiner et al. (eds.), *Advances in Evolutionary and Deterministic Methods for Design, Optimization and Control in Engineering and Sciences*, Computational Methods in Applied Sciences 36, DOI 10.1007/978-3-319-11541-2_29

instead of guaranteeing a minimum quality of service by completing the service within the customers time windows, companies try to service them as soon as possible within their corresponding time windows. Therefore we consider a bi-objective VRPTW where, in addition to the classical cost reduction, we consider a second minimization objective by measuring the quality of service through the average proportion on the time windows that customers have to wait for service.

VRP models can be found in the literature whose aims are to prioritize the satisfaction of customers needs. These models tackle the problem as an urgent supply of goods, that is, situations such as logistics under natural disasters, picking up children to go to school or maintenance services, all of which can be formulated with objectives and constraints that are similar to the proposed model. Among the models that can be found include the Multiple Travelling Repairmen Problem (M-TVRP) whose objective is to minimize total time delay or customer arrival [18]; The bus school routing problem [14] whose objectives, among others, is to minimize average distance travelled, or the average time used by students in arriving to their school; and the cumulative capacitated vehicle routing problem (CCVRP) a transportation problem which objective is to minimize the sum of arrival times at customers, instead of the classical route length, subject to vehicle capacity constraints. This type of problem is associated with the satisfaction of customer need, e.g. vital goods supply or rescue after a natural disaster [10, 31]. The orientation of these models does not emphasize the need for insertion of time window constraints are they are not considered, hence no references with these constraints appear in the literature. In these models, just as in the proposed model in this paper, the vehicles must arrive to all of the customers as soon as possible. In our model we specifically identify this objective, that is, to arrive as soon as possible within the established time window .

The approach considered for the problem is a bi-objective and discrete version of the Particle Swarm Optimization method. Particle Swarm Optimization (PSO) is a bio-inspired evolutionary metaheuristic proposed by Kennedy and Eberhart [19]. Similar to Genetic Algorithms, PSO is a population-based technique, inspired by the behavior of a population in the search for an optimum. PSO has been successfully applied to a wide range of industrial optimization problems due to its relatively easy implementation and high performance with low demand for computational resources. An extensive survey of PSO applications can be found in [15, 34].

Several adaptations of PSO that deal with discrete and multi-objective problems have been tested in the Literature. Particular domains with discrete space solutions, such as VRPTW, require to redefining how a particle moves in a discrete space and hence to redefining the mechanism for updating the velocity and position of particles. Kennedy and Eberhart proposed a discrete variant of the original PSO [20] limited to binary-valued solution elements. They interpreted changes of velocity in terms of probabilities by encoding the particle's position as a binary vector and using a stochastic velocity scheme. Since then, several authors have used this approach (e.g. [8, 28]). Other implementations of Discrete Particle Swarm Optimization (DPSO) have been introduced e.g. Yang et al. [43] that suggest a different way to update the velocity of particle. Al-kazemi and Mohan [4] use a method whose particles are influenced alternatively by the best position of their particular route and its neighbourhood.

Pampara et al. [32] use the angular modulation with four parameters in continuous PSO.

DPSO has also been applied to the VRP, however it has hardly ever been considered in the VRPTW case. Some references apply this metaheuristic to the VRP and its variants in [35]. Chen et al. [9] proposed a hybrid approach with Simulated Annealing (SA) to tackle the VRP. Ai and Kachitvichyanuku [2] solve the Capacitated VRP with two specific representations of solutions for the PSO, an extension of the same algorithms is applied to the VRPTW [1] and to the simultaneous Pick and Delivery VRP [3]. Zhen et al. [44] and Wang et al. [41] present other versions of the DPSO for Open VRP. In addition, Zhu et al. [45] show an improved adaptation of PSO for the VRPTW, a very basic implementation of the DPSO applied to some short made-up instances of the VRPTW . Sun et al. [38] developed a DPSO for this problem using a number of heuristics for the initialization of solutions and solution generations. Other recent versions of the DPSO for the VRPTW are proposed in [17, 23–26, 42]. One paper that stands out for its results is Marinakis et al. [26] who proposed another alternative to tackle the VRP which combine a DPSO with three other algorithms.

In multi-objective problems such as the proposed VRPTW it is necessary to compare the solutions in their different objectives. Several papers have suggested the adaptation of PSO to the Multiple Objective Optimization (MOO). Coello et al. [11] state that three main considerations must be taken into account when dealing with MOO using PSO:

1. The selection of the particles that will act as leaders.
2. The storage of non-dominated solutions found by the swarm.
3. The preservation of diversity within the swarm.

An important number of multi-objective PSO have been proposed in the literature with diverse approaches to tackle these considerations. In [5] and [36] a review of multi-objective PSO and its applications can also be found, including the VRPTW. In 2011, Li [22] proposed a multi-objective DPSO to solve a VRPTW. More recently others multi-objective DPSO that tackle the VRPTW, such as those proposed by Muñoz-Zavala et al. [30, 33], Shurog et al. [46] and Castro et al. [6, 7].

We consider a version known as Jumping Frog Optimization that can be used for this kind of discrete problems, the VRPTW. The Jumping Frog Optimization (JFO) approach proposed in [13, 27, 29] is based on the particles point of view instead of the solutions or particle's position. JFO have been applied to solve diverse optimization problems [12, 16, 37, 40] including the VRPTW [6, 7].

The remainder of this paper is organized as follows. Section 29.2 describes the formulation of the problem, the service oriented VRPTW. Section 29.3 describes the algorithm proposed JFO, a discrete variant of PSO which is used in this paper to tackle the problem. Next Sect. 29.4 analyses the performance of the proposed approach. Finally, in Sect. 29.5, we give some concluding remarks.

29.2 Formulation of the Problem

The VRPTW is given by a set of k vehicles to serve a set of n customers within given time windows, to find the set of corresponding routes that minimizes the operational costs. Each vehicle goes by a route that starts and ends at a depot visiting a number of nodes satisfying their demands. The assumptions of the VRPTW model are:

- Each vehicle is assigned to only one route.
- Each customer is visited by one and only one vehicle.
- Each route begins and ends at the depot.
- Each vehicle has a container with a capacity limitation and the total loading of each vehicle cannot exceed its capacity.
- Each customer is served within the time window.

We consider the following indices and model parameters:

- The vehicles indexes are: $k = 1, \ldots, m$;
- The customers indexes are: $i = 1, \ldots, n$;
- The depot index is $i = 0$;
- The amount of customer demand i is $q_i, i = 1, \ldots, n$;
- The capacity of vehicle k is $Q_k, k = 1, \ldots, m$;
- The cost (distance or time) of traveling from customer i (or depot 0) to j is $c_{ij}, i, j = 0, \ldots, n$;
- The unloading time at customer i is $u_i, i = 1, \ldots, n$;
- The time window of customer i is $[e_i, l_i]$; $i = 1, \ldots, n$; e_i and l_i are the respective earliest and latest time for serving customer i.

The decision variables are:

- Binary variables $x_{ij}^k, i, j \in [1, \ldots, n], k \in [1, \ldots, m]$, where $x_{ij}^k = 1$ if vehicle k goes from i to j and $x_{ij}^k = 0$ otherwise.
- Continuous variables r_i^k representing the vehicle k load when it reaches customer i. If vehicle k goes from i to j ($x_{ij}^k = 1$) then $r_j^k = r_i^k - q_i$.
- Continuous variables s_i^k representing the time when vehicle k starts to serve customer i. Analogously, if $x_{ij}^k = 1$ then $s_j^k = \max\{e_j, s_i^k + u_i + t_{ij}\}$.

The objectives are to optimize the operational costs and a measure of the quality of service based on the time that customers have to wait within their time windows for service; namely, we use the weighted average of these times. Then the three-indices formulation of the bi-objective extension of the VRPTW as an LP problem is as follows:

$$\text{Minimize} \sum_{k=1}^{m} \sum_{i=0}^{n} \sum_{j=0}^{n} c_{ij} x_{ij}^k \tag{29.1}$$

$$\text{Minimize} \frac{1}{n} \sum_{k=1}^{m} \sum_{i=0}^{n} \frac{s_{ik} - e_i}{l_i - e_i} \tag{29.2}$$

Subject to:

$$\sum_{k=1}^{m} \sum_{i=0}^{n} x_{ij}^k = 1, \quad j = 1, \ldots, n \tag{29.3}$$

$$\sum_{k=1}^{m} \sum_{i=0}^{n} x_{ji}^k = 1, \quad j = 1, \ldots, n \tag{29.4}$$

$$\sum_{i=0}^{n} x_{ij}^k = \sum_{i=0}^{n} x_{ji}^k = 1, \quad j = 1, \ldots, n, k = 1, \ldots, m \tag{29.5}$$

$$\sum_{i=1}^{n} x_{0i}^k = 1, \quad k = 1, \ldots, m \tag{29.6}$$

$$\sum_{i=1}^{n} x_{i0}^k = 1, \quad k = 1, \ldots, m \tag{29.7}$$

$$r_j^k - q_i - r_i^k \leq M(1 - x_{ij}^k), \quad i = 0, \ldots, n, \quad j = 1, \ldots, n, \quad k = 1, \ldots, m \tag{29.8}$$

$$r_0^k \leq Q_k, \quad k = 1, \ldots, m \tag{29.9}$$

$$s_i^k + t_{ij} + u_i - s_j^k \leq M'(1 - x_{ij}^k), \quad i, j = 1, \ldots, n, \quad k - 1, \ldots, m \tag{29.10}$$

$$e_i \leq s_i^k, s_i^k + u_i \leq l_i, \quad j = 1, \ldots, n, \quad k = 1, \ldots, m \tag{29.11}$$

$$x_{ij}^k \in \{0, 1\}, r_i^k \geq 0, s_i^k \geq 0, \quad k = 1, \ldots, m, \quad i, j = 0, \ldots, n \tag{29.12}$$

Equations (29.1) and (29.2) represent the objective functions in terms of costs and quality, respectively. Constraints (29.3) and (29.4) guarantee that only one route/vehicle enters and leaves from one node or that each customer is served exactly once. Constraints (29.5) establish the conditions to maintain continuity of the route, if a vehicle leaves a customer then it has reached it. Constraints (29.6) and (29.7) ensure that each vehicle leaves the depot and returns to it, thereby limiting vehicle use to one trip. Constraints (29.8) and (29.9) where M is a large scalar, establish the relationships between variables r_i^k and ensure that total customer demand in any route does not exceed the corresponding vehicle capacity Q_k. Constraints (29.10) state that a vehicle k cannot arrive at j before $s_i^k + t_{ij}$ if it is travelling from i

to j and guarantee subtours eliminations (M' is a large scalar). Constraint (29.11) ensure that time windows are observed. Finally (29.12) establish the conditions of the variables.

29.3 The Jumping Frog Optimization

Particle Swarm Optimization (PSO) is a nature-inspired metaheuristic that has been applied with success in many areas and appears to be a suitable approach for a wide range of optimization problems [21]. PSO is a population-based technique, inspired by the social behaviour of individuals (or particles) inside swarms in nature (for example, flocks of birds or schools of fish). It is easy to implement, requiring few parameter settings and computational memory. The original PSO algorithm can only optimize problems in which the elements of the solution are continuous real numbers since, in words of the inventors of PSO, it is not possible to "throw to fly" particles in a discrete space. Several modifications of the PSO algorithm for solving problems with discrete variables have been proposed in the literature. They are referred to as Discrete Particle Swarm Optimization (DPSO) methods [20].

The standard PSO considers a swarm S containing n particles ($S = 1, 2, \ldots, n$) in a d-dimensional continuous solution space. Each ith particle of the swarm has a position $x_i = (x_{i1}, x_{i2}, \ldots, x_{ij}, \ldots, x_{id})$, and a velocity $v_i = (v_{i1}, v_{i2}, \ldots, v_{ij}, \ldots, v_{id})$. The position x_i represents a solution to the problem, while the velocity v_i gives the rate of change for the position of particle i at the next iteration. Indeed, considering iteration k, the position of particle i is adjusted according to $x_i^k = x_i^{k-1} + v_i^k$.

Each particle i of the swarm communicates with a social environment or neighbourhood, $N(i) \subseteq S$, representing the group of particles with which it communicates, and which could change dynamically. In nature, a bird adjusts its position in order to find a better position, according to its own experience and the experience of its companions. In the same manner, considering iteration k of the PSO algorithm, each particle i updates its velocity reflecting the attractiveness of its best position so far (b_i) and the best position (g_i) of its social neighbourhood $N(i)$, according to the equation:

$$v_i^k = c_1 \xi v_i^{k-1} + c_2 \xi (b_i - x_i^{k-1}) + c_3 \xi (g_i - x_i^{k-1}) \tag{29.13}$$

The parameters c_i are positive constant weights representing the degrees of confidence of particle i in the different positions that influence its dynamics, while ξ refers to a random number with uniform distribution [0, 1] that is independently generated at each iteration and for each term.

The spirit of nature to deal with some real-life problems is often based on simple processes. Simple phenomena, easy ideas, sometimes works better than too elaborated and intricate ones. Trying to emulate this aspect of life, the *Jumping Frog*

Optimization (JFO) is a discrete PSO that keeps it its easy implementation and simplicity and can be applied to any combinatorial problem.

JFO considers a swarm S containing n particles ($S = 1, 2, \ldots, n$) whose positions x_i evolve in the solution space, jumping from one solution to another. The number of particles in the swarm is chosen, typical values are between 20 and 50. The position of a particle is encoded as a feasible solution to the problem. At each iteration, each particle has a random behaviour, or jumps to another solution in a manner guided by the effect of some attractors.

JFO considers three attractors for the movement of each particle i: its own best position to date (b_i), the best position of its social neighbourhood (g_i),and the best position to date obtained by all the particles, which is called the global best position (g^*). A jump approaching an attractor consists of changing the current solution by using some feature of the attractor. Each particle is further allowed to behave randomly and perform random jumps. A random jump consists of performing a random move on the solution. Algorithm 1 gives the main steps of the proposal.

The Discrete Particle Swarm Optimization named JFO has been applied to a multiobjective VRPTW problem in [6]. In this paper a large number of objectives is considered in order to show that incorporating some principles from multi-objective optimization allow particles to conduct a dynamic trade-off between objectives in order to reach feasibility. They show that without incorporating tailored heuristics or operators to tackle infeasibility, it is possible to evolve very poor infeasible route-plans to very good feasible ones using swarm intelligence.

In our implementation for the Service-Oriented VRPTW we consider the usual moves for VRPTW; i.e., to move one or more customers to other position of the route planning. The attractors are considered as the best particle taking into account, alternatively, each objective. A particle performs a jump towards the selected attractor by selecting a route from the selected attractor and applying a local search. The local search is based on the simplest move; i.e., to move one customer to other position of the route planning. We consider an archive consisting on a short set of disperse efficient solutions. The archive is updated taking into account the dominance relationship and the distance between the solutions in the archive.

Initially we generate a initial swarm by constructing feasible solutions at random. For each particle we keep the value of both objectives: the quality of service measure and the distance travelled. We get the average values of the two objectives in the initial swarm to normalize them, since the difference between the values of both objectives are quite large. From the initial we get a set of disperse good solutions as follows. First the two particles with best values for each one of the objectives are selected, and the distance between them, in the objective space, is computed. In order to get a reduced set of solutions uniformly scattered between them, the corresponding points in the objective space are obtained. The particles that are closest to each one of this point is used as initial archive set.

An attraction move of a particle consists of choosing a route of the attractor and insert it into the solution. The customers of this route are eliminated from the route that contains them in the solution. Then we explore the resulting particle solution

Algorithm 1: Bi-objective Discrete Particle Swarm Optimization for the Service-Oriented VRPTW

```
Input: VRPTW data
Output: Route Planning R;
Initialization:
- Set the size n_s of the swarm S;
begin
    - Generate the initial swarm S at random;
    - Update the vector A of the best positions;
    - Extract the best position g* in S;
    - Generate the social neighborhoods N_i ⊆ S;
    repeat
        for i = 1 to n_s do
            - Initialize the best social positions: g_i ← x_i;
        end
        for i = 1 to n_s do
            - Select ξ at random: ξ ← random(0, 1)
        end
        ;
        if ξ ∈ [0, c_1) then
            - attractor ← x_i;
        else if ξ ∈ [c_1, c_1 + c_2) then
            - attractor ← b_i;
        else if ξ ∈ [c_1 + c_2, c_1 + c_2 + c_3) then
            - attractor ← g_i;
        else if ξ ∈ [c_1 + c_2 + c_3, 1) then
            - attractor ← g*;
        - Move x_i: x_i ← Move(x_i, attractor);
        - Apply the local search: x_i ← LS(x_i);
        if x_i is better than b_i then
            - Update b_i ← x_i ;
        end
        if x_i ∈ N_j and x_i is better than g_j then
            - Update g_j ← x_i ;
        end
        if x_i is better than g* then
            - Update g* ← x_i ;
        end
        if not A ≺ x_i then
            - Update the archive A;
        end
        //x_i dominant
    until termination conditions;
end
```

to try to redistribute the customers of the smallest resulting routes in other routes in order to reduce the number of routes.

In order to decide if a new solution is included in the archive we proceed as follows. We traverse the archive checking if this solution is dominated for some solution already in the archive. In that case the new solution is discarded. Otherwise, we traverse again the archive looking for a solution dominated by the new one that will be replaced by the new solution. If the new solution dominates more than one solution in the archive, it replaces the nearest one (in the objective space). If the new solution is not dominated by any solution the archive and there is not solution in the archive that dominates the new one then we proceed as follows. We get the solution of the archive that is nearest to the new one. The new solution substitute the found solution in the archive if and only if the second nearest solution of the new solution is farther than the second nearest solution of the found solution.

The neighbourhoods are selected at random given their size that is stated to 1/5 the size of the swarm. The neighbourhoods are generated again each time the global best solution does not change from a generation to the next one.

29.4 Experimentation

This section describes the results from the computational experiments that were carried out in our study. The aim of the experiment is to evaluate the accuracy of the proposed JFO metaheuristic procedure, which is used to solve the bi-objective VRPTW oriented towards quality of service.

Two types of instances were used in the experiment for comparative purposes. One instance uses data from an actual planning case from a distribution company. The data provides the location of a set of customers which must be serviced on a specific day. We used available data from a randomly chosen set of 25, 50 and 75 customers, as well as their respective demand and time windows. The source of the second instance was the classic Solomon instances. Specifically we chose at random an instance from RCxxx, namely the RC207. The RCxxx instances combine the geographical distribution of the customers grouped in areas with customers located at random with a uniform distribution. In addition the RCxxx have wide large windows. Experiments used with these instances were carried out on 75 customers. Vehicle capacity was 1,000 load units.

The experimentation under the proposed JFO metaheuristic used different swarms and neighbourhood sizes. Specifically swarms of 10, 30 and 50 particles and 2, 4 and 6 particles for the neighbourhoods were used. In order to collect the results that used the JFO procedure for each one of the swarms and instances we compute 20, 50 and 70 iterations. In addition one instance and swarm were tested, specifically on a larger number of iterations if improvements were noticed, which will be discussed later on.

The JFO metaheuristic was implemented using a uniform combination of values c_i, with equal weights of 0.25 for each of the four types of movements (random, cognitive, social and global). That is we have considered that each movement has an equal probability during each one of the iterations.

Table 29.1 Computational results with real instances

Swarm-generations	Best		Average	
	TQs	TDist	TQs	TDist
25 Customers				
$S10 \triangleright NG20$	0.3729	579.0	0.3775	648.45
$S30 \triangleright NG50$	0.2815	526.2	0.2997	538.27
$S50 \triangleright NG70$	0.2294	422.7	0.3100	472.03
50 Customers				
$S10 \triangleright NG20$	0.4272	1140.3	0.4457	1267.18
$S30 \triangleright NG50$	0.3221	1081.1	0.3483	1253.39
$S50 \triangleright NG70$	0.3340	1036.2	0.3522	1234.31
75 Customers				
$S10 \triangleright NG20$	0.3805	1703.0	0.3932	1743.64
$S30 \triangleright NG50$	0.3542	1505.4	0.3660	1747.98
$S50 \triangleright NG70$	0.3361	1358.2	0.3501	1548.31

Table 29.2 Comparative results with real and Solomon instances

Swarm-generations	First iteration				Last iteration			
	Best		Average		Best		Average	
	TQs	TDist	TQs	TDist	TQs	TDist	TQs	TDist
Real instances	75 customers							
$S10 \triangleright NG20$	0.4432	1318.8	0.5039	1583.96	0.3805	1703.0	0.3932	1743.64
$S30 \triangleright NG50$	0.4802	1280.3	0.5137	1553.12	0.3324	1625.8	0.3660	1747.98
$S50 \triangleright NG70$	0.5232	1190.2	0.5175	1558.93	0.3126	1358.5	0.3501	1548.31
Solomon inst.	75 customers							
$S10 \triangleright NG20$	0.4852	1969.5	0.5439	2212.47	0.3742	2134.4	0.4273	2279.07
$S30 \triangleright NG50$	0.4764	1937.1	0.5513	2169.95	0.2977	2056.1	0.3444	2287.45
$S50 \triangleright NG70$	0.5264	1848.7	0.5481	2154.84	0.2252	2117.7	0.2733	2277.44

Table 29.1 shows a summary of the experimental results for each one of the actual instances with 25, 50 and 75 customers and with different swarm sizes 10, 30 and 50 respectively, and the total number of iterations on them 20, 50 and 70. The table shows the best solutions obtained with the values from the quality of service and the distance objectives. In addition to the average obtained values of these objectives are those for all particles of the swarm in the respective iteration.

Thus we carried out a comparison (Table 29.2) of the results from a real instance of 75 customers with the instance from the Solomon RC207 instance with the same amount of customers. In order to do so we used the same number of swarms and iterations as in the previous table and then compared the results from the first iteration with the results from the last iteration (producing the best solution).

Figures 29.1, 29.2 and 29.3 show the evolution of the solution archive for real instances for 25, 50 and 75 customers with swarm of 50 particles and 70 iterations.

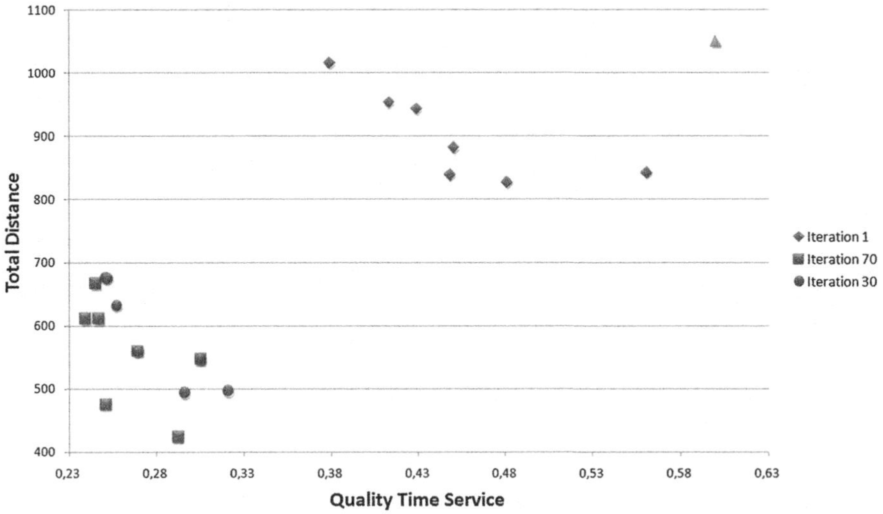

Fig. 29.1 Evolution of solution archive:real instance 25 customers, 50 particles and 70 iterations

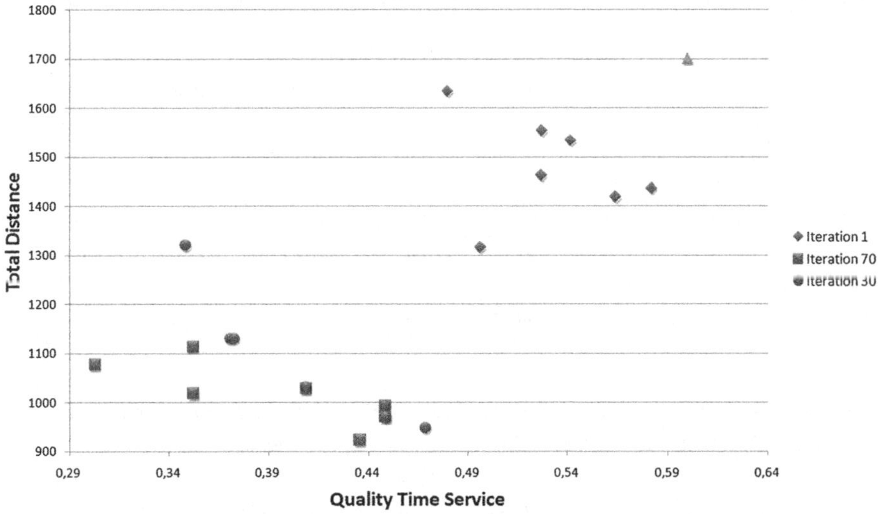

Fig. 29.2 Evolution of solution archive: real instance 50 customers, 50 particles and 70 iterations

Each figure represents the solution archive for iterations 1, 30 and 70. Improvement in the solution is clearly apparent as iterations increase.

An experiment using the actual instance of 50 customers and 50 swarms was carried out for the purpose of valuing the increase of iterations on the solution search. It was executed with 235 iterations. Table 29.3 shows the solution archive.

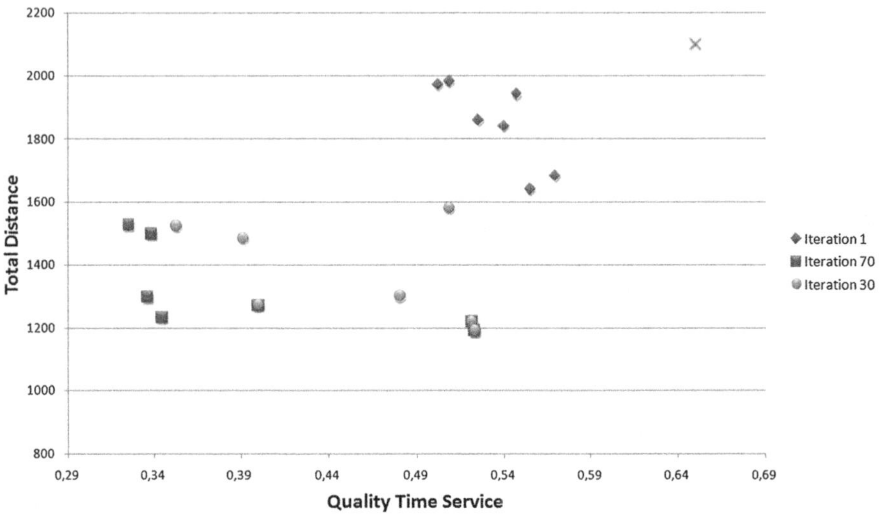

Fig. 29.3 Evolution of solution archive: real instance 75 customers, 50 particles and 70 iterations

Table 29.3 Current state of the solution archive with real instances of 50 customers 50 particles and 235 iterations

First Iteration		Iteration 70		Iteration 140		Iteration 235	
Quality	Distance	Quality	Distance	Quality	Distance	Quality	Distance
0.4799	1633.5	0.3040	1242.1	0.2955	1089.6	0.2955	1089.6
0.5146	1472.5	0.4157	1057.2	0.3533	1045.5	0.3248	956.1
0.5405	1530.2	0.3018	1293.5	0.2777	1277.7	0.2777	1277.7
0.5767	1307.4	0.4306	1039.2	0.4306	1039.2	0.3333	937.2
0.5950	1333.9	0.4333	1025.3	0.4333	1025.3	0.3037	1019.3
0.4834	1542.7	0.3706	1071.8	0.3248	1034.8	0.3030	977.9
0.4853	1100.0	0.4686	947.5	0.4686	947.5	0.3374	926.7

Table 29.4 Computational results with real instances of 50 customers 50 particles and 235 iterations

Real Inst.	First Iteration		Iteration 70				Iteration 235			
Objectives	Best	Average	Best	Average	Variation		Best	Average	Variation	
Quality	0.469	0.499	0.370	0.389	▽20.9	▽22.1	0.303	0.311	▽18.2	▽20.1
Distance	947.5	1278.7	1071.8	1096.6	△13.1	▽14.2	977.9	1026.4	▽8.8	▽6.4

Table 29.4 presents the results of the experimentation. Figure 29.4 is also included which displays the evolution of the solution archive by iteration.

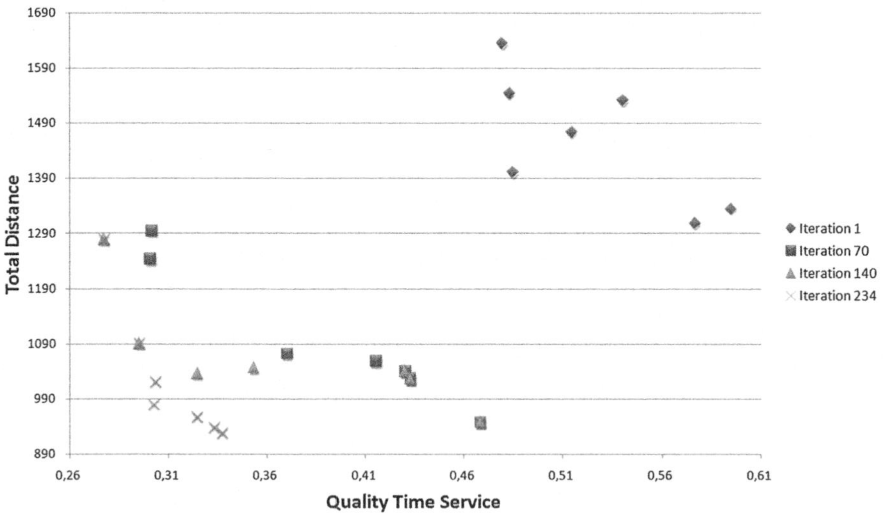

Fig. 29.4 Evolution of solution archive: real instance 75 customers, 50 particles and 235 iterations

29.5 Conclusions

This paper has tackled a new model of the bi-objective VRPTW which deals with seeking improvement of service quality. We approach this problem by not only looking at the objective of finding minimal distance routes but also consider factors such as maximizing customer satisfaction that is associated with the criteria of vehicles serving the customers as soon as possible within their time window. This paper provides the formulation for that model. A version of a PSO metaheuristic for discrete problems, JFO, is used to solve this problem.

The algorithm has been tested in the classical instances used for VRPTW and in a battery of instances obtained from real data of a distribution company. The experimental results reveal that the proposed procedure helps solve models with time windows and objectives which focus on the quality service. The results shows that the approach is suitable to offer the decision maker a reduced set of disperse non-dominated solutions. Future studies include improvements in the implemented metaheuristics by adjusting parameters.

References

1. Ai J, Kachitvichyanukul V (2009) A particle swarm optimisation for vehicle routing problem with time windows. Int J Oper Res 6(4):519–537
2. Ai TJ, Kachitvichyanukul V (2009) Particle swarm optimization and two solution representations for solving the capacitated vehicle routing problem. Comput Ind Eng 56(1):380–387

3. Ai TJ, Kachitvichyanukul V (2009) A particle swarm optimization for the vehicle routing problem with simultaneous pickup and delivery. Comput Oper Res 36:1693–1702
4. Al-kazemi B, Mohan CK (2000) Multi-phase discrete particle swarm optimization. In: FEA 2000: 4th international workshop on frontiers in evolutionary algorithms
5. Banks A, Vincent J, Anyakoha C (2008) A review of particle swarm optimization. part ii: hybridisation, combinatorial, multicriteria and constrained optimization, and indicative applications. Nat Comput 7:109–124
6. Castro J, Landa-Silva D, Perez J (2009) Exploring feasible and infeasible regions in the vehicle routing problem with time windows using a multi-objective particle swarm optimization approach. In: Krasnogor N, Melin-Batista M, Prez J, Moreno-Vega J, Pelta D (eds) Nature inspired cooperative strategies for optimization (NICSO 2008), vol 236. Studies in computational intelligence. Springer, Berlin, pp 103–114
7. Castro-Gutierrez J, Landa-Silva D, Moreno JA (2011) Nature of real-world multi-objective vehicle routing with evolutionary algorithms. In: Proceedings of the 2011 IEEE international conference on systems, man, and cybernetics (IEEE SMC 2011), IEEE Press, pp 257–264
8. Chang R, Lu C (2002) Feeder reconfiguration for load factor improvement. In: Power engineering society winter meeting, 2002, vol 2, pp 980–984. IEEE. doi:10.1109/PESW.2002.985152
9. Chen Al, Yang Gk, Wu Zm (2006) Hybrid discrete particle swarm optimization algorithm for capacitated vehicle routing problem. J Zhejiang Univ-Sci A 7(4): 607–614. doi:10.1631/jzus.2006.A0607
10. Chen P, Dong X, Niu Y (2012) An iterated local search algorithm for the cumulative capacitated vehicle routing problem. In: Tan H (ed) Technology for education and learning, vol 136, Advances in intelligent systems and computing. Springer, Berlin, pp 575–581
11. Coello CC, Lechuga GPMS (2004) Handling multiple objectives with particle swarm optimization. IEEE Tran Evol Comput 8(3):256–279
12. Consoli S, Moreno-Perez J, Darby-Dowman K, Mladenovic N (2010) Discrete particle swarm optimization for the minimum labelling steiner tree problem. Nat Comput 9(1):29–46
13. Consoli S, Perez JM, Darby-Dowman K, Mladenovic N (2008) Discrete particle swarm optimization for the minimum labelling steiner tree problem. In: Krasnogor N, Nicosia G, Pavone M, Pelta D (eds) Nature inspired cooperative strategies for optimization (NICSO 2007), vol 129. Studies in computational intelligence. Springer, Berlin, pp 313–322
14. Corberan A, Fernndez E, Laguna M (2002) Heuristic solutions to the problem of routing school buses with multiple objectives. J Oper Res Soc 53(4), 427–435. doi:10.1057/palgrave.jors.2601324
15. Eberhart RC, Shi Y (2001) Particle swarm optimization: developments, applications and resources. In: IEEE (ed) proceeding congress on evolutionary computation, vol 1, pp 81–86
16. Gomez-Gonzalez M, Lopez A, Jurado F (2013) Hybrid discrete PSO and OPF approach for optimization of biomass fueled micro-scale energy system. Energy conversion and management 65:539–545. doi:10.1016/j.enconman.2012.07.029
17. Gong YJ, Zhang J, Liu O, Huang RZ, Chung HSH, Shi YH (2012) Optimizing the vehicle routing problem with time windows: a discrete particle swarm optimization approach. IEEE Trans Syst Man Cybern Part C: Appl Rev 42(2):254–267
18. Jin-Kao H, Galinier P, Habib M (2000) Métaheuristiques pour l'optimisation combinatoire et l'affectation sous contraintes. Revue d'Intelligence Artificielle 13(2):283–324
19. Kennedy J, Eberhart R (1995) Particle swarm optimization. In: Proceedings of IEEE international conference on neural networks, vol 4, pp 1942–1948
20. Kennedy J, Eberhart R (1997) A discrete binary version of the particle swarm algorithm. In: IEEE international conference on systems, man and Cybernetics 1997. Computational cybernetics and simulation 1997, vol 5, pp 4104–4108. doi:10.1109/ICSMC.1997.637339
21. Kennedy J, Eberhart R (2001) Swarm intelligence. Morgan Kaufmann Publishers, San Francisco
22. Li J (2011) Study of a multi-objective vehicle routing problem with time window based on particle swarm optimization. In: 3rd international workshop on intelligent systems and applications (ISA) 2011, pp 1–4

23. Lin CT (2008) Using predicting particle swarm optimization to solve the vehicle routing problem with time windows. In: IEEE international conference on industrial engineering and engineering management (IEEM) 2008, pp 810–814
24. Liu X, Jiang W, Xie J (2009) Vehicle routing problem with time windows: a hybrid particle swarm optimization approach. In: 5th international conference on natural computation (ICNC) 2009, vol 4, pp 502–506
25. Mao Y, Deng Y (2010) Solving vehicle routing problem with time windows with hybrid evolutionary algorithm. In: 2nd WRI global congress on intelligent systems (GCIS) 2010, vol 1, pp 335–339
26. Marinakis Y, Marinaki M, Dounias G (2010) A hybrid particle swarm optimization algorithm for the vehicle routing problem. Eng Appl Artif Intell 23(4):463–472
27. Martinez FJ, Moreno JA (2008) Jumping frogs optimization: a new swarm method for discrete optimization. Technical Report 3, DEIOC. Universidad de La Laguna
28. Mohan CK, Al-kazemi BA (2001) Discrete particle swarm optimization. In: Proceedings of the workshop particle swarm optimization
29. Moreno-Perez J, Castro-Gutierrez J, Martinez-Garcia F, Melian B, Moreno-Vega J, Ramos J (2007) Discrete particle swarm optimization for the p-median problem. In: Proceedings of the 7th metaheuristics international conference. Montreal, Canada
30. Muoz-Zavala A, Hernndez-Aguirre A, Villa-Diharce E (2009) Particle evolutionary swarm multi-objective optimization for vehicle routing problem with time windows. In: Coello C, Dehuri S, Ghosh S (eds) Swarm intelligence for multi-objective problems in data mining, Studies in computational intelligence, vol 242, pp 233–257. Springer, Berlin
31. Ngueveu SU, Prins C, Wolfler-Calvo R (2010) An effective memetic algorithm for the cumulative capacitated vehicle routing problem. Comput Oper Res 37(11):1877–1885. doi:10.1016/j.cor.2009.06.014
32. Pampara G, Franken N, Engelbrecht A (2005) Combining particle swarm optimisation with angle modulation to solve binary problems. In: The 2005 IEEE congress on evolutionary computation, vol 1, pp 89–96. doi:10.1109/CEC.2005.1554671
33. Pengxin D, Shurong Z, Hongwei Z (2012) Improved particle swarm algorithm and its application in vehicle routing problem. In: Wu Y (ed) Software engineering and knowledge engineering: theory and practice, vol 114. Advances in intelligent and soft computing. Springer, Berlin, pp 269–274
34. Poli R (2008) Analysis of the publications on the applications of particle swarm optimisation. J Artif Evol App 2008:4:1–4:10. doi:10.1155/2008/685175
35. Potvin JY (2009) A review of bio-inspired algorithms for vehicle routing. In: Pereira F, Tavares J (eds) Bio-inspired algorithms for the vehicle routing problem, vol 161. Studies in computational intelligence. Springer, Berlin, pp 1–34
36. Reyes-Sierra M, Coello CAC (2006) Multi-objective particle swarm optimizers: a survey of the state-of-the-art. Int J Comput Intell Res 2(3):287–308
37. Seren C (2011) A hybrid jumping particle swarm optimization method for high dimensional unconstrained discrete problems. In: 2011 IEEE congress on evolutionary computation (CEC), pp 1649–1656. doi:10.1109/CEC.2011.5949813
38. Sun H, Wu B, Zhao YW, Wang W, Ma YL, Wang W (2004) Materials science forum. In: Ai X, Li J, Huang C (eds) Particle swarm optimization for vehicle routing problem with time windows (chap.), vol 471, pp 801–805
39. Toth P, Vigo D (eds) (2002) The vehicle routing problem, Monographs on discrete mathematics and applications, Society for industrial and applied mathematics publishing, vol 9
40. Wang H, Liu Y, Wu Z, Sun H, Zeng S, Kang L (2008) An improved particle swarm optimization with adaptive jumps. In: IEEE congress on evolutionary computation (CEC) 2008 (IEEE world congress on computational intelligence), pp 392–397. doi:10.1109/CEC.2008.4630827
41. Wang W, Wu B, Zhao YD, Feng P (2006) Particle swarm optimization for open vehicle routing problem. In: Huang DS, Li K, Irwing GW (eds) 2006 International conference on intelligen computing: part II (ICIC'06), Springer, Berlin, pp 999–1007

42. Wang Z, Li J, Fan J, Fan C (2010) Research on improved hybrid particle swarm optimization for vehicle routing problem with time windows. In: 2010 International conference on artificial intelligence and computational intelligence (AICI), vol 1, pp 179–183
43. Yang S, Wang M, Jiao L (2004) A quantum particle swarm optimization. In: Congress on evolutionary computation CEC2004, vol 1, pp 320–324. doi:10.1109/CEC.2004.1330874
44. Zhen T, Zhu Y, Zhang Q (2009) A particle swarm optimization algorithm for the open vehicle routing problem. In: International conference on environmental science and information application technology ESIAT 2009, vol 2, pp 560–563
45. Zhu Q, Qian L, Li Y, Zhu S (2006) An improved particle swarm optimization algorithm for vehicle routing problem with time windows. In: IEEE congress on evolutionary computation CEC 2006, pp 1386–1390
46. Zou S, Ding PX, Zhang HW (2011) The improvement of hybrid particle swarm algorithm and its application. Adv Mater Res 268:798–802

Chapter 30
An Approach for the Evaluation of Risk Impact of Changes Addressing Uncertainties in a Surveillance Requirement Optimization Context

Sebastian Martorell, Maryory Villamizar, Isabel Martón, Carmen Armero and Ana Sánchez

Abstract This paper presents an approach for the evaluation of risk impact of Surveillance Requirement changes addressing identification, treatment and analysis of uncertainties in an integrated manner, which is intended to be used in an optimization context. It is also presented an example of application of the methodology to study a SF change of the Reactor Protection System of a Nuclear Power Plant.

Keywords Licensing basis · Surveillance requirement · Uncertainity · Risk informed · Optimization · Nuclear power plant

S. Martorell (✉)
Departament d'enginyeria Química i Nuclear, Universitat Politècnica de València, Valencia, Spain
e-mail: smartore@iqn.upv.es

M. Villamizar
Industrial Engineering Department, Pontificia Bolivariana University-Bucaramanga, Bucaramanga, Colombia

I. Martón
Department of Chemical and Nuclear Engineering, Universitat Politècnica de València, Valencia, Spain

C. Armero
Departament d'Estadística i Investigació Operativa, Universitat de València, Valencia, Spain
e-mail: Carmen.Armero@uv.es

A. Sánchez
Departament d'Estadística i Investigació Operativa Aplicades i Qualitat, Universitat Politècnica de València, Valencia, Spain
e-mail: aisanche@eio.upv.es

D. Greiner et al. (eds.), *Advances in Evolutionary and Deterministic Methods for Design, Optimization and Control in Engineering and Sciences*, Computational Methods in Applied Sciences 36, DOI 10.1007/978-3-319-11541-2_30

30.1 Introduction

The Reactor Protection System (RPS) is a very important system for Nuclear Power Plant (NPP) safety, as it is responsible for shutting down the NPP by means of reactor trip in case of accidental conditions. So that, current regulation requires a very high reliability of the system in performing its vital and required safety function, i.e. reactor trip. The RPS is a sort of Safety Instrumented System (SIS) [1]. A SIS is defined as an "instrumented system used to implement one or more safety instrumented control functions. A SIS is composed of any combination of sensors, logic solver and final elements" [2]. The standard IEC 61508 requires every safety function to achieve a determined Safety Integrity Level (SIL). For low demand operating systems the SIL levels are defined in terms of average probability of failure on demand (PFD_{avg}, see Table 30.1).

Surveillance Requirements (SR) are part of Technical Specifications that are included into the Licensing Basis (LB) for operation of Nuclear Power Plants. Surveillance tests aim at limiting risk of undetected downtimes of safety related equipment by imposing equipment operability checks, which consist of testing of equipment operational parameters with established Surveillance Frequency (SF) and Test Strategy (TS).

Surveillance testing is executed periodically, and can be implemented using several strategies. The strategy establishes how the tests of the redundant components are scheduled with respect to one another. IEC 61508 (1998–2005) [3] defines proof test as a "periodic test performed to detect failures in a safety-related systems so that the system can be restored to an "as new" condition or as close as practical to this condition". This standard establishes the need of routine maintenance action in order to detect unrevealed failures, being proof test one of these activities. It thus has an important role in the achievement of safety integrity.

Surveillance requirements were established taking into account mainly deterministic criteria. A number of problems have been identified connected to SR that can jeopardize plant safety. The development of Probabilistic Risk Assessment (PRA) and its application since the early 80s to analyze SR changes has brought the opportunity to review SR consistency from a reliability-risk viewpoint, i.e. addressing the impact of the changes on plant safety on the basis of the reliability-risk information provided by the PRA, with particular attention to the role of the Surveillance Frequency included within Surveillance Requirements.

Table 30.1 SIL for low demand mode of operation [3]

SIL	PFD_{avg}
4	$\geq 10^{-5}$ to $<10^{-4}$
3	$\geq 10^{-4}$ to $<10^{-3}$
2	$\geq 10^{-3}$ to $<10^{-2}$
1	$\geq 10^{-2}$ to $<10^{-1}$

In this context, the NRC issued the first draft of Regulatory Guide RG 1.174 in 1998 [4], which remains a major milestone in the NRC initiative to risk-inform the regulations on changes to licensing basis (LB), which particularizes to analyze SR changes in RG 1.177 (1998) [5]. RG 1.174 and RG 1.177 require that all sources of uncertainty be indentified and analyzed such that their impacts are understood at the technical element level. It is only recently when some guidance is being proposed on the systematic treatment of uncertainties associated with the use of the PRA in risk-informed decision making of LB changes, NUREG-1855 [6] and EPRI-1016737 [7].

30.2 Types of Uncertainty

Uncertainties can be categorized as either random or epistemic. Random uncertainty reflects our inability to predict random observable events. Epistemic uncertainty represents our confidence in the model and in the numerical values of its parameters. This type of uncertainty is also called "state-of-knowledge" uncertainty or just "uncertainty". Epistemic uncertainties can be split into three categories [4]:

1. Parameter uncertainty is that which relates to the parameters of the PRA, given a choice of model. Even with a known model, the parameter values may still be unknown. Examples of parameter uncertainties include equipment failure rates, initiating-event frequencies, and human error probabilities.
2. Model uncertainty relates to the uncertainty in the assumptions made in the analysis and the models used. In many cases, there is limited knowledge and some disagreement on the proper model to represent a system. The result is that for a particular process, there are multiple competing models, each of which necessarily produces a different approximation of the same real-world system.
3. Completeness uncertainty represents the uncertainty due to the portion of risk that is not explicitly included or modeled in the PRA.

30.3 Framework for SR Change Optimization Addressing Uncertainties

Figure 30.1 shows the framework proposed in this paper for the optimization of Surveillance Requirements changes based on RAMS+C simulation and addressing treatment and analysis of uncertainties minimize in an integrated manner.

For sake of clarity in the sequel only a simplified A[U]+C problem is considered because of the generalization is straightforward. The general multi-objective optimization problem (MOP) can be formulated for the A[U]+C problem based on the tolerance interval method to minimize the vector of multi-objective uncertain functions, **f(x)**, subject to the vector of uncertain constraints, **g(x)** [8]:

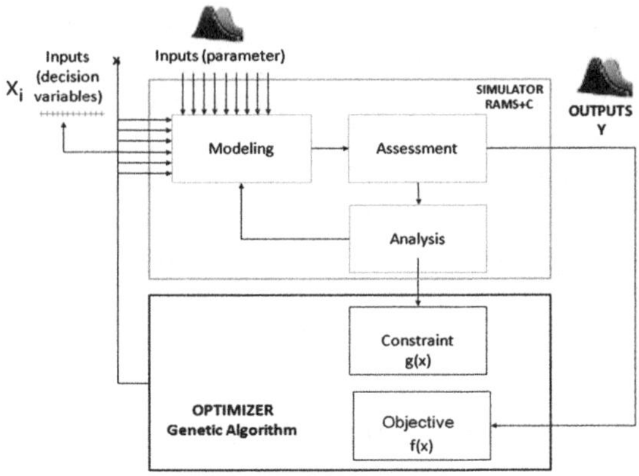

Fig. 30.1 Schematics of the RAMS+C based optimization problem under uncertain objective and constraints

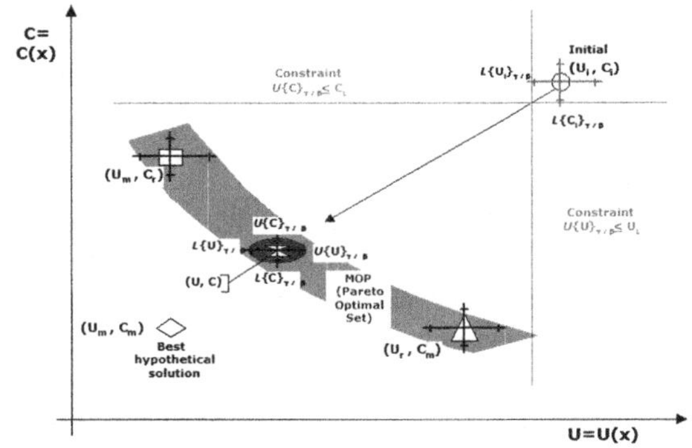

Fig. 30.2 Objective space of feasible solutions for MOP

$$\mathbf{f}(\mathbf{x}) = \{U\{U(\mathbf{x})\}_{\gamma/\beta}, U\{C(\mathbf{x})\}_{\gamma/\beta}\} \tag{30.1}$$

$$\mathbf{g}(\mathbf{x}) = \{U\{U(\mathbf{x})\}_{\gamma/\beta} \leq U_L, U\{C(\mathbf{x})\}_{\gamma/\beta} \leq C_L\} \tag{30.2}$$

Figure 30.2 shows the U–C plot of the possible solutions to the MOP considering herein uncertain objectives and constraints.

30.4 Evaluation of Risk Impact of SR Change Addressing Parameter and Model Uncertainties

A fundamental part of the framework proposed in the previous section is the evaluation/simulation of the RAMS+C impact of the Surveillance Requirement change used to formulate **f(x)** and **g(x)** functions.

A three steps based approach for the evaluation of risk impact of changes to Surveillance Requirements is proposed: modeling, assessment and analysis, which is based on the use of the PRA and includes identification of sources key to decision, treatment and analysis of uncertainties in an integrated manner. Risk evaluation is undestood herein with a general meaning of RAMS+C, while the example of application will focus on RPS system reliablity only.

While the analysis of parametric and model uncertainty is fairly mature, the analysis of completeness uncertainty cannot be handled in a similar formal manner. So that, only parameter and model uncertainties will be addressed in detail in the approach proposed in this paper. Guidance on addressing uncertainties from modeled and non-modeled risk contributors is given in NUREG-1855 [6] and EPRI-1026511 [9].

What concerns risk modeling, not only the usually addressed sources of uncertainty linked to PRA models and data but also the sources of model and parameter uncertainties associated with the assumptions in Surveillance Requirement change evaluation must be indentified, which will be classified in two main groups: parameter and model uncertainty [5].

Risk assessment is proposed by formulating the usual risk metrics for analyzing Surveillance Requirement changes in the literature [10–12]. It is proposed treatment of model and parameter uncertainties based on traditional sensitivity studies and uncertainty assessment respectively, the latter based of the probabilistic approach for uncertainty formulation and propagation by standard Monte Carlo Sampling (MCS) technique [13, 14].

Risk analysis is based on the well established comparison of the assessment of risk impact of the change including treatment of uncertainties with acceptance guidelines [15]. In addition, it is proposed also the use of both traditional importance measures and uncertainty importance measures (sensitivity analysis) in order to analyze respectively how basic events and parameter uncertainties influence the risk impact of the change proposed and its uncertainty [16, 17]. The former helps to identify the main risk contributors. Uncertainty importance measures help to identify which of uncertain parameters are most significant contributors to risk impact uncertainty and eventually the need of limiting the impact of the effect of particular uncertainties or of developing additional modeling and treatment of uncertainties (see Fig. 30.1).

A reasonable way of addressing epistemic uncertainties may consist of evolving from model to parametric uncertainties as much as possible, with an aim at allowing model refinement prior to use the results of the risk assessment to support quantification of objective and constraint functions. This paper focuses on demonstrating how the results of risk analysis, with application to study SR changes, can help in the achievement of such a model refinement (see internal loop of simulator in Fig. 30.1). By so doing, it is possible to build an optimization strategy based on the traditional use of MC simulation and GA optimization, i.e. MC-GA.

30.5 Case Study

The case study presents the results of the evaluation of the risk impact of a Surveillance Frequency (SF) change of circuit breakers of the Reactor Protection System of a PWR Nuclear Power Plant, which could be used to obtain a refined model for the MC-GA based SF optimization.

30.5.1 Problem Description

Figure 30.3 shows a schematic view of the RPS, which consists of two redundant and identical channels, A and B. The RPS can be actuated manually from the control room or automatically after a signal requiring reactor trip. This signal opens circuit breakers 52RTA and 52RTB. Manual or automatic actuation of the RPS deenergizes control rods, which are inserted in the reactor core by gravity. The by-pass breakers 52BYA and 52BYB allows testing of the main circuit breakers above using manual trip/rep.

Surveillance Requirements of the RPS establish a functional test of the redundant channels with a Surveillance Test Interval (TI), i.e. 1/SF, of two months and sequential testing. This means the first circuit breaker, e.g. 52RTA, is tested the first month and the second one, 52RTB, is tested the next month. The TI change consist of extending the current TI from 2 to 3 months; i.e. 1,440–2,160 h.

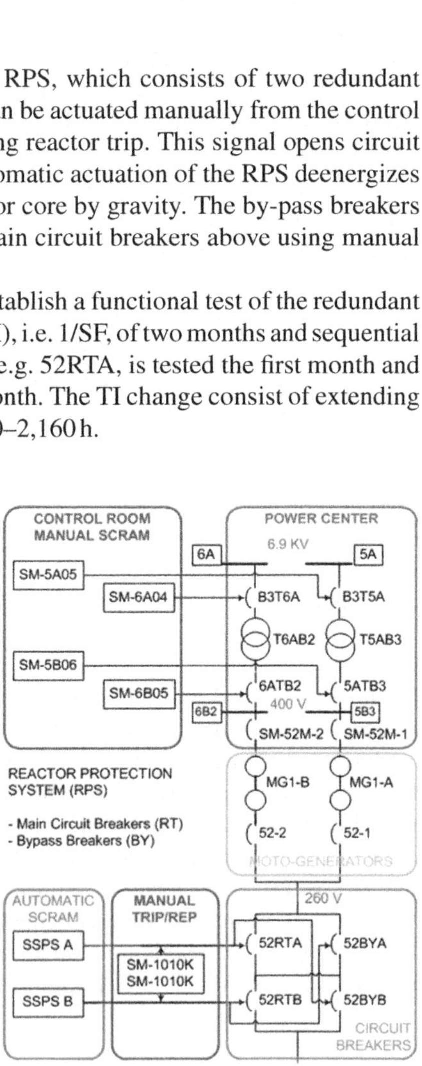

Fig. 30.3 Schematic view of the RPS

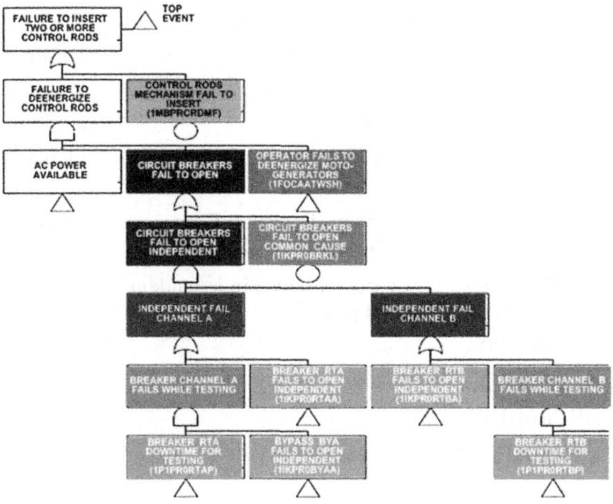

Fig. 30.4 Fault tree for RPS failure

30.5.2 Plant Specific PRA

Figure 30.4 shows a fault tree representing the failure of the RPS as top event included in the Level 1 PRA. Basic events representing control rods mechanism fail to insert, i.e. 1MBPRCRDMF, common cause failure (CCF) of the circuit breakers 52RTA and 52RTB, i.e. 1IKPR0BRKL, and human error to disconnect motor generators MG1-A y MG1-B, i.e. 1FOCAATWSH, are highlighted in addition to the AND gate representing occurrence of independent failures of RPS channels A and B.

Tables 30.2 and 30.3 present models and data used for modeling the basic events and their corresponding parameters in current PRA only for most important basic events belonging to the RPS that appear in the final Boolean equation after generating the minimal cut sets (MCS) using a cut-off criteria 10^{-12}.

30.5.3 Reliability Assessment of the RPS Before and After the SF Change as Compared to SIL Levels

Current regulation requires a very high reliability of the RPS in performing its vital safety function, i.e. reactor trip. Table 30.4 shows there is no significant reduction of the RPS reliability after the SF extension (base case), which remains very high. See also the results in Figs. 30.5 and 30.6. Table 30.4 summarizes also the results of a number of sensitivity studies performed in order to estimate the impact of model uncertainties on the assessment of risk impact of the SF change including parameter uncertainty as well (see also Fig. 30.5). In addition, comparing results in Table 30.4 (Fig. 30.5) with SIL requirements in Table 30.1, the highest SIL level 4 is achieved any case.

Table 30.2 More important basic events

Basic event	Description	BE model	Unavailability formula (u)
1MBPRCRDMF	Control rods mechanisms fail to insert	Standby	1/2*λ_R*MTBRT
1FOCAATWSH	Operator fails to deenergize generators	Demand	ρ_D
1IKPR0BRKL	Circuit breakers Common Cause fail to open (CCF)	Standby	1/2*λ_C*TI (sequential tests)
1IKPR0RTAA 1IKPR0RTBA 1IKPR0BYAA 1IKPR0BYBA	Circuit breaker independent fail to open	Standby	1/2*λ_I*TI
1FOPRATWSH	Operator fails to manually scram reactor	Demand	ρ_S
1P1PR0RTAP 1P1PR0RTBP	Downtime for testing	Test	τ/TI

Table 30.3 Data and parameters of basic events

Parameter	Description	Parameter type	Parameter/value
λ_R (h^{-1})	Control rods mechanisms failure rate	PDF	Ga(4.32 $\times 10^{-9}$; 0.49; 1.13 $\times 10^{8}$)
ρ_D (-)	Human error probability to deenergize generators	PDF	Ln(1.05 $\times 10^{-1}$; 5)
λ_C (h^{-1})	Circuit breakers Common Cause fail to open (CCF)	Standby	$\lambda_C = \beta^* \lambda_I$ Ln(1.21 $\times 10^{-7}$; 3)
β (-)	β -factor for Circuit breakers Common Cause failure rate	Constant	0.1277 (sequential tests)
λ_I (h^{-1})	Circuit breaker independent failure rate	PDF	Ga(9.47 $\times 10^{-7}$; 1.49; 1.57 $\times 10^{6}$)
$\rho_S(-)$	Human error probability to manually scam reactor	PDF	Ln(1.66 $\times 10^{-1}$; 5)
MTBRT (h)	Mean Time Between Reactor Trips	Constant	2160 (true trips) [Max TBRT = 13140]
TI (h)	Test Interval (1/SF)	Constant	1440 (current TI in TS)
$\tau(h)$	Test time	Constant	1.37

30.5.4 Measures of Importance of Basic Events and Parameters

Table 30.5 summarizes the results derived for the traditional importance measures for RPS Unavailability before the SF change (2 months) and ranked according to RAW.

Table 30.4 Results of quantification of unavailability for the base case and sensitivity studies

Case	Mean	5 %	95 %	Mean	5 %	95 %
Base case	1.38×10^{-5}	1.64×10^{-6}	3.97×10^{-5}	1.83×10^{-5}	2.19×10^{-6}	5.78×10^{-5}
Sensitivity						
MTBRT	3.74×10^{-5}	2.83×10^{-6}	1.18×10^{-4}	4.20×10^{-5}	3.66×10^{-6}	1.36×10^{-4}
FFNCT	1.38×10^{-5}	1.66×10^{-6}	3.56×10^{-5}	1.61×10^{-5}	1.92×10^{-6}	4.21×10^{-5}
HEP rookie	2.29×10^{-5}	2.15×10^{-6}	7.53×10^{-5}	3.20×10^{-5}	3.63×10^{-6}	1.122×10^{-4}
TS staggered	7.11×10^{-6}	5.31×10^{-7}	2.21×10^{-5}	8.33×10^{-6}	7.71×10^{-7}	2.60×10^{-5}

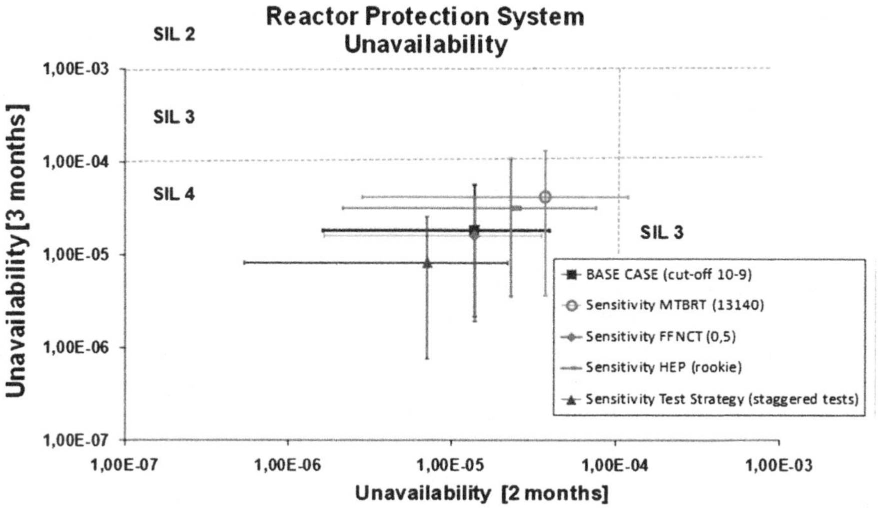

Fig. 30.5 Unavailability of RPS before and after SF change

Reference [18] proposes also the use of threshold values to determine the value of RAW below which we deem the basic event to be not risk-important. Threshold values of RAW are given for RPS Unavailability as compared to SIL levels using the following equations:

$$RAW_{U,SILx} = \frac{U_{SILx}}{U} \tag{30.3}$$

The following set of threshold RAW values for U is obtained using Eq. (30.3) and corresponding SIL levels:

$$RAW_{U,SILx} = \{7246; 724.6; 72.46; 7.26\} \tag{30.4}$$

Comparing RAW of basic events in Table 30.6 with threshold values in Eq. (30.4) it is found that uncertainty of basic events 1MBPRCRDMF and 1IKPR0BRKL is very

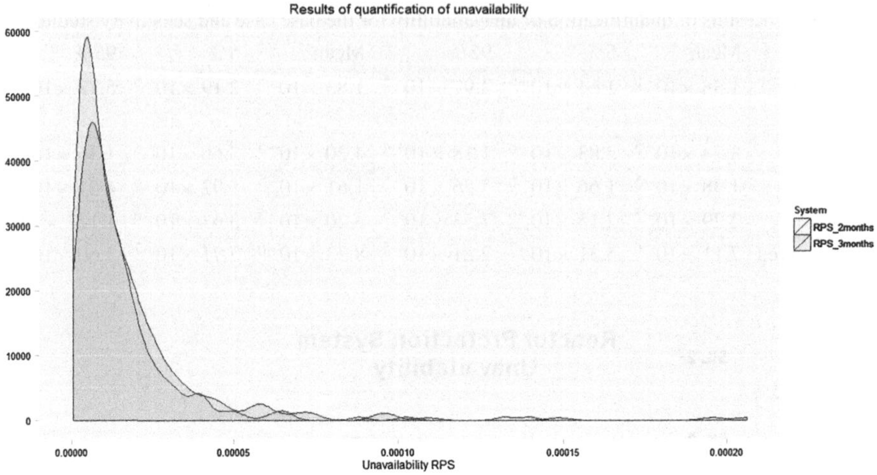

Fig. 30.6 PDF of unavailability RPS for the base case

significant, e.g. RAW beyond 7,300, which means RPS reliability could move from current SIL 4 to SIL 1. Table 30.6 summarizes the results derived for the uncertainty importance measures for RPS Unavailability before the SF change (2 months). In particular, Sobol and Spearman indices are presented. Both Spearmen and Sobol indices rank parameters in the same possition.

Table 30.6 results show, based on Sobol indices, the probability of fail to insert control rods mechanisms is the parameter uncertainty which influence the most uncertainty of RPS unavailability. Since the sum of all indices S_i is less than 1,

Table 30.5 Traditional importance measures of basic events (2 months)

Basic event	Unavailability	FV	RAW	RRW
1MBPRCRDMF	4.67×10^{-6}	3.38×10^{-1}	72330	1.51
1IKPR0BRKL	8.70×10^{-5}	6.61×10^{-1}	7596	2.95
1BLBC0G1CF	2.24×10^{-6}	1.41×10^{-5}	7.3	1
1FOCAATWSH	1.05×10^{-1}	6.63×10^{-1}	6.65	2.96

Table 30.6 Global sensitivity analysis (2 month)

Parameter	S_i	S_{Ti}	$S_{Ti} - S_i$	Spearman
ρ_D	0.612	0.601	−0.01	0.576
λ_R	0.186	0.188	0.002	0.566
λ_C	0.187	0.216	0.029	0.422
Total	0.980	–	–	

then, the model is non-additive and, therefore, there are interactions among input parameters above.

30.6 Concluding Remarks

This paper presents an approach for the evaluation of risk impact of Surveillance Requirement changes addressing identification, treatment and analysis of uncertainties in an integrated manner, which is intended to be used in an optimization context. The case study demonstrates how the results of risk analysis can help in the achievement of a model refinement by focussing on most significant risk contributors and uncertainties. By so doing, it may be possible to build an optimization strategy based on the use of MC simulation—GA optimization, by evolving first from model to parametric uncertainties accounting for the most relevant contributors.

Acknowledgments Authors are grateful to the Spanish Ministry of Science and Innovation for the financial support of this work (Research Project ENE2010-17449) and the Spanish Ministry of Economy and Competitiveness, Research Project MTM2013-42323.

References

1. Torres-Echeverria AC, Martorell S, Thompson HA (2009) Modelling and optimization of proof testing poli-cies for safety instrumented systems. Reliab Eng Syst Saf 94(4):838–854
2. International electrotechnical commission. IEC 61511 functional safety – safety instrumented systems for the process industry sector, Parts 1–3 (Switzerland, 2003)
3. International electrotechnical commission. IEC 61508. Functional safety of electrical/electronic/programmable electronic safety-related systems, Parts 1–7 (Geneva, Switzerland, 1998–2005)
4. RG 1.174. An approach for using probabilistic risk assessment in risk-informed decisions on plant-specific changes to the licensing basis, Revision 2 (USNRC, 2011)
5. RG 1.177. An approach for plant-specific, risk-informed decision making: technical specifications, Revision 1 (USNRC, 2011)
6. NUREG 1855. Guidance on the treatment of uncertainties associated with PRAs in risk-informed deci-sionmaking, Vol 1 (USNRC, 2009) (see also NUREG 1855 Draft Report, Revision 1, for Comment (2013)
7. EPRI-1016737. Treatment of parameter and model uncertainty for probabilistic risk assessments. EPRI (Palo Alto, CA, 2008)
8. Martorell S, Sánchez A, Carlos S (2007) A tolerance interval based approach to address uncertainty for RAMS+C optimization. Reliab Eng Syst Saf 92:408–422
9. EPRI-1026511. Practical guidance on the use of PRA in risk-informed applications with a focus on the treatment of uncertainty, Draft Report (2012)
10. NUREG/CR-6141. Handbook of methods for risk-based analyses of technical specifications (USNRC, 1995)
11. Kim IS, Martorell SA, Vesely WE, Samanta PK (1994) Risk analysis of surveillance requirements including their adverse effects. Reliab Eng Syst Saf 45(3):225–234
12. Kim IS, Martorell S, Vesely WE, Samanta PK (1992) Quantitative evaluation of surveillance test in-tervals including test-caused risks, NUREG/CR-5775 (BNLNUREG-52296, 1992)

13. Martorell S, Villamizar M, Villanueva JF, Serradell V, Sanchez AI (2009) Addressing uncertainties in risk-informed decision-making of changes to nuclear power plant technical specifications. In: Proceedings European Safety and Reliability Conference (ESREL 2009), pp 353–359
14. JCGM 101:2008. Evaluation of measurement data – supplement 1 to the guide to the expression of uncertainty in measurement – propagation of distributions using a Monte Carlo method (2008)
15. Caruso MA, Cheok MC, Cunningham MA, Holahan GM, King TL, Parry GW, Ramey-Smith AM, Rubin MP, Thadani AC (1999) An approach for using risk assessment in risk-informed decisions on plant-specific changes to the licensing basis. Reliab Eng Syst Saf 63:231–242
16. Aven T, Nkland TE (2010) On the use of uncertainty importance measures in reliability and risk analysis. Reliab Eng Syst Saf 95:127–133
17. Borgonovo E, Apostolakis GE, Tarantola S, Saltelli A (2003) Comparison of global sensitivity analysis techniques and importance measures in probabilistic safety assessment. Reliab Eng Syst Saf 79(2):175–185
18. Reinert JM, Apostolakis GE (2006) Including model uncertainty in risk-informed decision making. Ann Nucl Energy 33:354–369

Chapter 31
Scalable Deployment of Efficient Transportation Optimization for SMEs and Public Sector

Pekka Neittaanmäki and Tuukka Puranen

Abstract Transportation planning is central activity in logistic network design. In this study, we examine the deployment of optimization methodology to transportation planning. More specifically, we examine the adoption of system solving the well-known combinatorial optimization problem, the vehicle routing problem (VRP). Its application has resulted in efficiency gains in transportation logistics, but they have not been very widespread, and especially small-scale operators have not yet benefited from these systems. In this paper, we present a prospective case study on the issues during deployment of optimization, especially in the context of small and medium enterprises (SMEs). We propose a novel perspective to analyzing vehicle routing systems (VRSs), and complement the previous research on real-life aspects of commercial routing. In this study, we suggest a framework for analyzing VRS deployment, from a viewpoint frequently identified in enterprise architecture (EA) theory. To our knowledge, EA theory has not been applied to study the requirements of VRP solution methods. This new viewpoint allows us to identify new needs for widespread adoption of vehicle routing systems, and to derive additional requirements for the optimization methodology for SMEs. In practice, we identify several adoption barriers for VRSs and suggest potential strategies for lowering them.

Keywords Combinatorial optimization · Vehicle routing problem · Transportation planning · Metaheuristics · System deployment

31.1 Introduction

Transportation logistics is a major area of activity in the logistics field. Efficient transportation requires both efficient flow of material and people, as well as careful coordination of the entities performing the transportation. It is the issue of coordination that has especially been addressed by the academic logistic problem, the vehicle

---

P. Neittaanmäki (✉) · T. Puranen
Department of Mathematical Information Technology,
University of Jyväskylä, Jyväskylä, Finland
e-mail: pekka.neittaanmaki@jyu.fi

D. Greiner et al. (eds.), *Advances in Evolutionary and Deterministic Methods for Design, Optimization and Control in Engineering and Sciences*, Computational Methods in Applied Sciences 36, DOI 10.1007/978-3-319-11541-2_31

routing problem [1], in which the task is to simultaneously divide transportation activities to transporting entities and design optimal transportation routes for these entities. The goal is typically to minimize the number of entities (for example vehicles) and the total distance they travel.

Vehicle routing problem has several variants, each of which address a different set of properties encountered in real-life transportation planning. The most common variants include vehicle routing problem with time windows, pickup and delivery problem, multi-depot vehicle routing problem, and vehicle routing problem with backhauls. One of the main challenges in deploying optimization methodology to the VRP is this heterogeneity of these details. Operators have differing requirements [2], and different models typically require different solution methods to be solved efficiently and effectively [3].

Vehicle routing problems have been solved with a large array of optimization methods, ranging from exact methods [1] to metaheuristics [4] and hyperheuristic [5] methods. The most successful approaches to date have combined several metaheuristic components with a set of strong local search operators [6]. The downside of this diversity is the complexity of choosing efficient methods for the problem at hand [3].

A third challenge for a widespread utilization of optimization methodology is the fact that many logistic operators are relatively small. This results, due to the complexities mentioned, in an inability to acquire the necessary expertise to select, configure and deploy vehicle routing systems in general. In this paper, we attempt to provide a lightweight process for easier deployment of these methods, and discuss the implications of the process to the future of the optimization solution methodology.

The paper is structured as follows. In Sect. 31.2, we provide a context for the study, in Sect. 31.3, we describe briefly the research approach used, in Sect. 31.4 we propose a framework for examining the adoption of routing systems, and in Sect. 31.5 describe the key findings so far in the context of the proposed framework. Finally in Sect. 31.6, we conclude and suggest areas for further study.

31.2 Background and Contribution

The presented study is conducted as a part of a larger research project which aims to significantly lower the costs of deploying optimization in small and medium enterprises and public sector. There are two main hypotheses that form the basis of the project. Firstly, the utilization of optimization would provide value to SMEs and public sector institutions, and secondly, the observed lack of utilization is due to problems in deployment, and not in the optimization models and methods themselves.

The reasons for the above mentioned hypotheses are twofold: first, we have noted the success stories of the large-scale entities in deployment of optimization, indicating that the state of the art has advanced sufficiently for practical use, and second, the problems found in SMEs and public sector institutions exceed the efficient planning capabilities of the human dispatchers, i.e., they are non-trivial in most cases. In addition, our preliminary results indicate that there is a widespread demand for optimization solutions within SMEs, due to, for example, cost and environmental pressures.

It is still not clear which are the main reasons for the difficulty of the deployment in a scalable manner, and this paper presents the preliminary findings of our study. We have applied the process described in this paper in several cases, in organizations of different sizes and types. More specifically, we attempt to answer the following questions: why has commercial vehicle routing problem not been widely adopted, what are the barriers for the technology adoption, and especially, how do the findings affect the requirements for new vehicle routing problem models and solution methods. We propose a novel perspective to vehicle routing, and complement the already identified real-life aspects of commercial routing [7]. Our objective is to provide the SMEs with efficient tools for transportation planning in a scalable manner.

By *efficient* we mean accounting for the individual characteristics of the different cases well—unlike many of the current off-the-shelf routing systems which assume that one rich enough model and versatile enough algorithm set suffices. This has led to situation where systems either handle the simple cases (but as such do not provide enough value over the manual planning), or result in too complicated deployment and use in the complex cases. This has, effectively, lead to inefficient use of routing systems where most users do not benefit from the recent advances in the vehicle routing research. Thus to provide the efficiency needed by the heterogeneity of the SMEs, we derive additional requirements for the optimization methodology in this context.

To understand the *scalability* of the deployment, we identify several adoption barriers from the studied cases, and utilize a distinction frequently identified in EA frameworks: the domains—or layers—of business, data, and applications [8].[1] We build our deployment framework according to these three layers, categorize the barriers according to thess proposed framework, and suggest strategies for lowering them.

Previous studies has identified the need to consider commercial setting in general, and many of the publications on rich VRP variants have addressed the issue of modeling the aspects commonly found in real-life routing cases [7], as well as constructing systems capable of incorporating these aspects [9]. However to our knowledge, EA theory has not been applied to analyzing VRP solution methodology in this manner.

31.3 Research Method

The research is ongoing and is conducted as a prospective case study. More specifically, we offer optimization for deployment, pilot use, and operational use to several small and medium enterprises and public sector institutions and observe issues during the process.

[1] The *technology* domain is excluded at this stage of the study.

In practice, we have studied the adoption (or the reason for not adopting in some cases) of routing systems in 12 distinct cases at this point of the study. We refer to these cases with letters from A to L. The characteristics of the cases are briefly described as follows:

- **A**—A medium-sized enterprise operating nationally with more than hundred trucks for serving several distribution centers.
- **B**—A newspaper with internal transportation activity of less than ten trucks for daily deliveries of the publication on a regional level.
- **C**—A regional transportation company with less than 30 trucks for delivering goods and distributing free papers.
- **D**—A public sector entity with mission critical transportation tasks with several dozens of vehicles at a time.
- **E**—A public sector entity with 30 vehicles performing mission critical people transportation on a regional level.
- **F**—A regional transportation company with less than 20 trucks for delivering goods and packages.
- **G**—A food production establishment employing an internal fleet of less than 20 vehicles for transporting perishable goods with strict time limits.
- **H**—A medium sized municipality for providing school transportation service.
- **I**—A small municipality for providing school transportation service.
- **J**—A medium sized municipality providing school transportation service within both urban and peripheral areas.
- **K**—A medium sized nationally operating enterprise employing a fleet of 25 tanker trucks.
- **L**—A single vehicle courier service providing internal post delivery service for a municipality.

The cases were studied by analyzing the requirements, performing the deployment and observing and interviewing the case stakeholders, both end users and the individuals responsible for the procurement of ICT systems. From the studied cases we formulated a framework for analyzing the barriers of adoption and describe the identified barriers along with suggestions for lowering them. These are described in the two subsequent sections, respectively.

31.4 Proposed Framework

Although the stage of the research does not allow for quantitative analysis of the issues, we present the preliminary findings of the study with respect to the emerging process. One interesting finding is that the connection to the enterprise architecture theory also affects the optimization methodology design in a subtle manner, which we explain in detail shortly.

The proposed framework is designed to answer the necessary hypotheses for vehicle routing system adoption in a given case. The first hypothesis is the *hypothesis of value*, which assumes that the vehicle routing system can provide added value to

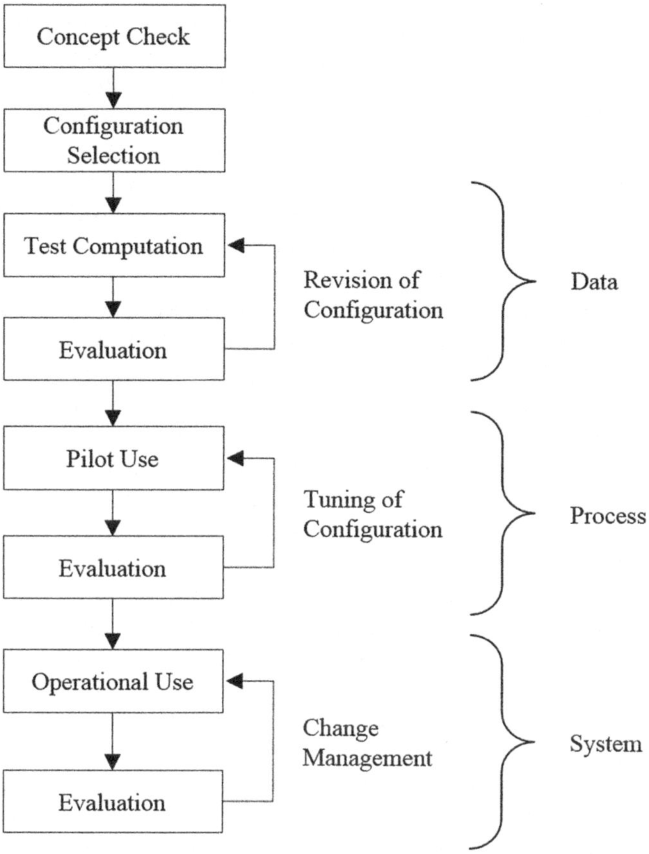

Fig. 31.1 The deployment process framework and the related aspects of the enterprise architecture layers

the operation of the enterprise. The second hypothesis is the *hypothesis of data*, which assumes that the data needed for the transportation optimization exists or can be generated in sufficient quality (EA data layer). The third hypothesis is the *hypothesis of process*, which assumes that the way of working does not change such that it undermines the other operations of the enterprise (EA business layer). The final hypothesis is the *hypothesis of system*, which assumes that the existing systems in the enterprise can be complemented by the new vehicle routing system (EA applications layer).

The deployment framework and process is outlined in Fig. 31.1. The process consists of the following steps. First a Concept Check is performed to ensure that the operator does actually require a VRP optimization solution (*hypothesis of value*). In Configuration Selection, the requirements of the operator are mapped with the aid of an optimization expert. Configuration is defined as the state of the variable settings of the system, including, most significantly, the optimization model and the optimization methods and their parameters. In Test Computation, the main aim is to ensure that the data required for the optimization process does exist and is valid (*hypothesis of data*). In addition, Evaluation is done for the selected configuration, based on the quality of the optimization results. After Pilot Use, the process of the operator has been adjusted to the use of automated transportation planning, and can be put into Operational Use (*hypothesis of process*). This finishes the deployment by integrating the other systems used by the operator with the optimization solution (*hypothesis of system*).

As we observe, there are three integration phases in the process: data, process, and system integration, which are also frequently identified in enterprise architecture theory.

Data required for optimization includes the necessary elements to construct the decision variables, constraints and objective of the optimization model. An important aspect is also the quality of data, more specifically, the data has to have a structure suitable for automated planning. One major issue has been data in free-form text where formal rules would be needed (e.g., incompatibility rules).

Process integration requires adjustments in the operational environment, some of which provide new opportunities for improvements. These include changes in planning frequencies, order processing capabilities, and the planning effort, all which, in many cases, affect the core of the operations of the logistic operator.

System integration consists of connecting the existing systems to the newly introduced planning tool. This includes introducing methods for data exchange and adjusting the necessary interfaces between the systems.

We have observed that the three phases of the process can be more easily managed during the deployment by keeping them separated. This also allows the operator to invest in the deployment gradually, as there is no need to start process integration if the data integration cannot be completed.

The successful deployment of the optimization system requires data, process, and system integration. However, all these phases require involvement of an optimization expert, which is prohibiting investment for SMEs. In order to make the deployment scalable, the process needs to be automated to be usable by a regular user. Note that this involves the change management during operations. In practice, changes in the requirements should be accommodated by the optimization model and methodology. This is a major requirement for the optimization methodology in the context of small and medium enterprises and public sector institutions.

31.5 Results

After formulating the deployment process framework, we attempted to identify the problems encountered in cases and pinpoint the exact phase of the encounter. This enables us to gather requirements and assumptions on not only the vehicle routing system itself, but also the optimization methods needed to solve the VRP instances in each case. We define barrier as *a reason that prevents the user to start using a routing system unless changes to the system are made or other necessary manual activities are performed.*We classified these reasons according to the phase of the deployment process it occurs.

31.5.1 Data Integration

We identified nine different data integration barriers for adopting the vehicle routing system. The most frequent problems concerned either the fact that some subset of the needed data was missing or there was not enough expertise to extract the data from existing systems. The following barriers were identified in the data integration phase:

- **D.1**—*Lack of data in digital form.* The data needed for building the optimization model is available, but is not in a digital form.
- **D.2**—*Missing structure of data.* The data needed for building the optimization model is available, but does not have a formal structure for automatic interpretation in the model building. Examples include descriptions of the compartment loading constraints in free-form text.
- **D.3**—*Missing task data elements.* The data needed to describe the tasks of the route planning has missing elements, such as missing or partial schedule information, e.g., time windows.
- **D.4**—*Missing resource data elements.* The data needed to describe the resources used on the routes, such as vehicles, has missing elements, such as speed profiles for different vehicle types.
- **D.5**—*Missing geographical information.* The data needed to describe the geographic area where the transportation takes place is missing necessary data, such as digital map or some key characteristics of the road network.
- **D.6**—*Missing cost structure data.* The data needed to formally evaluate the quality of the optimized plan is not available. Examples include hidden costs on loading and unloading vehicles and changes on the workflow on, e.g., warehouses due to changes in the routing procedures.
- **D.7**—*Inability to acquire data from existing system.* The data needed for modeling and solving the VRP is available, but it cannot be transferred from the existing system for utilization of optimization.
- **D.8**—*Inability to combine data from several existing systems.* The data needed for modeling and solving the VRP is available, but it is stored on several systems and not enough expertise is available to combine the data.
- **D.9**—*Low quality of existing data.* The data needed for modeling and solving the VRP is available, but the quality of the data is not sufficient. Examples include error in addresses, order quantities or schedule constraints.

When data integration barriers have been cleared, we may proceed to the process integration phase.

31.5.2 Process Integration

We identified six process integration barriers for adopting vehicle routing systems. Many of the barriers resulted from the physical reality that cannot be captured by the vehicle routing models, including complex changes in costs, procurement procedures and perceived quality by both the users and their customers.

- **P.1**—*Inability to describe the required operating process characteristics to the system due to lack of expertise.* The process of loading, packing, transporting and unloading involves operations whose cost, duration etc. depends on functions whose inclusion to the optimization model require more expertise than is available.
- **P.2**—*Prohibiting personnel role changes or other human resource issues.* The employment of a vehicle routing system results in changes in the roles of the personnel, such that the enterprise is unable to adapt to the changes. Examples include the inability to train, e.g., order management personnel to the increasingly centralized transportation planning.
- **P.3**—*Customer satisfaction decrease due to change in processes.* The employment of a vehicle routing system results in a perceived decrease in quality of service, such as differing driver visiting the customer in subsequent days.
- **P.4**—*Prohibiting changes in the physical operations.* The employment of a vehicle routing system results in prohibiting changes in the physical operations, such as need to manually reorder deliveries in warehouses as an additional process step.
- **P.5**—*Lack of resulting plan quality.* The plans produced by the vehicle routing system are not satisfying for the end user. Examples include underutilization of the fleet and obviously inefficient route sequences.
- **P.6**—*Opaqueness of the decision support leading distrust to plans.* The plans produced by vehicle routing system indicate efficiency, but the user is unable to evaluate the feasibility of the plans in reality.

After the changes resulting from the routing system adoption have been incorporated to the existing processes, the vehicle routing system can be integrated to the rest of the systems of the enterprise.

31.5.3 System Integration

In the system integration phase, the routing system is set to operational use. We identified three major barriers for adoption in this phase.

- **S.1**—*Inability to invest into a completely new system.* The system integration requires a prohibiting investment into a new system as a necessary supporting system is required before the vehicle routing system can be utilized.
- **S.2**—*Inability to integrate to existing systems.* The system integration to existing systems cannot be performed due to lack of resources or knowledge.

Table 31.1 Barriers by case

	A	B	C	D	E	F	G	H	I	J	K	L
D.1												•
D.2	–			•							•	
D.3				•	•		•				•	
D.4	•			•	•	•	•	•	•	•	•	
D.5				•								
D.6	–			+							+	
D.7							+					
D.8									•	•		
D.9	•											
P.1	–	•									+	•
P.2							+					
P.3					+		+	+	+	+		
P.4							+					
P.5	–		•			–	–		–	–	–	–
P.6				+	+							
S.1	+	–			•	•	•	•	•	•	+	•
S.2		–		•		+						+
S.3	–	–	–	–	–	–	–	–	–	–	–	–

Each row represents an identified barrier, and each column a studied case. Empty cell indicates that the barrier was not present in the case, + indicates that the barrier was identified as a potential problem, but has not yet confirmed as realized, • indicates that the barrier was realized, and – indicates that the barrier has not been assessed in the case

- **S.3**—*Inability to propagate changes to the system from the operational environment.* In practice, the operating environment of the enterprise changes frequently, and any planning tool needs to accommodate those changes. Examples include opening a new terminal and changing the cost structure of the transportation. The inability to incorporate these changes is a major concern as this is a hidden cost factor to the usefulness of the routing system, which is difficult to evaluate.

31.5.4 Barriers by Case

After identifying barriers for deployment, we identified the cases in where the barriers exist. The objective is to understand the implications of the barriers to different types of enterprises. The identified barriers by case are given in Table 31.1.

We can make several observations from the cases. First, in all cases, there is a realized barrier, and most of the cases have at least two active barriers. Most of the barriers are in the data integration phase, but almost all cases have also identified a barrier in the system integration. Process integration has the most variability between cases, and this may prove to be an interesting observation, although at this stage of the research it has to be yet verified.

On the data integration phase, only one case had no digital data available (D.1), and most data integration issues are related to resources, mainly vehicles (D.4), which are easier to generate than task data, as the data does not change frequently. Relatively few cases identified geographical information (D.5), cost structure (D.6), or existing data (D.7–D.9) as a barrier for data integration. On the process integration phase the identified potential problems relate most frequently decrease in customer satisfaction (P.3). In mission critical environments, the inability to trust the optimization results (P.6) was also identified as a possible barrier. As can be seen, the system integration has one major obstacle—inability to invest into completely new system in order to obtain a routing system (S.2)—in many of the cases. Note also that at this stage of the research, we have not been able to evaluate the change requirements during operational use (S.3).

31.5.5 Implications for Routing Models and Algorithms

From a data integration perspective, the most direct implication is the need for the algorithms to be able to accommodate manual changes by the user. This interactivity can help to adapt the system in situations where some of the data is not available or cannot be formalized to the system. From the modeling viewpoint, this means the ability to accept custom constraints during potentially interactive optimization.

From a process integration perspective, the ability to automatically select the algorithm parameters robustly to meet the differing needs of the different cases, is implied by the heterogeneity of the processes in the studied cases. This includes the ability to accommodate open planning in a continuous fashion, select algorithm parameters according to the optimization need and availability of the computational budget. In addition, the support for interactively performing manual changes, adjusting parameters and modifying constraints are likely to increase the planners' confidence in the produced plans.

From a system integration perspective, the main barrier in the studied cases is the inability to invest into a completely new system, which implies that a complementary subsystem or a service may be needed. This, in turn, implies a need for a case-specific routing system. However, scalability would require a generic routing system. To reconcile these requirements in the current context, we suggest using an optimization system that (1) uses an automatically adapting, e.g., hyper heuristic solution methodology and (2) complements the existing systems as a separate service.

31.6 Conclusions and Further Research

In this paper, we addressed the deployment of the vehicle routing problem optimization into operational use. We have suggested that the deployment of optimization to the SMEs, or the "long tail" of operators would yield large-scale benefits to society.

We argue, based on the preliminary findings in this study, that the self-adaptation and interactivity of the optimization methodology is necessary, not only from the viewpoint of the optimization results, but also that of large-scale system deployment and change management. In this light, priority should be given to methods that are capable of adapting to wide array of optimization problems, such as hyper heuristic solutions that combine the methodology from efficient metaheuristic components, instead of methods that improve a narrow set of results.

Although we are still at the early stage of the study, a number of patterns have emerged on the deployment of the optimization solutions. We certainly do not have solutions for each of the identified barriers, but we nevertheless feel that publishing the preliminary findings is beneficial for other researchers struggling to increase the adoption of routing systems. The next step of the study will be evaluation of the process after a large enough sample of operational use has been collected. In practice, we should verify that the VRP modeling and solution methodology approaches proposed here lower the identified barriers.

In addition, further studies are required especially in automation of the configuration selection and tuning from the viewpoint of the deployment process. This may have implications to the qualities required from optimization methods, which should be critically evaluated in the future. From the modeling viewpoint, interactive approaches may provide to be a fruitful direction, especially when formal data is not properly available. One option is to examine if it is possible to deduce and suggest missing constraints from the manual changes the users make to the solutions frequently.

Acknowledgments This research was funded by Tekes, the Finnish Funding Agency for Technology and Innovation. We thank Jouko Nieminen for reading drafts of this paper. The support is acknowledged with gratitude.

References

1. Toth P, Vigo D (eds) (2001) The vehicle routing problem. Society for Industrial and Applied Mathematics, Philadelphia
2. Drexl M (2011) Rich vehicle routing in theory and practice. Technical report, Johannes Gutenberg University Mainz
3. Sörensen K, Sevaux M, Schittekat P (2008) Multiple neighborhood search in commercial VRP packages: evolving towards self-adaptive methods. In: Adaptive and multilevel metaheuristics. Springer, pp 239–253

4. Puranen T (2011) Metaheuristics meet metamodels—a modeling language and a product line architecture for route optimization systems. Dissertation, University of Jyväskylä
5. Garrido P, Castro C (2009) Stable solving of CVRPs using hyperheuristics. In: Proceedings of the 11th annual conference on genetic and evolutionary computation. ACM, New York, pp 255–262
6. Vidal T, Crainic TG, Gendreau M, Prins C (2013) Heuristics for multi-attribute vehicle routing problems: a survey and synthesis. Eur J Oper Res 231(1):1–21
7. Hoff A, Andersson H, Christiansen M, Hasle G, Løkketangen A (2010) Industrial aspects and literature survey: fleet composition and routing. Comput Oper Res 37:2041–2061
8. The Open Group (2011) The open group architecture framework, TOGAF Version 9.1. http://www.opengroup.org/togaf Accessed 29 Sept 2013
9. Gacias B, Cegarra J, Lopez P (2009) An interdisciplinary method for a generic vehicle routing problem decision support system. In: Proceedings of the international conference on industrial engineering and systems management, Montreal, Canada, IEEE

Part VII
Engineering Design and Societal Applications: Electrical and Electronical Applications

Chapter 32
Estimation of the Electricity Demand of La Palma Island (Spain)

Begoña González, Antonio Pulido, Miguel Martínez and Gabriel Winter

Abstract Historical data of electricity demand in La Palma island (Spain) were collected and electricity demand estimates conducted by different organizations were sought. Some factors that could affect these data were studied and its predictions by the next years were looked for. The idea was to use these factors as explanatory variables in order to predict the values of electricity demand in the next years. Moreover, with the aim of minimizing the limitation of predicting the future based only on relationships between variables that occurred in the past, it has been considered the annual demand forecast for various scenarios, taking into account, for each of them, different variations of the explanatory variables. All that with the goal that the estimate band of the demand for each year includes the real future demand with high probability. This provided a prediction model that takes into account population and gross domestic product. Results and their graphical representation along with the other estimates found are presented. A similar approach was carried out to predict peak powers.

Keywords Prediction model · Electric demand · Peak powers · Flexible evolution algorithm · Robust design optimization

B. González (✉) · A. Pulido · M. Martínez · G. Winter
University Institute of Computational Engineering (SIANI),
Evolutionary Computation and Applications (CEANI),
Universidad de Las Palmas de Gran Canaria,
35017 Las Palmas de GC, Spain
e-mail: bgonzalez@iusiani.ulpgc.es

A. Pulido
e-mail: apulido@die.ulpgc.es

M. Martínez
e-mail: mmartinez@die.ulpgc.es

G. Winter
e-mail: gabw@step.es

D. Greiner et al. (eds.), *Advances in Evolutionary and Deterministic Methods for Design, Optimization and Control in Engineering and Sciences*, Computational Methods in Applied Sciences 36, DOI 10.1007/978-3-319-11541-2_32

32.1 Introduction

In this paper the current and projected supply and demand energy, along with energy needs in the 2025 time horizon, are tackled adopting a conservative prognosis. Such evolution should include significant increases in future energy demand, in response to growth in the high band, both the population and the activity of the different economic sectors: tourism, transport, etc. However it would be better to determine peak powers than demand energy, because they will indicate the requirements of power available at the thermal power station. Because of the fact that the electricity discharged into the grid at every instant must be equal to that consumed (because the electric parameters, both the frequency and the voltage, must be kept within a narrow range) it makes necessary to know the peak powers in order to determine the power that must produce the conventional generation park. It will be the peak power plus a specific reservation considering the possibility of failure of a group of generation and/or that another is in maintenance. Nowadays there are many unknowns for the immediate future. Rising oil prices and technical improvements could be the cause of slowing or increasing the sale and implantation of electric vehicles. That could convert the oil consumption into electrical one. The electricity self-generation in low or medium voltage is something unstoppable. Elements of electrical energy storage that may provide support at certain times, in order to smooth the curves of consumption and reduce the need for conventional power generating groups in peak times, are needed. Climate change and the rise in temperature in the Canary Islands, as well as the reduction of rainfall levels will lead to an increase in consumption. On the one hand by the needs of air conditioning and industrial refrigeration production, on the other hand by the great power consumption that presents the water cycle in the Canary Islands, where over half the population and the tourism sector demand large amount of desalinated water as well as purification and pumping requirements, resulting in over 10 % of the total demand. An improvement in process efficiency is imminent because new designs and an increasingly restrictive legislation. In fact a European target is to reach 20 % of efficiency in 2020. Smart electricity meters that differentiate the consumption prices according to the time interval at which it takes place are emerging. Also the rise of the price of electricity will cause a change in the behavior of consumers. The factor economic crisis is affecting economic activity very significantly, and therefore, this leads to a bias in the projected gross domestic product (GDP), which is greatly diminished compared to the boom years economy before 2008. It is also necessary to point out that the reference quantities of GDP from 2006, from which estimates are made, are not definitive, they are subject to revision. Thus the forecast for the next few years could be altered.

After this introduction, in Sect. 32.2 an assessment of the situation of energy demand from the information contained in the updated PECAN2006 (Canary Energy Plan 2006) is carried out and in Sect. 32.3 a prediction exercise is made in order to know what could be the annual electricity consumption and peak powers in the period 2012–2025.

32.2 Electricity Demand Past and Future

In the following, annual demand is considered in power station bars, i.e. the energy discharged in the grid from the conventional and renewable generators, not including direct consumption nor consumption supplied by own production. That is, the energy resulting from subtracting from the gross energy or the alternator terminals, auxiliary consumption of the various power stations and that generated by cogenerators, self-generators or renewable energy installations associated with consumption, for consumption of own premises (industry, hotels, homes, …). Therefore, the energy actually injected into the transporting electricity grid for the whole generation: power stations, cogeneration, renewable energy installations, etc.

In order to be specific about the evolution of electricity demand in La Palma island, different sources of demand estimation have been analyzed, among them the updated PECAN2006 from 2004 until the 2015 horizon. If estimates of the PECAN central scenario (scenario 2) made on the basis of an annual growth of regional GVA (Gross Value Added) of 2.8 % in 2005 and 2.9 % from 2006–2015, are compared with the real data of electricity demand on the island of La Palma between 2001 and 2011, provided by UNELCO-ENDESA (the only generator in ordinary regime in the Canary Islands) we can see that they are lower than demand data for the years 2004–2008, although the effect of the "economic crisis", which has been suffering since 2008, has not been taken into account in making these estimates. However,

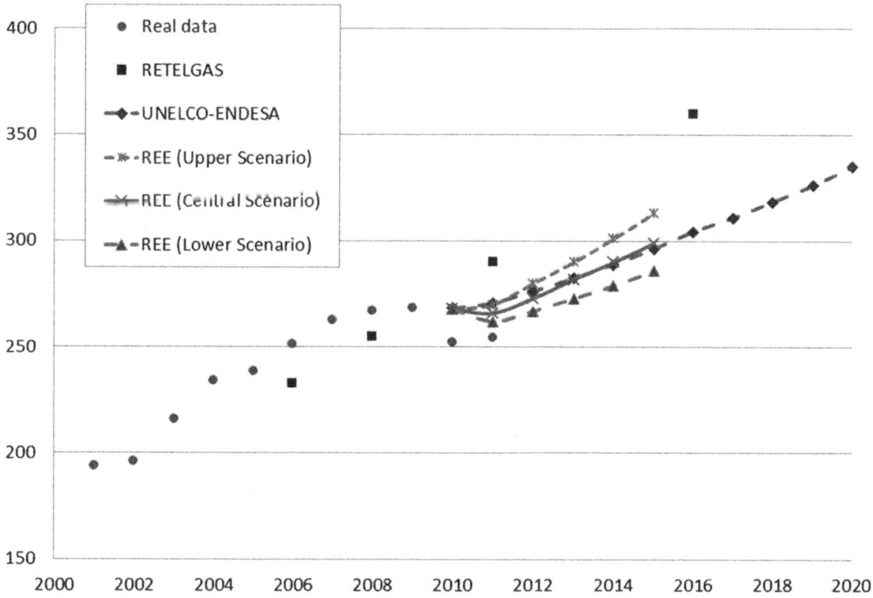

Fig. 32.1 Annual electricity demand in power station bars (GWh): historical data and forecasts by different sources

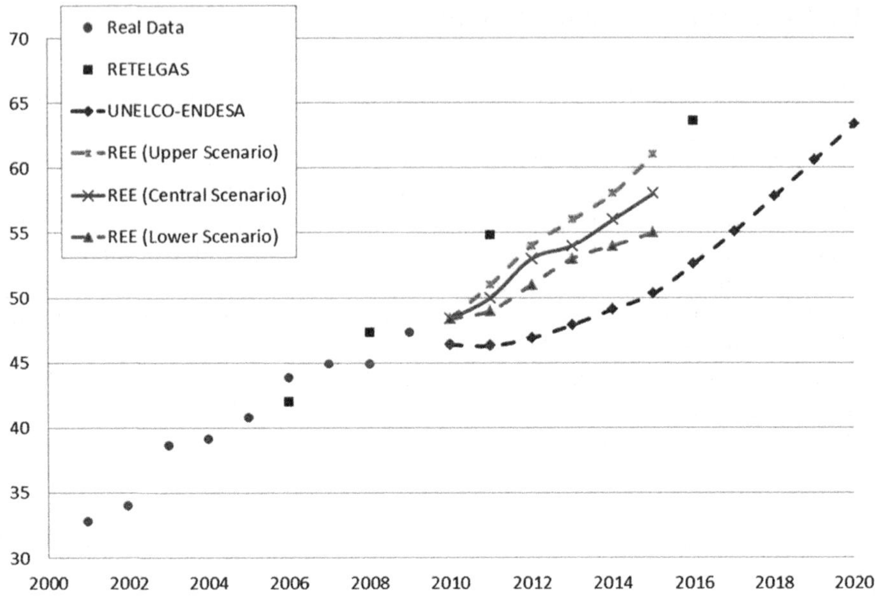

Fig. 32.2 Annual average hourly peak powers in power station bars (MW): historical data and forecasts by different sources

although the 2009 estimate still slightly underestimates the real demand, estimates from 2010 onwards overestimate it. Both historical data and forecasts of different annual electricity demand and peak power are shown in power station bars for the period 1998–2020 in Figs. 32.1 and 32.2, respectively. Estimates correspond to different companies predicting in different time intervals, although there is a significant interface between 2008 and 2015. Furthermore, historical data considered for the different forecasts not mach. In fact, while forecasts of PECAN (2004–2015) and RETELGAS (2006–2016) predate the economic crisis, those of UNELCO-ENDESA (2010–2020) and REE (2010–2015) are later. The update PECAN2006 assumed REE predictions for annual electricity demand and peak powers.

32.3 Electricity Demand and Peak Power Forecasting

To forecast demand over a time horizon, it must have historical values of explanatory variables and minimize the limitation of predicting the future based only on relationships between variables that occurred in the past.

This limitation is more severe in longer-term predictions. Therefore it has to be considered the annual demand forecast for various scenarios, taking into account, for each of them, different variations of the explanatory variables that contribute most to the demand. Considering likewise, if any, results predicted by different sources and with different methodologies, and contrasting the deviations of these results with reality occurred in recent years if possible. All that with the goal that the estimate band of the demand for each year includes the real future demand with high probability.

32.3.1 Model Selection

The process of selecting a model for predictive or explanatory purposes involves several procedures. One of them is to select the "best" subset of explanatory variables that explains the data in the simplest way. Data of electricity demand, population, GDP, consumption price index and electricity price were collected from 2001 to 2011 but so far, only long-term forecasts for population and GDP were found from official sources. For this reason only population and GDP were considered as explanatory variables.

When we looked for models to estimate load demand in the long term we found mainly econometric methods and methods based on neural networks [2–4]. From our point of view the advantage of the former is its transparency and explicit dependency from different explanatory variables. Figure 32.3 shows that there is a simple linear relationship between electricity demand (GWh) and both population and GDP (millions euros). The estimated simple linear regression functions are

$$\text{Electricity\_Demand} \approx -1,262.181 + 0.01748 \times \text{Population} \quad (32.1)$$

$$\text{Electricity\_Demand} \approx -198.7699 + 0.01083 \times \text{GDP} \quad (32.2)$$

An alternative approach is to consider a log–log model where the dependent variable as well as all explanatory ones are transformed to natural logarithms. In our

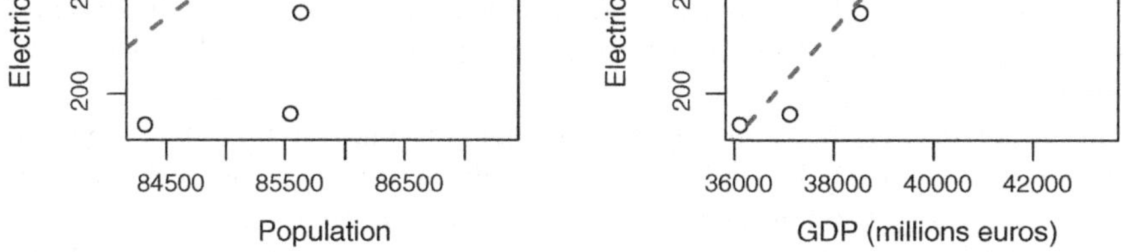

Fig. 32.3 Scatterplot of electricity demand (GWh) versus population and GDP (millions euros). In both cases, the *piecewise line* represents the fitted *regression line*

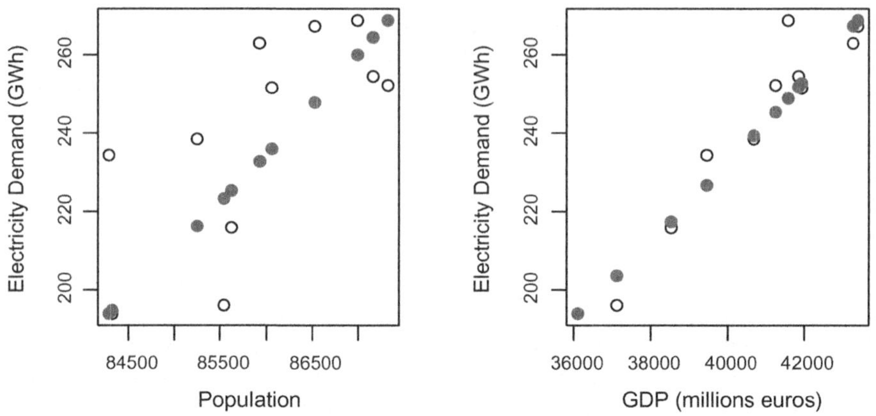

Fig. 32.4 Scatterplot of electricity demand (GWh) versus population and GDP (millions euros). In both cases, the *solid circles* represent the values fitted with the log–log model

case, as all the observations in the data set are positive the natural log transformation is applicable and the estimated simple linear regression functions are (see Fig. 32.4)

$$\ln(\text{Electricity\_Demand}) \approx -67.976 + 6.465 \times \ln(\text{Population}) \quad (32.3)$$

$$\ln(\text{Electricity\_Demand}) \approx -14.526 + 1.885 \times \ln(\text{GDP}) \quad (32.4)$$

These two functional forms give parameter estimates that have different economic interpretation. For example, the slope of the simple linear model gives us directly the change in electricity demand for a one-unit change in population or GDP and will vary depending on the data. In contrast, when a log–log model is considered the interpretation is given as an expected percentage change in electricity demand when population or GDP increases by some percentage. Such relationships are commonly referred to as elastic in econometrics, and the slope of the log–log model is referred to as an elasticity. According to Eqs. 32.3 and 32.4, to get the proportional change in electricity demand associated with a p percent increase in population or GDP, we have to calculate $\alpha = log[(100 + p)/100]$ and take $e^{6.465\alpha}$ or $e^{1.885\alpha}$, respectively. For example, a 1 % increase in population multiplies electricity demand by $e^{6.465 \times log(1.01)} = 1.0664$, i.e. increases the expected electricity demand by about 6.64 %, and a 1 % increase in GDP multiplies electricity demand by $e^{1.885 \times log(1.01)} = 1.0189$, i.e. increases the expected electricity demand by about 1.89 %.

Summarizing, elasticity measures how changing one variable affects others. If elasticity is equal to 0, demand is perfectly inelastic (i.e., demand does not change when explanatory variable changes). Values between zero and one indicate that demand is inelastic (this occurs when the percent change in demand is less than the percent change in explanatory variable). When elasticity equals one, demand is unit elastic (the percent change in demand is equal to the percent change in explanatory

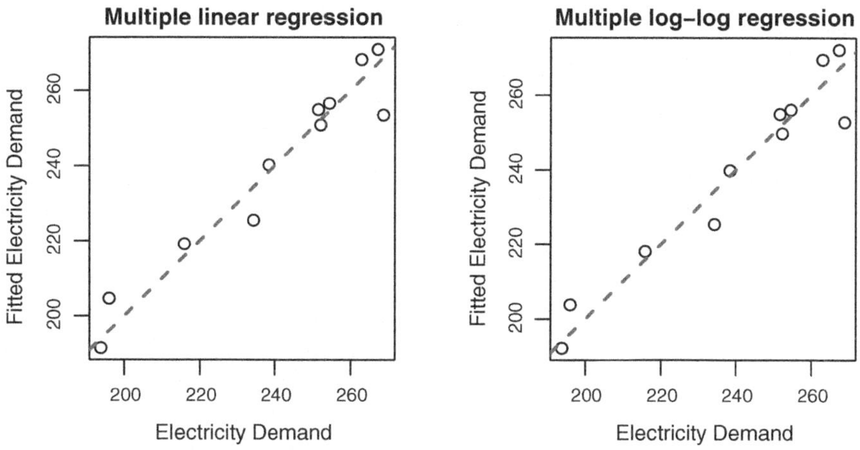

Fig. 32.5 Scatterplots of electricity demand data versus values fitted with multiple linear and log–log regression. In both cases, the *piecewise line* represents the line where the abscissa corresponds to the ordinate

variable). Finally, if the value is greater than one, demand is perfectly elastic (demand is affected to a greater degree by changes in explanatory variable).

Figure 32.5 shows scatterplots of electricity demand data versus values fitted with multiple linear and log–log regression:

$$\begin{aligned} \text{Electricity\_Demand} \approx & -369.323 \\ & + 0.0023 \times \text{Population} + 0.0102 \times \text{GDP} \qquad (32.5) \\ \ln(\text{Electricity\_Demand}) \approx & -20.768 \\ & + 0.627 \times \ln(\text{Population}) + 1.802 \times \ln(\text{GDP}) \\ & \qquad (32.6) \end{aligned}$$

According to Eq. 32.6, a 1 % increase in population multiplies electricity demand by $e^{0.627 \times log(1.01)} = 1.0063$, i.e. increases the expected electricity demand by about 0.63 %, and a 1 % increase in GDP multiplies electricity demand by $e^{1.802 \times log(1.01)} = 1.0181$, i.e. increases the expected electricity demand by about 1.81 %.

32.3.2 Population

For estimation of the evolution of the population on the island of La Palma in the period 2012–2025 it was considered the projection of the population of Spain in the mid-term (2010–2020) of the Statistics National Institute (INE), according to which the Canarias population will go from growing 23.92 % in the period between 2000 and 2011 to grow just 2.71 % between 2011 and 2020. Although in the case of the island of La Palma, the population would decrease (see Table 32.1).

Table 32.1 Projection of the population on the island of La Palma

2001 (Datum)	2011 (Datum)	2015	2020	2025
84,319	87,163	86,649	86,285	86,135

Table 32.2 Gross domestic product at market prices of Canary Islands (period 2004–2011)

	Canary Islands	Spain
2001/2000	4.9	3.6
2002/2001	2.8	2.7
2003/2002	3.8	3.1
2004/2003	2.4	3.3
2005/2004	3.1	3.6
2006/2005	3.1	4.0
2007/2006	3.2	3.6
2008/2007	0.3	0.9
2009/2008	−4.2	−3.7
2010/2009	−0.8	−0.1
2011/2010	2.1	0.7

The projected evolution by the INE is based on the classical method of components, which part of the resident population and the observed data for each of the basic demographic components of mortality, fertility and migration, obtaining the evolution of the population under some assumptions on the becoming of these three phenomena.

32.3.3 Gross Domestic Product at Market Prices

In terms of Canarian economic growth from 2004 to present (period 2004–2011), Table 32.2 shows annual growth rates of real GDP at market prices based on official data from the Regional Accounting of Spain by INE. Clarify that the values for the three latest years are still provisional in greater or lesser degree and could be revised.

With respect to forecasts of GDP growth (Deputy Ministry of Economy, and Economic Affairs with the EU. Ministry of Economy and Treasure) for the period 2010–2013 (see Table 32.3) it is necessary to clarify that the calculation of these forecasts has been made in the absence of additional financial shocks, and in a context of uncertainty both over the intensity of the crisis effects and over the extension in time and magnitude of the economic recovery. That is why these forecasts could be revised in shorter time periods than usual in another circumstances. Anyway, we considered GDP data for the 2010 and 2011.

Regarding long-term forecasts of Canary's GDP for the period 2014–2020, in line with the observations made in the mid-term scenario, to comment that economic

Table 32.3 Gross domestic product forecasts (Canary Islands)

2010	2011	2012	2013
−0.4	0.4–0.9	1.1–1.6	1.8–2.3

uncertainty levels are amplified. That is why it has chosen to establish annual average growth for the whole period. It has established three growth scenarios. First, it established a central scenario in which the upside and downside risks are balanced. Secondly it considers a downside scenario in which the hypothesis is an output from the crisis more slow than anticipated, and with potential shocks in the price of raw materials. The opposite hypothesis would be good as basis for the estimate of the third scenario, named at random. These long-term forecasts for the period 2014–2020 are 2.5, 1.5 and 3.0 for central, downside and upside scenarios, respectively.

32.3.4 Evolutionary Algorithm

A Genetic Algorithm (GA) has been used as an optimizing tool, with selection and mutation associated with an expert system, called Flexible Evolution Algorithm (FEA) [5]. The FEA is subdivided into several functions, called engines. These subroutines have been designed to group the diverse actions that are to be executed during the optimization depending on their objectives. In this way, all the learning tasks will be clustered in a learning engine, and something similar will happen with all the selection schemes, sampling strategies or decision mechanisms. A general scheme of the FEA can be seen in Fig. 32.6. Starting with the initial population IP, and for all the iterations of the population obtained earlier, we evaluate the fitness function of each candidate solution and from here, and for each generation, the Decision Engine acts over all the different stages of the algorithm. So the decision engine will decide what kind of learning, selection and/or sampling strategy will be used in every generation until the stop criterion is reached, which is also determined by the Decision Engine. The Learning Engine stores everything that could be useful afterwards, such as information about the variables or statistics. The intention is to use this information to learn about the process and even to establish rules that could be fruitful and will be included in the decision engine afterwards. The Selection Engine chooses which solutions are to be sampled and the sampling method that will be used, whereas the sampling engine carries out the mutation and crossover processes over the variables. Finally, the Filtering Engine removes the possible errors in the solutions before entering in the next iteration.

Main ingredients of the implementation of the FEA are: a Dynamic Structure of Operators (DSO), an Enlargement of the Genetic Code (EGC) of individuals and the use of a Central Control Mechanism (CCM). In order to obtain an efficient DSO two unique classes of operators have been defined: Selection and Sampling. Most existing crossover and mutation operators have been included in the Sampling class and the definition of two additional characteristics of all members of this class was also advisable: the Nature and the Range. The DSO enables the use of any

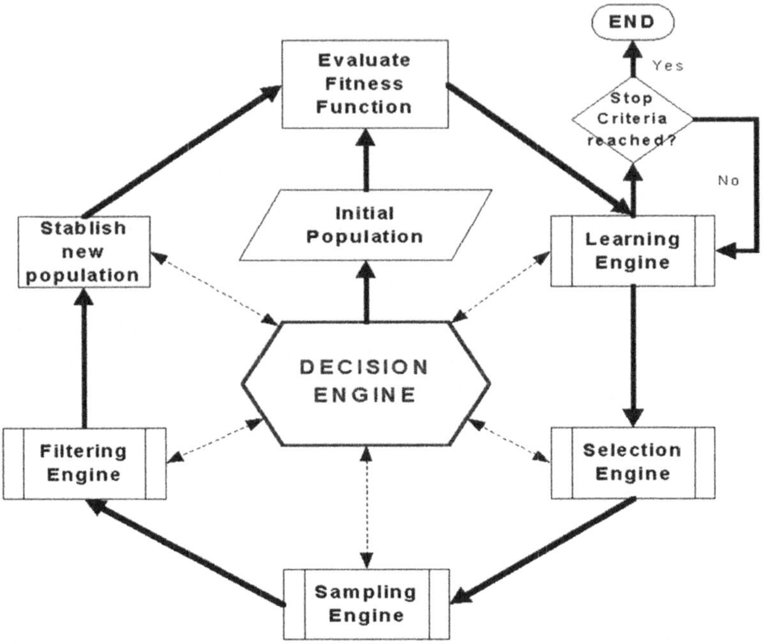

Fig. 32.6 General scheme of the flexible evolution algorithm (FEA)

of the operators at each step along every optimization run depending on operator's previous contribution to the common task (to get the optimum). The genetic code of the individuals must be enlarged (EGC) in order to include useful information for the process control included in the CCM. After several implementations, the identification of the sampling method used to obtain each variable of the individuals has proved to be the most useful information, but only when a simple Probabilistic Control Mechanism (PCM) based on rules IFTHEN- ELSE is used as a CCM. The joint use of the DSO, the EGC and the CCM has permitted the elimination of the crossover and mutation probabilities. The PCM is responsible for reaching a trade-off between the exploration and the exploitation of the search space, which is equivalent to achieving a competitive and cooperative balance among the sampling operators.

32.3.5 Robust Design Optimization

Robust Design Optimization (RDO) is a method of minimizing the effect of input parameter uncertainty on the solution without eliminating the causes. There are different approaches to perform robust optimization in practice [1]. We had opted for a mean-variance robustness approach that consist in consider as FEA fitness function the following one:

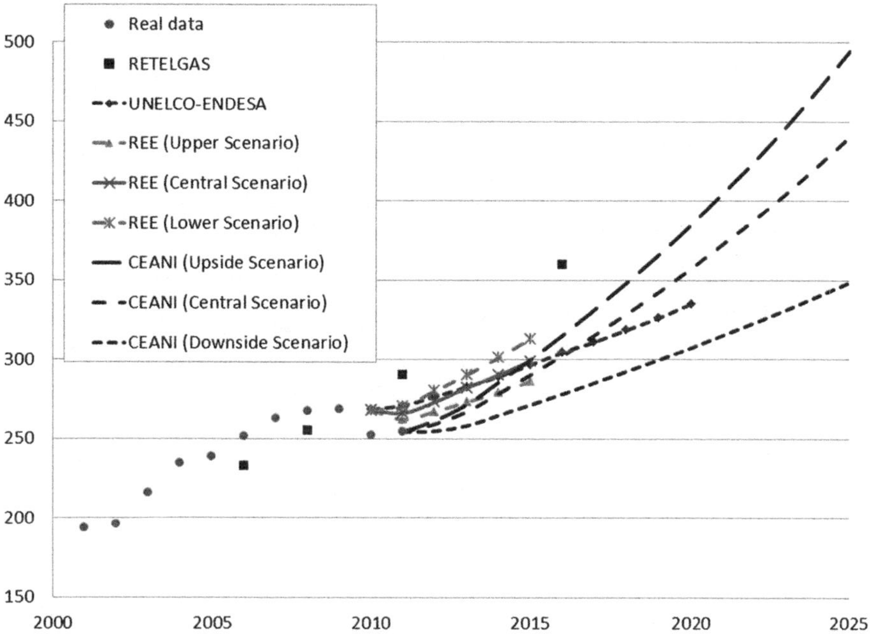

Fig. 32.7 Annual electricity demand in power station bars (GWh): historical data and forecasts by different sources including our result (CEANI forecasts)

$$f(\mathbf{x}) = \mu(f^*(\mathbf{x})) + W \times \sigma^2(f^*(\mathbf{x})), \tag{32.7}$$

where $\mathbf{x}$ represents an individual of any population of the FEA, $f^*(\mathbf{x})$ represents a set of different values of the real fitness function evaluated at the individual $\mathbf{x}$ if the input parameters with uncertainty change a little bit, W is a real number (weight), μ denotes the mean and σ^2 the variance.

32.3.6 Log–Log Econometric Model

Estimates of the annual electricity demand and annual average hourly peak power (in power station bars) were calculated from a log–log econometric model where population and GDP at market prices are the explanatory variables (see Figs. 32.7 and 32.8):

$$\log(y) = a + b \times \log(population) + c \times \log(GPD), \tag{32.8}$$

The a, b and c parameters were fitted by least squares using the evolutionary algorithm described in Sect. 32.3.4 with robust design optimization (see Sect. 32.3.5). The y parameter represents both annual electricity demand and annual average hourly

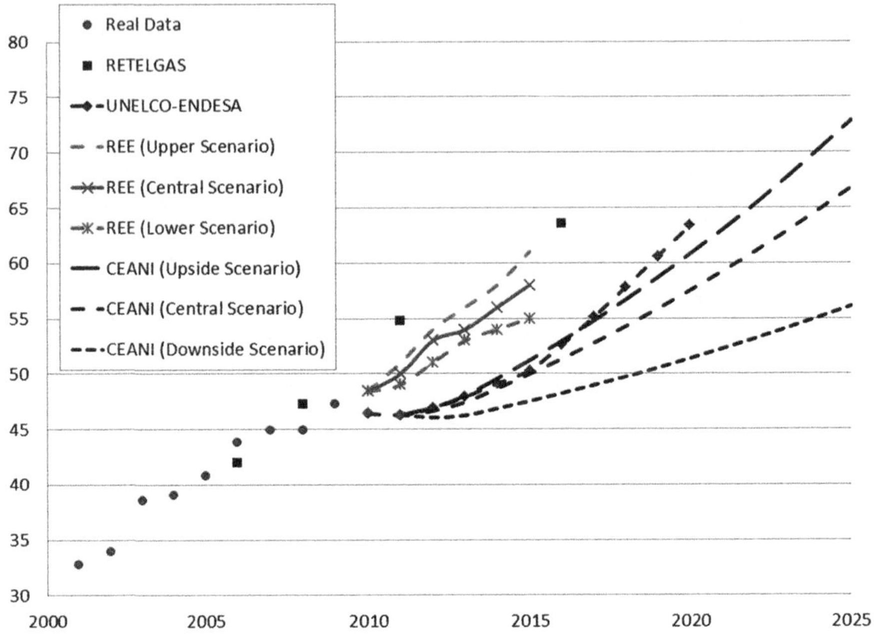

Fig. 32.8 Annual average hourly peak powers in power station bars (MW): historical data and forecasts by different sources including our result (CEANI forecasts)

peak power. The prevailing uncertainty around some explanatory variables (GDP at market prices) justifies the incorporation of robust design to FEA in order to obtain a, b and c parameters of the model considered. Figure 32.9 shows scatterplots of both electricity demand data and average hourly peak powers data versus the values fitted with the log–log econometrics models obtained:

$$\ln(\text{Electricity\_Demand}) \approx -15.526 + 0.308 \times \ln(\text{Population}) + 1.698 \times \ln(\text{GDP}) \tag{32.9}$$

$$\ln(\text{Hourly\_Peak\_Powers}) \approx -54.265 + 3.951 \times \ln(\text{Population}) + 1.272 \times \ln(\text{GDP}) \tag{32.10}$$

According to Eq. 32.9, a 1 % increase in population multiplies electricity demand by $e^{0.308 \times log(1.01)} = 1.0031$, i.e. increases the expected electricity demand by about 0.31 %, and a 1 % increase in GDP multiplies electricity demand by $e^{1.698 \times log(1.01)} = 1.0170$, i.e. increases the expected electricity demand by about 1.7 %.

According to Eq. 32.10, a 1 % increase in population multiplies average hourly peak powers by $e^{3.951 \times log(1.01)} = 1.0401$, i.e. increases the expected average hourly

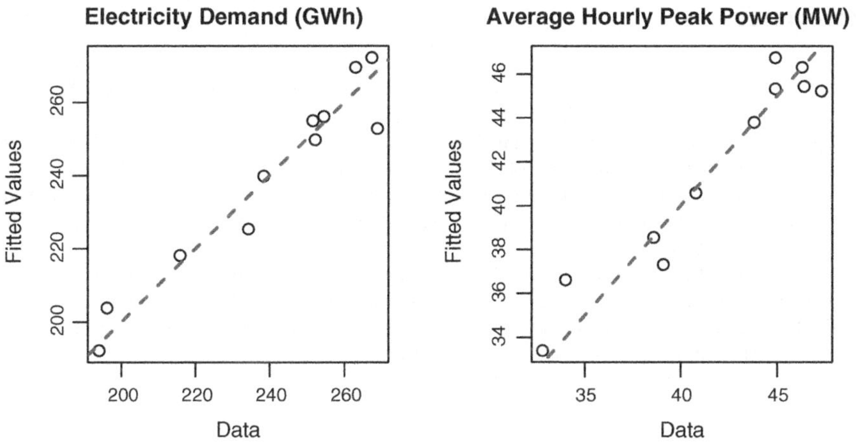

Fig. 32.9 Scatterplots of both electricity demand data and average hourly peak powers data versus the values fitted with the obtained log–log econometrics models. In both cases, the *piecewise line* represents the line where the abscissa corresponds to the ordinate

peak powers by about 4.01 %, and a 1 % increase in GDP multiplies average hourly peak powers by $e^{1.272 \times log(1.01)} = 1.0127$, i.e. increases the expected average hourly peak powers by about 1.27 %.

32.4 Conclusions

This work arose from the need of an energy plan for each island of the Canary Islands. Because this plan would have to anticipate the amount of land required for energetic facilities through 2025, electricity demand projection during that period of time was necessary.

For economic reasons the islands must be self-sufficient in the largest extent possible because energy cannot come from abroad. So if forecasts are not as accurate as possible, energy self-sufficiency may be compromised. When we looked for models to estimate load demand in the long term we found mainly econometric methods and methods based on neural networks. From our point of view the advantage of the former is its transparency and explicit dependency from different explanatory variables.

The current economic situation and job insecurity precariousness as well as price increases in energy can bring great changes in the use of electricity in coming years. Therefore we consider necessary to incorporate these variables into the model. And now we are working in that sense.

References

1. Beyer HG, Sendhoff B (2007) Robust optimization a comprehensive survey. Comput Meth Appl Mech Engrg 196:3190–3218
2. Ghods L, Kalantar M (2011) Different methods of long-term electric load demand forecasting; a comprehensive review. Iran J Electr Electron Eng 7(4):249–259
3. NZIER (2009) Review of electricity demand forecast model. Report commissioned by the New Zealand electricity commission. http://www.electricitycommission.govt.nz/opdev/modelling/gpas/2005SOO/Demand/Review
4. NZIER (2011) Electricity demand model review. Report to Transpower. https://www.transpower.co.nz/sites/default/files/plain-page/attachments/nzier-review-of-transpower-forecast.pdf
5. Winter G et al (2005) Flexible evolutionary algorithms: cooperation and competition among real-coded evolutionary operators. Soft Comp 9(4):299–323

Chapter 33
Optimization of the Dimensionless Model of an Electrostatic Microswitch Based on AMGA Algorithm

Jorge Santana-Cabrera, José Miguel Monzón-Verona, Francisco Jorge Santana-Martín, Santiago García-Alonso and Juan Antonio Montiel-Nelson

Abstract In this paper a micro genetic algorithm for multi-objective optimization (AMGA) is used to minimize the number of function evaluations of the dimensionless model of an electrostatic microswitch. A non-dimensional dynamic model is proposed, and three objective functions are defined: the closing dimensionless time of the first impact, the maximum dimensionless speed and the maximum dimensionless displacement of the first impact. This work has been carried out using dimensional analysis. Results demonstrate an interesting methodology based on AMGA for optimizing the closing time and displacement of the first impact in a microswitch.

Keywords Electrostatic microswitch · Dimensional analysis · Multi-objective optimization

J. Santana-Cabrera (✉) · J.M. Monzón-Verona · F.J. Santana-Martín · S. García-Alonso
J.A. Montiel-Nelson
Institute for Applied Microelectronics (IUMA),
University of Las Palmas de Gran Canaria (ULPGC), 35015 Las Palmas de GC, Spain
e-mail: jscabrera@iuma.ulpgc.es

S. García-Alonso · J.A. Montiel-Nelson
Department of Electronic Engineering and Automation (DIEA),
University of Las Palmas de Gran Canaria (ULPGC), 35015 Las Palmas de GC, Spain
e-mail: sgarcia@diea.ulpgc.es

J.A. Montiel-Nelson
e-mail: montiel@iuma.ulpgc.es

J.M. Monzón-Verona · F.J. Santana-Martín
Department of Electrical Engineering (DIE),
University of Las Palmas de Gran Canaria (ULPGC), 35015 Las Palmas de GC, Spain
e-mail: jmonzon@die.ulpgc.es

F.J. Santana-Martín
e-mail: fsantana@die.ulpgc.es

D. Greiner et al. (eds.), *Advances in Evolutionary and Deterministic Methods for Design, Optimization and Control in Engineering and Sciences*, Computational Methods in Applied Sciences 36, DOI 10.1007/978-3-319-11541-2_33

33.1 Introduction

MEMS electrostatic microswitches possess high insulation of the electromechanical switches and ultra-low losses. They also have low power consumption and small size and the advantages of low cost of solid state relays manufactured with microelectronic technology. Besides, electrostatic microswitches operate in a large range of frequencies. These properties make possible the massive application of these MEMS devices to wide technology fields, in particular, to the telecommunication industry, to wireless devices such as microswitches for antennas and switches for reception-transmission, among others.

One of the main subjects in resistive microswitch design is related to the interaction between the tip and the substrate, and the damage accumulated produced by the bouncing of the tip on the substrate. It is well known that the tip of a resistance microswitch bounces several times on the substrate before reaching a permanent contact [1].

We have used dimensional analysis for measured performance of some dimensionless parameters. The use of a dimensionless model is a valuable procedure used to study engineering problems [2]. By applying Buckingham [3] theorem, dimensionless parameters are obtained.

Π Buckingham theorem applied to dimensionless analysis establishes that, an equation with a number of variables related between them that defines a physical problem is reduced to another similar dimensionless equation but with a lower number of variables. A dimensionless parameter consists in a group of variables joined in a way that the dimensionless expression is the unit. The number of dimensionless sets for a particular problem is equal to the difference between the total number of variables minus the number of fundamental dimensions.

Our contribution in this work consists in designing a microswitch with the following goal: minimize the time of the first contact tip-substrate to increment the working frequency and, at the same time, minimize the maximum velocity and the oscillation of the first bouncing to decrease the number of bounces.

These goals are in contradiction because if we decrease the closing time, then, the velocity of the first impact increases and, hence, there is a greater bouncing, see Fig. 33.1. This fact indicates that we are dealing with a multiobjective optimization problem.

This paper is organized as follows. Section 33.2 presents the theoretical fundaments of the optimization multiobjective of a microswitch. Section 33.3 provides the results and discussion. Finally, in Sect. 33.4 conclusions are presented.

33.2 Theoretical Fundaments of the Optimization Multiobjective of a Microswitch

In general, the optimization multiobjective problem is defined as follows: find the vector $\overrightarrow{x}^* = [x_1^*, x_2^*, \ldots, x_n^*]^T$ that satisfaces the m restrictions of the inequalities

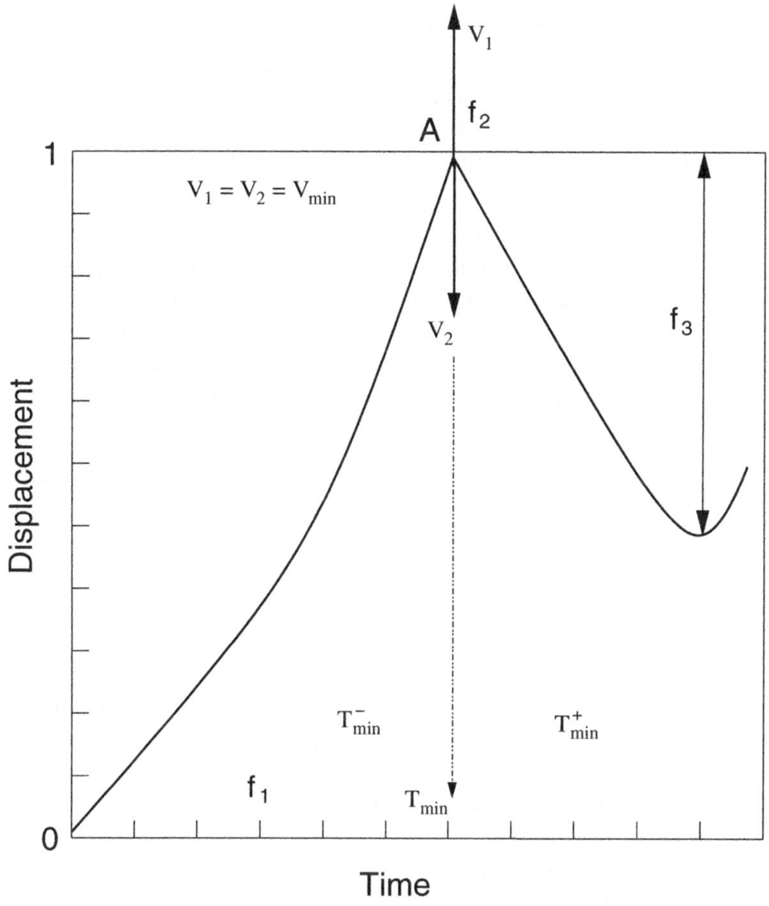

Fig. 33.1 Dimensionless analysis of the velocity for $A_1 = 0.1$

$$g_i(\vec{x}) \geq 0 \qquad i = 1,\ 2, \ldots, m$$

the p restrictions of the equalities

$$h_j(\vec{x}) = 0 \qquad j = 1,\ 2, \ldots, p \tag{33.1}$$

and optimizes the vector of functions

$$\mathbf{f}(\vec{x}) = [f_1(\vec{x}),\ f_2(\vec{x}), \ldots, f_k(\vec{x})]^T \tag{33.2}$$

where $\mathbf{x} = [x_1,\ x_2, \ldots, x_r]^T$ is the vector of variable decision.

Multiobjective optimization problems require three basic elements:

- A conflicting trade-off between two or more objective functions. They are a quantitative measure of the system to be optimized.
- The variables that affect the objective functions.
- The restrictions: a set of relations, equations and inequalities that some variables must satisfy.

For obtain an optimal design of the microswitch the dinamic behavior of the system is modelled and analyzed. A cantilever beam represents a basic physical model of microswitch. We have analyzed this beam through a lumped parameter model of mass-spring-friction and two state variables (velocity and position). In this model, we stress the interaction between the tip of the beam and the substrate that has been modeled introducing a "Lennard–Jones" force [1]. In our work, see Fig. 33.1, the three objective functions are:

- $f_1(\overrightarrow{x})$: The closing time—to increase the working frequency of the microswitch.
- $f_2(\overrightarrow{x})$: The first impact velocity.
- $f_3(\overrightarrow{x})$: The first impact displacement.

We have identified six dimensionless parameters. In first place, A_1, which is a proportion between the impeller electrostatic force and the elastic force associated to the cantilever beam. Second, the quality factor Q which is inversely proportional to friction coefficient.

From the dynamical analysis, we have obtained two performance related to the time domain, the velocity and the position. From this performance we obtain important data for the design of the microswitch: the time $T_{\min}$ to establish the first contact tip/substrate, the maximum velocity for the first contact, $\tilde{v}_{\max}$, and the maximum elongation after the first impact, $\tilde{r}_{\max}$.

The MEMS switch is modeled as a one-degree of freedom system which is the position of the tip of the cantilever r. It consists on a mass m, initially placed at a distance g_0 from the substrate, a spring with elastic constant k, and a dashpot with damping coefficient b. Thus, the motion of the system is described by the classical second order linear Eq. 33.3, where F_{EL} is the electrostatic actuation force and F_{LJ} the Lennard-Jones force that provides the mechanical interaction between two facing surfaces. F_{EL} and F_{LJ} is expressed as 33.4 and 33.5, respectively. In this work, the area of interaction has been assumed as $A = 100\ \mu\text{m}^2$, $C_1 = 10^{-20}\,\text{Nm}$ and $C_2 = 10^{-80}\,\text{Nm}$ [1].

$$m\frac{d^2 r}{dt^2} + b\frac{dr}{dt} + k\,r = F_{EL} + F_{LJ} \tag{33.3}$$

$$F_{EL} = \frac{\frac{1}{2}\varepsilon_0 A_0 V^2}{\left(g_0 + \frac{d_\varepsilon}{\varepsilon_r} - r\right)^2} \tag{33.4}$$

$$F_{LJ} = \frac{C_1 A}{(g_0 - r)^3} - \frac{C_2 A}{(g_0 - r)^9} \tag{33.5}$$

The dimensionless equations of the dynamic model are:

$$\frac{d\tilde{v}_e}{d\tilde{t}} = \left[\frac{A_1}{(1 + A_4 - \tilde{r})^2}\left[1 + A_5(1 - \tilde{r})\right] - \frac{\tilde{v}_e}{Q} - \tilde{r} + \frac{A_2}{(1 - \tilde{r})^3} - \frac{A_3}{(1 - \tilde{r})^9}\right] \tag{33.6}$$

$$\frac{d\tilde{r}_e}{d\tilde{t}} = \tilde{v}_e \tag{33.7}$$

where A_1, A_2, A_3, A_4 and A_5 are non dimensional parameters and Q is the quality factor.

A_1 and Q are expressed as follows:

$$A_1 = \frac{\varepsilon_0 A_0 V^2}{2k g_0^3} \tag{33.8}$$

and

$$Q = \frac{\sqrt{km}}{b} \tag{33.9}$$

33.3 Results and Discussion

The proposed optimization problem is described as follows

$$min \quad (f_1(\overrightarrow{x}),\ f_2(\overrightarrow{x}),\ f_3(\overrightarrow{x})) \tag{33.10}$$

where

$$\overrightarrow{x} = (A_1, Q) \tag{33.11}$$

is the decision variable vector that corresponds to the dimensionless parameters A_1 and Q described previously.

$f_1(\overrightarrow{x})$ corresponds to the first dimensionless bounce time. At this instant the microswitch contacts the substrate for first time, $f_1(\overrightarrow{x}) = T_{\min}$, see Fig. 33.1.

$f_2(\overrightarrow{x})$ is $\tilde{v}_{e_{\max}}$, which is the velocity for the first bounce at $T_{\min}$ value; in this time instant, the velocity V_1 for $T_{\min}^{-}$ is equal to V_2 at $T_{\min}^{+}$, see Fig. 33.1. Note that the velocity senses are opposite.

$f_3(\overrightarrow{x})$ is $\tilde{r}_{\max}$ which is the dimensionless maximum position of the first bounce and correspond to a $V = 0$ velocity.

The instants $T_{\min}$ and $T_{\max}$ are determined in the execution time of the numerical solution. For this reason, we have analyzed two consecutive instants of the simulation, comparing the value of the variables $\tilde{r}_e$ and $\tilde{v}_e$.

Table 33.1 Execution time of the experiments

Population	Generation	Evaluations	Time (s)
80	10	800	2,405
80	50	4,000	20,566
80	62	4,960	25,568
100	62	6,200	31,996
60	200	12,000	62,245

Differential Eqs. 33.6 and 33.7 have been solved using a RKF with fixed step programmed in C++ [4] wich uses the algorithm AMGAII-Archive-based Microgenetic Algorithm. We have chosen a real codification for each chromosome composed by the design variables A_1 and Q. In general, the precision is better than the binary codification [5]. This is improved adding more bits, but it increases the simulation time. The used time step Δ was 10^{-6} s and is constant during the whole simulation.

All the simulation results were obtained in a SunFire X2200, with 2 CPU AMD Opteron 2214 Dual Core (2,2Ghz), 4 GB RAM, using the O. S. Red Hat Enterprise Linux Server 5.3.

We have solved this problem based on the concept of Pareto optimal solution applying genetic multiobjective algorithms. The numerical solution of the objective functions is highly time consuming, see Table 33.1. Hence, we have reduced the number of evaluations of the objective functions.

We propose the use of Archieve based Micro Genetic Algorithm (AMGA) [6] method because this algorithm generates a small number of new solutions in each evaluation, improving the total time for the evaluation [7].

Taking into account the first group of results of the dimensionless dynamic model, we conclude that the most important dimensionless parameters are A_1 and Q. A_1 represents the quotient between the minimum electrostatic force and the maximum elastic force and Q represents the quotient between the maximum energy stored in the spring and mass, and the friction looses.

Figure 33.2 illustrates the Pareto optimum with the objective functions closing time for the first impact versus the maximum velocity in the first impact. Figure 33.3 shows the Pareto optimum and represents the three objective functions mentioned in the previous section [6]. Figures 33.4 and 33.5 represent the influence of A_1 in the closing time and velocity of the first impact. As A_1 increases the closing time decreases and the closing speed increases.

Figure 33.6 shows the influence of the dimensionless parameter Q in the maximum oscillation of the first impact. As Q increases the closing position increases.

In this work, we have developed a methodology based on AMGA for optimizing the closing time and displacement of the first impact in a microswitch. From Figs. 33.2, 33.3, 33.4, 33.5 and 33.6, the relationship between dimensionless parameters A_1 and Q and closing time and displacement is illustrated.

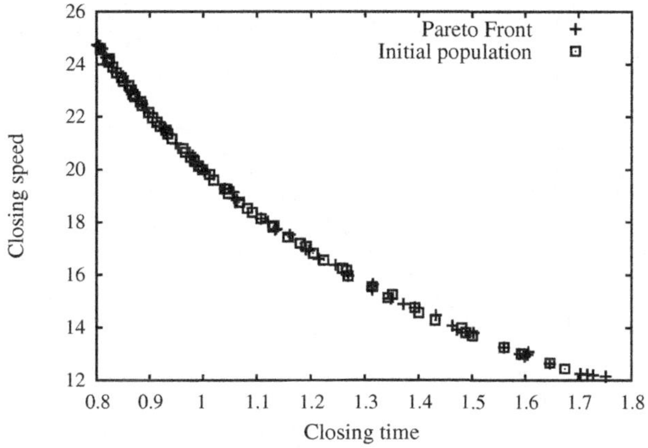

Fig. 33.2 Pareto solution time versus speed

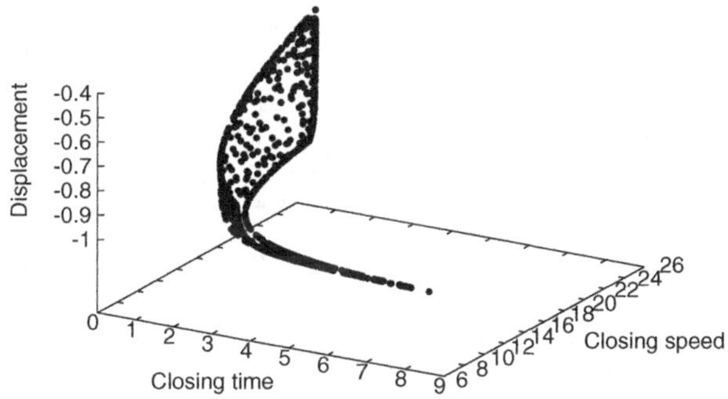

Fig. 33.3 3D Pareto AMGA method first impact

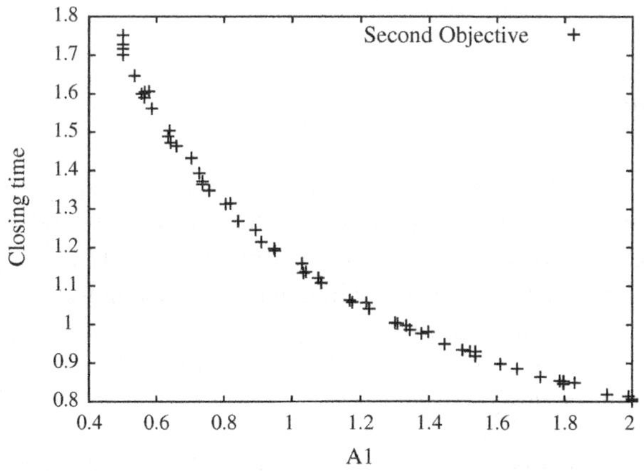

Fig. 33.4 Dimensionless time versus A1

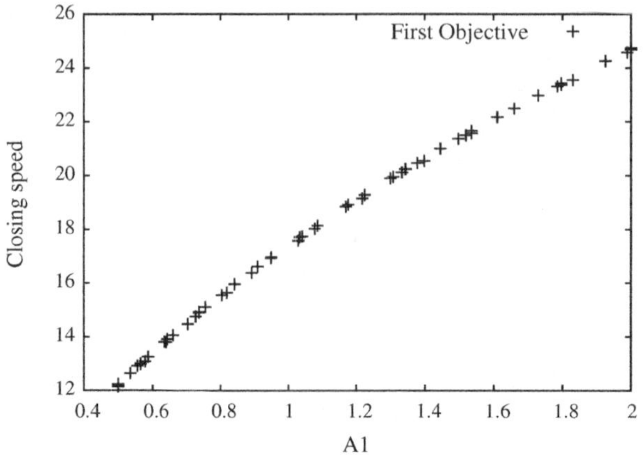

Fig. 33.5 Dimensionless speed versus A1

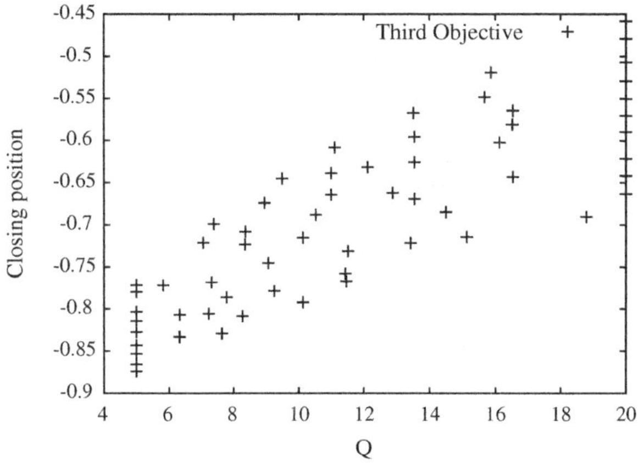

Fig. 33.6 Dimensionless speed versus Q

33.4 Conclusions

Based on a dimensionless model of a microswitch for the closing time and displacement, we have analyzed the sensitivity of this Micro Electro Mechanical System—MEMS—to the variation of the dimensionless parameters of the model. In this way, we predict the dynamic behavior of a microswitch. By using multiobjective Genetic Algorithms we have optimized the bouncing of the microswitch which is one of the major designing concern. We have used the AMGA multi-objective GA for obtaining the Pareto front of the design space.

Acknowledgments This work was funded by project BATTLEWISE—TEC2011-29148-C02-01—of the Spanish Ministry of Economy and Competitiveness.

References

1. Granaldi A, Decuzzi P (2006) The dynamic response of resistive microswitches: switching time and bouncing. J Micromech Microeng 16(7):1108
2. Vasca F, Verghese GC (1999) Adimensional models and participation factors for the analysis of induction motor dynamics. In: Industrial electronics, 1999. ISIE '99. Proceedings of the IEEE international symposium on, vol 2, pp 480–485
3. Misic T, Najdanovic-Lukic M, Nesic L (2010) Dimensional analysis in physics and the buckingham theorem. Eur J Phys 31(4):893
4. Nicolet A, Delince F (1996) Implicit runge-kutta methods for transient magnetic field computation. IEEE Trans Magn 32(3):1405–1408
5. Michalewicz Z (1996) Genetic algorithms + data structures = evolution programs, 3rd edn. Springer, London
6. Tiwari S, Koch P, Fadel G, Deb K (2008) Amga: an archive-based micro genetic algorithm for multi-objective optimization. In: Proceedings of the 10th annual conference on genetic and evolutionary computation, GECCO '08, ACM, New York, pp 729–736
7. Monzon-Verona JM, Garcia-Alonso S, Sosa J, Montiel-Nelson JA (2013) Multiobjective genetic algorithms applied to low power pressure microsensor design. Eng Comput 30(8):1128–1146

Chapter 34
Generation of New Detection Codes for GPS Satellites Using NSGA-II

J. Sosa, Tomás Bautista, Daniel Alcaraz, S. García-Alonso and Juan A. Montiel-Nelson

Abstract In this paper we obtain new detection codes, to determine whether a GPS satellite in particular is visible, using NSGA-II as multi-objective optimization engine. Our approach takes into consideration the length of the code and the sampling frequency in comparison with other approaches found in the literature that fix those design parameters. The obtained new detection codes produce an improvement of the 19 % in terms of CPU execution time. Results demonstrate that both design parameters must be taken in consideration to obtain high quality detection codes.

Keywords Genetic algorithms · GNSS · Gold codes · Low computational effort · Multi-objective optimization

34.1 Introduction

Nowadays, the detection of GPS signals for performing location tasks is one of the most commonly demanded applications [1]. In particular, the fast expansion of the mobile telephony, the increasing of the CPU capabilities and the reduced

---

J. Sosa (✉) · T. Bautista · D. Alcaraz · S. García-Alonso · J.A. Montiel-Nelson
Department of Electronic Engineering and Automation (DIEA),
Institute for Applied Microelectronics (IUMA),
University of Las Palmas de Gran Canaria (ULPGC),
35015 Las Palmas de Gran Canaria, Spain
e-mail: jsosa@iuma.ulpgc.es

T. Bautista
e-mail: bautista@iuma.ulpgc.es

D. Alcaraz
e-mail: dalcaraz@iuma.ulpgc.es

S. García-Alonso
e-mail: sgarcia@iuma.ulpgc.es

J.A. Montiel-Nelson
e-mail: montiel@iuma.ulpgc.es

D. Greiner et al. (eds.), *Advances in Evolutionary and Deterministic Methods for Design, Optimization and Control in Engineering and Sciences*, Computational Methods in Applied Sciences 36, DOI 10.1007/978-3-319-11541-2_34

battery charging times provide a wide market where GPS applications are potentially explored [2, 3]. However, in this scenario, since the GPS receiver just becomes one application running concurrently with some others in a single device, the requirements of low computational effort and low power consumption are mandatory [4]. Reducing the computational effort to determine whether a GPS satellite is visible or not is a hot topic in this research area [5–7]. In the existing literature three basic approaches are presented to cope with this problem. The first one consists in reusing most of the computation with additional hardware. This approach is called *split-sum methodology* [8]. Other authors propose to obtain a single detection code that allows to know if more than one satellite is visible or not [9]. Finally, in another approach [10] authors present a methodology to obtain GPS detection codes of 341 bits achieving a lower computational effort. In this paper we explore the idea of obtaining reduced length detection codes for GPS satellites presented in [10] using a multi-objective approach and we introduce as new optimization variables the length of the reduced code and the sampling frequency.

34.2 Problem Definition

In order to detect whether a satellite is visible or not, receivers compare the incoming GPS identifiers (at a frequency L1 of 1,575.42 MHz), with all the possible GPS satellite identifiers [11]. A satellite identifier is a Pseudo Random Number (PRN). Basically, a PRN is an array of binary digits where each digit is called *chip*. The length of the PRN array is 1,023 chips [12]. Each satellite has assigned an unique PRN as identifier. Every satellite transmits its own PRN identifier every millisecond. The comparison function is as follows:

$$Comp(PRN^{sat}, D^{rx}) = [a_1, a_2, \ldots, a_L] \qquad (34.1)$$

$$a_i = \sum_{j=1}^{L} PRN^{sat}_{mod_L(i+j)} * D^{rx}_i$$

where L is the length of the PRN, that is, 1,023 chips for each GPS satellite. D^{rx} is the incoming radio frequency data that is acquired by the GPS analog front-end receiver. PRN^{sat} is the PRN identification of the GPS satellite (*sat*). This function takes in consideration all the possible alignments between the incoming GPS identification data and the compared PRN. Therefore, this is the reason to obtain an array of values and not only a single value. The size of array $Comp(PRN^{sat}, D^{rx})$ is L, one value for each possible alignment between the incoming data and the compared PRN.

Following the theory, when a satellite is visible the array $Comp(PRN^{sat}, D^{rx})$ has an unique maximum. This maximum value is called Detection Peak (D_P). The location/index where the D_P is placed in the comparison array is called code-phase. The code-phase determines the starting chip of the PRN sequence. The other values in this array are lower than the D_P value. These lower values are called Noise (N). If the compared satellite (PRN^{sat}) is not visible, all values in array $Comp(PRN^{sat}, D^{rx})$ are noise.

Nowadays, in the literature there exist multiple approaches to implement the detection Eq. 34.1. For instance, it is quite easy to translate this Eq. 34.1 from the time domain to the frequency domain and use the Fast Fourier Transform (FFT) and its inverse (IFFT) to obtain the same results. However, in practical GPS receivers there exist only two basic detection techniques [13]. Their main key in comparison with other approaches are their implementation simplicity and the required computational effort.

In one of these approaches the L1 incoming signal is oversampled, that is, every chip of the PRN is sampled more than once, so the D_P grows with the increasing sampling frequency. In terms of Eq. 34.1, implementing the oversampling only requires to set the correct value to L.

The other solution, instead of increasing the sampling frequency, increases the sampling period. As a consequence, the recorded incoming GPS data contains more than one complete PRN sequence. Therefore, the Detection Peak increases its value in proportion to the increment of the sampling period, that is, the signal recording time is increased.

Stepping up the sampling frequency and/or the sampling period increases the sensibility of a GPS receiver when the GPS signal-to-noise ratio is too low. However, ordinary applications like open-sky navigation systems, i.e., typical GPS receiver for car tracking, only takes one of both methodologies with reduced increasing factors.

In this research, we propose to obtain new detection codes to determine whether a satellite is visible or not. The main feature of those new codes is their reduced length in comparison with the original PRN. Proposed new codes require lower computational effort than the traditional PRN [15]. Moreover, we are based on the approach presented in [10] where authors introduce a novel methodology to obtain new PRN detection codes using GA. In this previous approach, the research is focused in determining what kind of multi/single objective algorithm is more suitable for this type of application. However, in order to narrow the search space of the problem, authors fix the length of the new detection codes to a submultiple of the original length. This submultiple is 341 chips ($341 \times 3 = 1{,}023$).

Our proposal in this work is redefining the problem presented in [10] with two new variables to optimize. The first is the length of the detection code and the second is the sampling ratio. We define the sampling ratio as the size of the new detection code divided by 1,023 chips. Since we reduce the sampling ratio below the unity, we will use the term dropping ratio as a more adequate concept or definition. The approach presented in [10] has a length of 341 and a dropping ratio of 3 (that is, to take 1 sample and drop 2 for every 3 samples of the incoming signal). This results in a sampling ratio of 1/3 as shown in Fig. 34.1.

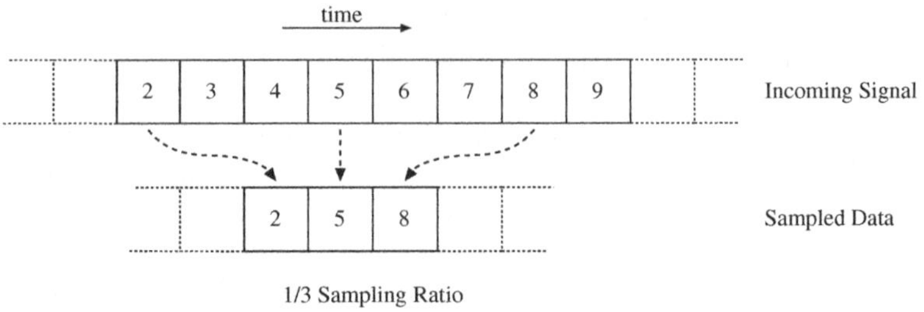

Fig. 34.1 Example of 1/3 sampling ratio

34.3 Problem Codification

Based on the approaches found in the literature, we choose the NSGA-II [14] as GA multi-objective optimization engine. The codification for the individual is as follows. Each individual represents a new detection code. The individual is made by an array of Boolean values (see Fig. 34.2). The status of each Boolean value can be only a logic zero and a logic one. The unknown or error states are forbidden in this engine. The length of the individual array determines the length of the new detection code. Finally, each element of the individual array corresponds with an element of the new detection code; this makes that the index is the same in both arrays.

Unsurprisingly, the cost function is basically the Eq. 34.1. As before mentioned, this cost function is an array of comparison values. However we are not interested in all these values but our attention is focused on the Detection Peak and the Noise. Moreover, we know that the dropping ratio determines the total number of Detection Peaks in the comparison array. For example, if the length of the new detection code is set to 341 and the dropping ratio is set to 3, then there exist 3 different Detection Peaks. So the comparison array contains as many Detection Peaks as set the dropping ratio. Other values on this comparison array are Noise values.

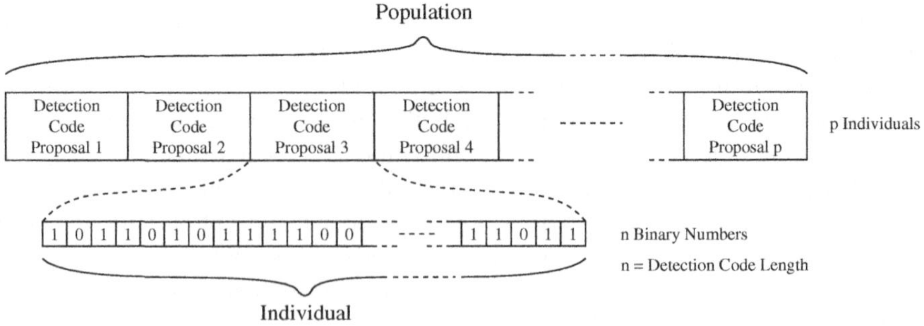

Fig. 34.2 Problem codification

We are interested in obtaining new reduced GPS detection codes with a low Noise and high Detection Peaks. In order to obtain those values, after evaluating each individual with Eq. 34.1, the evaluation function extracts the minimum Detection Peak and the maximum Noise from the computed comparison array.

34.4 Experiments

We assume that the search space of this problem grows with the length of the new detection codes and dropping ratio used. If we provide all the degrees of freedom on all our problem variables, the required computational effort is increased in great matter. The magnitude of this computational effort produces non-practical CPU execution times. In this sense, in order to obtain the solutions in a practical CPU time, we include the following rule:

$$|Proposed\ length\ \times\ Dropping\ Ratio - 1{,}023| < \xi \tag{34.2}$$

where ξ is the maximum allowed deviation between the original PRN and new proposed PRN.

Table 34.1 presents the NSGA-II optimization engine setup. We define two objective functions to optimize. The first one is to maximize D_P (minimize $-D_P$) and the second is to minimize the Noise (N). The NSGA-II optimization engine is controlled by a program written in C that proposes lengths and sizes randomly using the rule introduced in Eq. 34.2.

In other words, our application (C program) proposes different lengths and dropping ratios and the NSGA-II optimization engine obtains new detection codes. That is, the application starts with a proposal of a new length and dropping ratio to obtain a new detection code. Then the NSGA-II optimization engine takes the control and produces a Pareto-Front curve for those parameters. Once the optimization is finished, our application extracts the best solution of the Pareto-Front. This procedure starts over and over with several lengths and dropping ratios. In addition, after extracting the best solution our application evaluates the convergence of the proposed

Table 34.1 NSGA-II setup in our experiments

Parameter	Value
Num. objectives	2
OBJ_1	$-D_P$
OBJ_2	N
Population	40
Generations	10 k/100 k
Crossover	0.8
Mutation	1/PRN length
Seed	Random/uniform distribution

parameters. If the application determines that the solution can be refined/improved the NSGA-II is executed again with more generations (from 10 k to 100 k).

In order to measure the quality of new detection codes we introduce some concepts. We define the Detection Gap (D_G) as follows:

$$D_G = min(D_P) - max(N) \tag{34.3}$$

This means that the Detection Gap is the distance between the minimum Detection Peak ($min(D_p)$) and the maximum Noise ($max(N)$). Bigger Detection Gaps give better detection codes.

We also need to evaluate the required mathematical operations. For this purpose, we label the number of required multiplication/addition operations to execute Eq. (34.1) as CPU Operations (CPU_O). In this case, fewer CPU Operations require lower resources for hardware implementation.

Moreover, we define the CPU Effort (CPU_E) as the CPU Operations per Detection Gap, that is:

$$CPU_E = \frac{CPU_O}{D_G} = \frac{num.\ required\ mul/add}{min(D_P) - max(N)} \tag{34.4}$$

In our experiments we compare our proposal with the traditional methodology [15]. This traditional methodology consists of implementing Eq. (34.1) directly without any of our proposed improvements.

Table 34.2 presents some results of our application when we look for reduced codes of GPS satellite ID 1. In this experiment, we set ξ to a maximum of 400 chips. The first and second columns of the table contain the proposed code length and the dropping ratio. The number of generations are shown in the third column (in times of $1k$ generations). The fourth column gives the value of the rule defined in Eq. (34.2). The following two columns exhibit the best optimized minimum Detection Peak ($min(D_P)$) and maximum Noise ($max(N)$). The seventh column indicates the Detection Gap (D_G). The following column measures the CPU Operations (CPU_O). Ninth column evaluates the CPU Effort CPU_E. Finally, the last column presents the difference (CPU_{Diff}) between our proposal and the traditional methodology for the CPU Effort ($CPU_E(our) - CPU_E(traditional)$). In addition, please note that the last two rows in this Table 34.2 show the best solution obtained in [10] and also using the traditional methodology, respectively.

As expected, results from Table 34.2 are better as greater is the total number of generations in terms of maximum Noise and/or minimum Detection Peak. Last column presents a comparison between the traditional methodology and our new reduced detection code. The comparison is done in terms of required computational effort and detection gap to determine whether a satellite is visible or not. The negative values in this column denote better solutions than using the traditional methodology.

From Table 34.2, we observed that, for instance, the combination 320 × 3 (length × dropping ratio) with 100 k generations and 212 × 3 with 10 k generations have similar values in last column, −73 and −76, respectively. The first one has the double

Table 34.2 NSGA-II experiments for Satellite ID 1 and $\xi < 400$

Code length	Dropping ratio	Generations ($\times$ 1 k)	Rule value	$min(D_P)$	$max(N)$	D_G	CPU_O	CPU_E	CPU_{Diff}
384	2	10	768	206	64	142	147,456	1038.4	−59
448	2	10	896	236	66	170	200,704	1180.6	84
480	2	10	960	250	70	180	230,400	1280.0	183
512	2	10	1,024	276	72	204	262,144	1285.0	188
212	3	10	636	90	46	44	44,944	1021.5	−76
288	3	10	864	116	54	62	82,944	1337.8	241
304	3	10	912	126	56	70	92,416	1320.2	223
320	3	10	960	134	56	78	102,400	1312.8	216
336	3	10	1,008	140	60	80	112,896	1411.2	314
341	3	10	1,023	141	59	82	116,281	1418.1	321
256	4	10	1,024	90	52	38	65,536	1724.6	628
512	2	100	1,024	276	70	206	262,144	1272.5	176
288	3	100	864	136	54	82	82,944	1011.5	−85
304	3	100	912	148	56	92	92,416	1004.5	−92
320	3	100	960	156	56	100	102,400	1024.0	−73
341	3	100	1,023	177	57	120	116,281	969.0	−128
256	4	100	1,024	98	52	46	65,536	1424.7	328
341	3	40	1,023	177	89	88	116,281	1321.4	224[a]
1,023	1	–	1,023	1,023	69	954	1,046,529	1097.0	0[b]

$min(D_P)$ minimum Detection Peak, $max(N)$ maximum noise, D_G Detection Gap, CPU_O CPU Operations, CPU_E CPU Effort, CPU_{Diff} CPU Effort Difference
[a]Data from [10]
[b]Traditional methodology [15]

of detection gap than the second one, but the second proposal requires a half of the CPU Operations to evaluate Eq. (34.1). Therefore, this comparison demonstrates that there exist several combinations of code lengths and dropping ratios that have similar ratios of CPU effort and detection gap.

The results obtained with our application, as shown on Table 34.2, are always better than the presented in a previous work [10] in terms of maximum Noise. In case of the Detection Peaks our approach obtains at least the same or better values than those in the referred work [10].

Table 34.3 presents the summary of the results for all checked lengths. The first column shows the PRN length. The second, the third and the fourth columns provide the maximum, minimum and average computational effort difference (CPU difference) between our proposal presented in this document and the traditional methodology. The last column in this table gives the obtained improvement in percentage. A negative value here means that our proposal is better than the traditional methodology in the case referred to.

Table 34.3 Obtained results from experiments for all satellites

PRN length	CPU_{Diff}			Improvement (%)
	Max	Min	Avg	
256	114.26	−159.25	−44.38	4.05
264	27.13	−339.43	−202.94	18.50
272	−40.09	−342.06	−208.95	19.05
280	−65.42	−328.37	−170.86	15.57
288	55.00	−267.56	−173.64	15.83
296	26.28	−270.43	−135.58	12.36
304	303.24	−241.30	−116.91	10.66
312	−38.91	−227.86	−115.25	10.51
320	40.78	−229.20	−98.23	8.95
328	47.51	−215.16	−80.58	7.35
341	114.26	−159.25	−44.38	4.05

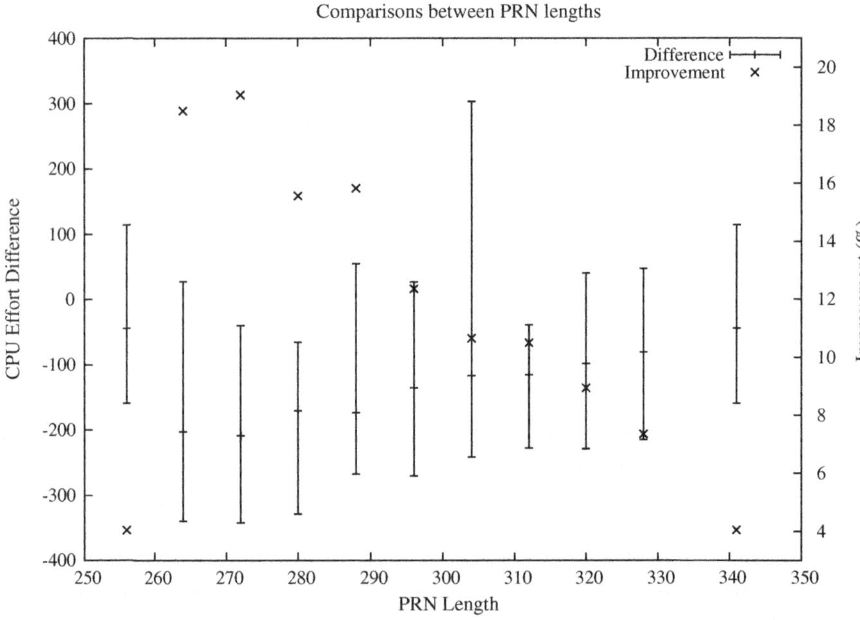

Fig. 34.3 Results comparison in terms of PRN length and CPU effort improvement

Figure 34.3 display graphically the data on Table 34.3. On the left vertical axis is represented the computational effort difference units. In addition, on the right vertical axis the average improvement in percentage is shown. The horizontal axis gives the detection code length (proposed PRN length). In this Fig. 34.3, each vertical bar represents the maximum, minimum and average obtained improvement (left vertical axis). Finally, the × symbol identifies the improvement in percentage (right vertical axis).

The ideal goal in our application is to obtain a set of detection codes where the obtained improvement is equal for all satellites. In addition, due to hardware restrictions, all detection codes must have with the same length. These are moved to Fig. 34.3 looking for a bar (set of detection codes with same length) and its maximum and minimum as close as possible to the bottom of the figure. In this sense, there are two solutions very close with similar results, lengths 264 and 272, where the improvement in average is 18.50 and 19.05 % respectively. The advantage of length 272 set is not only the better percentage, but also this set provides improvements in all its detection codes (maximum, minimum and average improvements are negative values).

34.5 Conclusions

In this paper a methodology is proposed to take into account new design parameters to obtain reduced GPS detection codes to determine whether a satellite is visible or not. Those new design parameters are the length of the reduced detection code and the dropping ratio. New optimal detection codes are obtained using NSGA-II as optimization engine. In addition, we present a new metric to evaluate the performance of new detection codes in terms of required CPU effort. Results demonstrate that new detection codes exist that exhibit both similar or better performance in terms of CPU effort, detection gap, code length and/or dropping ratio.

Acknowledgments This work is patent pending and was funded under project BATTLEWISE (TEC2011-29148-C02-01) of the Ministry of Economy and Competitiveness.

References

1. USAF Navstar GPS (2003) Where am I? Are we there yet? Air Space Power J 6(2):182–197
2. Kaplan E, Hegarty C (1996) Understanding GPS principles and applications. Artech House, Norwood
3. Duncan MJ, Badland HM, Mummery WK (2009) Applying GPS to enhance understanding of transport-related physical activity. J Sci Med Sport 12(5):549–556
4. Prasad R, Ruggieri M (2005) Applied satellite navigation using GPS, GALILEO, and augmentation systems. Artech House, Norwood
5. Borre K (2006) A software defined gps and galileo receiver: a single-frequency approach. Birkhuser, Boston
6. Hyoungmin S, Haeyoung J, Changdon K (2007) A new GNSS signal acquisition algorithm based on cross-correlation sequence with reduced signal-receiving time. In: International conference on control, automation and systems (ICCAS'07), pp 2563–2567
7. Chih-Hung W, Wei-Han S (2011) A study on GPS GDOP approximation using support-vector machines. IEEE Trans Instrum Meas 60(1):137–145
8. Gunawardena S, van Graas F (2006) Split-sum correlator simplifies range computations in GPS receiver. Electron Lett 42(25):1469–1471
9. Jan S-S, Lin Y-C (2009) A new multi-C/A code acquisition method for GPS. GPS Solutions 13(4):293–303

10. Sosa J, Montiel-Nelson JA, Nooshabadi S (2011) Low power GPS pseudo random numbers using genetic algorithms. In: Evolutionary and deterministic methods for design, optimisation and control with applications to industrial and societal problems (EUROGEN11), Capua, Italy
11. Lee S-W, Kim J, Jeong M-S, Lee YJ (2011) Monitoring atomic clocks on board GNSS satellites. Adv Space Res 47(10):1654–1663
12. Michalski A, Czajewski J (2004) The accuracy of the global positioning systems. IEEE Instrum Meas Mag 7(1):56–60
13. Hamza G, Motawie I (2009) Implementation of a complete GPS receiver using Simulink. IEEE Circuits Syst Mag 9(4):43–51
14. Deb K, Pratap A, Agarwal S, Meyarivan T (2002) A fast and elitist multiobjective genetic algorithm: NSGA-II. IEEE Trans Evol Comp 6:182–197
15. Dunn MJ (2012) Navstar GPS space segment/navigation user interfaces (IS-GPS-200). U.S. Coast Guard Navigation Center

Author Index

D. Greiner et al. (eds.), *Advances in Evolutionary and Deterministic Methods for Design, Optimization and Control in Engineering and Sciences*, Computational Methods in Applied Sciences 36, DOI 10.1007/978-3-319-11541-2

The manufacturer's authorised representative in the EU is Springer Nature Customer Service Centre GmbH, Europaplatz 3, 69115 Heidelberg, Germany. If you have any concerns regarding our products, please contact ProductSafety@springernature.com

Printed and bound by CPI Group (UK) Ltd, Croydon, CR0 4YY
30/09/2026
02225570-0018